# ECOLOGY: AN EVOLUTIONARY APPROACH

**J. MERRITT EMLEN**
Indiana University

# ECOLOGY: AN EVOLUTIONARY APPROACH

**ADDISON-WESLEY PUBLISHING COMPANY**
Reading, Massachusetts
Menlo Park, California · London · Amsterdam · Don Mills, Ontario · Sydney

This book is in the
**ADDISON-WESLEY SERIES IN THE LIFE SCIENCES**

*Consulting Editor:* JOHNS HOPKINS, III

*Second printing, May 1977*

ISBN 0-201-01894-2
ABCDEFGHIJ-HC-7987

*To John T. Emlen and Ernest Merritt,*
*great men, great scientists*

# Preface

Never before in our history has it been so vital that we understand and intelligently oversee the use of our environment. We need to know in detail the effects on biological productivity and relationships of adding chemical effluents to our streams, lakes, and oceans; we need to know the effects of diminishing numbers of species on the stability of remaining populations, including our own; we need to understand how our own evolutionary heritage predilects us to certain behavior patterns. The most urgently needed knowledge perhaps involves the relationships within whole biological communities, and it is precisely these relationships about which, unfortunately, we probably know the least.

The reason we know so little of biological communities is fairly straightforward. For one thing, communities are terribly complex. But it is more than that. The understanding of biological principles comes through the meticulous building of knowledge and statement of laws, starting from the simplest axioms. Such axioms are few in biology and take one of two forms. They are either the laws of chemistry and physics, or the laws of evolutionary change. Ecology, the study of organisms in their environment, starts from both.

Limnologists, for example, base their conclusions largely on the first, while evolutionists rely on the second. Accordingly, practitioners of these two disciplines may look on the same ecological questions from very different viewpoints. To a limnologist a creature encysts, perhaps, because of pH, temperature, or chemical changes in the environment. To the evolutionist the creature encysts because, in the past, those ancestors which possessed genetic material predilecting them to encyst under the appropriate conditions more often survived to pass on that genetic material than individuals which possessed other genes. The approach of the limnologist is mechanistic, or *proximate*; that of the evolutionist *ultimate*. Both approaches are equally valid and, in concert, very fruitful. But the second—and to a large degree also the first—approach applies to individual adaptations to the environment. To apply it to biological communities without falling into the dangers of argument by analogy, we must first look at individual adaptation, then apply what we learn of individual adaptation to population changes and interactions, and finally apply this last knowledge to communities. Community ecology is the field of ecology farthest removed from the basic axioms available, and hence is least thoroughly understood.

Most of what we know of ecology is of an empirical nature. In the past, faced with the vast complexity of the biological world, classifying and describing was the common approach. In fact, until a considerable amount of this had occurred, it was impossible for any sort of theoretical framework to be erected. It was not until the middle of the present century that the basic laws of natural selection and the accumulation of ecological facts became sufficient for the formulation of general hypotheses to become commonplace. At about the same time computers became generally available and two groups split off from the group of classical ecologists.

The first were the evolutionary ecologists, and the second the so-called systems analysts. Both, adhering to new philosophies and using new mathematical tools, have tended to look down on those engaged in the old "woodsy lore" ecology. On the other hand, the classicists view their upstart offspring as sloppy and out of touch with biological reality, Many systems analysts attack the evolutionists for the same reason. These views are, to some extent, justified, but are also generally based on prejudice, for often the classical ecologist remains so because he is without the mathematical training to join the new schools, and those engaged in evolutionary theory or systems analysis avoid the field (and sometimes the laboratory, too) out of ignorance of nature, or laziness.

There is nothing backward or intellectually unstimulating about gathering data. In fact it is undoubtedly the most demanding of ecological work. On the other hand there is nothing nebulous (except in the case of sloppy work) about theory in general and the evolutionary approach in particular. Both theoretician and descriptive ecologist should remember that the final arbiters of a theory's value are descriptive data.

Many people contributed to the preparation of this book. In particular I wish to thank James Crow and John T. Emlen of the University of Wisconsin; Robert Sokal and George Williams at the State University of New York at Stony Brook; Gordon Orians and Robert Paine, University of Washington; Stephen Emlen, Cornell University; John Vandermeer, University of Michigan, and Henry Horn, Princeton University. Others contributing significantly to the ideas, information and writing of the book are Gay Reuhl Emlen, Douglas Futuyma, Joel Sohn, and Charles Taylor.

*Bloomington, Indiana*                                                    J. M. E.
*August* 1972

# Introduction

Several ecology texts have been written, but few have attempted to approach ecology from an evolutionary viewpoint except at the most elementary level. Many students scarcely know that an evolutionary–theoretical framework for ecology exists, and many others are predilected by training to view such a framework with suspicion. In addition there are many people in such fields as mathematics, physics, or engineering who may be naturally drawn to evolutionary ecology because of its mathematical overtones, but who will never join the field because of the need to wade through the plethora of elementary background definitions and facts. It is my contention that much of this background material is, in fact, unnecessary to an understanding of ecological principles and that the serious student will be more likely to delve into ecology after he has been shown a logical framework on which to hang his ideas.

This book, then, is designed for advanced students, both undergraduate and graduate, in the sciences. Section I assumes a basic knowledge of Mendelian genetics but, except for this one necessary retreat, is designed to be read by anyone with an elementary college background in mathematics. Those with no knowledge of biology may, of course, have difficulty in some places, while some of the passages included to alleviate these difficulties will strike trained ecologists as somewhat trivial. I have tried to hold such "inconsistency" in level to a minimum, but the book's goals require a degree of compromise in this respect.

Ecology may be thought of as the study of the relationships between organisms and their environment. The environment is considered in its broadest form as a collection of physical and biotic factors in the form of weather, terrain, prey, predators, competitors, etc., and the social structures within which organisms make their homes. To build an evolutionary–theoretical framework for ecology, one must start with the axioms of evolutionary theory, and, from these axioms, form hypotheses about the nature of individual plants and animals. These hypotheses, along with descriptive formulas, may then be applied to the investigation of populations, and, finally, to ecological communities.

A thorough understanding of the basic axioms of evolution—principally the laws of natural selection—is absolutely essential to such an approach. Some of the dangers of naively applying the concept of natural selection are obvious. We presume an existing trait to be advantageous because it has evolved and then, in retrospect, explain its evolution on the basis of some imagined advantage.

The imagined advantage (and "advantage" must be precisely defined) is often unmeasurable, rendering the above explanations circular.

There is another need to understand the axioms. It is easy to fall into a trap set by the following statement of evolutionary law: "natural selection acts to increase fitness." As will be pointed out later, this statement must be strongly qualified. Furthermore, it does not follow from the statement that natural selection maximizes fitness, and the relation between "fitness," "adaptedness," and "evolutionary advantage" is not at all clear. Many authors have failed to note these facts, often with unfortunate results. To avoid such pitfalls of ignorance, this book presents the axioms of evolution in some detail in Section I. Following the hierarchy of levels in the building of a theoretical framework, Section II deals with individual adaptations, ecology of the individual. Section III covers the ecology of populations, and Section IV discusses the ecology of communities.

# Contents

SECTION I  MECHANISMS OF EVOLUTION

**Chapter 1  Population Genetics**

**Chapter 2  The Genetics of Natural Selection**

**Chapter 3  Special Considerations in Evolutionary Theory**

**Chapter 4  Selection in Heterogeneous Environments**

# Section I
# Mechanisms of
# Evolution

# 1
# Population Genetics

Ecology is in large part the study of the adaptation of organisms to their environment. It is therefore hardly surprising that the theory of natural selection should play an important role in the growth of ecology. It is true that an almost total ignorance of evolutionary mechanisms need not deter an ecologist from making educated guesses about the adaptive value of any given trait. Such guesses, however, may lead this ecologist into a most flagrant sort of circular reasoning: He explains the existence of an observed trait on the basis of some benefit he thinks it conveys on its bearer, but deduces its beneficial nature from the fact that it exists and has, therefore, presumably evolved. A careful consideration of mechanisms by which a given trait could have evolved greatly decreases the chances of circular reasoning and ultimately supplies the ecologist with a greater understanding of the adaptive significance of any given trait. In addition, a good understanding of evolutionary mechanisms will often enable an ecologist to infer the circumstances causing and surrounding the development of a given trait and indirectly explain or predict the existence of additional, associated traits. Obviously, if one is to deal creatively with ecological theory some knowledge of evolutionary mechanisms is indispensable.

Many organisms are haploid or polyploid, sexually or asexually reproducing. Sex and reproductive status may be determined genetically in some species, somatically in others. Some genetic traits are carried on autosomes, others on sex chromosomes. To cover all these (and other) aspects of genetics would be literally impossible in the space of four chapters. For this reason, we restrict ourselves to a consideration of diploid organisms and ignore cases of sex-linked traits. For the interested reader, arguments analogous to those given in this text, but applying to other, special cases are usually easily derived and in general can be found in genetics texts (see Crow and Kimura, 1970).

With the exception of genes located on the sex chromosomes, every gene locus in a diploid individual is represented by two genes, one deriving from the individual's mother, the other from its father. Thus, for each gene locus, every individual may be thought of as the combination of two genes drawn from a pool of parental genes, one gene representing the male contribution, one the female contribution. Random mating in the parental generation is analogous to random selection of genes from this *gene pool*. Nonrandom mating corresponds to a biased selection of genes.

1

## 1.1 THE HARDY–WEINBERG LAW

Most genes are characterized by several allelic forms. For example, the $A$, $B$, $O$ blood type locus in man is represented by at least three. A few, superficially, may be represented by only two (male pattern baldness in man, for example), but on closer examination generally turn out to be multi-allelic. It is often convenient, however, to collapse the set of all alleles into two subsets. Because assortment of alleles in meiosis is independent, the two subsets also assort independently and may be treated as simple alleles. For simplicity we shall limit ourselves to this simplified approach and shall denote the two allelic forms (subsets of alleles) for a given locus by the letters $A_1$, $A_2$. In this simple case, any diploid population can be characterized at a given locus by the three genotypes, $A_1 A_1$, $A_1 A_2$, $A_2 A_2$. The first and last genotypes are referred to as *homozygous*, the middle, containing both alleles, as *heterozygous*. The individuals carrying them are called *homozygotes* and *heterozygotes*, respectively. If the gene pool representing this locus for the population consists of $A_1$ genes with frequency $p$, and $A_2$ genes with frequency $q = 1 - p$, it is obvious that, given random mating (random drawing from the gene pool), the probability of drawing two $A_1$ genes is $p^2$, one $A_1$ and one $A_2$, in either order, $pq + qp = 2pq$, and two $A_2$ genes, $q^2$. These probabilities represent the chances of the formation of $A_1 A_1$, $A_1 A_2$, and $A_2 A_2$ individuals, respectively, and thus the frequencies of such individuals in the next or $F_1$ generation. Now, if the $F_1$ generation consists of $n$ individuals, then the number of $A_1 A_1$, $A_1 A_2$, and $A_2 A_2$ individuals is, respectively, $p^2 n$, $2pqn$, $q^2 n$, and since $A_1 A_1$ individuals carry two $A_1$ genes, $A_1 A_2$ individuals carry one $A_1$ gene, and $A_2 A_2$ individuals carry none, the total number of $A_1$ genes carried by individuals of the $F_1$ generation must be

$$2(p^2 n) + 1(2pqn) + 0(q^2 n) = 2pn.$$

Similarly, the number of $A_2$ genes is $2qn$. The proportion of the $F_1$ generation gene pool consisting of $A_1$ genes is then

$$\frac{2pn}{2pn + 2qn} = p,$$

which is identical to the corresponding proportion in the parental generation. Thus gene frequencies $(p, q)$ and, consequently, genotype frequencies $(p^2, 2pq, q^2)$ do not change from generation to generation in a randomly breeding population. Furthermore, the subsequently invariant genotypic frequencies appear immediately, in the first generation following random breeding. This law of constancy in gene and genotype frequency is known as the *Hardy–Weinberg law*.

As any ecologist knows, nature is vastly complex and any model so simple as the Hardy–Weinberg law is almost certain to be a simplification of the real world. There are, in fact, beyond the assumption of random breeding stated above, several further assumptions behind the derivation of the law. The Hardy–Weinberg law describes the statistically most probable outcome of random breeding.

For example, the offspring of two parents of genotype $A_1A_1$ and $A_1A_2$ are most likely to be $A_1A_1$ and $A_1A_2$, occurring in equal numbers. The assumption that a $50 : 50$ ratio of offspring types really occurs, however, is fairly accurate only when dealing with average figures based on large numbers of matings. A second assumption implicit in the Hardy–Weinberg argument is that there is no migration between the population and others near it. A third assumption is that there is no changing of genes from one allelic form to another through mutation. Finally it is assumed that individuals of all genotypes contribute equal numbers of offspring to the next generation. In so far as different genotypes, as reflected in different phenotypes, differ in their mortality rates and fecundity, this last assumption clearly is untenable. Thus the usefulness of the Hardy–Weinberg law obviously lies not in its accuracy but in its incompleteness, for it tells us what would happen in a statistically ideal situation where there is no mutation or migration, and where survival and reproductive success are equal for all genotypes. Where real populations deviate from the Hardy–Weinberg prediction, it is fair to surmise that these ideal conditions do not exist.

## 1.2 STATISTICAL CONSIDERATIONS

### 1.2.1 Random Genetic Drift

The statistical aspects of the gene locus can be viewed in either of two ways: as a matter of sampling error in a single population, or in terms of inbreeding. We shall discuss each of these in order.

Consider three animal populations, constant in size, of 2, 20, and 200 individuals. For the sake of simplicity let the gene frequencies at a given locus be $p = q = 0.5$. If the populations are to be truly randomly breeding—i.e., if any individual is equally likely to mate with any other—we must remove the restriction of separate sexes. When this is done we see that reproduction in the three populations can be thought of as the drawing from a gene pool of, respectively, 4, 40, and 400 genes for every gene locus. In all cases the most likely (and the average, or expected) fraction of $A_1$ genes drawn is one-half. In reality, however, it may be something quite different. For example, the probability that in the first population $p$ will, in fact, turn out to be as far from the expected 0.5 as 0.75 or even farther, is

$$\binom{4}{3}\left(\frac{1}{2}\right)^4 + \binom{4}{4}\left(\frac{1}{2}\right)^4 = 0.31.*$$

In the other two populations, the corresponding probabilities are

$$\sum_{x=30}^{40}\binom{40}{x}\left(\frac{1}{2}\right)^{40} = 0.0032, \qquad \sum_{x=300}^{400}\binom{400}{x}\left(\frac{1}{2}\right)^{400} \to 0.$$

Thus, in large populations, the actual gene frequency may be very close to that

---

* See Appendix 1.

expected, but in progressively smaller populations the probability of a proportionately large error increases. The expected error is easy to calculate. If one
were to draw one gene (or gamete) at a time from the gene pool, only one of two
gene frequencies, 0 or 1, could be observed. The deviations from the mean ($p$)
are thus $|1 - p|$, which occurs whenever an $A_1$ gene is drawn (with frequency $p$),
and $|0 - p|$, with frequency $q$. The mean error per drawing then, is

$$\pm (p|1 - p| + q|0 - p|) = \pm 2pq.$$

Since the total number of genes drawn in a population of $n$ individuals is $2n$, the
mean error in gene frequency over all drawings ($\overline{\Delta p}$) is

$$\overline{\Delta p} = 2pq/2n = pq/n. \tag{1.1a}$$

Obviously, the size of the expected error decreases with population size. The
variance in error can be found also. The variance in error for one drawing is

$$(p|1 - p|^2 + q|0 - p|^2) = pq^2 + qp^2 = pq.$$

The variance in error for the whole population, which is the variance in error for
the average drawing, is thus

$$\text{Var}(\Delta p) = pq/2n. \tag{1.1b}$$

In every generation there is a chance of sampling error which, on average, changes
$p$ by $\pm pq/n$, a deviation with a variance of $pq/2n$. There is always a chance
that the error will be sufficient to completely eliminate one allele from the
population.

Needless to say, if chance alone causes a detectable drift in gene frequencies
from the expected in any given generation, the accumulation of this *genetic drift*
over several generations may be considerable. Thus small populations of one species
that have been isolated for several generations may well show differential drift
in genetic composition and thus also in phenotypic makeup. Ford (1964) points
out, as have others, including Wright (1948), that even very small amounts of
immigration and emigration or natural selection tend to mask the process of genetic
drift and may render it an impotent and therefore unimportant factor in genetic
changes in most populations.

Nevertheless, even if the principle of genetic drift is ignored as a direct
influence on genetic change, there are indirect and related effects which should
be pointed out, lest the reader become too enamored of that *meilleur des mondes
possibles*, the world of the arithmetic mean. The existence of random genetic
changes in small pocket populations of many species may effect differential
survival of local groups and thus affect the dynamics of dispersion and the exact
direction of evolution (see Chapters 3 and 4). Random genetic changes may
result in the accumulation of new genes and recombinations, and thereby
increase the probability of the spread of new genotypes. Since any local
population may be considered a statistical sample of the whole population, there
is a chance that small, local populations will differ widely in their phenotypic

makeup. Perhaps for this reason baboon troops in Africa, which number between 9 and 185 individuals (average 36 to 42, DeVore and Washburn, 1963) and are essentially discrete populations, often differ in such characteristics as hair color, tail length, and facial structure (DeVore and Washburn, 1963).

## 1.2.2 The Founder Effect

When a small portion of a total population disperses to and colonizes a new area, the gene pool of this founder group may differ considerably from the gene pool which gave it birth. The individuals representing a recently settled island species, for example, may differ from their mainland congenors. The sampling error generating these genetic differences between a parent population and a small offshoot group finding a new population has been named the *founder principle*, and is thoroughly discussed by Ford (1964).

Differences between groups of the same species which exist due to the founder effect may be subsequently exaggerated. Since the gene pools of the two groups are initially different, natural selection must operate on different raw materials with different potentialities for change. Selection, then, unless effectively opposed by migration, should result in greater differences between the groups. A laboratory verification of this prediction is given by Dobzhansky and Spassky (1962). These workers followed two series of populations of the fruit fly, *Drosophila pseudoobscura*, one derived from a single pair of flies whose offspring were all heterozygous for the genes (gene arrangements, actually) "Pike's peak" and "arrowhead," the other started with a series of crosses of several strains, all offspring, again, being heterozygous. Both series, then, began with individuals identical with respect to the above-mentioned genes, but the first involved populations much more similar with respect to other genetic material than the second. Over several generations, as expected, the variability in the frequency of Pike's peak and arrowhead between the populations rose consistently higher in the second than in the first series, showing that the direction of genetic change varies with the structure of the rest of the gene pool.

Scattered over parts of Colorado, New Mexico, and Arizona lives a tufted-eared squirrel known as Abert's squirrel (*Sciurus aberti*), and within the range of this species, but isolated from it by geological factors, is the north rim of the Grand Canyon, the Kaibab plateau. It is hypothesized that sometime in the recent past a founder group of Abert's squirrels managed somehow to reach and colonize the plateau. This hypothesis is based on the fact that squirrels very similar to *S. aberti* now inhabit the area. Since that part of the Kaibab plateau inhabited by the related squirrels closely resembles the surrounding habitat of Abert's squirrel (Ponderosa pine forest) there seems no *a priori* reason to expect natural selection to have operated differently on the founding population and the parent population.

Nevertheless, the animals comprising the two populations differ quite markedly in one aspect of their morphology, the former being entirely brown and black, and the latter similar but with a spectacular white tail. The differences are considered sufficient by some to justify categorization of the two populations as separate species. The large number of Kaibab squirrels, and the fact that these squirrels differ consistently from the parental stock, indicates that natural selection has been operating. It also shows that the differences between the two types represent more than the effects of genetic drift or the founder effect without subsequent evolution. One suspects that differential evolution has resulted from similar selective pressures operating on genetic stock that differed originally because of the founder effect.

### 1.2.3 Inbreeding

Inbreeding, or the breeding of related individuals, leads to important changes in the genetic structure of populations due to genetic drift. Consider a randomly breeding population of size $n$ and follow the fate of one replicating gene in one individual in a generation we shall define as number 0. In the next generation, if all genes replicate with equal success, the gene in question and its replicates make up a fraction $1/(2n)$ of all genes in the population. Thus the probability that one such gene will unite with another is $1/(2n)$. When such an event occurs, the resulting individual is homozygous for genes which are *identical by descent* from a single ancestral gene (Cotterman, 1940; Malecot, 1948, cited in Crow and Kimura, 1970). Note that the proportion of individuals which are identically, as opposed to nonidentically, homozygous for that gene by virtue of descent from generation 0, remains, for all subsequent generations, $1/(2n)$.

Suppose we consider the replicates of that original gene as it appears in generation 1. A proportion $1/(2n)$ of the population in all succeeding generations will be identically homozygous by virtue of descent from this generation also, and the same is true of all possible ancestral generations. An individual may be identically homozygous with respect to descent from any one of several generations. After one generation, a proportion $[1-1/(2n)]$ of the population of homozygotes is still not identically homozygous, but after the next generation there is a $1/(2n)$ chance that individuals may be identically homozygous by descent from that generation, so that the proportion of the population still untouched is

$$\left(1 - \frac{1}{2n}\right)\left(1 - \frac{1}{2n}\right) = \left(1 - \frac{1}{2n}\right)^2.$$

After $t$ generations, the corresponding expression is $[1 - 1/(2n)]^t$. One minus this value, or the fraction of homozygotes that are identically homozygous, is called the *inbreeding coefficient*, $F$:

$$F(t) = 1 - [1 - (1/2n)]^t. \tag{1.2}$$

In species with separate sexes, the coupling of two identical genes cannot occur until the second generation (since an individual cannot breed with itself) so that one generation is lost and the expression becomes $1 - [1 - 1/(2n)]^{t-1}$.

We now digress a moment and reconsider the variance of the sampling error in breeding populations $(\mathrm{Var}\,(\Delta p) = pq/2n)$. We may picture the calculation of this variance by examining a parent population with gene frequencies $p_{(0)}$ and $q_{(0)}$, and the set of all possible offspring populations (Fig. 1.1a). The variance of $\Delta p$ is thus $\overline{(p_{(1)} - p_{(0)})^2}$. But since the average value of $p_{(1)}$ must be $p_{(0)}$, we can write:

$$\mathrm{Var}\,(\Delta p) = \overline{(p_{(1)} - \overline{p_{(1)}})^2} = \mathrm{Var}\,(p),$$

where $\mathrm{Var}\,(p)$ is the variance of $p$ in the offspring generation.

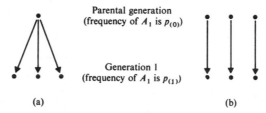

(a)                                                    (b)

**Figure 1.1**

Now consider a very large group of individuals comprising many sub-populations of identical size and gene frequency (Fig. 1.1b). Clearly, the variance in $p$ between offspring populations here is the same as that in Fig. 1.1(a). Thus if we picture a system of subpopulations of size $n$, initially all of the same gene frequency, after one generation the variance in $p$ is $\mathrm{Var}\,(p) = \mathrm{Var}\,(\Delta p) = pq/2n$. After two generations the expected variance in gene frequency, where $p_{(2)ij}$ is the frequency of the $A_1$ allele in population $j$ in the second generation, rising from population $i$ in the first generation and $p_{(2)i} = \overline{p_{(2)ij}}$, averaged over all $j$ (see Fig. 1.2), is given by

$$E\,(\mathrm{Var}\,p_{(2)}) = E\,\overline{(p_{(2)ij} - \overline{p_{(2)ij}})^2} = E\,(\overline{p_{(2)ij}^2} - \overline{p_{(2)ij}}^2)$$

$$= E\,(\overline{p_{(2)ij}^2} - p_{(0)}^2) = E\,[(\overline{p_{(2)ij}^2} - \overline{p_{(2)i}}^2) + (\overline{p_{(2)i}}^2 - p_{(0)}^2)]$$

$$= E\,(\overline{p_{(2)ij}^2} - \overline{p_{(2)i}}^2) + E\,(\overline{p_{(1)}^2} - p_{(0)}^2)$$

$$= E\,(\mathrm{Var}\,p_{(2)i}) + \mathrm{Var}\,p_{(1)}$$

$$= \frac{\overline{p_{(1)}q_{(1)}}}{2n} + \mathrm{Var}\,p_{(1)}.$$

Similar reasoning shows that, in the general case,

$$E\,(\mathrm{Var}\,p_{(t)}) = \frac{\overline{p_{(t-1)}q_{(t-1)}}}{2n} + (\mathrm{Var}\,p_{(t-1)}). \qquad (1.3)$$

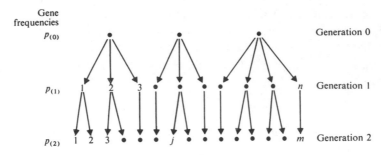

**Figure 1.2**

But

$$\text{Var } p_{(t-1)} = \overline{\left(p_{(t-1)} - \overline{p_{(t-1)}}\right)^2} = \overline{\left(p_{(t-1)} - p_{(0)}\right)^2}$$

$$= \overline{\left(p_{(t-1)}^2 - 2p_{(t-1)}p_{(0)} + p_{(0)}^2\right)}$$

$$= \overline{\left(p_{(t-1)} - p_{(t-1)}q_{(t-1)}\right)} - 2\overline{p_{(t-1)}}p_{(0)} + p_{(0)}^2$$

$$= p_{(0)} - \overline{p_{(t-1)}q_{(t-1)}} - 2p_{(0)}^2 + p_{(0)}^2$$

$$= p_{(0)} - \overline{p_{(t-1)}q_{(t-1)}} - \left(p_{(0)} - p_{(0)}q_{(0)}\right)$$

$$= p_{(0)}q_{(0)} - \overline{p_{(t-1)}q_{(t-1)}}.$$

Hence

$$\frac{\overline{p_{(t-1)}q_{(t-1)}}}{2n} = \frac{p_{(0)}q_{(0)} - \text{Var } p_{(t-1)}}{2n}. \tag{1.4}$$

Combining Eqs. (1.3) and (1.4) then gives

$$E\left(\text{Var } p_{(t)}\right) = \frac{p_{(0)}q_{(0)}}{2n} + \left(1 - \frac{1}{2n}\right)\left(\text{Var } p_{(t-1)}\right),$$

and expanding this expression, we have

$$E\left(\text{Var } p_{(t)}\right) = p_{(0)}q_{(0)}\left[1 - \left(1 - \frac{1}{2n}\right)^t\right]. \tag{1.5}$$

By comparing this with Eq. (1.2), we see that

$$E\left(\text{Var } p_{(t)}\right) = p_{(0)}q_{(0)}F_t. \tag{1.6}$$

Thus inbreeding and genetic drift are directly related.

The usefulness of Eqs. (1.5) and (1.6) can be seen by writing

$$E\left(\text{Var } p_{(t)}\right) = E\left(\overline{p_{(t)}^2} - \overline{p_{(t)}}^2\right) = E\left(\overline{p_{(t)}^2} - p_{(0)}^2\right) = p_{(0)}q_{(0)}F_t = \overline{p_{(t)}}\,\overline{q_{(t)}}F_t,$$

whence the average frequency of the $A_1A_1$ genotype over the whole population,

$\overline{p_{(t)}}^2$, is given by $\overline{p_{(t)}}^2 + \overline{p_{(t)}q_{(t)}}F_t$. Similar arguments can be made for $A_1A_2$ and $A_2A_2$. The results are shown below:

$$\overline{p_{(t)}^2} = \overline{p_{(t)}}^2 + \overline{p_{(t)}q_{(t)}}F_t,$$
$$\overline{2p_{(t)}q_{(t)}} = 2\overline{p_{(t)}q_{(t)}}(1 - F_t), \qquad (1.7)$$
$$\overline{q_{(t)}^2} = \overline{q_{(t)}}^2 + \overline{p_{(t)}q_{(t)}}F_t.$$

The meaning of the above equations is simply that where inbreeding occurs, the Hardy–Weinberg law no longer holds. There is, instead, a gradual increase in the frequency of homozygotes throughout the population. The reason for this can be appreciated in a less quantitative sense by noting that in a population made up of many small subpopulations, genetic drift or inbreeding may lead to the extinction of one or the other allele in any given subpopulation. Once the gene is gone it cannot return (we have assumed no mutations). As more and more subpopulations lose one of the alleles, it is obvious that the population as a whole must become more homozygous. No true population is divided into discrete units of equal size, or, for that matter, even discrete units, for there is always some genetic interchange between subgroups unless geographic isolation occurs. Thus the equations above may overemphasize the effects of inbreeding. However, no population is truly randomly breeding, so that the tendency for increase in homozygosity is virtually always present.

It must be stressed that we have considered the statistical aspects of gene change as if no other factors impinged on gene frequency.

## 1.3 MUTATION AND MIGRATION

Besides neglecting the statistical considerations discussed above, the Hardy–Weinberg law also assumes an absence of mutation and migration between subpopulations. Mutations, however, do occur, and may be of several types: unidirectional ($A_1$ to $A_2$), or bidirectional ($A_1 \rightleftarrows A_2$), point genetic (one locus), or chromosomal.

Chromosomal mutations, or *aberrations*, will be covered in the next chapter. For the moment we concern ourselves with point-genetic mutation and denote the probability that an $A_i$ allele mutates to the $A_j$ form in a given generation by $\mu_{ij}$. Since, in a population of $n$ individuals, there are $2np$ $A_1$ genes and $2nq$ $A_2$ genes, the expected number of genetic changes from $A_1$ to $A_2$ must be $2np\mu_{12}$ and from $A_2$ to $A_1$, $2nq\mu_{21}$. The expected net change in the number of $A_1$ genes is thus $2n(q\mu_{21} - p\mu_{12})$, and the expected change in $p$ is

$$\overline{\Delta p} = q\mu_{21} - p\mu_{12}. \qquad (1.8)$$

If no other factors influence gene frequency change, equilibrium will occur, solving the above equation, when

$$p = \frac{\mu_{21}}{\mu_{12} + \mu_{21}}.$$

It is unlikely that this equilibrium will ever be realized. Most mutation rates are on the order of $10^{-6}$, so that a value of $p = 10^{-5}$ is probably generous, and this cause of change in frequency must contest with other causes such as natural selection or genetic drift. The expected change in gene frequency due to drift was given as $pq/n$, which, for $p = 0.5$, for example, is $0.25/n$. Changes due to mutation approach this magnitude only in populations for which $0.25/n$ is roughly equal to $10^{-5}$—that is, when $n$ exceeds about 25,000 individuals.

Where migration between adjacent populations exists, we can write the following: Let $n_i$ be the number of individuals in the $i$th population, and $m_{ij}$ be the fraction of individuals in population $i$ moving to population $j$ over a specified period of time (say one generation). Then, if $p_i'$ is the expected frequency of the $A_1$ allele in the $i$th population after migration has occurred,

$$p_i' = \frac{2n_i p_i + \sum_{j \neq i}(2n_j p_j m_{ji} - 2n_i p_i m_{ij})}{2n_i + 2\sum_{j \neq i}(n_j m_{ji} - n_i m_{ij})}.$$

For two populations,

$$p_i' = \frac{2n_1 p_1 + 2n_2 p_2 m_{21} - 2n_1 p_1 m_{12}}{2n_1 + 2n_2 m_{21} - 2n_1 m_{12}} = \frac{n_1 p_1(1 - m_{12}) + n_2 p_2 m_{21}}{n_1(1 - m_{12}) + n_2 m_{21}}.$$

Thus

$$\overline{\Delta p_1} = p_1' - p_1 = \frac{n_1 p_1(1 - m_{12}) + n_2 p_2 m_{21}}{n_1(1 - m_{12}) + n_2 m_{21}} - \frac{n_1 p_1(1 - m_{12}) + n_2 p_1 m_{21}}{n_1(1 - m_{12}) + n_2 m_{21}}$$

$$= \frac{n_2 m_{21}(p_2 - p_1)}{n_1(1 - m_{12}) + n_2 m_{21}}. \tag{1.9}$$

Equilibrium occurs when $\overline{\Delta p_1} = 0$: when $p_1 = p_2$.

If adjacent populations don't possess equal gene frequencies for all loci, they are thus never in equilibrium. The above discussion only shows the averaging effect of migration. Real populations diverge for reasons of drift, different mutation rates, differential survival of different genotypes in the different populations, or differential tendency to migration among different genotypes.

## 1.4 MATING PREFERENCES

Breeding within a population is not always random and in many and perhaps even most species takes on rather definite, predictable patterns (also see Chapter 3). At least three possible situations exist.

**1.** Individuals choose as mates other individuals who possess either similar or opposite genotypes with respect to some locus. Imprinting is a phenomenon in which offspring develop psychological fixations (usually) on their parent(s), and in birds and mammals is known to affect later sexual behavior and preferences. It is entirely possible that imprinting results in a tendency for young animals to seek mates resembling their parents (Warriner *et al.*, 1963). The genetic effects of such mate selection can be seen as follows: Suppose that of the three genotypes displayed at a given locus, all individuals of one genotype mate only with others of the same genotype. The $A_1A_1 \times A_1A_1$ breedings will yield only $A_1A_1$ offspring, the $A_2A_2 \times A_2A_2$ all $A_2A_2$ offspring, and the $A_1A_2 \times A_1A_2$, one-quarter $A_1A_1$, one-quarter $A_2A_2$, and only one-half $A_1A_2$. The frequency of heterozygotes is decreased each generation.    The consequence of strict positive assortative mating is isolation of two homozygous lines and a subsequent trend toward speciation.    Usually, though, this type of mating is far from absolute and should result simply in a depressed frequency of heterozygotes.    For further discussion of this matter see Mainardi *et al.* (1965), and Kalmus and Maynard Smith (1967). The commonness of this form of *positive assortative* mating is not known, nor are the evolutionary reasons for its existence (but see Chapter 4, Section 2.2, for a discussion of the advantages of inbreeding).

2.    Mating may be frequency dependent.    That is, the frequency with which an individual of one genotype successfully mates may be dependent on the relative frequency of that genotype in the population.    Ehrman (1966, 1967, 1970) has discovered that in *Drosophila* a female may choose males possessing genotypes of low frequency.    The result of such behavior is the prevention of the loss of any genotype with respect to whatever locus is affected.

3.    The third and best-understood possibility involves the preferential choice of certain phenotypes, whether or not they resemble the individual choosing, and regardless of their frequencies.    This situation leads to differential mating success of different genotypes, and is discussed in Chapter 3 under intersexual selection.

## 1.5 DIFFERENTIAL MORTALITY AND FECUNDITY

The constant-gene-frequency law of Hardy and Weinberg rests also on an assumption that all genotypes have equal survival and reproductive rates. Abundant evidence to the contrary will be offered throughout the rest of this book; one rather simple example is given below.    Karn and Penrose (1951) have shown that, if neonatal mortality in humans is plotted against weight at birth, with weight along the abscissa, the result is a trough-shaped curve showing minimal death frequency at about eight pounds.

If individuals carrying one of two genes have higher survival and/or higher reproductive success than individuals not carrying the gene, it is clear that the frequency of the responsible gene will increase over that of its allele in subsequent generations.    This is equivalent to saying that individuals carrying the gene will become relatively more frequent.

One of the most beautiful and best-studied examples of ongoing change in gene frequency due to differential survival of genotypes is that involving the peppered moth (*Biston betularia*) in England.    In the first half of the nineteenth century only pale, speckled individuals were known.    In 1848, however, a jet black form was described near Manchester.    Subsequently, more black individuals appeared and by the middle of the twentieth century the dark form easily outnumbered the original pale form throughout most of England. The explanation for this rather abrupt change in color morph (on an evolutionary time scale) seems to lie in the increasing amount of industrial air pollution which first appeared early in the eighteen-hundreds. Prior to the Industrial Revolution the peppered moth customarily spent the day perched on the trunks of trees, where its pale-speckled wings blended with the lichens growing on the tree bark. Its coloration could be considered *cryptic*, or protective, an adaptation for escaping the attention of predators. With the appearance of industrial soot, however, the tree trunks in such areas as Manchester became blackened, the lichens died, and *Biston betularia* contrasted vividly with its background. At this point individuals carrying the genotype for the black morph acquired a selective advantage and their frequency in the total population rapidly increased.

Kettlewell (1955, 1956), on the basis of marked-recaptured individuals, estimated that in polluted areas near Birmingham, the pale morph suffered twice the mortality of the dark morph. Direct observations showed birds taking 43 of the pale, but only 15 of the dark form when the two morphs were released in equal numbers. Near Dorset, in one of the few accessible but unpolluted areas of the English countryside, Kettlewell found, as might be expected, that the trees still support clean lichens and that the dark form of the peppered moth is at a disadvantage compared with the pale form. Here birds were observed to devour 164 dark-colored and only 26 pale-colored moths when equal numbers of each were made available (Kettlewell, 1956). While there are a few areas where the pale-colored moths are still at a selective advantage, there seems to be a general darkening trend over most of the country. Genetic studies have shown the presence of at least two distinct black forms (Sheppard, 1960), each controlled by a different gene. By 1960 the two forms had spread to the point where they naturally interbred. In addition to the peppered moth, roughly 50 other English moth species were also gradually turning black (Kettlewell, 1955). For further discussion and references, see Kettlewell (1961), Clarke and Sheppard (1966), and Creed (1971).

### 1.5.1 The Laws of Natural Selection

Natural selection operates through the influence of the environment (both internal and external) on the reproductive success of individuals. All else being equal, the genotypes of any individuals carrying genes or gene combinations which increase their fecundity (number of offspring produced) increase in frequency with time. We say that selection *favors* the phenotype with higher fecundity, that this phenotype is at a *selective advantage*. Consider a population consisting of gaudy and cryptic individuals. If the latter suffer lower predation due to their coloration, more will survive, and an individual of this morph will be more likely to reproduce than a gaudy individual. Cryptic individuals then will, on average, produce a disproportionate number of offspring; they will be favored by selection.

A measure of the expected, ultimate success of a genotype, known as its *fitness*, is given by the ratio of its numbers in one generation to its numbers in the preceding generation. Clearly, three factors are involved in fitness: the frequency of mating combinations giving rise to the genotype, the subsequent fecundity of such pairs, and the probability of survival to reproductive age of the resulting offspring. The first of these three factors depends on mating preferences of different phenotypes. Preferences have already been briefly mentioned and will be discussed more thoroughly in Chapter 3.

Meantime, consider any population whose members breed only once in their lifetimes and in which no mating preferences exist. In addition, let us suppose that differences in fecundity between different genotype pairs is small, so that the frequency of genotypes among offspring approaches a Hardy–Weinberg equilibrium. Fitness, in this somewhat simplified view of the real world, is essentially a measure of survival to adulthood, and is the ratio of offspring of the two generations. If fecundity varies significantly, the picture does not change qualitatively, but the calculations presented below are made considerably more cumbersome.

Consider now a single gene locus for which we can define three genotypes, $A_1A_1$, $A_1A_2$, and $A_2A_2$. Suppose that the proportion of the $A_1$ gene in the gene pool just prior to breeding is $p$. Since differences in fecundity are assumed to be small, the frequencies of offspring genotypes at the time of zygote formation are $p^2$, $2pq$, $q^2$. If we now denote the fitnesses of the three genotypes by $W_1$, $W_2$, and $W_3$, it is clear that, at reproductive age, the ratio of genotypes in the $F_1$ generation is $p^2W_1 : 2pqW_2 : q^2W_3$. The actual numbers of animals representing each genotype at this time can be found simply by multiplying the above values by the population size. That is, the actual numbers are $np^2W_1$, $2pqnW_2$, $nq^2W_3$, where $n$ is the number of individuals in the parental generation. Since the $A_1A_1$ genotype consists of two $A_1$ genes, the $A_1A_2$ genotype of one, and the $A_2A_2$ of none, the number of $A_1$ genes present after one generation of selection must be

$$2(np^2W_1) + 1(2pqnW_2) + 0(nq^2W_3) = 2np(pW_1 + qW_2).$$

Similarly, the number of $A_2$ genes is

$$0(np^2W_1) + 1(2pqnW_2) + 2(nq^2W_3) = 2nq(pW_2 + qW_3).$$

The total number of genes is

$$2np(pW_1 + qW_2) + 2qn(pW_2 + qW_3) = 2n(p^2W_1 + 2pqW_2 + q^2W_3).$$

Thus $p'$, the new frequency of the $A_1$ gene, is, after one generation of selection,

$$p' = \frac{p(pW_1 + qW_2)}{p^2W_1 + 2pqW_2 + q^2W_3}.$$

But the denominator in the above expression is simply the average value of $W$, ($\overline{W}$), over the whole population. Thus

$$p' = \frac{p(pW_1 + qW_2)}{\overline{W}}. \tag{1.10}$$

The above results are summarized in Table 1.1.

**Table 1.1.** Gene frequency change under natural selection

| Genotype | $A_1A_1$ | $A_1A_2$ | $A_2A_2$ |
|---|---|---|---|
| Number at birth | $p^2n$ | $2pqn$ | $q^2n$ |
| Number at birth in next generation | $p^2nW_1$ | $2pqnW_2$ | $q^2nW_3$ |
| Frequency of $A_1$ gene after one generation of selection | $\dfrac{2(p^2nW_1) + 1(2pqnW_2)}{[2(p^2nW_1) + 1(2pqnW_2)] + [1(2pqnW_2) + 2(q^2nW_3)]}$ | | |
| | $= \dfrac{p(pW_1 + qW_2)}{\overline{W}}$ | | |

The expected change in frequency of the $A_1$ gene over one generation is thus

$$\overline{\Delta p} = p' - p = \frac{(p^2W_1 + pqW_2) - p\overline{W}}{\overline{W}}$$

$$= \frac{(p^2W_1 + pqW_2) - (p^3W_1 + 2p^2qW_2 + pq^2W_3)}{\overline{W}}$$

$$= \frac{p^2W_1(1 - p) + pqW_2(1 - p) - p^2qW_2 - pq^2W_3}{\overline{W}}$$

$$= \frac{pq}{\overline{W}}[p(W_1 - W_2) + q(W_2 - W_3)]. \tag{1.11}$$

If we examine the expression for average fitness, we see that its partial derivative with respect to $p$ can be given by

$$\frac{\partial\overline{W}}{\partial p} = \frac{\partial}{\partial p}(p^2W_1 + 2pqW_2 + q^2W_3) = 2[p(W_1 - W_2) + q(W_2 - W_3)].$$

When we substitute this into Eq. (1.11), we obtain the expression (after Wright, 1937)

$$\overline{\Delta p} = \frac{pq}{2\overline{W}}\frac{\partial\overline{W}}{\partial p} = \frac{pq}{2}\frac{\partial \ln \overline{W}}{\partial p} : (\ln \overline{W} = \log_e\overline{W}). \tag{1.12a}$$

Since $\overline{\Delta p}$ represents the change in gene frequency over one generation, it may be thought of as a *rate* of change in frequency of the $A_1$ gene, so that Eq. (1.12a) can also be written, where $\overline{\Delta p/\Delta t}$ is now the average rate of change per time and $T$ is the generation time,

$$\overline{\Delta p/\Delta t} = \frac{pq}{2T}\frac{\partial \ln \overline{W}}{\partial p}. \tag{1.12b}$$

Where $T$ varies with genotype, it is convenient to incorporate it into the expressions for fitness:

$$W_1{}^{1/T_1}, \quad W_2{}^{1/T_2}, \quad W_3{}^{1/T_3}.$$

Thus Eq. (1.12b) can be written

$$\overline{\Delta p/\Delta t} = \frac{pq}{2\overline{W}^{1/T}}\frac{\partial \overline{W}^{1/T}}{\partial p} = \frac{pq}{2}\frac{\partial \ln \overline{W^{1/T}}}{\partial p}.$$

Referring to Eq. (1.12b), note that when fitness is increased by an increase in $p$ ($\partial \overline{W}/\partial p$ is positive), then $\Delta p$ is also positive, and that when $\overline{W}$ is decreased by an increase in $p$ ($\partial \overline{W}/\partial p$ is negative), $\Delta p$ is negative. What this means is that natural selection will always change gene frequency in such a manner as to increase fitness. This conclusion is basic to the whole theory of natural selection, and we shall return to it repeatedly throughout this book.

### 1.5.2 The Rate of Gene Frequency Change

In Eqs. (1.11) and (1.12), we have simple expressions describing the rate at which the frequency of a gene increases in a population. (Note that these expressions do not take into consideration the effects of inbreeding, mutation, or migration.) The parameters are $\partial \ln \overline{W}/\partial p$, $pq/2$, and generation time $T$. We shall consider each of these parameters in order.

When one says that the rate at which a gene spreads is proportional to $\partial \ln \overline{W}/\partial p$, this is equivalent to saying that when large changes in fitness occur with small changes in $p$, selection operates rapidly. If we remember that selection operates on genotypes through individuals, that $\overline{W}$, and thus $\ln \overline{W}$, in reality refers to the fitness of the average genotype, and that different values of $p$ at the individual level must mean different genotypes, this is also equivalent to saying that selection operates more rapidly at a single gene locus as differences in fitness between genotypes increase. Thus a gene which conveys a 20% advantage to its bearer spreads more rapidly than one which conveys a 10% advantage.

When either $p$ or $q$ are small, $pq/2$ is small, so that selection occurs slowly. This means that the rate at which a gene increases in frequency is slow initially, increases at middle frequencies, then slows again, frequency approaching 1.0 as an asymptote.

Finally, the rate of selection depends on the generation time $T$. This dependency perhaps can best be clarified by noting that the rate of natural selection varies directly with the difference in rate at which the various genotypes contribute genes to subsequent generations. If one genotype has one-half the generation time of another—that is, if it breeds twice as often as the other—all else being equal, it will pass genes twice as fast and eventually come to dominate the population.

Before we leave Eqs. (1.10) through (1.12) and their implications, let us recall again that these equations are approximations only. The assumptions on which their derivation is based—that mating preferences do not exist and that differences in fecundity are small (as well as the assumption that breeding occurs only once in a lifetime)—are not always true. Prior to the nineteen-forties, it was generally believed that the selective advantage of one genotype over another rarely exceeded 1%. It was with this understanding that fitness values were treated as if invariant with $p$. We now know that differences in fitness may exceed 25% (Ford, 1964).

As an example, Kettlewell found the melanic (black) form of *Biston betularia* to have about a 30% advantage over the pale form in the Manchester district. Thus in some cases, $W_1$, $W_2$, and $W_3$ must be rewritten as functions of $p$ so that the equations become only rough approximations to the actual genetic events taking place in a population. However, large differences in fitness do not affect the equations if they are due strictly to differential survival (as is probably the case with *Biston betularia*). Furthermore, it seems unlikely on theoretical grounds (see Crow and Kimura, 1970, for a review discussion on the cost of selection) that such large selective advantages are more than a rare phenomenon (but see Discussion I). Thus in all but extraordinary cases Eqs. (1.10) through (1.12) should describe reality to a close approximation.

### 1.5.3 Genetic Equilibrium

We now consider the conditions for selective equilibrium, that is, the conditions under which $\Delta p$ in Eq. (1.12a) or (1.12b) is zero. From the equations it is clear that $\Delta p$ is zero when $p = 0$. This situation occurs when $W_3 > W_2 \geqslant W_1$ or $W_3 \geqslant W_2 > W_1$ so that $A_2$ is favored and $A_1$ eliminated. $\Delta p$ is also zero when $p = 1$ ($q = 0$), corresponding to the situation in which $W_1 > W_2 \geqslant W_3$, or $W_1 \geqslant W_2 > W_3$. The complete establishment of one allele at the expense of others is known as *fixation*; the gene frequency is said to be *fixed*. When $W_2$ is less than either $W_1$ or $W_3$, $\overline{W}$ reaches its maximum value when $p = 0$ or 1, so that in this case also, selection results in fixation. The only remaining case is that in which $W_2$ exceeds both $W_1$ and $W_3$ ($\overline{W}$ is maximized when $p$ lies between 0 and 1) and $p(W_1 - W_2) + q(W_2 - W_3) = 0$. Since $q = 1 - p$, the appropriate solution is given by

$$\hat{p} = \frac{W_2 - W_3}{2W_2 - W_1 - W_3}, \qquad W_2 > W_1, W_3, \qquad (1.13)$$

where $\hat{p}$ represents the equilibrium value of $p$.

In conclusion, let us say that selection leads to fixation unless the fitness of the heterozygote is greater than that of either homozygote. In this case, gene frequency reaches a stable equilibrium in which both alleles are present and the frequency of the $A_1$ gene is that given in Eq. (1.13). Superior fitness of the heterozygote is termed *heterosis*, the locus is referred to as *heterotic*, and the importance of the phenomenon can hardly be overemphasized.

There are a number of well-established cases of heterosis, the classic example being that of the sickle-cell locus in man. Here, homozygosity for one allele (*SS*) results in individuals with normal, healthy red blood cells, while homozygosity of the other allele (*ss*) results in red blood cells incapable of transporting adequate amounts of oxygen; this usually causes death to the victim prior to reproductive age.

If the story were as simple as this, it is obvious that the lethal gene would very quickly be eliminated by natural selection. The situation is not so simple, however: The heterozygote condition results in almost-normal red blood cells and, in addition, conveys on its bearer a certain degree of resistance to malaria. In temperate zones such a resistance would hardly be important, since malaria is uncommon and the heterozygote would be essentially identical to the normal homozygote in terms of effective phenotype. Here the sickle-cell gene would certainly never increase in frequency and, indeed, should disappear. People of temperate climates rarely possess it. In tropical areas, however, the heterozygote advantage is obvious, and in parts of tropical East Africa, individuals carrying the sickle-cell trait exist with a frequency of up to 40%. Allison (1955) has shown that roughly three out of four individuals homozygous for the lethal gene die before reaching reproductive maturity, and that the heterozygote advantage over the normal homozygous state is of the order of 25%.

His conclusions were reached in the following way. Examination of blood samples indicated, as mentioned, that about 0.40 of all adult individuals carried the *s* gene. Of these, it was estimated that 2.9% were homozygous for the gene. Thus, the frequency of *ss* has to be about $0.029 \times 0.40 = 0.012$, leaving 0.388 of the population heterozygous. The frequency of the sickle-cell gene, *q*, must then be

$$\frac{2(0.012) + 1(0.388)}{2} = 0.206.$$

If this is indeed the correct frequency, then, assuming random breeding, the frequency of normal homozygotes, in the absence of selection, should be $p^2 = (1 - 0.206)^2 = 0.630$. Heterozygotes should comprise a fraction $2pq = 2(0.206)(1 - 0.206) = 0.327$ of the population, the remaining $q^2 = (0.206)^2 = 0.042$ consisting of the sickle-cell individuals. The results are shown in Table 1.2. It is easy to see that only about one in four *ss* individuals survive, and that the selective advantage of the heterozygote is

$$\frac{1.19 - 0.95}{0.95} = 25\%.$$

**Table 1.2.** Gene frequencies and calculated fitness values for the sickle-cell trait

|  | Before selection (Hardy–Weinberg) | After selection (observed) | Relative fitness |
|---|---|---|---|
|  | (a) | (b) | (b/a) |
| $SS$ | 0.630 | 0.600 | 0.95 |
| $Ss$ | 0.327 | 0.388 | 1.19 |
| $ss$ | 0.042 | 0.012 | 0.29 |

## 1.6 SYNTHESIS

The effects on gene frequency of all factors so far considered are summarized in Table 1.3. The changes are taken to be over one generation.

**Table 1.3.** Review of the causes of gene frequency change

| Drift | $\overline{\Delta p} = \pm pq/n$     Var $(\Delta p) = pq/2n$ | (Eqs. 1.1a, b) |
|---|---|---|
| Mutation | $\overline{\Delta p} = q\mu_{21} - p\mu_{12}$ | (Eq. 1.8) |
| Migration | $\overline{\Delta p} = [n_2 m_{21}(p_2 - p_1)]/n_1(1 - m_{12}) + n_2 m_{21}$ | (Eq. 1.9) |
| Mate selection | May or may not affect $p$; always affects genotype frequency | |
| Survival and fecundity | $\overline{\Delta p} = \dfrac{pq}{\overline{W}}[(p(W_1 - W_2) + q(W_2 - W_3)]$ | (Eq. 1.11) |

**1.** Where only mutation and migration operate, it is clear that gene frequency will change in such a manner that $p$ approaches $\mu_{21}/(\mu_{21}+\mu_{12})$ in both populations. If the values of $\mu$ differ in the two populations, migration will never equalize gene frequencies. However, the differences will be negligible unless migration is virtually nonexistent or the difference in mutation rate unusually large.

**2.** Suppose selection acts to bring $p$ to some value $p_s$. Then if $p_s$ is not equal to $\mu_{21}/(\mu_{21}+\mu_{12})$, mutation will constantly tend to change gene frequency away from its selective equilibrium value. The resulting equilibrium may be found by writing $\overline{\Delta p}_{(\text{selection})} + \overline{\Delta p}_{(\text{mutation})} = 0$. If $p_s = 0$ or 1, mutation may act as a block to complete fixation, for there may be back-mutations to the deleterious allele. Consider, for example, a series of loci which are essentially fixed except for back-mutation. The probability that any given individual will possess a deleterious gene by virtue of back-mutation at a given locus may be very small (say $10^{-5}$). But if 10,000 loci are considered, the expected number of deleterious genes per individual is 0.1. If that 1 gene in every 10 individuals has only a small effect, there will be little change in $\overline{W}$ due to its presence. On the other hand, if its effect is lethal, there may be a significant drop in average fitness (to 90%) as a result.

**3.**  Where selection in two adjacent areas differs, migration may have a profound effect on equilibrium gene frequences.  Suppose, for example, that $n_1$ and $n_2$ are constant so that (Eq. 1.9) $n_2 m_{21} = n_1 m_{12}$.  Then at equilibrium ($W$'s refer to population 1):

$$\overline{\Delta p}_{1(\text{migration})} + \overline{\Delta p}_{1(\text{selection})} = \frac{m_{12} n_2}{n_1} (p_2 - p_1)$$

$$+ \frac{p_1 q_1}{\overline{W}} [p_1(W_1 - W_2) + q_1(W_2 - W_3)] = 0.$$

Again it is clear that a selective equilibrium will be changed due to gene flow between the two populations.  The equilibrium gene frequency in population 1 ($p_1$) depends on the rate of selection, and thus on

$$\frac{\partial \ln \overline{W}}{\partial p_1},$$

as well as the migration rates and gene frequency in the other population.

In conclusion, let us say that the factors affecting the frequency of alleles at a gene locus are several.  A rigorous simultaneous treatment of all such factors is extremely complicated.  Thus, in order to evaluate these factors, we had to examine each separately.  The resulting picture is one of selection first changing gene frequency toward some value which maximizes average fitness and then opposing genetic drift, mutation, and migration so that frequency remains as close to its optimum value as possible.  The bulk of selective gene frequency change in populations is merely the change necessary to maintain the status quo.

# 2
# The Genetics of Natural Selection

## 2.1 LINKAGE

In order to treat gene loci as we have in Chapter 1 it is necessary to make the assumption that the alleles at one locus assort, or are parceled out to gametes, independently. If one locus is in some way bound to another, then selection on one may disrupt the randomness with which the other is drawn from the gene pool. Picture a two-locus system with alleles $A_1$ and $A_2$ at one locus, $B_1$, $B_2$ at the other. Four types of gametes ($A_1B_1$, $A_1B_2$, $A_2B_1$, $A_2B_2$) are possible. We denote the respective frequencies by $r$, $s$, $t$, $u$. Now if the arguments given in Chapter 1 are to apply, these loci must assort independently. This condition is clearly met if the ratio of $A_1$ to $A_2$ alleles in a randomly drawn sample is the same whether the associated allele is $B_1$ or $B_2$. That is, assortment is independent if and only if $r/t = s/u$, or $ru - st = 0$.

Where gene loci occur on separate chromosomes this situation usually prevails, but when the loci share the same chromosome (are *linked*) this is not always the case. During meiosis, paired chromosomes do break and occasionally rejoin with the homologous partner. In such a case two linked genes may be separated. This process of *crossing over*, however, is not sufficient to account for independent behavior of the two loci unless breaks occurring in the chromosomal segment between them are very frequent. To allow for linkage we denote the value of $ru - st$ by $d$, and consider what happens to $d$ over one generation. If, in the population in question, the proportion of chromosomes that break between the $A$ and $B$ loci and cross over is $c$, then the following is true: Every time $A_1B_1$ and $A_2B_2$ chromosomes pair, $cA_1B_2$ and $cA_2B_1$ chromosomes result. Since the frequencies of $A_1B_1$ and $A_2B_2$ chromosomes are $r$, $u$, respectively, the probability with which they come together in a randomly breeding population is $ru$. Thus the expected proportion of $A_1B_1$'s and $A_2B_2$'s lost due to crossing over each generation per chromosome is $c(ru)$. On the other hand, each time a crossover occurs between $A_1B_2$ and $A_2B_1$ chromosomes (which join with probability $st$), $c(st)$ $A_1B_1$ and $A_2B_2$ chromosomes are gained. Thus, denoting the appropriate genotypic frequencies after one generation with a prime, we can write

$$r' = r - c(ru) + c(st) = r - c(ru - st) = r - cd,$$

$$u' = u - c(ru) + c(st) = u - c(ru - st) = u - cd,$$

$$s' = s + c(ru) - c(st) = s + c(ru - st) = s + cd,$$

$$t' = t + c(ru) - c(st) = t + c(ru - st) = t + cd.$$

The new value of $d$ ($d'$) is given by

$$d' = r'u' - s't' = (r - cd)(u - cd) - (s + cd)(t + cd)$$

$$= ru - rcd - ucd + c^2d^2 - st - scd - tcd - c^2d^2$$

$$= d(1 - rc - uc - sc - tc) = d[1 - c(r + u + s + t)]$$

$$= d(1 - c).$$

Thus, every generation, the value of $d$ is reduced by a fraction, $c$, and after sufficient time approaches zero. When this occurs, the criterion for independent consideration of the two loci ($ru - st = 0$) is met. We say that the loci are in *linkage equilibrium*. Note that where the two loci are far enough apart on a chromosome to allow a very high frequency of chromosome breakage between them, equilibrium may be reached very quickly ($c$ is large). Where the loci are closely linked, approach to equilibrium is very slow.

Consider now the interaction between this trend toward linkage equilibrium and natural selection. We suppose, for example, that selection pushes both loci to fixation. In this case only one chromosome type remains and there can no longer be crossovers resulting in new gene combinations (*recombinations*). When, however, a heterotic situation prevails, the selected frequency of *AB* chromosomes may differ from that expected at linkage equilibrium. The resulting balance of the opposing forces is a *linkage disequilibrium*. The problem can be quantitatively formulated as follows. We construct a mating diagram (Table 2.1) in which the

**Table 2.1.** Fitness values for a two-locus system

| Frequency of parental types | $r$ | $s$ | $t$ | $u$ |
| --- | --- | --- | --- | --- |
|  | $A_1B_1$ | $A_1B_2$ | $A_2B_1$ | $A_2B_2$ |
| $r$   $A_1B_1$ | $W_{11}$ | $W_{12}$ | $W_{13}$ | $W_{14}$ |
| $s$   $A_1B_2$ | $W_{21}$ | $W_{22}$ | $W_{23}$ | $W_{24}$ |
| $t$   $A_2B_1$ | $W_{31}$ | $W_{32}$ | $W_{33}$ | $W_{34}$ |
| $u$   $A_2B_2$ | $W_{41}$ | $W_{42}$ | $W_{43}$ | $W_{44}$ |

fitness of the various chromosome combinations are given. The frequencies of the matings in a randomly breeding population are given by the appropriate product of chromosome frequencies. Suppose we ignore maternal effects so that $W_{ij} = W_{ji}$. Then, following the same basic procedure as in Chapter 1, we can write, where $\overline{W}$ is the fitness averaged over the entire population,

$$r = \frac{2(r^2) + 1(rs) + 1(rt) + 1(ru) + 1(sr) + 1(tr) + 1(ur)}{2}$$

$$= \frac{2r^2 + 2rs + 2rt + 2ru}{2}$$

$$= \frac{2r(r + s + t + u)}{2}$$

(before selection and crossing over), and

$$r' = \frac{(r^2 W_{11} + rs W_{12} + rt W_{13} + ru W_{14}) - c(ru W_{14} - st W_{23})}{\overline{W}}$$

$$= \frac{r\overline{W}_r - c(ru W_{14} - st W_{23})}{\overline{W}}$$

(after selection and crossing over).  We can thus write

$$\Delta r = r' - r = \frac{r(\overline{W}_r - \overline{W})}{\overline{W}} - \frac{cQ}{\overline{W}}. \tag{2.1a}$$

where $Q = ru W_{14} - st W_{23}$.

By similar arguments, it is easily shown that

$$\Delta s = \frac{s(\overline{W}_s - \overline{W})}{\overline{W}} + \frac{cQ}{\overline{W}},$$

$$\Delta t = \frac{t(\overline{W}_t - \overline{W})}{\overline{W}} + \frac{cQ}{\overline{W}} \tag{2.1b}$$

$$\Delta u = \frac{u(\overline{W}_u - \overline{W})}{\overline{W}} - \frac{cQ}{\overline{W}}.$$

(See Felsenstein, 1965, and Kimura, 1965.)

Only if $r(\overline{W}_r - \overline{W})$, $s(\overline{W}_s - \overline{W})$, $t(\overline{W}_t - \overline{W})$, $u(\overline{W}_u - \overline{W})$, and $d$ $(=ru - st)$ are all simultaneously zero can both selective and linkage equilibrium occur. Clearly this occurs with fixation, but it is highly unlikely otherwise. Note, then, that we must proceed with caution when considering the effects of selection on gene frequency, for linkage equilibrium cannot always be assumed.

## 2.2 CHROMOSOMAL ABERRATIONS

We have seen that natural selection acts to modify or to maintain gene frequencies (in the face of crossing over, mutation, and migration). Selection also acts to change the frequencies of various chromosomal aberrations. For example, consider a case of linkage disequilibrium. Let us suppose that the $A_1$ gene is more fit in the presence of $B_1$ and that $A_2$ is more fit in the presence of $B_2$. Then selection acts to increase the frequencies of $A_1B_1(r)$ and $A_2B_2(u)$ over those of $A_1B_2(s)$ and $A_2B_1(t)$. Once $ru$ rises above $st$, crossing over will oppose selection. Clearly, if selection can eliminate crossing over, then it is free to bring chromosome frequencies to an equilibrium value representing maximum fitness. That is, selection will increase fitness if it can block crossovers. At least one means of accomplishing this exists. Breaks may occur in a chromosome on both sides of the segment containing the two loci, $A$ and $B$. If this segment rotates 180 degrees before reuniting with the remainder of the chromosome, an *inversion* is formed. Because of the aligning of homologous chromosomes at meiosis, the possibility of recombination in the next generation is destroyed (see Fig. 2.1). So long as heterosis with respect to gene combinations exists, crossing over prevents selection from attaining maximum fitness and the presence of inversions is favored (see Fisher, 1958). Note, however, that so long as normal and inverted chromosomes do not differ in their contributions to an individual's fitness, they should occur in equal frequencies at equilibrium. Inversions effectively block crossovers only when paired with normal homologs and the maximum frequency of such pairings occurs when normal and inverted species are equal in number.

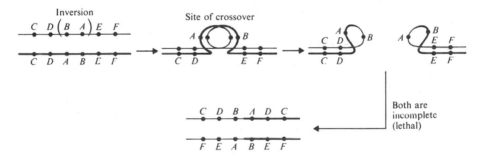

Fig. 2.1. The effect of a crossover on gamete viability.

Another means of blocking, or slowing, the rate of crossing over is to bring the loci under consideration closer together on their chromosome. As they become closer there is less chromosomal material between them and so less chance that a break will occur in that space. The deletion of chromosomal segments or their repositioning can accomplish this end and may often be selected for unless it adversely affects fitness for other reasons. Of course, if all crossing over were blocked, there would be no opportunity for mutually beneficial genes to come together unless already linked. Furthermore, it would be impossible to achieve the recombination of those genetic innovations that arise through mutation and which have the potential of changing the course of selection toward higher fitness. It would not be advantageous to eliminate crossing over completely, for clearly it is often beneficial.

Suppose that two genes, each beneficial, occur on separate chromosomes. Obviously, an individual will have a better chance of passing both genes on to its offspring if the two can become linked in some way. Linkage of these two genes would be particularly adaptive if they were interdependent in their effects. Chromosome breakage with consequent joining of the originally separate segments with the two genes (*translocation*) then may be adaptive. Those individuals inheriting the translocation will be favored by selection.

Genes which affect more than one trait are known as *pleiotropic*. Such genes, if they convey both beneficial and deleterious effects, present a dilemma to the population which cannot be resolved with chromosomal inversions or translocations. Here *modifier* or *epistatic* genes may be selected for. These are closely associated with those genes under consideration (*major genes*) in their effects but not necessarily linked with them. Such genes act to modify the actions of other genes and may be selected to diminish or even totally negate the deleterious effects of one gene while enhancing its beneficial effects. Action of an epistatic gene may be direct and specific (perhaps at the level of transcription) or indirect (the advantage or disadvantage of a trait determined at one locus may be altered by behavioral responses to the environment which are mediated, in turn, through other loci). Note that the alleles at locus *B* in the previous discussion are epistatic to the effects of the alleles at locus *A* and vice versa.

## 2.3 GENE COMPLEXES

It should be clear that in addition to changes in the frequencies of individual genes or chromosomes, equally important results of natural selection are the linking of interdependent and modifier genes, the unlinking of genes whose effects are mutually depressing on fitness, and the coevolution of genes and their modifiers. P. M. Sheppard (1960) put it this way: "Because genes can alter each others' effects, an organism, to survive, must have genes which interact appropriately to produce a balanced integrated system, and such an assemblage of genes is called a *gene complex*."

The evolution of gene complexes leads us to two very interesting considerations. First, suppose the environment of a given population changes in such a way that a certain trait ceases to have selective value, while other traits, controlled by genes in the same complex with those affecting the first, continue to be selected. A genetic change which is adaptive for one trait has only a very small probability of also being adaptive for other, linked traits, and it follows that new, selected innovations affecting the latter traits are usually detrimental to the first. In fact selection affecting the latter traits in the absence of selection for the first will generally result in the deterioration of the first. Since virtually all traits are controlled by genes in gene complexes, it follows that relaxation of selection for a trait usually leads to its degeneration. In addition, a useless trait may be an energy drain or an additional possible site of infection. For example, cave-dwelling fish which have no use for vision gradually lose their eyesight. The eyes become sunken and skin grows over the sockets. In addition, the protective and/or communicative value of skin color no longer is of value and there is a loss of pigmentation (Sadoglu, 1967).

A corollary of the genetic degeneration of unselected traits is that strong selection for a trait or set of traits may result in genetic changes so rapid that genetic compensations for coupled traits cannot keep pace. Under such circumstances, the coupled traits are bound to deteriorate (Lerner and Dempster, 1951). Mather (1953) has accumulated supporting data for this corollary by subjecting the fruit fly, *Drosophila melanogaster*, to strong artificial selection for perfect bilateral symmetry. Such selection should result in new gene frequencies and arrangements in the gene complexes involved because selection compensates for the "epigenetic violence" (Mather, 1953) done to the organisms. It follows, then, that relaxed selection for symmetry will lead to the gradual restoration of well-being with respect to other affected traits and, because of the experimentally induced changes in the gene complexes, result in even more asymmetry than that originally existing. This is exactly what Mather found.

A second consequence of the evolution of gene complexes can be examined as follows: Suppose that a given major gene is beneficial, its allele deleterious. The two homozygote genotypes by definition then have different fitness values. What of the heterozygote? Obviously any modifier genes which tend to mask the effect of the deleterious allele or enhance the effect of the advantageous allele will increase the fitness of the heterozygote. In short, selection should cause the heterozygote to resemble more and more closely the homozygote with the higher fitness. Such a mechanism has been invoked as an explanation for the evolutionary origin of the phenomenon known as *genetic dominance* (Fisher, 1928, 1958).

Much criticism has been leveled at Fisher's theory of dominance. Wright (1934) feels that the intensity of selection acting on modifiers is simply too small to bring about the evolution of dominance, and adds his own theory. Haldane (1939) expands this theory: We assume that the enzyme produced by a gene in a population exists in sufficient quantity to carry out its function properly so that any greater amount would not benefit the organism. Thus if a mutant allele which increased the amount of enzyme appeared, it would not be at a selective advantage. If the new allele triggered the production of less enzyme, however, it would be at a disadvantage. Furthermore, in the latter case, the organism would function normally if heterozygous, but if homozygous would produce less enzyme, because the new allele (the deleterious allele) would be recessive. Regardless of the mechanism involved in the evolution of dominance, pleiotropic genes are in general dominant for their beneficial effects and recessive for their deleterious effects (Sheppard, 1960). This fact has an interesting and extremely important consequence for gene complexes.

When a complex of genes becomes very closely linked so that crossing over between its members essentially ceases, the complex may be treated as a single locus and is known as a *supergene*. Since many (perhaps all) genes are pleiotropic, it should not be difficult in any complex to find one with several effects on its bearer. In fact, in a complex we are very likely to find at least two genes for which the homozygous forms $A_1A_1$ and $B_1B_1$ are beneficial with respect to some effects and deleterious with respect to others; the homozygous forms $A_2A_2$, $B_2B_2$ are less beneficial for the first and less deleterious for the latter. An animal of genotype $A_1B_1.../A_1B_1...$ will thus be superior in some respects to an individual of genotype $A_2B_2.../A_2B_2...$, but inferior in others. If dominance can be evolved, however, all genes will tend to show dominance for their respective beneficial effects. Heterozygous individuals, $A_1B_1.../A_2B_2...$, will therefore possess the beneficial traits of both homozygotes and show superiority (higher fitness) over both. Thus if pleiotropy is common, it follows that supergenes are often heterotic. (We note here that heterosis is often applied to any type of hybrid vigor. In this text we stay with the definition given in 1.5.3: Heterosis applies to loci with respect to which heterozygotes are more fit than homozygotes.) In addition, the notion of evolution of dominance can be applied to the modifier genes themselves. There is selective pressure toward the suppression by certain genes of the deleterious effects of other, linked genes. If the suppressor genes are pleiotropic and therefore heterotic, then in the event that such epistasis occurs, heterosis of the supergenes containing the suppressor genes will also occur. Heterosis involving supergenes may provide for the indefinite maintenance of several alleles at any given (super) locus and is an important mechanism in maintaining genetic diversity in populations.

This raises an important point. A gene "locus" is defined in terms of the variability in observable criteria caused by genetic material located there. To a behaviorist, a locus might be that site of all adjacent genetic material acting on maze-learning ability. To a *Drosophila* geneticist, a locus might be that point on a given chromosome which houses the genetic material affecting eye color. To a biochemist a locus might determine an enzyme species or even a polypeptide or amino acid. "Locus" may refer to virtually any continuous (or perhaps even discontinuous) strand of genetic material and what to one worker is one trait may to another be many. There is no qualitative difference between a gene and a gene complex. The dominance argument might best be summarized by the comment that heterosis should in general be more likely to occur when the locus in question contains larger amounts of genetic material.

Finally, there is no reason why the above argument for heterosis cannot be applied to pleiotropic genes (as opposed to supergenes). That is, since pleiotropic genes are generally dominant for their beneficial effects, heterosis of such genes should be quite common. A variation of this argument was first put forward by Jones (1917; see also East, 1936). It has recently been suggested (Lewontin, 1967) that heterosis may be an extremely common phenomenon, and findings of numerous isozyme polymorphisms (variety of maintained forms of proteins) in natural populations (for example, Lewontin and Hubby, 1966) support this view at least in part. It is possible, though, that much of this heterozygosity is due to alleles neutral, or nearly so, in their effects on fitness (Arnheim and Taylor, 1969; King and Jukes, 1969).

## 2.4 METRIC TRAITS

Whereas a number of traits such as eye color in *Drosophila*, "pygmy" in house mice, male pattern baldness in man, or seed color in peas are all-or-none phenomena, most traits are continuous. Height, weight, or tail length, for example, are not discrete characters, but take on a continuous range of values. Such traits are measurable and are known as metric characters. They are controlled not by a single gene locus, but many, and the genes at these loci are called *polygenes*.

To examine the role of polygenes in an organism consider the simplest of all possible cases, in which there is no linkage or epistasis and in which the effects of all of $m$ loci of either of two allelic forms, designated by subscripts 1, 2, are equal and additive. Thus, for example, if $A_1$, $B_1$, $C_1$, etc., each contribute one centimeter to tail length, and $A_2$, $B_2$, $C_2$, etc., each contribute two centimeters, an individual of genotype $A_1 B_1 \ldots / A_1 B_1 \ldots$ will have a tail of $2m$ cm and an individual of genotype $A_2 B_2 \ldots / A_2 B_2 \ldots$ will have a $4m$ cm tail. An individual with genotype $A_1 B_1 \ldots / A_2 B_2 \ldots$ displays a tail $3m$ cm long.

At the $i$th locus let the frequency of the subscript-1 gene be $p_i$. A gene pool of $n$ individuals in our example can then be thought of as occupying a series of $m$ boxes, each with two allelic forms of which the subscript-1 form is represented in the $i$th box with frequency $p_i$. The expected number of subscript-1 genes drawn is

$$np_1 + np_2 + \cdots + np_m = nm\bar{p},$$

where $\bar{p}$ is the arithmetic mean of the $p_i$'s, and the frequency of subscript-1 genes in the population is $nm\bar{p}/nm = \bar{p}$. A simple application of the binomial distribution function then gives the frequencies of individuals with $k$ subscript-1 genes:

$$\text{Frequency of individuals with } k \text{ subscript-1 genes} = \binom{2m}{k} \bar{p}^k (1 - \bar{p})^{2m-k}.$$

A distribution curve drawn from the above expression for $m = 10$ and $\bar{p} = 0.4$ is illustrated in Fig. 2.2 and the tail size $(= 1 \cdot k + 2(20 - k) = 40 - k)$ is shown, with $k$, on the abscissa.

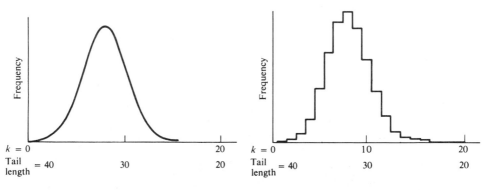

**Fig. 2.2.** An example of the variation in a metric trait with change in genotype.

It is obvious that as the number of polygenes involved in the determination of a trait increases, the distribution more and more closely approaches a continuous rather than a stepped curve. It is also obvious that environmental influences on genic expression will tend to round off the steps in a distribution, making it continuous. Thus a combination of many polygenes, and variable environmental influences produce the continuous distributions in many metric traits observed in nature.

## 2.5 QUANTITATIVE GENETICS

### 2.5.1 Basic Concepts

The effects of selection on metric characters can be visualized graphically (Fig. 2.3). The distribution of a trait is shown, the shaded area representing those extremes which are selected against. As selection eliminates one extreme, the alleles affecting the trait in that direction will gradually decrease in frequency and the distribution curve will move toward the other extreme. When selection against both extremes is the same, both allelic forms are selected against equally and the population gene frequency comes to equilibrium. Note that genes at nonheterotic loci will gradually be brought to fixation by selection.

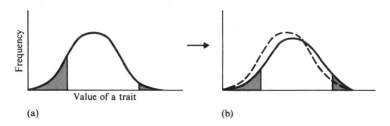

**Fig. 2.3.** The effect of selection on a metric trait.

As more and more loci become fixed, the variance in a metric trait due to genetic factors must decrease and approach zero. Those loci which are heterotic, on the other hand, will not become fixed; the genetic variance will not approach zero. Since heterotic equilibria are stable, natural selection can no longer act on them once they are attained unless fitness of the various genotypes is altered. This is true even though there is still clearly some trait variability due to genotype differences.

The trait variance due to genetic differences can be broken down into two components. If we ignore the existence of all genetic dominance so that the effects on the trait of the various alleles are additive, we can calculate what is known as the *additive genetic variance* ($V_A$). Dominance genetic variance ($V_D$) is the correction, which arises due to the existence of genetic dominance, and which, when added to the additive variance, gives the total genetic variance ($V_G$).

Now, heterosis occurs strictly because the heterozygote possesses a phenotype which is more fit than either homozygote. Clearly this cannot be if there is no dominance since, in this case, the heterozygote should be intermediate in fitness between the homozygotes. The heterotic maintenance of genetic equilibrium, then, is associated with genetic variance attributable strictly to dominance effects. The additive genetic component should, as is the case of nonheterotic genes, tend toward zero. But, as stated above, selection cannot change equilibrium gene frequencies. Thus selection is neither affecting nor being affected by $V_D$. Selection acts in response to additive genetic variance and, in the process, decreases additive genetic variance. The variability in size of a metric trait, then, decreases under the influence of natural selection until the variance in the additive genetic component reaches zero. Some variability remains as a result of dominance and environmental factors. There are special cases (Warburton, 1967) in which the above argument does not hold but these need not concern us here.

The matter of selection for metric traits can be more rigorously approached. Let us first consider one locus and denote the average values of the metric trait for individuals of genotype $A_2A_2$, $A_1A_2$, $A_1A_1$ by $x, y, z$. A good example is that of the trait "pygmy" in laboratory mice (King, 1950, 1955). Here, the normal or *wild type* individual averages 14 g in weight while the heterozygote averages 12 g and the homozygote pygmy weighs only 6 g. In this case, $x$ is 6 g, $y$ is 12 g, and $z$ is 14 g. Note that there is a certain amount of dominance (measured by $y - [(x + z)/2]$ but not complete dominance.

Suppose that $y$ (the size of the heterozygote) had been greater than either $x$ or $z$. Then we would say that there is *overdominance*. If, in this hypothetical case, large size were advantageous, then overdominance would lead to heterosis. In the present case of pygmy mice, heterosis would occur only if intermediate-sized mice ($y$) were more fit than large or small. Overdominance is equivalent to heterosis when the character being measured is fitness.

Suppose that the pygmy gene is present with a frequency of 0.2. Then the relative frequencies of the genotypes are $q^2 = 0.04$, $2pq = 0.32$, $p^2 = 0.64$, and the size $M$ of the average mouse in the population is

$$M = (0.04)(6) + (0.32)(12) + (0.64)(14) = 13.04 \text{ g}.$$

The general expression can be written

$$M = p^2z + 2pqy + q^2x. \tag{2.2}$$

With respect to the locus in question, each genotype may now be measured by the degree to which it deviates, on the average, from the population mean. The deviations are known as the *genetic values* of the genotypes (see Table 2.2). Clearly, if there is no epistasis, the total genetic value of an individual may be found by adding the genetic values of its genotypes at all loci. Thus, where $G$ is the total genetic value, and $G_i$ that at the $i$th locus,

$$G = \sum_i G_i.$$

Table 2.2

| Genotype | Size | Population mean size | Genetic value ($G$) | |
|---|---|---|---|---|
| $A_1A_1$ | $z$ | $p^2z + 2pqy + q^2x$ | $z - M = q[(1 + p)z - 2py - qx]$ | |
| $A_1A_2$ | $y$ | $p^2z + 2pqy + q^2x$ | $y - M = (p^2 + q^2)y - p^2z - q^2x$ | |
| $A_2A_2$ | $x$ | $p^2z + 2pqy + q^2x$ | $x - M = -p[pz + 2qy - (1 + q)x]$ | |

The variance in a trait due to genetic differences in a population (genetic variance $= V_G$) can be found very simply:

$$V_G = \sum_i V_G = \sum_i \overline{(G_{ij} - \bar{G}_i)^2},$$

where $G_{ij}$ is the genetic value of the $j$th genotype at the $i$th locus. But since $\bar{G}_i$, measured as a deviation from the population mean is, by definition, zero, this becomes

$$V_G = \sum_i \overline{G_{ij}^2} = \sum_i \{p_i^2[q(1 + p_i)z_i - 2p_iy_i - q_ix_i]^2$$

$$+ 2p_iq_i[-p_i^2z_i + (p_i^2 + q_i^2)y_i - q_i^2x_i]^2 + p_i^2[-p_i(p_iz_i + 2q_iy_i - (1 + q_i)x_i]^2\},$$

where $p_i$ is the frequency of the $A_1$ gene and $x_i$, $y_i$, $z_i$ are the appropriate values of the trait at the $i$th locus. After much tedium, this reduces to

$$V_G = 2p_iq_i[p_i(z_i - y_i) + q_i(y_i - x_i)]^2 + 4p_i^2q_i^2\left(y_i - \frac{x_i + z_i}{2}\right)^2. \tag{2.3}$$

The term on the left is usually written $2p_iq_i\alpha_i^2$. The term $y_i - [(x_i + z_i)/2]$ is, as mentioned above, a measure of dominance. We will now show that the left term in the equation represents the additive variance and thus that the right term is the dominance genetic variance.

Of all individuals carrying the $A_1$ gene at the $i$th locus, a proportion $p_i$ will carry another $A_1$ gene and a proportion $q_i$ will carry the $A_2$ gene. Therefore, the average genetic value of all individuals carrying the $A_1$ gene at this locus must be $p_iz_i + q_iy_i$. Thus, on average, the value of the $A_1$ gene is given by $p_iz_i + q_iy_i - M$. This is written $\alpha_{A_1i}$ and is, conceptually, the difference in genetic value of individuals carrying the $A_1$ gene at the $i$th locus and the average individual in the population as a whole.

$$\begin{aligned}
\alpha_{A_1i} &= (p_iz_i + q_iy_i) - (p_i^2z_i + 2p_iq_iy_i + q_i^2x_i) \\
&= q_i(p_i(z_i - y_i) + q_i(y_i - x_i)) = q_i\alpha_i, \\
\alpha_{A_2i} &= (p_iy_i + q_ix_i) - (p_i^2z_i + 2p_iq_iy_i + q_i^2x_i) \\
&= -p_i(p_i(z_i - y_i) + q_i(y_i - x_i)) = -p_i\alpha_i.
\end{aligned} \tag{2.4}$$

If the values of the $A_1$ and $A_2$ alleles are additive, therefore, we obtain the results listed in Table 2.3.

**Table 2.3**

| Genotype | Additive genetic value ($A$) |
|----------|------------------------------|
| $A_1A_1$ | $\alpha_{A_1} + \alpha_{A_1} = 2q\alpha$ |
| $A_1A_2$ | $\alpha_{A_1} + \alpha_{A_2} = (q - p)\alpha$ |
| $A_2A_2$ | $\alpha_{A_2} + \alpha_{A_2} = -2p\alpha$ |

$$\bar{A} = p^2(2q\alpha) + 2pq(q - p)\alpha + q^2(-2p\alpha) = 2p^2q\alpha + 2pq^2\alpha - 2p^2q\alpha - 2pq^2\alpha = 0$$

But the genetic ($G$) as opposed to the additive ($A$) genetic values are, as defined, $z - M$, $y - M$, $x - M$. Thus we have expressions for the genetic values and the additive genetic values of genotypes. The difference is the dominant genetic value, $D$:

$$D = G - A, \quad \text{or} \quad G = A + D.$$

We are now in a position to calculate the additive genetic variance. As before, let subscript $i$ denote the $i$th locus. Then, from the values in Table 2.3,

$$V_A = \sum_i \{p_i^2(2q_i\alpha_i - \bar{A}_i)^2 + 2p_iq_i[(q_i - p_i)\alpha_i - \bar{A}_i]^2 + q_i^2(-2p_i\alpha_i - \bar{A}_i)^2\}$$

$$= \sum_i \{p_i^2(2q_i\alpha_i)^2 + 2p_iq_i[(q_i - p_i)\alpha_i]^2 + q_i^2(-2p_i\alpha_i)^2\}$$

$$= \sum_i [(4p_i^2q_i^2\alpha_i^2) + (2p_iq_i^3\alpha_i^2 - 4p_i^2q_i^2\alpha_i^2 + 2p_i^3q_i\alpha_i^2) + (4p_i^2q_i^2\alpha_i^2)]$$

$$= \sum_i 2p_iq_i\alpha_i^2(2p_iq_i + q_i^2 + p_i^2)$$

$$= \sum_i 2p_iq_i\alpha_i^2.$$

We write

$$V_A = \sum_i 2p_iq_i\alpha_i^2 = \sum_i 2p_iq_i[p_i(z_i - y_i) + q_i(y_i - x_i)]^2. \tag{2.5}$$

(See Eq. 1.11.)

### 2.5.2 The Rate of Evolution

We have already noted that selection acts on the additive genetic variance, and in Eq. (2.5) we have an expression for that variance.

Suppose $M$ denotes the population mean value of some metric trait. The quantities of interest are $\Delta M$ and $\Delta V_A$ per generation. We first note that for small $\Delta p_i$, $\Delta M$ approaches

$$\sum_{i \in \mathcal{M}} \frac{\partial M}{\partial p_i} \Delta p_i,$$

where $\mathcal{M}$ is the set of all loci with genes affecting $M$, $p_i$ is the frequency of one of two alleles at the $i$th locus, and the loci in $\mathcal{M}$ are in linkage equilibrium and not mutually epistatic. From Eq. (1.12a) we know that $\Delta p_i$ is given by

$$\Delta p_i = \frac{p_i q_i}{2\overline{W}} \frac{\partial \overline{W}}{\partial p_i} \,.$$

Substituting, we find

$$\Delta M \rightarrow \sum_{i \in \mathcal{M}} \frac{p_i q_i}{2\overline{W}} \frac{\partial \overline{W}}{\partial p_i} \frac{\partial M}{\partial p_i} \,,$$

and if the genes associated with $\mathcal{M}$ are not pleiotropic (so that

$$\frac{\partial \overline{W}}{\partial p_i} = \frac{\partial \overline{W}}{\partial M} \frac{\partial M}{\partial p_i} \Big) ,$$

$$\Delta M \rightarrow \sum_{i \in \mathcal{M}} \frac{p_i q_i}{2\overline{W}} \frac{\partial \overline{W}}{\partial M} \left( \frac{\partial M}{\partial p_i} \right)^2 .$$

But from Eq. (2.2), $M$ is given by $(p_i^2 z_i + 2p_i q_i y_i + q_i^2 x_i)$ for any $i$ so that

$$\frac{\partial M}{\partial p_i} = \frac{\partial}{\partial p_i} (p_i^2 z_i + 2p_i q_i y_i + q_i^2 x_i) = 2[p_i(z_i - y_i) + q_i(y_i - x_i)].$$

This, in turn, is equal (see Eqs. 2.4) to $2\alpha_i$. Thus

$$\Delta M \rightarrow \sum_i \frac{p_i q_i (2\alpha_i)^2}{2\overline{W}} \frac{\partial \overline{W}}{\partial M} \,.$$

But $2p_i q_i \alpha_i^2$ is the additive genetic variance with respect to $M$. Thus

$$\Delta M \rightarrow \frac{V_A(M)}{\overline{W}} \frac{\partial \overline{W}}{\partial M} \,; \tag{2.6}$$

$\partial \overline{W}/\partial M$ is the amount by which a change in the value of $M$ affects the value of $\overline{W}$.

Because of the simplifying assumptions made in deriving this simple expression, it is useful only for qualitative predictions. It is clear that the evolution of a trait will be slow if there is little variability in its expression in the population or if the existing variability is largely nonadditive (i.e. is due to heterosis, epistasis, or environmental influences). In addition, evolution will be slow if changes in the value of the trait have little effect on fitness.

A special case of the above is that for which the trait in question is itself the fitness of the average individual in a population. In this case the assumptions of no epistasis or pleiotropy are not necessary, $\partial \overline{W}/\partial M$ is, of course, 1.0, and

$$\Delta \overline{W} = V_A/\overline{W}. \tag{2.7}$$

Here $V_A$ is the additive genetic variance in fitness. This is a form of the *fundamental theorem of natural selection* first postulated by Fisher (1958).

To explore the rate at which the additive genetic variance changes with selection, we first examine the special case depicted in Eq. (2.7). Let $Z_{ij} = W_{ij} - \overline{W} = W_{ij} - 1$ when $\overline{W} = 1$, where $W_{ij}$ is the fitness of the $j$th genotype at the $i$th locus. Then, if $f_{ij}$ is the frequency with which the $j$th genotype appears at the $i$th locus, we can write

$$V_G = \Sigma f_{ij}(W_{ij} - \overline{W})^2 = \Sigma f_{ij}Z_{ij}^2,$$

where the sum is over all $i$ and $j$. The additive genetic variance is $V_G - V_D$. But after one generation, the new value of $f_{ij}$ is $f_{ij}' = f_{ij}W_{ij} = f_{ij}(Z_{ij} + 1)$. So

$$V_G' = \Sigma f_{ij}(Z_{ij} + 1)Z_{ij}^2 - [\Sigma f_{ij}(Z_{ij} + 1)Z_{ij}]^2$$

$$= \Sigma f_{ij}Z_{ij}^3 + \Sigma f_{ij}Z_{ij}^2 - (\Sigma f_{ij}Z_{ij}^2 + \Sigma f_{ij}Z_{ij})^2.$$

But from the above, $\Sigma f_{ij}Z_{ij}^2 = V_G$, while $\Sigma f_{ij}Z_{ij} = \overline{Z} = 0$. Thus

$$V_G' = \Sigma f_{ij}Z_{ij}^3 + V_G - V_G^2, \quad \text{and} \quad V_A' = \Sigma f_{ij}Z_{ij}^3 + V_G - V_G^2 - V_D'.$$

If the change in gene frequency is not very great so that $p_iq_i$ for all $i$ changes very little, then the dominance correction will be nearly the same for both generations (see Eq. 2.3) so that

$$\Delta V_A = V_A' - V_A \rightarrow \Sigma f_{ij}Z_{ij}^3 - V_G^2. \tag{2.8}$$

The quantity $\Sigma f_{ij}Z_{ij}^3$ is the third moment of fitness about the mean and is the measure of skewness in the distribution of genotype fitnesses in a population. When this distribution is most negatively skewed, selection will most strongly reduce the additive genetic variance (Warburton, 1967). The reader is referred to Kimura (1958) for a derivation of the equivalent of Eq. (2.8) for the continuous case.

With respect to the diminishing of the additive genetic variance in some trait of value $M$, we can write (for very small $\Delta M$)

$$\Delta \overline{W} \rightarrow \left(\frac{\partial \overline{W}}{\partial M}\right) \Delta M, \quad \text{so} \quad \frac{\Delta \overline{W}}{\Delta M} \rightarrow \frac{\partial \overline{W}}{\partial M}.$$

But (Eqs. 2.6, 2.7)

$$V_A(M) \rightarrow \frac{\Delta M \overline{W}}{\partial \overline{W}/\partial M} = \frac{\Delta M V_A}{(\partial \overline{W}/\partial M)\, \Delta \overline{W}} = V_A \frac{1}{(\partial \overline{W}/\partial M)(\Delta \overline{W}/\Delta M)}$$

$$\rightarrow \frac{V_A}{(\partial \overline{W}/\partial M)^2}.$$

Thus so long as $\partial \overline{W}/\partial M$ does not change significantly over one generation, we can write

$$\Delta V_A(M) \approx \frac{\Delta V_A}{(\partial \overline{W}/\partial M)^2}, \quad \text{or} \quad \Delta V_A \approx \Delta V_A(M)\left(\frac{\partial \overline{W}}{\partial M}\right)^2.$$

Obviously, the same arguments with respect to skewness and size of $V_A$, $V_D$ apply here as for the case of fitness variance. In addition, $V_A(M)$ is decreased more rapidly when $(\partial \overline{W}/\partial M)$ is small; that is, when a change in $M$ has only a small effect on fitness, it takes a large change in $V_A(M)$ to make a small change in $V_A(W)$.

### 2.5.3 Heritability

Selection is strong and therefore very effectively opposes back mutation, drift, and migration on traits $(M)$ for which $\partial \overline{W}/\partial M$ is large. Selection on such traits raises fitness and lowers $V_A(M)$ more than selection acting on traits of less importance. That is, variation in traits strongly correlated with fitness should be due mostly to dominance or environmental effects. This appears to be the case. For example, in cattle the proportion of variability in amount of white spotting in Friesians that is additive-genetic is 0.95, while the similar proportion for conception rate (clearly related in a more direct way to fitness) is only 0.01. In pigs the corresponding figure for thickness of back fat is 0.55, for body length, 0.50, but for litter size, 0.15. In mice, the figure is 0.60 for tail length at six weeks, 0.35 for body size at six weeks, and just 0.15 for litter size (cited in Falconer, 1960).

Since $V_A(M)$ is clearly the critical factor (along with $\partial \overline{W}/\partial M$) in determining rates of selection, it would be very valuable to be able to measure it. There are a number of ways in which this is done; one of the simpler techniques is shown below. Suppose we choose a large number of family groups and in each group measure the phenotypic value of whatever we are interested in for one parent and the average offspring. The phenotypic value of the parent is given by $A + D + E$, where the symbols stand, respectively, for additive, dominance, and environmental components. If breeding is random and the one parent is $A_1 A_1$ at some locus, then its offspring will each possess at least one $A_1$ gene, the other being drawn randomly from the gene pool. The deviation in value of these offspring from the population mean is thus $\alpha_{A_1}$ (see Table 2.3 and preceding discussion) which is one-half the additive genetic value of the parent. If the parent is heterozygous, one-half of the offspring will carry one of the alleles, the other half the other allele, and the genetic value of the two groups of offspring will be $\alpha_{A_1}$, $\alpha_{A_2}$. The average offspring will have a value $G = \frac{1}{2}(\alpha_{A_1} + \alpha_{A_2})$ which, again, is one-half the additive value of the parent. Thus, where the value of the parent is $(A_p + D_p) + E_p$, the value of its average offspring is $(\frac{1}{2}A_p) + E_o$. The subscripts $p$, $o$ stand for parent and offspring. Where there are $n$ such family pairings measured, we can write (since the values are all deviations from the mean)

Cov (phenotypes of parent, average offspring)

$$= \text{Cov}\,(p, \bar{o})$$

$$= \frac{1}{n} \sum (\tfrac{1}{2}A_p + E_o)(A_p + D_p + E_p)$$

$$= \frac{1}{n} \sum \frac{A_p^{\,2}}{2} + \frac{1}{n} \sum \frac{A_p D_p}{2} + \frac{1}{n} \sum \frac{A_p E_p}{2} + \frac{1}{n} \sum A_p E_o + \frac{1}{n} \sum D_p E_o + \frac{1}{n} \sum E_o E_p.$$

But the additive genetic and dominance genetic values are uncorrelated with each other (that is the possession of a given $A_p$ indicates nothing of the corresponding $D_p$) and the environment. Therefore, all but the first and last terms of the above expression are zero. Furthermore, if we suppose that essentially none of the environmental variance is due to mean differences between families (or design our experiment so that this is so), then there is, by definition, no correlation between $E_o$ and $E_p$ so that the last term also becomes zero. We are left with

$$\text{Cov}\,(p, \bar{o}) = \frac{1}{n}\sum \frac{A_p{}^2}{2} = \tfrac{1}{2}V_A. \tag{2.9}$$

The fraction of phenotypic variance which is due to additive genetic factors is the *heritability* (heritability $= h^2 = V_A/V$), and measures the proportion of the variability in a trait on which natural selection acts. It is clearly a measure of the extent to which differences in traits are inherited, hence the name heritability.

## 2.6 SYNTHESIS AND OVERVIEW

The adaptation of an organism to its environment involves the selective process on a very large number of loci. There are myriads of simultaneous selective pressures acting to change or maintain everything from body size to color to hormone sensitivity thresholds to nerve pathways to the subtleties of the efficient use of allotted time and energy. The difficult fact to grasp is that selection can apparently work effectively on many loci and chromosomes simultaneously. This is particularly hard to imagine when we note that unless a population is extremely large, a small selective advantage may be negated by chance coappearance of the responsible gene with deleterious genes at other loci, for, after all, fitness of a genotype at some locus is the mean value of all fitnesses of that genotype in combination with all possible genotypes at other loci. Clearly, change in gene frequencies is not necessarily a steady process toward increased fitness, with no backsteps.

Let us examine the question of gene substitution when fitness differences are very small. Suppose, for example, that the normal (wild-type) gene at a locus displays complete dominance, and let the initial frequency of the new gene be $q$. Then, after one generation, the frequency of the normal gene, $p_{(1)}$, is

$$p_{(0)} \left( \frac{p_{(0)}W_1 + q_{(0)}W_2}{\overline{W}} \right) = \frac{p_{(0)}(p_{(0)} + q_{(0)})W_1}{\overline{W}} = \frac{p_{(0)}W_1}{\overline{W}}. \quad \text{(See Eq. 1.10.)}$$

If the population is stable, $\overline{W}$ by definition is 1, so that we can write $p_{(1)} \approx p_{(0)}W_1$. After $n$ generations, $p_{(n)} \approx p_{(0)}W_1{}^n$, and the frequency of the new gene will be $1 - p_{(0)}W_1{}^n$. Suppose $W_1 = 0.99$ while $\overline{W} = 1$ and $q_{(0)} \to 0$. Then the number of generations for 90% fixation is $n$, such that $0.90 \approx 1 - (1)(0.99)^n$. This gives a value for $n$ of roughly 229 generations. Where fitness differences are larger, selection proceeds more rapidly. Thus it is clear that even for genes conveying very slight selective advantages, gene substitution may occur over geologically very short periods of time. It is only shorter-term climatological and ecological changes (along with mutation, migration, etc.) that prohibit animals from becoming really finely attuned to their environment.

For natural selection to act there must be additive genetic variance. Thus the matter of the maintenance of genetic variability becomes important. A certain amount of variability occurs due to mutation and chromosomal aberrations. Where selection acts differently on adjacent populations migration acts to maintain diversity. Even if selection pressures are similar in two populations there may be different genetic strategies followed if only for fortuitous reasons. Where dominance selection makes supergenes heterotic, the introduction of one superallele into populations selecting for others leads to the maintenance of several forms which, in the absence of migration, might not be found. Gene complexes of this sort contain much potential variability tied up because of close linkage. But the rare recombinations within loci slowly release this variability for natural selection to act on. Suppose that a population suddenly begins growing rapidly. (Dispersal into a new environment without competitors might result in such growth.) Under such circumstances the ascendancy of advantageous genes needn't occur at the expense of normally lethal genes, and the latter will be eliminated more slowly. In fact, if the lethal genes were initially in equilibrium frequency, the population growth would result in a relative weakening of the selection pressures against them and a consequent rise in their frequency. The increase in these usually rare genes now allows for their incorporation into the other genetic material in ways which may, occasionally, prove advantageous. Thus the growth phase in a fluctuating population is one of gaining genetic innovations. The following decline occurs in response to increased selection pressures and acts to weed out those innovations that are not beneficial. The result is that selected changes in traits may be more rapid in fluctuating than in stable populations (Ford, 1964).

It is important to note in this argument on population fluctuation that innovations appear and are selected against with more strength than in stable populations only with respect to whatever factors are implicated in the fluctuations. If, for example, the population rise is due to a drop in mortality, selection for fecundity is not necessarily relaxed (in fact the ability to pass genes rapidly becomes more directly dependent on fecundity and may result in stronger selection on correlated traits). When mortality rises and sends the population into decline, there is not necessarily an increase in selection pressure on fecundity or any trait not associated with mortality. Thus when we speak of faster rates of evolution in fluctuating populations we must be certain to note which traits are responsible for the control of population size.

Because not all individuals in a population possess the optimal (highest fitness) genotype, a real population will suffer more deaths and fewer births than one composed of all optimal genotypes. Where $\overline{W}_A$ and $W_A$ (max) are the additive components of, respectively, the population mean and population maximum fitnesses, and $n$ is the number of individuals in a population, $n(W_A(\max) - \overline{W}_A)$ gives the effective number of deaths implicated in the action of natural selection on that population. We call such "deaths" *selective deaths*. Those differences in mortality and reproductive failure due to factors other than genetic differences play no role in natural selection.

# 3
# Special Considerations in Evolutionary Theory

One often encounters the expression that if something is good for the species (or population) it will evolve. Since what is good for the species may, in turn, be detrimental to its individual members, the above sentiment shows admirable trust in the inherent goodness of nature toward populations. But it reflects more a romantic fallacy than clear logic. Natural selection as we have so far explored it acts on genotypes through the differential success of individuals, not species or populations, so that it is the disadvantage or advantage to individuals that determines the course of evolution. Nevertheless, in many species there appear to exist traits which, although of benefit to the population, are of apparent disadvantage to the individual possessing them (so-called *altruistic traits*). Examples include the sharing of food, warning signals which may advertise the signaler's presence to predators, and cooperative effort such as is found in social insect colonies.

Many moths promptly die after egg laying, an act hardly benefiting the deceased, but perhaps beneficial to their offspring, because fewer moths attract fewer predators. Many male insects and spiders allow themselves to be devoured by their mates after copulation. This sacrifice on the part of the male assures an extra energy supply to the female who, in turn, can utilize the energy to lay more eggs. But the act would certainly not appear to benefit the males. Female rhesus monkeys have been reported to protect half-sisters from attacks by angry males (Washburn *et al.*, 1965).

Many examples of "altruism" may in fact be selfish—an individual aids another in return for later favors (Trivers, 1971). Other examples seem to be genuinely unselfish.

## 3.1 GROUP SELECTION

In addition to the above and many other examples there are a number of traits which work to the detriment of their owners' selfish interests apparently toward the purpose of population or species survival. Picture a set of circumstances in which a population has a tendency to fluctuate wildly. Such populations, it can be argued, will eventually oscillate to a low point from which they cannot recover and hence will become extinct. Populations which persist, then, must possess some sort of homeostatic mechanism which dampens fluctuations while other populations die out. The genes responsible for their success will, by virtue of their persistence, spread until all populations are of this sort (Dunbar, 1960). A central factor is "voluntary" cutback of reproductive potential. It may be argued that, at least in some circumstances, natural selection acts on predators to render them more and more efficient in their hunting until eventually they are capable of wiping out their own food supply. As in the above case, such populations will become extinct and will be replaced by other populations with more "prudent" predators that hold down their individual genetic contribution to future generations in the interests of their own and the group's future success. Selection of this sort would operate through the group rather than the individual; it is known as *group selection*.

There is little doubt that group selection operates in nature. Whether group selection might act in opposition to natural selection and come out ahead is quite a different matter. In order to explore this possibility we divide group selection into two parts. The first deals with groups that are small and largely isolated from one another genetically, and the second deals with larger, more diffuse groups.

The first type of group selection was examined by Wright (1945, 1956a, 1956b) and later by others including Slobodkin (1953), who set forth the conditions under which it might successfully act in the interest of traits opposed by selection at the individual level. Since it involves competition between small, semi-isolated groups (demes), it is best referred to as *interdeme selection*. The mechanism by which it operates is as follows.

Consider a gene which, if sufficiently frequent, decreases mortality or increases reproductive success in the deme in which it is found. The critical frequency may be achieved initially through mutation and drift. A deme possessing such a critical frequency of the gene grows and divides into daughter groups more rapidly than other demes. In fact, if the total population over all demes remains unchanged, the daughter groups, at least some of which possess the gene in more than the critical frequency, must arise at the expense of other demes possessing the gene at lower frequencies. The result is an increase in the frequency of the gene. If local extinctions are rapid enough among those demes possessing the gene in low frequency, a trait may conceivably spread even if opposed within the groups by natural selection acting through individual survival and reproduction.

There are a number of animal species which display dispersion patterns suitable for the operation of strong interdeme selection. Anderson (1964) reports a population of house mice (*Mus musculus*) which was divided into several small, isolated units living in a barn. Hershkowitz (1962) notes that small mammals are often rather sedentary in nature and occupy small areas as "pocket" populations. It seems probable that isolated pocket populations are common and perhaps the rule among small rodents. Thus, since mortality is rather high in mice, group extinction rates may under certain circumstances also be high, and it is conceivable that mice experience interdeme selection strong enough to occasionally oppose, with success, natural selection acting through individuals. Many birds also may have population structures conducive to strong interdeme selection. It is reported that birds of several species return to the same nesting areas year after year, thus perhaps forming genetically isolated groups at the nesting season. Whether the possibility of extinction of an entire nesting area is high enough to enable group advantages to override personal disadvantages is questionable. Reproductive isolation of small groups may occur in many reptiles and amphibians which home to specific breeding sites on specific shores of specific ponds. The frog *Rana clamitans*, for example, shows breeding site fidelity and homes accurately from distances as far as 500 yards (Oldham, 1967). Fish show specific breeding site attachments also. Wright and Hasler (1967) have suggested that in the white bass, *Roccus chrysops*, in Wisconsin, breeding site attachment may result in effective genetic isolation between breeding groups. Amphibian, reptile, and fish populations are capable of rapid expansion, often suffer rather extraordinary mortality (much of which may be nonselective—see 2.6) when very young, and are thus prime candidates for strong interdeme selection.

If interdeme selection is to overshadow natural selection at the individual level, it must, by definition, proceed at a faster pace. From Chapter 2 we borrow the expression for the change in a character of size $M$ (in subpopulation $i$ of a large population of isolates) over one generation:

$$\Delta M_i = \frac{V_A(M)_i}{\overline{W}_i}\left(\frac{\partial \overline{W}}{\partial M}\right)_i.$$

Over many demes that rate of change in $M$ is thus

$$\Delta M = \overline{\frac{V_A(M)}{\overline{W}}\frac{\partial \overline{W}}{\partial M}} = \overline{V_A(M)\left(\frac{1}{\overline{W}}\right)\left(\frac{\partial \overline{W}}{\partial M}\right)},$$

and if $V_A(M)$ and $(1/\overline{W})/(\partial \overline{W}/\partial M)$ are uncorrelated,

$$\Delta M = \overline{V_A(M)}\,\overline{\left(\frac{1}{\overline{W}}\right)\left(\frac{\partial \overline{W}}{\partial M}\right)} = \overline{V_A(M)}\,B_{(\text{ind})}, \tag{3.1}$$

where $B_{(\text{ind})}$ is a measure of the importance of the trait to individuals. In the case of interdeme selection, the same expression holds except that now the appropriate measure of variance is the total genetic variance of mean values of $M$ among the demes, $V_G(\overline{M})$.

$$\Delta M = V_G(\overline{M})B_{(\text{deme})}, \tag{3.2}$$

where now $B_{(\text{deme})}$ is a measure of the importance of the trait to demes. If the generation times, $T_{(\text{ind})}$ and $T_{(\text{deme})}$, are incorporated, the average rate of change in $M$ per unit time over the whole collection of demes is

$$\frac{\Delta M}{\Delta t} = \frac{\overline{V_A(M)}B_{(\text{ind})}}{T_{(\text{ind})}} + \frac{V_G(\overline{M})B_{(\text{deme})}}{T_{(\text{deme})}} \tag{3.3}$$

(See also Crow, 1955; Crow and Kimura, 1970). But

$$\overline{V_A(M)} = \frac{1}{N}\sum_i^N V_A(M_i) \quad \text{and} \quad V_A(\overline{M}) = V\left(\frac{1}{N}\sum_i^N M_i\right) = \frac{1}{N^2}\sum_i^N V_A(M_i),$$

where $N$ is the number of demes. Thus $\overline{V_A(M)}$ will exceed $V_A(\overline{M})$, and unless $V_D(\overline{M})$ is considerable, $\overline{V_A(M)}$ will be larger than $V_G(\overline{M})$. Furthermore, demes are composed of individuals so that the rate of deme multiplication cannot be faster than the average rate of individual multiplication. That is, demes spawn new demes at intervals at least as great as those at which individuals produce new individuals ($T_{(\text{deme})} \geqslant T_{(\text{ind})}$). On the basis of these facts it is easy to see that interdeme selection is unlikely to be a significant force in relation to individual selection. Finally, only very small amounts of migration destroy the nature of a deme as a discrete unit of selection, thus diminishing still further the relative significance of group selection. Only when $B_{(\text{deme})}$ is considerably larger than $B_{(\text{ind})}$ is it likely that interdeme selection becomes more important than selection at the individual level.

A special case, and exception, occurs in apomictic species (those breeding asexually). For example, if the demes involved are actually clones (all members are genetically identical), then there is no selection at the individual level ($V_A = 0$) and no question but that interdeme selection is the dominant force. It is, in fact, no different in this case from individual selection (Haldane, 1932). It is the only form of selection in such genera as *Volvox*, *Physalia* (Siphonophora) except when individual differences arise within the colony due to genetic mutation.

The second form of group selection—involving larger groups—suffers all the difficulties besetting interdeme selection but to an even greater degree. As the population size of the subgroups acting as units of selection increases, $T_{(\text{deme})}$ becomes increasingly greater than $T_{(\text{ind})}$. Simultaneously, the difference between $\overline{V_A(M)}$ and $V_A(\overline{M})$ increases as well. The requirements for the successful opposition of individual selection by group selection soon become incredibly stringent.

In spite of the difficulties mentioned above, group selection in larger groups has a number of strong advocates. In particular, Wynne-Edwards (1962) has invoked group selection to explain a large number of phenomena. He reasons that a population that grows too large will run out of food or some other resource, an event obviously detrimental to its chances for survival. Thus individuals have acquired, through group selection, *epideictic behavior* which serves to pass information allowing other individuals to assess the size of their population, and tendencies to cut back voluntarily in their reproductive activities (and associated behavior) when the assessed population size becomes too large. The "prudent predator" is an example of an animal which stays its own efforts in reproduction (a natural result of eating less) to avoid increasing its population and thus overeating its prey. Territoriality, which limits reproduction by excluding individuals unable to acquire breeding areas, is also cited as an example of population control evolved through group selection, an example of epideictic behavior.

Most and probably all phenomena discussed by Wynne-Edwards can also be explained through natural or kin selection (see below). However, it is easy and often very dangerously misleading to argue after the fact; it is possible to find some imagined selective advantage for almost any trait. This, then, is not a valid reason for rejecting Wynne-Edwards' approach in all cases, as some have done. It is reasonable to conclude that interdeme selection seldom (and large-group selection virtually never) results in the evolution of altruistic traits. But it is probably correct to say also that interdeme selection, and to a smaller extent large-group selection, will enhance or oppose selection at the individual level and encourage or discourage traits selectively neutral or nearly neutral at the individual level.

## 3.2  KIN SELECTION

Selection acts on gene frequencies. Thus if the death of a carrier of a given gene results in the survival of enough other individuals carrying the gene who would otherwise die, there will be an increase in the number of that gene in the population. A trait which is detrimental to its owner may still be selected for if it is sufficiently beneficial to individuals associated with the owner. Such traits are called *altruistic*, and one mode of selection responsible for their existence is known as *kin selection* (Haldane, 1955; Fisher, 1958; Hamilton, 1963; 1964a and b; Maynard Smith, 1964). Let us examine kin selection closely.

We can picture a population as being, at any instant in time, subdivided into a number of nonoverlapping subgroups of $n + 1$ interacting individuals. Suppose that in some of these groups at least one member is an altruist acting in response to a given situation. We let the ratio of altruism-producing (altruist) genes carried by the altruist to those carried by each of the $n$ recipients of the altruism, averaged over one generation, be $1/\alpha$, and denote by $\Delta \overline{W}_A$ the average loss in fitness to altruists arising from a generation of altruistic acts. The term $\Delta \overline{W}_R$ will represent the corresponding gain to each of the recipients. There will also be a loss in fitness to all genetic lines due to experiences of their members in situations calling for altruistic behavior, but in which no altruist is forthcoming. Call this loss, averaged over a generation, $\Delta \overline{W}$. Now let $p$ be the frequency of the altruist gene and assume that groups change in composition so that mating is random over the population as a whole. Call $\theta$ the mean proportion of groups containing an altruist, and suppose that the altruist gene is recessive so that all altruists are homozygous. Then the expected number of altruist genes gained per group, over one generation, due to experiences in which an altruist is present, is

$$[2n\alpha \, \Delta \overline{W}_R - 2 \, \Delta \overline{W}_A]\theta,$$

and, since the proportion of genes which are altruist in groups lacking an altruist is $p$, the loss of these genes due to experiences in the absence of altruism is

$$[2p(n + 1) \, \Delta \overline{W}](1 - \theta).$$

The net gain is

$$2[\theta(n\alpha \, \Delta \overline{W}_R - \Delta \overline{W}_A) - (1 - \theta)p(n + 1) \, \Delta \overline{W}].$$

The net gain, per group, in both alleles is

$$2[\theta(n\alpha \, \Delta \overline{W}_R - \Delta \overline{W}_A) - (1 - \theta)p(n + 1) \, \Delta \overline{W}]$$
$$+ 2[\theta(1 - \alpha)n \, \Delta \overline{W}_R - (1 - p)(1 - \theta)(n + 1) \, \Delta \overline{W}]$$
$$= 2[\theta(n \, \Delta \overline{W}_R - \Delta \overline{W}_A) - (1 - \theta)(n + 1) \, \Delta \overline{W}].$$

The new frequency of the altruist gene, $p'$, is thus given by

$$p' = \frac{2(n+1)p + 2[\theta(n\alpha\,\Delta\overline{W}_R - \Delta\overline{W}_A) - (1-\theta)\,\Delta\overline{W}p(n+1)]}{2(n+1) + 2[\theta(n\,\Delta\overline{W}_R - \Delta\overline{W}_A) - (1-\theta)\,\Delta\overline{W}(n+1)]}$$

and

$$\Delta p = p' - p = \frac{\theta}{(n+1)\overline{W}}[n\,\Delta\overline{W}_R(\alpha - p) - \Delta\overline{W}_A(1-p)]. \tag{3.4}$$

But, where altruists are homozygous, $\alpha$ is the probability that either allele is identical by descent with one of the altruist's plus the probability that this is not so but that the allele is altruist nevertheless. Where the proportion of genes in the population identical by descent to the altruist's genes is $r$, then: $\alpha = r + (1 - r)p$. Substituting this into Eq. (3.4), we get

$$\Delta p = \frac{\theta}{(n+1)\overline{W}}\left(n\,\Delta\overline{W}_R r(1-p) - \Delta\overline{W}_A(1-p)\right) = \frac{\theta(1-p)}{(n+1)\overline{W}}\left(n\,\Delta\overline{W}_R r - \Delta\overline{W}_A\right). \tag{3.5}$$

Thus the altruist gene will spread if and only if $r > \Delta\overline{W}_A/n\,\Delta\overline{W}_R$. The value $r$ is clearly a measure of the relatedness of members of the altruist-containing group (known as the *coefficient of relationship*); hence the name kin selection. The above conditions apply when the altruist trait is recessive, the usual situation with new genes, and are slightly relaxed as dominance increases. For more information see Hamilton (1963, 1964a). We now look at some possible examples.

**1.** The value of $r$ between parent and offspring is $\frac{1}{2}$. Initially, with the first appearance of a gene for altruism, it is unlikely that both parents will be altruists so that the effects of the actions of only one need concern us. For the altruistic trait in question to spread, $r\,(=\frac{1}{2})$ must exceed $\Delta\overline{W}_A/n\,\Delta\overline{W}_R$. Suppose $n = 1$. Then if the beneficial effect on the young of some altruistic parental act increases its fitness—taking into consideration the possible consequent death of its parent—by at least twice the amount by which the act lowers the parent's fitness, on average, the altruistic trait will be selected. Where $n = 2$, the needed change in fitness ratio is 1.0.

If a parent has the proper ecology and physiology to benefit its offspring greatly without risking its own life, this condition is most likely to be met. Consequently many large species or predaceous species, which by virtue of their size or aggressiveness are capable of defending their young, do display parental care. While other considerations are also pertinent to the existence of parental care, the following examples are nonetheless suggestive. Octopi exhibit parental care while slow-moving mollusks generally don't. Many arthropods carry their eggs about with them and, in general, the larger and/or more active species such as lobsters, crayfish, and many spiders later care for the hatchlings. Smaller arthropods and less active or sessile species, such as amphipods and isopods, also carry their eggs about but generally desert them at hatching. Among snakes, guarding of young is most evolved in such groups as the boids (large and strong) and the vipers (poisonous) (Neill, 1964). If the parents are capable of escape when an indefensible situation arises, the opportunity to abandon protection of the young alleviates the deleterious side effects of altruism on the altruist. Birds, which are capable of flight (and flightless birds which have no natural predators), defend their young while turtles don't. In birds, with the exception of brood parasites, at least one parent inevitably cares for the young. This care takes the form of feeding, often active defense in the form of mobbing (see Chapter 5), and occasionally distraction displays such as the broken wing act of the killdeer, *Charadrius vociferus*. It is interesting that birds have evolved the ability to assess the benefits and dis-advantages of parental care and will desert if the nest is badly disturbed. If disturb-ance occurs early in the nesting cycle, a bird may desert and re-nest with a minimum loss in time and energy. If the disturbance occurs later, much will be lost and the disturbance necessary to force desertion should, theoretically, be greater. It appears that in fact desertions in some bird species may be less common during the nestling than during the egg-incubation stage. Young (1963) found that among 858 eggs and 444 nestling redwing blackbirds, *Agalaius phoeniceus*, 59 eggs (6.9%) and only 4 (0.9%) nestlings died due to desertion. In the yellow-headed blackbird, *Xanthocephalus xanthocephalus*, the corresponding figures were 8.2% and 0.0%.

Mammals, since the young are dependent on their mother's milk, always show maternal care, though seldom paternal care.

**2.** The evolution of hawk alarm calls in birds must, initially, have depended on the relatedness of individuals on a nesting ground (Maynard Smith, 1965). The same is true of "helpers" in birds (Skutch, 1961; Brown, 1970) who aid their neighbors in the care of young or in other ways. Whether birds nesting in the same area are closely related is not known, but the fact that many apparently return to the same general, and often specific, areas to nest year after year suggests that this is a strong possibility. It is instructive that when family units of the chaffinch, *Fringilla coelebs*, break up in the winter with the consequent drop in the value of *r* in any group of individuals, there is no longer any use of hawk alarm calls (Marler, 1956). This suggests that, in some circumstances at least, altruism may be directed toward closely but not distantly related individuals. The advantage, through kin selection, of cooperation is inevitably opposed by the advantage, through individual selection, of selfishness. Since the balance between these advantages will vary with the degree of genetic similarity (and thus strength of kin selection) between individuals, such discriminatory behavior is clearly advantageous. In the redwing blackbird, and European titmice, *Parus sp.*, the call is used throughout the winter but the winter flocks of titmice are amalgamations of family units and the red-wing is probably large enough that $\Delta \overline{W}_A$ is insignificant (Orians and Christman, 1968). In the redwing, moreover, the survival of the polygamous male is less important to its offspring than that of the female since the male does not con-tribute as much as the female in care of the young; the female's survival is vital to the young. Only the male voices the hawk alarm call in this species. In tricolor blackbirds, *Agalaius tricolor*, the colonial nesting habit squeezes many pairs together and undoubtedly raises the potential risks of an alarm call. This species has no hawk alarm call (Orians and Christman, 1968).

**3.** A particularly good example of the effects of kin selection has been described by Hamilton (1964b) for the case of many social insects. Recall that in honey bees, for example, males are haploid, the result of unfertilized eggs, while females are diploid. If we visualize a somewhat simplified system in which $A_1$ represents an altruist gene, $A_2$ its allele, and suppose that the queen bee is inseminated by only one male, then the following is true. Initially, when the altruist gene first appears in the population, it will occur either in the male or in a heterozygous form in the queen. The chance that a gene of very low frequency—which by definition it must be—occurs in homozygous form is negligible. Thus two equally likely mating schemes are possible: $A_1 \times A_2 A_2$, and $A_2 \times A_1 A_2$. If one male offspring possesses the altruist gene, one-half of his brothers must possess it. If a female possesses it, an expected $\frac{3}{4}$ of her sisters must carry it. Males, then, have a coefficient of relationship of $\frac{1}{2}$ (Crozier, 1970). For females, $r = \frac{3}{4}$, and females will be related to their daughters by a factor $r = \frac{1}{2}$ (if a queen possesses the gene, chances are one in two that it will be passed to any daughter). Given the choice of laying her own eggs or tending to the development of the queen's eggs (her sisters), then, a female should choose the latter since her relation to her sisters is closer than her relation to her own potential offspring. Thus females become workers rather than queens. The obvious question then arises: Why, eventually, do females begin to behave as queens? The answer is that as soon as the cost (in time and energy) of caring for one more sister becomes $1\frac{1}{2}$ times as great as the cost of producing one's own young, the latter behavior is favored. Selection has apparently favored a trigger mechanism which, when a hive becomes overcrowded, results in a change in worker behavior such that some of the developing eggs they tend will develop into queens.

**4.** Hamilton also discusses some of the ramifications of reproductive mode on kin selection. Species which reproduce vegetatively produce clones whose members are genetically identical. Only mutation results in genetic divergence between members. Thus the coefficient of relationship of the members is 1.0 and the condition, $r > \Delta \overline{W}_A / n \Delta \overline{W}_R$ is more likely to be met for behavioral traits in such animals than in sexually reproducing species. Such altruistic traits as cooperative division of labor are consequently more likely to evolve in the former than the latter. Thus vegetatively reproducing cells of species such as *Volvox* form colonies while sexually reproducing cells such as *Chlamydomonas*, most ciliates, and flagellates are solitary creatures. Many of the Hydrozoa, which breed asexually, form colonies (for example, *Tubularia*), the most impressive cases belonging to the Siphonophora: individual polyps each perform functions vital to the survival of the whole colony and are interdependent for their survival. Without the intercellular cooperation resulting from the action of kin selection on vegetatively reproducing cells, the evolution of the Metazoa might never have occurred.

A number of additional considerations involving kin selection deserve brief comment. First, kin selection need not operate solely on altruistic traits. In fact, it is probable that most traits benefiting an individual's relatives are also beneficial to the individual. Kin selection is most interesting with respect to altruism, however, because its action regarding self-advantageous traits cannot be distinguished from that of natural selection in the usual sense. In addition to the evolution of altruism, consider a "selfish" trait which benefits its bearer at the expense of relatives. Kin selection will overbalance natural selection for the selfish trait, resulting in its decline, if $r$ is sufficiently large. It is "fit" to be selfish with non-relatives, but with relatives, only sparingly so.

The initiation and rate of kin selection for altruistic traits is correlated with the relatedness of interacting individuals. In addition, the action of kin selection depends upon the relative ages of interacting animals. If an individual's relatives are all past their reproductive ages, the value accruing from altruism is zero, for the genes saved will never be replicated and passed on. On the other hand, if an individual is himself past reproductive age, the potential loss of genes due to his altruistic character is unimportant because they, likewise, will not be passed. In the former case altruism could never evolve, but in the latter would evolve with virtual certainty. Selfishness should evolve in the former, but not the latter situation. Altruism is most likely to evolve or to be preferentially shown toward those individuals with the greatest expected reproductive contribution in their futures, i.e. toward youngsters. Altruism is most likely to be shown by individuals with the least reproductive potential, i.e. the older individuals. We can easily verify this by observing the interactions of parents and children in man and any number of other species with parental care.

Finally, it is important to note that altruism can evolve between individuals totally unrelated ($r = 0$) when it aids individuals related to either or both. For example, altruism between mates often evolves due to its effect on their offspring. Protection of, or cooperation with, a mate may result in greater reproductive success.

## 3.3 THE RELATION OF KIN AND GROUP SELECTION

Although kin selection involves competition between discrete groups in a population it is not necessarily the same process as interdeme selection. In the latter, groups must be permanent; that is, they cannot mingle or exchange members with others. In the case of kin selection the discrete groups may be temporary units. However, inbreeding within a deme leads to relatedness of group members. Thus when isolated demes exist and are closely knit in the sense that an altruistic act by one member affects all others, then interdeme selection becomes a special case of kin selection.

A note of warning needs to be added. Both group selection and kin selection are discussed furiously at present by ecologists of differing viewpoints, and the result is a gradually emerging consensus on a subject for which there are few data. Are hawk alarm calls really altruistic? It is generally agreed that they are, or at least were so in the past, but this is not really known to be true. Similarly, most ecologists would deny that the migration of young animals from the home area of their parents is altruistic; it is believed that the young are either driven out actively or that it is selectively advantageous to them, as individuals, to leave. But is this true? The theories of group and kin selection are well enough developed for use as predictors and they are capable of generating new problems and ideas—and as such are worthwhile theoretical constructs—but only when they can be applied in specific cases. Unfortunately, it is difficult to separate them from the laws of natural selection acting on individuals except in cases of altruism and, except in the trivial cases regarding parental care or parental death, no one has ever collected the necessary data to prove the existence of a truly noncultural altruistic trait.

## 3.4 SEXUAL SELECTION

### 3.4.1 Intrasexual Selection

The role of social pressures in effecting genetic change was first recognized and supported by Charles Darwin in *The Descent of Man and Selection in Relation to Sex* (1898). In this book, Darwin notes that secondary sex characteristics often differ greatly between male and female and that these differences far exceed variation within either sex. There follows a marvelously detailed collection of examples and a discussion remarkable for its insight considering the paucity of genetic information available at the time.

A number of cases of sexual dimorphism are due to the action of natural selection acting on survivorship, and fecundity, rather than social (sexual) success. Thus, for example, the large complex antennae of male saturnid and other moths have evolved not as benefits in social encounters, but as supersensitive receptors for picking up the scent of female-produced sex attractants. Similarly, the clasping structures found commonly in fish and other animals are used as mechanical aids in sexual behavior. Both of these traits may be said to have evolved through the agency of natural selection on fecundity. On the other hand, many sexual characteristics can be explained only on the basis of some social benefit accruing to their owner. Such benefits may be of two types. Either they provide their bearer with superior competitive powers against other individuals of the same sex, thereby placing it in a better situation for indirectly attracting and holding a mate, or they endow it with increased "sex appeal." There is no shortage of examples for either case. As an example of a trait increasing competitive prowess (evolved through *intrasexual selection*), we can mention the antlers of deer. Antlers are used in contests of strength (or bluff) apparently related to mate attraction. Along with such special combat structures, males of many species have increased both in size and aggressiveness. For example, it is a general although not universal rule among both birds and mammals as well as many other groups, both vertebrate and invertebrate, that the male is larger than the female. It is a basic tenet of animal behavior that males are generally more aggressive than females (and that male hormone treatment can increase aggressiveness). Male baboons, for example, outweigh their females by about $2\frac{1}{2}$:1, and the behavior of at least the sexually active males is in striking contrast to that of the more docile females. There is little question that the same disparities in size and temperament hold in man also.

A behavior pattern often strongly related to sex is territoriality. The males of many bird species defend territories and in at least some cases this behavior is directly related to mating. In most others, the territory is indirectly a factor in mating. The female may choose that male with the largest or the "best" territory in order to find adequate food or shelter for her young. There is some evidence (see Robel, 1967, for example) of a positive correlation between territory size and mating success, but obviously size is only one aspect of territorial quality.

The selective pressure on males to successfully defend territories is nowhere better illustrated than in the Australian magpie *Gymnorhina tibicen* (Carrick, 1963). In this species a small number of males with territories make up semipermanent groups with their associated females. Their defended areas comprise the country best suited to magpie nesting and leave subordinate individuals with only marginal habitat. Outcast individuals form less permanent groups and display little breeding activity. First-year, inexperienced males do not breed but, with other subordinate males, bide their time and watch for the death of an established male to provide an opening into the inner group. Only about one-fourth of all males breed in any given year.

In the absence of territorial behavior, male assertiveness is often exemplified in the form of dominance hierarchies in which the highest-ranking males do most of the breeding. In chickens, dominant cocks fertilize more eggs than lower-ranking birds (Guhl and Warren, 1946). Baboons and rhesus monkeys are other examples, although in the latter species greater gene passage by the highest-ranking males has been questioned (Conoway and Koford, 1963).

Together with the evolution of territoriality and fighting ability must have come selective pressures for increased communicative ability. To be able to bluff one's opponent saves much time, energy, and injury. Sexual selection has thus favored traits tending to allow individuals to intimidate others more successfully. To fake a larger size or aggressiveness than one is actually endowed with is clearly one possibility, and we find many cases in which animals raise their hair (dogs, for example) or feathers to give a glorified impression of size. The remarkable chest expansion of which two teenage toughs are capable in a face-off is another example of size bluffing. Specific acts implying imminence of attack—that is the *intention movements* preceding such acts—also take on communicative significance and are selected in their own right as bluffing mechanisms. The threat postures of most birds are believed to have derived from such intention movements as are the various ornaments such as red epaulets in the redwing blackbird or crests in the stellars jay, *Cyanocitta stelleri*, which emphasize the displays. Sudden raising of hair or feathers, as well as giving the impression of size, may mimic the appearance of sudden approach. All these characteristics are meaningless without the social context and cannot be explained on the basis of natural selection acting on survivorship or fecundity.

### 3.4.2 Intersexual Selection

In addition to intrasexual selection, in which fighting ability or ease of intimidation is selected, there is another form of sexual selection in which traits are selected because the opposite sex finds them attractive. This is known as intersexual selection. To defend the existence of such a selective mechanism one must first demonstrate that mate preference exists. This is a subject for which, unfortunately, little evidence has been gathered. In fact, subjectively, many species appear to be quite unselective in their choosing of a mate. Where the subject has been specifically examined, however, mate preferences have been found. For example, preference in the laboratory has been demonstrated in such divergent types as dogs (Beach and LeBoef, 1967), rhesus monkeys (Herbert, 1968), *Mus musculus* (Mainardi *et al.*, 1965), pigeons (Warriner *et al.*, 1963), and *Drosophila* (Merrill, 1960; Ehrman, 1966, 1967, 1970). The differential mating success of individuals in such nonterritorial species as baboons and man certainly reflects preference in mate choice. It seems reasonable, therefore, to expect intersexual selection in at least a number of species and perhaps many.

There are several socially selected traits for which a specific tag of intrasexual or intersexual is too crude. For example, the musical trilling of the wings during flight of the male broad-winged hummingbird (*Selasphorus platycercus*) is certainly social in function and may be both assertive and sexually stimulating. The crests of many birds and the dewlaps and bright colors of many lizards may likewise serve both functions, as the ritual pecking of hollow logs by woodpeckers almost certainly does. Song has a social function but its purpose varies greatly. In zebra finches, *Poephila guttata*, its function appears entirely sexual (Morris, 1954), whereas in cliff swallows, *Petrochelidon pyrrhonota*, it is aggressive in function (Emlen, 1954). In the hawfinch, *Coccothraustes coccothraustes*, it appears to be totally unrelated to pair formation (Mountfort, 1956).

On the other hand there are many sexual differences which cannot easily be explained as anything but the result of intersexual selection. To mention a few: Many male beetles possess magnificent horns and projections which appear not to be used in combat. Many of these beetles seem not to compete actively with each other for mates. Increases in size of body and mandible in males as opposed to females are not necessarily found. The horns appear purely ornamental and are probably sexual attractants. Many insects (for example, most cockroach species) produce substances with aphrodisiac effects. Flies, butterflies, fiddler crabs, spiders, and many others perform often elaborate courtship displays to attract their mates, such magnificent performers as birds of paradise being perhaps the best known, popularly, and most spectacular examples. Some birds build ornamented structures, or bowers, to "impress" the female, and others such as the long-billed marsh wren (*Telmatodytes palustris*) build nests which the females "examine" before deciding on a mate (Verner, 1963a, b). Male great-tailed grackles (*Cassidix mexicanus*) have evolved huge, unwieldy tails which have been shown to be selectively disadvantageous other than in the social context and which appear not to be used in intrasexual conflicts (Selander, 1965).

The variety of sexual attractants is nothing short of fantastic, and one may legitimately ask what there may be about some particular trait that makes it attractive. Why is it that jewel fish find red coloration stimulating while ducks generally are excited by head turning and bobbing? Why do peacocks and grackles exaggerate the tail or tail coverts while chickens display combs and wattles? Why does the cuttlefish (*Sepia*) find chromatophore-induced zebra stripes sexually desirable while lycosid spiders prefer ornamented legs and lizards pushups? To a certain extent the traits which have evolved and become exaggerated through inter-sexual selection may have been randomly determined. Because of a peculiar quirk in behavior brought about genetically, pre-peahens may quite by chance have found a long tail covert, rather than some other trait, exciting. Such preference may originally have resulted from pleiotropic effects of genes selected for some other purpose. Once such a preference occurred throughout a population, however, it might have continued to evolve in a self-accelerating manner. First, males with long tail coverts would have produced more offspring, and long tail coverts would thus have increased in frequency. Second, a female who produced daughters which chose, in turn, long tail coverts in males (thereby increasing the covert length and consequent chances of her own male offsprings' mating success) would have contributed more to the ancestry of future generations than a female who produced indiscriminate daughters. Thus genes favoring female selection of males with long tail coverts would have been favored. Both selective processes would have inter-acted and led to exaggeration in tail covert length. Note that selection in this case is acting on one generation through its effects on the next—kin selection. In general, as in the case of the great-tailed grackle, such self-accelerated selection will proceed until balanced by natural selection acting on increased male survivorship.

More tangible and possibly more important factors also appear to determine the direction of intersexual selection. In species that are dimorphic due to intra-sexual selection, those traits so selected may be indications of genetic superiority. For example, superior size and aggressiveness may be indicative of a superior ability to protect the female (for example, in some cervids, bovids, man) or to provide for her and her young (man), as well as a superior ability to intimidate other males. In species where interference from other individuals of the same species is a possibility (colonial and many territorial birds) the advantages of possessing a mate with good fighting and intimidating ability are obvious. In general, then, we might expect females to choose males who most graphically display their maleness. Any sign of social superiority and any emphasis of sexual differentiation arising from intrasexual selection in the male is likely to be seized upon by the female and thus selected independently of its original importance. Of course, all characters selected with respect to intimidating ability do not necessarily become secondary sexual stimulants. It appears, for example, that the epaulet of the redwing blackbird is used solely to display dominance in a male's territory. The female subsequently chooses a territory, not a male directly (G. H. Orians, personal communication; D. G. Smith, in press). Data gathered by Smith show that of males whose epaulets had been blackened with dye at the time of pair formation, 17 of 47 retained territories, while of control birds, handled but not dyed, 37 of 40 (with four cases in question) retained territories. The difference is significant at the 0.01 level

$(\chi^2 = 7.103)$. Noble and Bradley (1933) believe that the erection of crests and dewlaps, as well as the ubiquitous pushups observed in lizards, have no intersexual function at all. Nevertheless, it is hardly surprising that many traits appear to have both intrasexual and intersexual value.

Another determinant of mate preference may be waste. A male which can expend the time and energy and afford the risks involved in bright colors, complex, often unwieldy structures, and elaborate displays, thereby advertises his physical prowess to the female. As a symbol of superiority the development of any wasteful or potentially dangerous characteristic may have selective value, but only if the female recognizes the correlation between extravagance and its permissibility and only so long as the male avoids overextension. Once the selective process has begun, however, it may self-accelerate to the point where overextension is common. The bright colors of some lizards which display in open areas make them very vulnerable to predation, and one may picture lesser suitors, colors subdued, waiting in cover for their own chance to display as the supply of males rapidly turns over.

Anyone who has ever watched a peacock displaying and noted the apparent, almost painful intensity of purpose cannot fail to be impressed by the usual blasé reaction of the female. Clearly, such a state of affairs is not conducive to mating and any action designed to startle the female or catch her attention should be advantageous. Whether secondary sex characters ever originate as attention-getting devices is not known but such a consideration undoubtedly plays a role in their later evolution. The shaking of the tail coverts by the peacock would seem to have such distractive value. It also may be that distraction is the prime selective force behind the brilliant colors found in so many male insects, fish, amphibians, and birds.

Certainly displays cater to that sensory mode which can best be counted on to pick up the sexual message. Species with strongly developed olfactory senses will most strongly advertise their sexual intentions by means of odors. Visual and auditory animals such as birds rely on visual display when in open areas and song when in areas of dense cover. This is strikingly demonstrated by hummingbirds which display vocally in the latter but exhibit elaborate iridescent flight courtship in open habitats (Sibley, 1957).

Sexual selection acts to modify characteristics in other ways than selection for survival and fecundity. Since the latter acts to increase fitness in the nonsocial context, it follows that sexual selection must nearly always result in lower fitness values respecting all considerations other than breeding success and its associated social phenomena. Thus boat-tailed grackles display a balanced fledgling sex ratio but an adult ratio favoring females 2.42 to 1.0 (Selander, 1965). The advantages of social success have clearly taken a toll in terms of increased male mortality. In other species, bright colors may have the disadvantage of drawing predators. One way around the difficulties posed by sexually selected traits, of course, is to lose them after the breeding season. Thus the male three-spined stickleback (*Gasterosteus aculeatus*), which in the spring and summer possesses metallic blue eyes, a blue-gray back, and a red throat, fades perceptibly upon completion of his sexual activities. In North America, many wood warblers and ducks undergo a post-breeding-season molt from which they emerge dull copies of their former brilliance. It must be pointed out, though, that loss of colors does not always happen even when it would appear to be easily accomplished. For example, the redwing blackbird undergoes a post-season molt but does not lose its red epaulets. Similarly, hummingbirds are brilliant throughout the year. One may argue that the disadvantage stemming from possession of such markings is very slight or occurs only on the breeding grounds, but such arguments seem more desperate than valid. A more reasonable explanation lies in the fact that such birds as redwings and hummingbirds are brilliantly red, blue, or dark green, colors which appear drab to mammalian predators without color vision. Unless avian predation (color vision is universal in raptors) occurs there seems no immediate need, with respect to predation, to become drab (D. G. Smith, personal communication).

Hamilton and Barth (1962) have suggested an alternative explanation for the loss of bright coloration in the winter. They suggest that intrasexually evolved characteristics have aggressive overtones and that the advantages of strifeless winter living have selected against the retention of such traits. In general terms, certain traits are lost during much of the year because natural selection overbalances sexual selection. The traits reappear at other times of year because at these times sexual selection is stronger than natural selection. Phenotypic plasticity has evolved allowing adaptation to both sets of circumstances (see 4.3).

There are many examples of traits which may be lost and then regained, such as antlers of deer, ischial swelling and coloration in monkeys, and tail feathers in wydahs. Also many cases are found in which sexual characteristics are not developed until individuals are old enough to have a reasonably good chance of mating success. The selective question posed is: Does this individual have a good enough chance of success to make the time and energy expenditure of courtship and the risks inherent in sexual markings worthwhile? If not, the individual may retain juvenile characteristics into its adult life. The males of many bird species such as gulls retain juvenile coloring until their second year, although physiologically they are quite capable of sexual activity in their first.

### 3.4.3 Sexual Dimorphism and Species Isolation

Sexual selection could not operate at all were it not for the fact that individuals—particularly males—compete for mates. In view of this fact, one might expect to find sexual dimorphism most pronounced in those species which, because of polygamous or promiscuous mating systems, must impress members of the other sex the most often or strenuously. The promiscuous primates such as macaques and baboons, and the promiscuous cervids (deer) are quite strikingly dimorphic in size or shape in contrast to such monogamous creatures as cats and dogs (with the exception of the lion and the uproarious exceptions of domesticated cats and dogs). It has been suggested (Skutch, 1940) that birds with long pair bonds are generally less dimorphic than species with short pair bonds. Hawks, sparrows, thrushes, and even the wood warblers show less sexual dimorphism than do the polygamous or promiscuous grackles, redwing, and yellow-headed blackbirds. It cannot be overemphasized, however, that such generalizations linking mating behavior and dimorphism are not universal. Sexual selection may result in differences much less severe or obvious to humans than striking contrasts in size or color. The polygamous long-billed marsh wren, *Telmatodytes palustris*, for example, shows no obvious morphological sexual differences although male and female behavior patterns in the nesting season are quite distinct (Verner, 1963a, b). Also, at least some birds with long pair bonds, for example the cardinal (*Richmondena cardinalis*), may be dimorphic. Furthermore, the situation described, while generally valid for birds and mammals, as well as reptiles, amphibians, and some invertebrates, emphatically does not hold true for fishes. In fact, here promiscuity is generally related to monomorphism. The situation will be discussed further in Chapter 6.

One generalization that can be made is that sperm costs considerably less to produce than yolky eggs or maternal care. In species which provide for their young's welfare by laying large, food-rich eggs or spending time and energy in their care, the female must obviously exercise some judgment in her choosing of the right mate. The male can hardly lose by spreading sperm everywhere possible, but the female can be saddled with the job of raising infertile or inferior young. Mate selection, then, falls primarily to the female. This is the reason why sexual selection, both intrasexual and intersexual, operates primarily on the male. This is less true in monogamous species where less mate advertising is needed and where the male often aids in care of the young. In such cases, generally, either both or neither of the sexes display obvious secondary sex characteristics.

The form and intensity of sexual differences owe themselves not only to the criteria discussed above but also to the proximity of related species. It is obviously disadvantageous for a female to produce eggs, and possibly spend much time and energy caring for them, if they are fathered by another species. Usually, such hybrids are sterile if they are ever conceived at all. It thus becomes advantageous for the female to choose that male which is most obviously of the proper species. Such choice is selected for and, in turn, produces sexual selective pressures on the males of the two species to diverge in appearance. Sympatric tree shrew species seem to have solved the problem by diverging in size and copulatory posture (Sorenson and Conoway, 1966). Fruit flies, *Drosophila sp.*, avoid mistaken mating through divergent courtship displays and Dobzhansky and Koller (1939) report that, as one would expect, the differences between two species are more pronounced where they overlap in range than where they are *allopatric* (nonoverlapping). Ehrman (1965) demonstrated the same phenomenon for two races of *Drosophila paulistorum*. Ducks of the genus *Anus* hybridize frequently, and an even greater frequency may be prevented by bright, divergent coloration. On the island of Hawaii the mallard, *A. platyrhynchus*, is the only member of the genus present, and here the males have lost their bright colors (Sibley, 1961). A quick glance through Peterson's *A Field Guide to the Western Birds* (1941) reveals a number of groups in which females of closely related species are virtually indistinguishable. It is instructive to examine the males of these species. Among the hummingbirds, the female Costa's (*Calypte costae*) and the female black-chinned (*Archilochus alexandri*) hummingbird are impossible to separate in the field. Similarly the female rufous (*Selasphorus rufus*) and female Allen's (*Calypte anna*) hummingbird are almost identical. There seems no *a priori* reason why the males should be different if the females are not. The ranges of the first two species, however—respectively southern California, southern Utah, southern Nevada, and Arizona; Mexico, southern California, Arizona, New Mexico, and west Texas—are strongly overlapping and the males strikingly different. The ranges of the latter two—Oregon, southwest Montana, and Canada; California and Arizona—are nonoverlapping and the males virtually identical. The females of the bullocks (*Icterus bullocki*) and hooded (*Icterus cucullatus*) oriole are extremely similar in appearance. The ranges of the two species overlap and the males are very different. The females of the California (*Carpodacus pupureus*) and Cassin's (*Carpodacus cassini*) purple finch are very similar. Their ranges overlap only in the coastal mountain range of California and here they are partly isolated by altitude, the Cassin's purple finch preferring higher areas. The males are very similar. The male of the common redwing blackbird is black with a red epaulet bordered below with yellow. In central California it overlaps with the tricolored redwing, almost identical in appearance but with a white rather than yellow border. In the area of overlap, the common redwing has lost the yellow border and thus has diverged in appearance from its relative.

## 3.5. SEXUALITY

### 3.5.1 The Evolution of Sex

The most striking effects of sexuality seem to be of negative adaptive value. Among small invertebrates, for example, the necessary time involved in sexual activities slows the reproductive process and in species which reproduce very often and rapidly this time lost may significantly alter fitness. We might expect this to be true, for example, in protozoans and rotifers, where the former generally reproduce by fission unless environmental conditions dictate sexual behavior (deterioration of the environment, aging of individuals, shortage of food), and many species of the latter reproduce by parthenogenesis (females reproduce diploid young with male fertilization). In species where rapid multiplication is important, especially during periods of dispersal, asexual reproduction is expected and found (Stalker, 1956). Among many other animals, especially vertebrates, sexual behavior requires a considerable expenditure of time and energy in such activities as courtship and territorial defense. In addition, these activities may require considerable time in open, unsheltered areas where the risk of predation is great. Finally, and perhaps most important, a sexually reproducing individual passes on only one-half of its genetic complement to each offspring.

The usual answer to the question of "why sex?" is that sexual reproduction allows for genetic recombination. In the long term, the chance for the simultaneous incorporation of new, beneficial mutations is increased (in large populations) by sexual reproduction (Muller, 1932; Crow and Kimura, 1965). A mechanism for the selection in one generation of increased fitness or variance for another far in the future, though, is rather hard to imagine. Of course, any mechanism which allows independent selection of linked genes—that is, linkage equilibrium—will be advantageous and clearly the recombination made possible through sexual reproduction will act toward this end. Maynard Smith (1968), however, following an argument by Sturtevant and Mather (1938), shows that if an asexual, haploid population displays independent assortment it may maintain it. Suppose that genotypes $ab$, $Ab$, $aB$, and $AB$ exist and that the presence of an $A$ allele raises fitness by $1 + k_A$, the $B$ allele by $1 + k_B$. Then, where the effects of alleles are multiplicative, relative fitnesses can be given by $1, 1 + k_A, 1 + k_B, (1 + k_A)(1 + k_B)$. If the initial frequencies are $r$, $s$, $t$, $u$, then after selection

$$r'u' - s't' = (1)(1 + k_A)(1 + k_B)ru - (1 + k_A)(1 + k_B)st$$

$$= (1 + k_A)(1 + k_B)(ru - st).$$

Thus, if $ru - st$ is initially zero, it will remain so. Where this is the case, sexual reproduction will be of no advantage in speeding selection. Of course, the effects of the gene substitutions need not be multiplicative, in which case sex is still at least a small aid in promoting selection. But can this aid overbalance the rather formidable disadvantages to the existence of sex? Maynard Smith continues his argument by noting that if adjacent populations undergo differential selection so that, say, $AB$ increases in one, $ab$ in the other, after which the two populations are brought together somehow, perhaps in a third habitat, $ru$ will be much greater than $st$. In this case the advantage of recombination through sexual reproduction is tremendous. Another advantage to the sexual scheme of reproduction is that recombination through sexual union of gametes can bring together "rare advantageous genes that occur in separate individuals, thus hastening incorporation of rare mutant combinations" (Crow and Kimura, 1969). The validity of these arguments clearly depends on the action of group selection. Williams (unpublished manuscript) has other suggestions. Consider a population of individuals some of which reproduce asexually and the rest of which reproduce sexually. The variety of genotypes produced per parent is greater among the sexually reproducing line. Hence such parents produce young with a greater spread in relative probability of surviving to reproduce (viability) (see Fig. 3.1). Suppose now that selection disposes of all individuals except those of relative viability greater than $V_1$. Roughly one-half of the offspring of both types of parents survive to reproduce. But the sexually reproducing parent is passing on only one-half as many genes per offspring as its asexual competitors and asexuality is favored. If selection is more severe, and allows only individuals of relative viability greater than $V_2$ to survive, more than twice as many sexual as asexual offspring survive. In this case there is no doubt that sexuality is favored by selection. Where mortality is low or nonselective (different genotypes do not differ in their respective mortality rates) and fecundity is not highly variable, sexuality will not be likely to evolve. Most of the evolutionary history of life, though, has involved organisms with enormous reproductive capacities and very high mortalities. Unless the bulk of these deaths were nonselective (a matter of some controversy) then sex should have been the generally expected trend (as it apparently has been). Where these conditions of stringent selection have been lost some species have secondarily become asexual. Most of these apomicts are found in the lower phylogenetic levels, in such phyla as Cnidaria or Turbellaria. A few higher invertebrates, such as many of the rotifers and arthropods, have secondarily reverted to obligate parthenogeny, and this has apparently occurred even among vertebrates. Maslin (1962, 1971), Lowe and Wright (1966), and Kluge and Eckardt (1969) have reported purely parthenogenic species in both teiid and iguanid lizards.

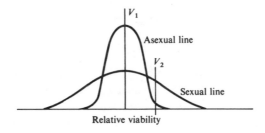

**Fig. 3.1.** Viability of offspring of sexual and asexual lines.

In species for which environmental conditions remain relatively constant but then undergo sudden and rapid deterioration, we might expect facultative sexuality to occur (Williams, unpublished manuscript). For example, among many freshwater rotifers and cladocerans, low mortality, the deletion by one-half of the genome, and the time required for sexual behavior make the sexual mode of reproduction disadvantageous during the spring and summer months. However, when conditions change in the fall and winter and mortality rises, these species suddenly change to sexuality. Williams suggests that large species, as for example in the classes Aves and Mammalia, have low enough fecundity and mortality rates that they might be expected to be asexual. Nevertheless, the high degree of physiological and anatomical adaptation to sexuality and the dependence in some species at least on parental care by both parents may make reversion to asexuality a very difficult process, one which would take a very long time to occur if, in fact, it could occur.

### 3.5.2 The Number of Sexes

Virtually all animal phyla have sexually reproducing species. Quite obviously, then, sex must have some *raison d'être*. Why, with the exception of some protozoans, are all species represented by only two sexes? Two possible schemes involving more than two sexes can be envisioned. In the first, the joining of several different sexes would be required for reproduction. In the second, the joining of any two of several sexes would be sufficient. The first scheme is clearly less advantageous than the second for the simple reason that simultaneous meetings of several individuals of several different sexes is less likely to occur and more difficult to accomplish than the meeting of just two individuals of different sexes. The second scheme seems eminently reasonable. The advantages and disadvantages would be identical to those suggested for a two-sex system except that the time required to find an appropriate mate would be cut (Siegel, 1955). Obviously, an animal living in a very sparsely populated environment would find himself mating successfully more often if mating could occur (say) with five out of every six individuals encountered (six sexes) rather than one out of every two (two sexes). In fact the chances of successfully mating over any given period of time will be increased fivefold. There seems to be virtually nothing disadvantageous in such a system; nevertheless there is at least one rather clear reason for its rarity. Each possible couplet of sexes will produce young with a given relative average fitness. Unless these relative values fluctuate over time periods longer than generation time, or the chances of mating in the absence of many sexes becomes extremely small (i.e. in very sparse, non-motile species), selection will eliminate the possibility of all mating combinations except that one producing the most fit offspring. In other words, all but two sexes will be eliminated through competition. It is noteworthy that in the entire animal kingdom the only species with multiple sexes (mating types) are protozoans, animals with little motility and, in relation to their size, often highly rarefied populations. *Paramecium*, *Oxytricha*, and *Euplotes*, for example, all have species with multiple mating types. *Euplotes cristatus* demonstrates six mating types, all of which are capable of conjugation with the other five and all of which occur together in one area in the Mediterranean (Wichterman, 1967). It is obvious but not particularly disturbing to the above hypothesis that phylogenetically more advanced species with low motility often live in sparse populations yet display no more than two sexes. It seems unlikely that a species with anatomy, physiology, morphology, and behavior highly evolved for two-sex reproduction should revert to a multisex system, and the rarity of the latter system would seem to preclude the possibility that it is the ancestral condition in a recent taxon of this sort.

A rather peculiar form of sexual behavior occurs in vertebrates, insects, and possibly other groups. Haskins *et al.* (1960) report several species, among them a beetle, *Ptinus latro*, to be sexually reproducing but with no indication of paternal inheritance. *Ptinus latro*, an all-female species, overlaps in its range with another beetle, *P. hirtellus*, and mates with males of the latter species. The male gamete, however, acts only to trigger embryonic development and is not incorporated into the genetic material of the egg. *Mollienisia formosa*, a fish, behaves in a similar way, mating with males of congeneric species but displaying neither males nor paternal inheritance (Hubbs and Hubbs, 1932; Haskins *et al.*, 1960; Darnell *et al.*, 1965). The evolutionary explanation of this phenomenon is unknown.

# 4
# Selection in Heterogeneous Environments

The components of fitness, fecundity, mortality rate, and mating success are dependent not only on genotype but on environmental conditions: That genotype conveying maximum fitness under one set of circumstances may be quite different from another genotype maximizing fitness under other circumstances. Thus, as environmental conditions change over space (along ecological gradients, or between patches of differing habitat), selective pressures favor different genes and gene frequencies. Environments may also vary over time, making adaptation to a variety of situations mandatory for survival. Where a population encounters living conditions which vary in time or space, its environment is said to be heterogeneous. It is the effects of heterogeneity on fitness and the direction of evolution that concerns us in this chapter.

## 4.1 ECOLOGICAL GRADIENTS

The world is full of ecological gradients. For example, average temperatures change with latitude both on land, in lakes, and at sea. Temperatures also change with altitude of land and depth of water. Variability in temperature generally increases as one moves from tropical to temperate regions or from coastal to inland areas. Moisture gradients, as well as soil and mineral gradients, occur across the North American great plains and other areas and, at more local levels, gradients may occur in the form of anything from ground slope to wind speed. In addition, there are gradients in the relative proportions of different habitat types in space or situations in time. The proportion of space covered with snow or the proportion of the year in which temperatures fall below freezing is high in Arctic regions, decreases toward temperate areas, and reaches zero in the tropics.

Along those gradients in which change is effected over very short distances it is possible for individuals to make behavioral adjustments. The wood ants, *Formica polyctena* and *F. lugubris*, for example, prefer temperatures of about 22°C in the early spring, and temperatures upward to 32°C by the end of April. Movements along steep temperature gradients within their colonies allow these ants to realize their preferences (Herter, 1924, 1925, cited in Sudd, 1967).

The study of *preferenda* in animals is an old one and many examples could be cited. At the moment, however, we are concerned with the effects of gradients on genetic change over distance. Butterflies of the genus *Colias* have been described as being more heavily pigmented on the undersides of their wings in northern as opposed to more tropical latitudes, and in high as opposed to low altitudes.  These differences hold both among and between species and suggest the existence of differential selection.  Since butterflies are heliothermic (control their body temperatures by absorbing solar heat), and since dark colors absorb more heat than light colors, it has been suggested that the coldness associated with high latitudes and altitudes increases the advantage of dark (light-absorbing) coloration.  Watt (1968) has used thermistor probes to show that in fact dark-colored individuals heat up more rapidly than lighter individuals.  Another example of phenotypic change over an ecologic gradient is found in the song sparrow (*Melospiza melodia*).  As we move northward from this species' southern breeding range in southern California to its northern limits in Alaska, we find it beginning to breed at progressively later dates (Johnston, 1954).  The reason clearly relates to the later appearance of vegetation and hence insect food.

The blue/snow goose, which shows a gradient in frequency of the two (blue or white) morphs from north to south in North America, has been studied exhaustively by Cooch (1961, 1963) and Cooke and Cooch (1968).  Here, as one might expect, the white morph which is protectively colored against snow is most common in the far north part of the breeding range, but becomes less common toward the south.  A warming trend in the eastern Canadian Arctic has resulted in an increase in the blue phase in recent years.

The classic generalizations concerning ecological adaptation across gradients are the rules of Bergmann, Allen, and Gloger.  The arguments behind the first two generalizations run as follows.  The amount of heat held by an individual is proportional to its volume, while the rate at which heat is lost is proportional to its surface area.  Thus in homeotherms, all else being equal, it is best in cold climates to minimize the ratio of surface area to volume.  This can be done in either or both of two ways: increase body size or decrease the length of protruding structures.  Thus Bergmann's law asserts that, all else being equal, tropical animals should be smaller than those living in the Arctic, and Allen's rule states that extremities in Arctic species should be shorter than those in individuals inhabiting warmer areas.  To an extent these rules are true but, as with most ecological generalizations, they are not universal.  First, all else is not equal.  Animals living in cold climates may possess heavier fur.  In the case of man, Scholander (1955) suggests that Eskimos may actually be warmer under their parkas than those tropical people who go naked.  Yet Allen's rule applies: limbs and fingers are shorter in the Arctic (Scholander's argument is contested by Newman, 1966).  Scholander (1955) raises more serious objections to Bergmann's and Allen's rules when he points out that tiny birds with skinny legs, cranes and herons with long extremities, and such minute homeotherms as shrews live abundantly in the far north.  Furthermore, in the puma, raccoon, and otter, individuals generally are larger in warmer parts of the species range.  Other objections can be raised.  While the rules may be generally true for gradients between Arctic and temperate regions, there are reasons to believe them to be less reliable and perhaps even reversed along temperate-tropical gradients. For example, in diurnal species living on tropical deserts or savannas it is possible that large body size would help limit the acquisition of unwanted radiant heat.

Gloger's rule states that colors will be darker in warm climates and lighter in cold ones. While it is true that dark colors blend better (to the interests of both prey and predator) in temperate or tropical forests and white blends better against snow, there are also reasons for discounting the general validity of this rule. Dark colors absorb heat and in species for which temperature control is critical (for example in small mammals or *Colias* butterflies) and for those which do not need the ability to blend into a background (large, herding herbivores such as musk oxen), the relative value of a white coat may be overbalanced by the heat value of a dark coat. Small mammals in general and musk oxen in particular are dark colored in the Arctic.

Individuals which differ in appearance or life pattern at different points in an ecological gradient due to genetic differences are called *ecotypes*. The term, although applying to differences in butterfly pigmentation, size of body and extremities, etc., is generally applied to physiological differences. This is perhaps due to the fact that most studies of ecotypes have concerned adaptations to soil nutrients or reproductive adaptations to latitudinal changes in light-dark cycles in plants. One important example of physiological-ecotypic variation in animals should be mentioned. Bullock (1955) has noted that although the activity rates of warm water marine invertebrates slows considerably if they are taken into cold water, the individuals normally inhabiting cold waters are not significantly different in activity rates from those living in warm areas. Apparently natural selection has favored physiological temperature optima which correspond to the prevailing temperature conditions and which change with latitude.

Let us examine the genetics behind the adaptive clines that accompany ecological gradients. To take a somewhat simplified case, suppose that differing conditions along a gradient result in opposite selection pressures at each end for some particular gene locus. Then the highest fitness value at one end will occur with genotype $A_1A_1$ while at the other end $A_2A_2$ will be superior. If we consider only this one locus for the moment, it is clear that either one homozygote or the other, or the heterozygote must display maximum fitness between the ends of the gradient, because no other genetic configuration is possible. If we suppose that $A_1A_1$ (or $A_2A_2$) is superior in the gradient center and that $A_1A_2$ is never optimal, then $A_1A_2$ individuals will be selected against everywhere. However, only very small migratory movements across the gradient will be needed to maintain both alleles at all points on the gradient. The frequency of the $A_1$ gene will range from nearly one at one end of the gradient to nearly zero at the other. Suppose that $A_1A_2$ is optimal somewhere along the gradient. This is perhaps more reasonable (in this simplified case) since $A_1A_2$ is likely to be intermediate in its effects between $A_1A_1$ and $A_2A_2$ and may then be expected to be adaptive in conditions inter-mediate between those at the gradient ends. In such a case heterosis occurs over part of the gradient and assures the maintenance of both alleles. Again, the frequency of the $A_1$ gene will range from nearly one at one end of the gradient to nearly zero at the other.

The second case above lends itself to further analysis. First, fitness is a function of both gene frequency and environmental parameters. Thus, in terms of the distance from one end of a gradient to the other we can, as a simplification, write the fitness of $A_1A_1$ at point $D$ where $D$ is the fraction of the distance along the gradient, as $xD$. The fitness of $A_2A_2$ can be written $x(1 - D)$. The fitness of $A_1A_2$ we write as $xh$ (where $h$ may be a function of $D$).

Within the part of the gradient where heterosis occurs, that is, where $xh > xD$, $x(1 - D)$, the equilibrium frequency of $A_1$ is given by

$$\hat{p} = \frac{W_2 - W_3}{2W_2 - W_1 - W_3} = \frac{xh - x(1 - D)}{2xh - xD - x(1 - D)} = \frac{h + D - 1}{2h - 1}$$

(see Eq. 1.13). A plot of $\hat{p}$ on the ordinate against $D$ on the abscissa is given in Fig. 4.1.

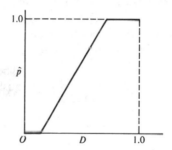

**Fig. 4.1.** Change in equilibrium gene frequency over a gradient.

On the basis of this simplified picture, we should expect monomorphic populations at either end of an ecological gradient, and trimorphic populations ($A_1A_1$, $A_1A_2$, $A_2A_2$) occurring over a portion of the gradient, the frequency of the morphs varying continuously with distance from either end.

To what extent continuous clines in phenotype occur across ecological gradients in nature is difficult to say, for even abrupt steps in a phenotypic cline may be softened by phenotypic mixture due to migration and thus give the appearance of continuity. What is clear, however, is that in many species, even those with rather high motility, remarkably abrupt steps in phenotypic clines do occur. This has been noticed particularly in plant species (for example, see Jain and Bradshaw, 1966). On the most northern part of one of the Shetland islands, the melanic form of the moth *Amathes glareosa* occurs with a frequency of 97%. The frequency drops to 1% on the most southern part of the island only 70 miles away, and a large part of the drop (35% to 13%) occurs over an area only eight miles in width (Kettlewell and Berry, 1969). Such abrupt changes are reasonable on this particular island which is characterized by very steep ecological gradients.

However, sometimes steep clines occur in the apparent absence of sharp ecological gradients. An excellent case is that described by Dowdeswell and Ford (1953) and Ford (1964) for the meadow brown butterfly, *Maniola jurtina*, on the English mainland. This species displays on the underside of its hind wings a row of spots which may number anywhere from zero to five. Most individuals seem to possess either no spots or two spots. All populations examined are found to be either unimodal for no spots or bimodal for zero and two spots. Tables 4.1 and 4.2 show some of the results obtained by these authors, the areas named in order along an east to west gradient across northern (Table 4.1) and southern (Table 4.2) England. Five years after these observations were made in 1952 the situation was reexamined. It was found that the distribution of the two population types had changed and that the steps in the cline separating them had moved several miles (Creed *et al.*, 1959). The pattern of abrupt changes in population type, however, was the same as previously observed, as it continued to be in 1958 through 1960 (Creed *et al.*, 1963). In one area, the separating point was marked only by a hedge on one side of which occurred a unimodal, on the other a bimodal population. The sharp demarcation was blunted only by local mixing of the two populations as their members flew freely back and forth across the hedge. No obvious ecological difference was apparent between the environments on the two sides of the hedge. What apparently is happening is that at certain points along the east–west gradient some unknown but critical ecological value is reached which demands differential selection on its two sides—

**Table 4.1.** Distribution of *Maniola jurtina* in northern England

| East | Number of spots | | | | | | Area collected | |
|---|---|---|---|---|---|---|---|---|
| | 0 | 1 | 2 | 3 | 4 | 5 | | |
| | 36 | 9 | 5 | 3 | 0 | 0 | Okehampton | Unimodal at zero spots |
| | 21 | 12 | 3 | 0 | 0 | 0 | Lydford | |
| | 10 | 2 | 2 | 0 | 0 | 0 | Holsworthy | |
| | 28 | 17 | 21 | 7 | 0 | 0 | Lewannick | Bimodal at zero and two spots |
| | 30 | 16 | 18 | 7 | 1 | 0 | Lanivet | |
| West | | | | | | | | |

**Table 4.2.** Distribution of *Maniola jurtina* in southern England

| East | Number of spots | | | | | | Area collected | |
|---|---|---|---|---|---|---|---|---|
| | 0 | 1 | 2 | 3 | 4 | 5 | | |
| | 47 | 31 | 10 | 4 | 1 | 0 | Newton Abbot | Unimodal at zero spots |
| | 15 | 8 | 1 | 0 | 0 | 0 | Noss Mayo | |
| | 34 | 14 | 4 | 2 | 0 | 0 | Plymstock | |
| | 47 | 22 | 31 | 4 | 1 | 0 | Plymouth | Bimodal at zero and two spots |
| | 21 | 14 | 18 | 3 | 0 | 0 | Roborough | |
| | 39 | 24 | 29 | 7 | 0 | 0 | Foeck, Falmouth | |
| West | | | | | | | | |

selection which is strong enough to result in abrupt phenotypic changes in spite of rather free migration. But unless there exists some very steep ecological gradient totally unobserved by man, such a conclusion seems unreasonable.

Suppose that no such sharp gradient exists. Can the jump in phenotype still be explained? One possible explanation is that dominance for $A_1$ changes to dominance for $A_2$ at some critical point. This explanation is probably not valid, though, since it assumes very strong selection for dominance (which probably never occurs). Clarke (1966) has treated this sort of problem as follows.

Assume first the validity of the model depicted in Fig. 4.1. Then suppose that there exists a modifier gene, $B$, with dominance over its allele, $b$, and suppose that in the presence of $B$, the fitnesses of $A_1A_1$, $A_1A_2$, $A_2A_2$ are changed to $xD + xr$, $xh + xs$, $x(1 - D) + xt$. Under the influence of the modifier gene the equilibrium frequency of $A_1$ is given by

$$\hat{p} = \frac{x(h + s) - x(1 - D + t)}{2x(h + s) - x(D + r) - x(1 - D + t)} = \frac{h + s + D - t - 1}{2h + 2s - r - t - 1}.$$

Whereas the slope of the line in Fig. 4.1 (corresponding to genotype $bb$ in this case) is given by $d\hat{p}/dD = 1/(2h - 1)$, the slope of the corresponding line for the $B$-genotype is $d\hat{p}/dD = 1/(2h + 2s - r - t - 1)$. Both lines are shown in Fig. 4.2. In 4.2(a), $r + t - 2s > 0$, in 4.2(b), $r + t - 2s < 0$. The intersection of the two lines can be found by setting the alternate $\hat{p}$ values equal to each other. The solution is given by

$$D = \frac{ht - hr + r - s}{r + t - 2s}, \qquad \hat{p} = \frac{t - s}{r + t - 2s}.$$

We now digress a moment and note that $B$ will be selected when $W_{B-} > W_{bb}$.

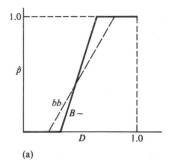 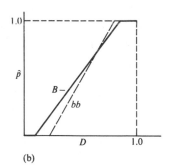

(a)                                          (b)

**Fig. 4.2.** Change in equilibrium gene frequency over a gradient.

That is, when

$$p^2(D + r) + 2pq(h + s) + q^2(1 - D + t) > p^2D + 2pqh + q^2(1 - D)$$

or, more simply, when

$$p^2r + 2pqs + q^2t > 0.$$

Substituting $1 - p$ for $q$ and solving the quadratic inequality, we see that $B$ is favored when

$$p \gtrless \frac{(t - s) + \sqrt{s^2 - rt}}{r + t - 2s} : (r + t - 2s) \gtrless 0.$$

Considering the first case $(r + t - 2s) > 0$, we see that the critical value of $p$ is reached at

$$\frac{t - s + \sqrt{s^2 - rt}}{r + t - 2s},$$

which is larger than $(t - s)/(r + t - 2s)$, the value of $\hat{p}$ at the intersection of the two lines in Fig. 4.2(a). Thus, as $D$ increases, $\hat{p}$ increases along the dashed line in Fig. 4.2(a) until the critical value of $D$ is reached. At this point, $p$ jumps to the solid line, resulting in a picture such as that shown in Fig. 4.3(a). In the case of $r + t - 2s < 0$, the critical value of $p$ is likewise

$$\frac{t - s + \sqrt{s^2 - rt}}{r + t - 2s},$$

but since $r + t - 2s$ is negative, this occurs to the left of the point of intersection of the two lines. Here $\hat{p}$ again follows the dashed line (in Fig. 4.2b this time) and jumps to the solid line before the two meet (Fig. 4.3b). In both cases a jump in the adaptive cline, in the same direction as the continuous cline preceding and following the jump, is noted though no such jump occurs in the ecological gradient.

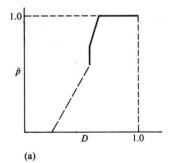 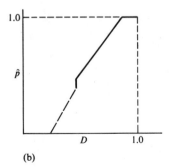

(a)                          (b)

**Fig. 4.3.** Change in equilibrium gene frequency over a gradient.

The sharpness of the step in the phenotypic cline depends on the degree of migration across the critical point in the ecological gradient. This is because migration blunts the abruptness through mixing the adjacent populations and, through interbreeding, slows the selection of $B$ on one side of the critical point and $b$ on the other. Clearly, the extent to which such steps occur, as well as the clarity of their demarcation, will depend on the mobility of the species in question and thus, indirectly, on their size and mode of life. An important point is that once a step in the cline is established, so that a large change in gene frequency exists over a small distance, the likelihood is increased that the critical $p$-value for other modifier genes occurs at the same value of $D$. Thus a whole series of modifiers may exist, each resulting in a gene frequency jump at the same point on the gradient, each of which acts with the others to make selective pressures on the major gene very strongly different on either side of the critical point in space, even though no jump in the ecological gradient exists.

## 4.2 PATCHY ENVIRONMENTS

Environmental change does not always take the form of gradients. A much more prevalent situation is the patchy, or *mosaic* environment. Patchiness is recognized easily by man in the form of fields interspersed with woodlots, woods broken by bracken openings, rock piles, squirrel middens or patches of vaccinium in boreal forest, and many other forms. Not so obvious to us are patterns of changing soil types or microclimate, although these may be of vital importance to the survival of a large variety of organisms. Does a given species show specialized adaptation and habitat preference for one patch type in its mosaic world, or generalized adaptation and tolerance for several? Perhaps the organism changes its preference and distribution between patch types with the seasons, or possesses phenotypic plasticity allowing it to express adaptations for whichever patch it happens to be inhabiting at any given time. At any rate, it should be obvious that natural selection may operate in different directions in different, adjacent patches. A simple example is given in the work of Struhsaker (1968). The Hawaiian intertidal snail, *Littorina picta*, possesses varying amounts of shell sculpturing. Highly sculptured shells will offer greater resistance to water pressure and therefore be washed ashore or out to sea more readily. Consequently, it is to be expected that the equilibrium between selective forces acting to promote and discourage shell sculpturing will shift toward more ornate shells in areas of weak wave force, and toward simple shells in areas of strong wave force. Where the intertidal bench slope is steep, wave force is greatest and sculpturing, as expected, least developed. Regression of larval sculpturing on that of the parents (see Chapter 2) showed that shell sculpturing is at least partly genetically determined.

     The European garden snail, *Cepaea nemoralis*, is a polymorphic species with respect to a number of characteristics including shell banding pattern and shell background color. The variety with respect to both of these features seems to be very stable and has persisted since pre-neolithic times at least (Cain and Sheppard, 1954). Background color, determined by a single locus, may be either yellow (which appears green when the living animal occupies the shell) or pink (which appears brown). Brown shells may also occur because of gene action at another locus. The primary predator of these snails is the song thrush (*Turdus philomelos*) which, as a visual predator, might be expected to take those individuals that stand out most against their environmental background because of color. This is indeed the case. Sheppard (1951) reports that shell fragments of predated individuals are, proportionately to the number of living snails, mostly of that color least suited to a given area. Yellow (green) shells survive best in gardens and hedgerows, pink (brown) in forest and other dark-background areas. In April, before leaves have appeared and the environment is dark colored, yellow shells made up 42% and 43% respectively of the shell fragments in two areas. In early May, against a slightly more leafy background, the frequencies fell to 22% and 27%. Later in May, only 9% and 15% of those snails destroyed possessed yellow shells. This predation may be an important factor in selection for shell color, as is made clear when we note that thrush predation alone may account for mortality rates of up to 85.5% per year (Cain and Sheppard, 1954). Thus *Cepaea* appears to have adapted to an environment made up of green and brown patches by matching its shell color to its local background and thereby becoming polymorphic. For further information and discussion on *Cepaea*, see Cain *et al.*, 1960; Ford 1964.

## 4.2.1 The Concept of Grain

What constitutes a patch will vary from species to species, and the effective size of a recognized patch varies with the size and motility of an animal. A rough classification recognizes three basic concepts: A patch may be so small with respect to the habits of an animal that it is not even noticed. Thus a cow may be unaware of a small clump (read "patch") of clover among the grass it is busy devouring; it is immaterial to a sleeping moose that the earth beneath its left front hoof contains significantly more soil arthropods than that beneath its right hind knee. A species traversing such *fine-grained* patches does not, and cannot, change its response to match its microhabitat of the moment. Natural selection must act as if the environment were a mean of all the patches encountered.

At the other extreme, patches may be such that an organism can spend its whole life in one type or another. In this case there will generally be a superior type of patch into which individuals will preferentially move if possible (superior implying higher fitness). In such *coarse-grained* environments three situations exist. Species that do not broadcast their gametes will produce offspring which continue to inhabit the parental patch type. Clearly, adaptation will favor specialization to one patch type. The world is full of species which inhabit isolated island patches. Pikas, *Ochotona princeps*, for example, live in fell fields but never venture much beyond the boundaries of these rocky patches into the surrounding subalpine forest or meadow.

Species which tend to wander about but do not randomly disperse their young may choose patch types and will ultimately inhabit the one patch type to which they are best suited. When high population densities lower the desirability

of these patches the individuals will move into next best living quarters. In this
case selection may act divergently, at least on some generations, favoring adaptation to all of two or more different patch types. *Disruptive selection* of this sort,
particularly if migration between patch types is slight, may result in divergence of
the populations into separate morphs. Migration and subsequent interbreeding
between populations then result in a mixture of morphs in each. Examples of
specialized subpopulations of this sort are also in abundance. Blair (1947), for
instance, describes pocket populations in New Mexico of the cactus mouse,
*Peromyscus eremicus*, in which pelage color varies from place to place in accordance
with soil color.

The third situation involving coarse-grained environments is that in which
local populations disperse their gametes randomly so that offspring are forced to
live in whatever patch type they happen to settle. This description may be somewhat more extreme than anything found in nature—even barnacle larvae show
some tendency to choose the substrate on which they settle, and wind-dispersed
insects have the capability of correcting a bad situation by walking or flying
about—but it will do as an approximation to the real behavior of many species.
The nature of adaptation in such species is not immediately obvious.

Of course, most environments are neither strictly fine grained nor coarse
grained, but are in between. A wolf pack may, of necessity, traverse both forest
and tundra on a given night. A sparrow, to survive, may have to forage both
on the open ground, in short, dense grass, and on shrubs. Any number of encounters
with other individuals or objects constitute patches in a sense. In none of these
cases do the patches blend into each other imperceptibly and in none of these
cases are the patches such that whole lives may be lived in them. None are,
strictly speaking, either fine or coarse grained.

## 4.2.2 The Fitness Set and the Adaptive Function

While most environments are intermediate between fine and coarse grained,
it is instructive to look at these two extremes as a beginning to an understanding
of fitness in a heterogeneous environment. We do not, remember, have any
simple arguments relating to specialization or generalization with respect to the
different patches in fine-grained environments (adaptation to the mean patch
may imply a jack-of-all-trades approach or a master-of-one-take-your-chances-
in-the-others approach). The same is true of coarse-grained environments where
the inhabitants broadcast their gametes randomly, and in environments of intermediate grain. Under what circumstances will selection favor specialization
to one patch type, and under what circumstances will it favor generalized adaptation
to several or all?

Consider the fitnesses in each of two patch types of the set of all realizable
genotypes in a population. A genotype predisposing a phenotype optimally
constructed to withstand the rigors of one patch type will undoubtedly have
accomplished that optimum at the expense of fitness in the other patch type.
For example, the black swallowtail butterfly, *Papilio polyxenes*, may pupate

either on tree trunks or on the undersides of leaves.  Detection of the pupae by predators is reduced if those pupae on tree trunks are brown.  But brown coloration is disadvantageous against leaves.  To become specialized to leaves (green) an individual must sacrifice the trait (brown coloration) which adapts it to tree trunks.  An intermediate phenotype, meaning a different and probably intermediate genotype, of brownish-green color would probably be of intermediate fitness in both patch types.  The situation may be shown graphically as in Fig. 4.4 (Levins, 1968).  Here $A$ denotes the relative fitness in patch one, $W_1$ (along the ordinate), and patch two, $W_2$ (along the abscissa), for an individual specialized to patch one, and $B$ denotes the same for a specialist in patch two.  Where the two patch types are very similar, so that individuals may make significant adjustments physiologically or behaviorally, a small genetic change improving fitness in one patch type should not greatly lower fitness in the other.  Thus the set of genotypes intermediate between $A$ and $B$ form a convex *fitness set* (shaded area, Fig. 4.4a).  In situations where the patches are very dissimilar with respect to the individuals' abilities to adjust somatically, a change in genotype increasing survival in one patch may result in a drastic loss of survival ability in the other.  Then the fitness set, enclosed by $OAB$, is concave (shaded area, Fig. 4.4b).  Note that while no genotype is capable of the fitness values indicated by the dashed line in Fig. 4.4(b), the values do describe the average fitness of a population consisting of varying proportions of $A$ and $B$.

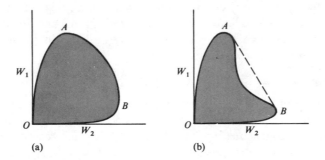

**Fig. 4.4.** Fitness sets (a) similar patches, (b) very different patches.

We now define what Levins (1968) refers to as the *adaptive function*.  Consider first the case of a population for which the two patch types under consideration constitute a fine-grained environment.  As stated earlier, selection must respond to the mean fitness over all patch types so that if a proportion $P$ of an individual's time is spent in patch type one and the remainder in patch type two, $\overline{W} = PW_1 + (1 - P)W_2$.  Writing this in the form

$$W_1 = \frac{1}{P}[\overline{W} - (1 - P)W_2], \tag{4.1}$$

it is clear that Eq. (4.1) defines a family of lines each describing a given value of $\overline{W}$. That is, the "adaptive function" defines a set of fitness isoclines. In Fig. 4.5 these have been superimposed on the fitness sets of Fig. 4.4. Since selection acts to increase fitness, it effectively chooses that genotype (available genotypes are by definition contained in the fitness set) with the highest fitness, i.e. that point on the fitness set tangent to the highest fitness isocline. In Fig. 4.5(a), this is given by $X$; in 4.5(b), by $X_1 = A$. Note, however, that increasing and maximizing fitness are two different things. If all initial genotypes are below and to the right of $X_e$ then $X_2$ denotes the genotype toward which selection acts. If all initial genotypes are above and to the left of $X_e$ then $X_1$ is favored.

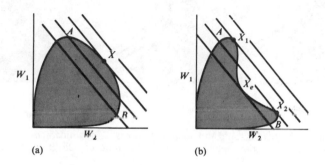

**Fig. 4.5.** Superimposed fitness sets and adaptive functions (a) similar, fine-grained patches, (b) very different, fine-grained patches.

Two general conclusions can be drawn from the above arguments. First, it is clear from Fig. 4.5 that when patch types are similar with respect to the ability of individuals to somatically adjust (the fitness set is convex) a jack-of-all-trades is favored. When patch types differ sufficiently (the fitness set becomes concave) selection favors specialization to either one patch type or the other. In the first case, we say the population is generalized, in the second case, specialized.

Suppose now that the two patch types in question constitute a coarse-grained environment and that gametes are randomly broadcast. We let $P$ be the proportion of generations which occupy patch type one; the remaining generations occupy the second patch type. The proportion $P$ will be roughly equal to the proportion of individuals in patch type one at any given time. Then, since $\overline{W}$ is the ratio of individuals in one generation to that in the previous generation, the ratio of individuals after $n$ generations to the initial number is given by $\overline{W}^n$. But of these $n$ generations, $Pn$ occur in patch type one and $(1 - P)n$ occur in patch type two. Thus $\overline{W}^n = W_1^{Pn} \cdot W_2^{(1-P)n}$, and $\overline{W} = W_1^P \cdot W_2^{(1-P)}$. Rewriting, we have

$$W_1 = (\overline{W}/W_2^{1-P})^{1/P}. \tag{4.2}$$

Figure 4.6 shows this adaptive function superimposed on three possible fitness sets. The point $X$ in 4.6(a) is the same as point $X$ in 4.5(a); $X_1$ and $X_2$ in 4.6(c) are the same as $X_1$ and $X_2$ in 4.5(b). About the only difference here is that a fitness set must become considerably more concave before selection favors specialization. The real interest in these figures involves the points $X'$. We noted above that the points along the dashed line joining $A$ and $B$ represented possible average fitness values not for individual genotypes but mixtures of genotypes. Thus the points $X'$ represent the maximum fitnesses realizable by populations and exceed the maximum fitnesses realizable by any genotype. That is, these points represent optimal population structure, the population configuration favored by group selection (see 4.2.3).

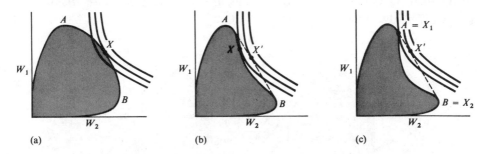

**Fig. 4.6.** Superimposed fitness sets and adaptive functions (a) similar, coarse-grained patches, (b) less similar, coarse-grained patches, (c) very different, coarse-grained patches.

Real-life situations are intermediate between those depicted in Figs. 4.5 and 4.6, so that the general conclusions common to both extreme cases will apply also to the real world. Inasmuch as real adaptive functions will usually be somewhat curved, the group selection statement also applies to the real world.

Of course, larger individuals and more highly mobile individuals divide their environments into larger and larger patches, so that what are fine-grained differences to one species become coarse-grained to another. The clover clump which is fine-grained to the cow is certainly coarse-grained to an aphid. But we have seen that specialization to one or several patch types is usually the result of selection in coarse-grained environments unless the species in question broadcasts its gametes randomly. Thus small and relatively immobile creatures should be more likely to show a variety of morphs than large, mobile species.

### 4.2.3 Genetic Mechanisms

At any given moment, most populations will comprise a range of genotypes. If all genotypes lie to one side or the other of $X_e$ it is then clear what will subsequently happen (unless selection acts in huge jumps, or, what is equivalent, very few loci are involved, in which case the value of $X$ with the highest fitness is chosen). But what if genotypes occur to both sides of $X_e$? We can easily see that selection will favor individuals of both extremes—those close to $A$ and those close to $B$—over intermediate forms, but what will the outcome of this disruptive selection be? A number of laboratory experiments have been done in an effort to explore the effects of disruptive selection; these may be useful as a guide to answering the above questions.  Thoday (1958b) kept cultures of *Drosophila melanogaster* which he subjected to artificial disruptive selection for sternopleural chaetae (bristle) number.  Three sets of experiments were run.  In each of two experiments, four cultures were maintained, two with flies selected for high bristle number, and two with flies selected for low bristle number.  In one experiment, the flies with the highest and lowest bristle number were chosen from all four cultures and opposites mated (negative assortative mating with free gene flow between high and low populations).  In the second the procedure was the same except that there was positive assortative mating (like with like).  In the third, positive assortative mating was coupled with stabilizing selection (selection against the extremes in bristle number).  As one might expect, the first and third experiments yielded very little change in variability of bristle number, while variability increased considerably in the second.  Thoday, in effect, produced a *polymorphic* population (one with a variety of morphs) by simulating disruptive selection in a population where positive assortative mating occurred.

In later experiments, Thoday and Gibson (1962) found in one strain of *D. melanogaster*, after ten generations of disruptive selection, that reproductive isolation had occurred.  Apparently, the low viability of intermediate phenotypes, due to the experimenters' selection, had favored the evolution of positive assortative mating.  Clearly if disruptive selection leads to reproductive isolation, and, in the presence of reproductive isolation, leads to polymorphism, then it is possible (Figs. 4.5, 4.6) that concave fitness sets may lead to several rather than just one specialized phenotype.

While the experiments of Thoday and Gibson are suggestive, they are controversial (Scharloo, 1964; Scharloo *et al.*, 1967; Thoday and Gibson, 1970, 1971; Scharloo, 1971).  Other workers have tried and failed to get reproductive isolation under similar circumstances.  It appears that Thoday and Gibson may have gotten the results in ten generations that would normally take far longer.  Nevertheless the hypothesis that disruptive selection may result in polymorphism seems borne out.

Chabora (1968) has found that even in the absence of positive assortative mating, disruptive selection on bristle number in *D. melanogaster* led to the divergence of two lines after only five generations (one strain), ten generations (three more strains), and twenty seven generations (fifth of eight strains tested).

Note that to a less obvious extent inbreeding, like positive assortative mating, will allow the evolution of more than one morph.  Thus where fitness sets are concave, inbreeding may be favored through either kin selection or group selection.

A species in which disruptive selection has produced two or more morphs will continue to evolve in such a manner that modifier genes maximize the fitness of each morph.  Each morph will evolve a gene complex regulating its character.  The genes responsible for determining which morph is to develop are known as *switch genes*.

We now examine a somewhat modified version of a model proposed by Maynard Smith (1962).  We denote by $A_1A_1$, $A_1A_2$, $A_2A_2$ the three genotypes possible at a locus with two allelic forms and suppose that all three genotypes move freely over a patchy environment with two patch types, showing no genotype-specific preferences for patch type.  Thus gene frequencies are the same in both of the two patch types.  Suppose now that the fitness of $A_1A_1$ and $A_1A_2$ can be given by $W_A + K$ in patch type one, $W_A$ in patch type two, the fitness of $A_2A_2$ by $W_a$ in patch type one, $W_a + k$ in patch type two.  This system of complete dominance reflects the expected behavior of a switch gene.  From the equations in Chapter 1 we now see that the change in the frequency ($p$) of gene $A_1$ in patch types one and two will be

$$\Delta p_{(1)} = \frac{pq}{\overline{W}}\left[p(W_1 - W_2) + q(W_2 - W_3)\right] = \frac{pq[p(0) + q(K)]}{1 + p(1 + q)K} = \frac{pq^2K}{1 + (1 - q^2)K},$$

$$\Delta p_{(2)} = \frac{pq[p(0) + q(-k)]}{1 + q^2k} = \frac{-pq^2k}{1 + q^2k}. \tag{4.3}$$

The net change in $p$ is given by $P\Delta p_{(1)} + (1 - P)\Delta p_{(2)}$, where $P$ is the proportion of individuals in patch type one, equal in value to the proportion of time the average individual spends in the first patch type.  Now both $A_1$ and $A_2$ will be maintained in the population if, when $p$ approaches unity, $|P\,\Delta p_{(1)}| < |(1 - P)\,\Delta p_{(2)}|$, and if when $p$ approaches zero, the opposite is true.  These conditions are equivalent to

$$P\frac{pq^2K}{1 + K} < \frac{pq^2k}{1}(1 - P) \quad \text{or} \quad \frac{P}{1 - P} < \frac{k}{K}(1 + K),$$

$$P\frac{pq^2K}{1} > \frac{pq^2k}{1 + k}(1 - P) \quad \text{or} \quad \frac{P}{1 - P} > \frac{k}{K(1 + k)}. \tag{4.4}$$

Thus it is possible, for at least some range of $P$, that selection will favor the existence of a switch gene mechanism promoting the evolution of two different morphs under disruptive selection, even in a freely breeding population.  The frequency of the switch gene alleles and thus the frequency of the two morphs will depend on the interaction of the selection depicted in Eq. (4.4) and the group selection operating to bring the population to its optimum state ($X'$ in Figs. 4.5 and 4.6).  In the case of the garden snail, *Cepaea*, Ford (1964) believes shell color to be determined by a heterotic switch gene.

What all this means is that when the fitness set is concave (fine-grained environment) or more concave than the adaptive function (coarse- and inter-mediate-grained environment), disruptive selection may exist and may lead to the coexistence, in a freely breeding population (or an inbreeding or positive assortative mating population) of several morphs.  That is, selection favors polymorphism. The equilibrium frequency of the various morphs will reflect their fitnesses in the different patches and also (except in a strictly fine-grained environment) the effects of group selection acting to maximize fitness of the population itself.

   Note that the evolution of polymorphism, through natural selection operating at the level of the individual, may result in an *approach* to the population optimum $(X')$ favored by group selection.

### 4.2.4 Gradients in Patch Frequency

If the fitness set is highly convex, slight changes in $P$ move the point of tangency with the adaptive function only slightly (see Fig. 4.7a). As environments diverge (the fitness set flattens), the same changes in $P$ result in far greater changes in optimal phenotype (Fig. 4.7b). Thus if an ecological gradient consists of changing proportions of patch types (as is the case with forest and grassland in the transition zone between United States eastern deciduous forest and the long-grass prairie, for example), genotype and, presumably, phenotype may either change only slightly or, at some point, take a sudden jump. Although the ultimate reason is totally different from that proposed by Clarke (1966), and the mechanism is not explicit, this is a situation similar to the step clines discussed previously.

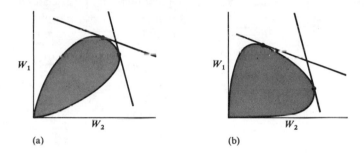

**Fig. 4.7.** Fitness sets and adaptive functions for two values of $P$ (a) similar patches, (b) very different patches.

   The above arguments are summarized in Table 4.3.

## 4.3 FLUCTUATING ENVIRONMENTS

Temporal differences of a fluctuating nature may also be thought of as coarse or fine grained. Where individuals can leave an area during bad times the environment is clearly fine grained; where they are forced to remain in one place, the environment is coarse grained. Grain in the temporal sense and grain in the spatial sense are not independent, since an ability to escape, as above, indicates potentially coarse-grained differences between the area evacuated and the area to which the escape is made. Thus as winter approaches and food supply becomes scarce, many species, particularly birds, leave one area for another. By their adaptive response of escape the two potentially coarse-grained areas visited are effectively rendered intermediate-grained with respect to survival, feeding, and other factors.

**Table 4.3.** Summary of arguments: Adaptation and population structure in spatially heterogeneous environments

| Grain environments | Fine-grained or coarse-grained environments | | Coarse-grained environments with gametes broadcast randomly |
|---|---|---|---|
|  | Similar | Quite different |  |
| Optimal phenotype | Generalized | Specialized | Specialized |
| Population structure | Monomorphic | Monomorphic or polymorphic | Pockets of monomorphic specialists |
| Changes over a gradient of patch-type frequencies | Gradual clines in morph | Discrete jumps from one monomorphic population to another or change in relative frequencies of morphs in a polymorphic population | Gradual change in the relative frequencies of specialized morphs |

## 4.3.1 Rates of Fluctuation

Suppose that temporal change is very slow so that changes of the sort necessary to define a fitness set occur only over several generations. Then suppose, as a simple approximation, that the fitness of a population falls off with the square of the difference, in some measure, between that phenotype ($P^*$) optimal at a given time and the actual realized phenotype ($P$) at that time. Then we can write the expected average fitness over time

$$E(\overline{W}) = \overline{W}_{\max} - \overline{(P - P^*)}^2.$$

This can be expanded to read

$$E(\overline{W}) = \overline{W}_{\max} - [\overline{(P^2)} - 2\overline{(PP^*)} + \overline{(P^{*2})}]$$

$$= \overline{W}_{\max} - [(\overline{P^2} - \overline{P}^2 + \overline{P}^2) - 2(\overline{PP^*} - \overline{P}\,\overline{P^*} + \overline{P}\,\overline{P^*})$$

$$+ (\overline{P^{*2}} - \overline{P^*}^2 + \overline{P^*}^2)]$$

$$= \overline{W}_{\max} - \{[\operatorname{Var} P + \overline{P}^2] - 2[\operatorname{Cov}(P, P^*) - \overline{P}\,\overline{P^*}] + [\operatorname{Var} P^* + \overline{P^*}^2]\}$$

$$= \overline{W}_{\max} - (\overline{P} - \overline{P^*})^2 - [\operatorname{Var}(P) + \operatorname{Var}(P^*) - 2\operatorname{Cov}(P, P^*)].$$

As is intuitively obvious, variability in the environment (and thus variability in optimal phenotype, $P^*$) reduces fitness. Variability in realized phenotype also reduces fitness unless that is compensated for by its correlation with $P^*$. Thus if the genetic system responds accurately and quickly enough that $2\,\mathrm{Cov}\,(P, P^*) >$ $\mathrm{Var}\,(P)$, then it is beneficial for a population to respond genetically to changes in its environment. Where the genetic system is sufficiently rigid $[2\,\mathrm{Cov}\,(P, P^*) <$ $\mathrm{Var}\,(P)]$ a constant genotype is best. Constancy means lack of selective response, which implies both low additive genetic variance and consequently lowered phenotypic variance at any time. Genetic tracking of environmental changes requires increased $V_A$ which means the existence of individuals with nonoptimal genotypes and lowered average fitness for the population. Thus successful response to change raises $\overline{W}$ but depends on increased genetic variance which lowers $\overline{W}$. Levins (1964) has shown that genetic tracking results in a net increase in fitness over genetic constancy only when the environmental changes are autocorrelated (i.e. predictable) to a sufficient degree. Predictability of change, theoretically at least, allows for a higher covariance of $P$ and $P^*$ than would otherwise be attainable. Whether populations can avoid genetic change is another question. Probably it is possible to buffer the genetic system against all but very long-term and/or very strong changes in selective pressure, but not completely.

More rapid fluctuations in the environment can be treated with fitness sets and adaptive functions, the diagrams in Figs. 4.5 and 4.6 still being suitable for our purposes. Where changes are slight so that the fitness set is convex— or only slightly concave—optimal strategy calls for a generalized phenotype, capable of handling both situations. Where changes are sufficiently dramatic, optimal strategy calls for specialization. There is no reason to believe that natural selection will not act toward these ends. Let us examine the situation more closely. Changes occurring gradually, several times over an individual's life span, are generally responded to physiologically. For example, a man moving to an area of higher altitude finds his red blood cell count climbing. The response is reversible. Such animals as ptarmigan or weasels show body changes in the color of new feathers or hair following a molt, appearing brown in summer and white in winter. Long-term behavioral responses also categorize this type of response to change. The pika's hay-pile building and use for food in winter is an example, as is the ability of animals to change their foraging techniques seasonally in response to changes in prey densities and distribution. These are all examples of generalists' responses. Specialists respond in more dramatic ways, almost in a sense escaping one set of environmental conditions. For example, species as divergent as protozoans, rotifers, flatworms, and insects may form cysts in response to environmental deterioration. These cysts are resistant forms which represent a state of suspended animation. The animal is in a sense both adapting to and escaping from its environment. It cannot, however, be considered a generalist in a sense of displaying even remotely equal activity under different circumstances. Less extreme cases involve the migration, aestivation, and hibernation of animals. A meadow mouse (*Microtus*) which remains active all winter under the snow is a generalist, while its summer companion and hibernating winter drop-out, the jumping mouse (*Zapus*), is a specialist.

Still shorter-term changes in environment are generally met with behavioral (and, secondarily, rapid physiological) changes. Rises in temperature bring appropriate behavioral responses in ectotherms. Salamanders may burrow more deeply into the soil or frogs move to more moist areas. Attacking and fleeing are responses to rapidly changing situations. As opposed to these generalist type responses there exist specialized, pseudo-escape responses. Hummingbirds, because of their tremendously great metabolic needs (very high surface area to volume ratio) cannot maintain themselves for any great length of time without feeding. Thus at night they become torpid, lowering their basal metabolism rate below that maintained under normal, waking conditions. Finally, virtually all of the higher vertebrates as well as some other groups sleep at times during the day. Whether sleep is a semi-escape mechanism from certain conditions, or whether it is a physiologically necessary act which occurs during the more un-favorable circumstances faced by the species at certain times, is immaterial. Diurnal species such as birds or primates are daytime specialists just as flying squirrels are nighttime specialists.

In conclusion then, long term changes in the environment should be genetically tracked—that is, populations should respond genetically to the changes—when the autocorrelation of successive events is sufficiently high. Data on this matter are lacking. It is also undoubtedly true that long-term migrations may result from such changes. Changes which occur several times over the life span of an individual are usually responded to physiologically, or with migration. When the fitness set is convex, the individual should become generalized, meaning that the ability to physiologically adapt to changing conditions allows the con-tinuation of normal activities. If the fitness set is concave, specialization results, meaning that physiological adaptation no longer allows for normal activities in all circumstances, but allows for great efficiency under some conditions, and some form of inactivity and escape during others. Very rapid changes in situation generally require behavioral responses (although such rapid responses as brady-cardy are physiological).

Note that there is a close correlation between the nature of response to temporal and spatial changes. In a patchy environment approaching a fine-grained nature, movements from patch to patch constitute rapid changes of environment over time and will thus be accompanied by behavioral responses. Movements over less fine-grained environments, in which an animal spends sufficiently long enough periods of time, will be accompanied by physiological adjustments (human responses to several days in a new altitude or in a new time zone, for example). In coarse-grained environments, adjustments are more apt to be genetic, corresponding to long-term fluctuations of the environment in time. Often the different modes of adjustment complement one another. The butterfly *Colias eurytheme* in the San Joaquin valley of California is dimorphic. White morphs predominate in the early morning and evening and yellows are relatively abundant at midday, indicating different behavioral responses to a rapidly changing environment by the two morphs. It seems that the white form is more fit in cool or less bright situations and the yellow in warmer, brighter situations. (See the discussion of *Colias* under 4.1.) This hypothesis is borne out by the fact that northern populations, in cooler, less bright surroundings, display greater relative numbers of the white morph (Hovanitz, 1953). This population structure is consistent with the view that fluctuations in temperature or light intensity over periods of perhaps several years have presented the butterfly with a concave fitness set. Subsequent disruptive selection may have resulted in dimorphic populations in which the frequency of the morphs varied over gradients in the proportion of cool and warm years (north to south). Evolution of the gene complexes affecting the two morphs is reflected in the behavioral response to more rapid environmental change of the same nature. Concave fitness sets have promoted behavioral specialization to one set of conditions and partial escape from the other.

### 4.3.2 Heterosis and Polymorphism

Maynard Smith's model for the maintenance of a switch gene will not work in environments changing in time. There is, however, another approach that does work (and is also applicable to patchy environments). Denote the fitnesses of $A_1A_1$, $A_1A_2$, and $A_2A_2$ as shown below.

|  | $A_1A_1$ | $A_1A_2$ | $A_2A_2$ |
|---|---|---|---|
| Fitness in situation 1 | $W_1$ | $W_1 + a$ | $W_1 + b$ |
| Fitness in situation 2 | $W_2 + B$ | $W_2 + A$ | $W_2$ |

If a proportion $P$ of the time is spent under conditions 1, then

$$\overline{W}_{A_1A_1} = PW_1 + (1 - P)(W_2 + B)$$

$$\overline{W}_{A_1A_2} = P(W_1 + a) + (1 - P)(W_2 + A)$$

$$\overline{W}_{A_2A_2} = P(W_1 + b) + (1 - P)W_2.$$

Heterosis occurs when $\overline{W}_{A_1A_2} > \overline{W}_{A_1A_1}, \overline{W}_{A_2A_2}$. That is, when

$$P > \frac{(B - A)}{a + (B - A)}, \qquad \frac{A}{A + (b - a)}. \tag{4.5}$$

When fluctuating selective pressures occur, this condition is not necessarily met. Furthermore, if the selective pressures are too great or if the periodicity of the fluctuations is too long, selection may result in fixation before the reverse trend gets started. Where fluctuations are short and not too strongly marked, however, it is possible that they result in a kind of heterosis and thereby act to maintain a polymorphism. Short-term fluctuations may accomplish this end through another means: temporal isolation. For example, the earthworm, *Allolobophora chorotica*, has two morphs which are described as green and pink.

Aviary experiments showed green to have an advantage in terms of protective coloration (although in the absence of predation, pink displayed a higher fitness). In dry situations, however, when background color is no longer green, pink appears to be more fit. Thus fluctuations between wet and dry seasons (short term) favor first the green then the pink forms. The two morphs probably have different optimal reproductive periods or locations due to this fluctuation, and it is possible that reproductive isolation might result. Satchell (1967) has shown that, in fact, the hybrids of the two morphs have low viability, indicating that partial isolation has occurred for some time.

Table 4.4 summarizes the arguments pertaining to temporal heterogeneity.

## 4.4 MARGINAL POPULATIONS

The definition of "species" is in no sense settled and the usual meaning, defended by Mayr, of a set of interbreeding individuals is rather vehemently objected to in some quarters. For present purposes a working definition is offered. The reader who would rather not think of species in the sense presented can, if he likes, substitute some other word to fit the definition. The word "species" here represents a set of individuals descended from a single freely interbreeding population and retaining an arbitrary degree of phenetic similarity to the parent population. The degree of similarity may be whatever we wish it to be, but in the usual sense of species it is stretched to the point at which extreme samples can no longer produce viable offspring, or slightly farther.

Any such species displays a *range* or a collection of areas in which it lives. If the total number of individuals in the species remains roughly constant over time, then $\overline{W}$ must be roughly one over the whole range. No restriction, however, is placed on fitness at any point in the range. We will define *range center* as that (or those) areas in the range where fitness is high for a given population density, and *periphery* as those areas in which fitness is low given the same population density. Clearly, if individuals are capable of moving about they will tend to move toward areas of high fitness where their reproductive success will be greatest. However, as their numbers increase in one area they may begin to eat out the food supply or attract predators with the result that fitness drops (see Chapter 10). Population density will vary such that, at equilibrium, fitness is constant over virtually the entire range. However, beyond the limits of the area to which adaptation is possible, the few wandering individuals will show a fitness of less than one. Obviously, population maintenance in these extreme peripheral areas is possible only when a net migration occurs from those areas where $\overline{W} > 1$ to those in which $\overline{W} < 1$. Thus there comes a point at which further range extension would place individuals into such "rigorous" conditions with respect to their adaptive potential that even migration cannot maintain their numbers. This point represents the limits of the range and, by definition, determines the geographic limits of the species.

**Table 4.4.** Summary of arguments: Adaptation and population structure in temporally heterogeneous environments

| | Long term | | Shorter term | | Rapid (fluctuations) | |
|---|---|---|---|---|---|---|
| | conditions differ | | conditions differ | | conditions differ | |
| | greatly | little | greatly | little | greatly | little |
| Nature of response | Genetic. If autocorrelation between successive events is low, none (ideally) | | Physiological or behavioral | | Behavioral | |
| Phenotype | Monomorphic and specialized* | | Monomorphic and specialized (escape through migration, hibernation, etc.) | Monomorphic and generalized | Monomorphic and specialized (escape), or polymorphic | Monomorphic and generalized |

\* Specialized, generalized refer to a situation in time, ignoring spatial heterogeneity.

Populations inhabiting peripheral areas, near the limits of the range, are called *marginal* populations.

## 4.4.1 Selection Differences in Central and Marginal Populations

Selection pressures on peripheral populations are likely to be somewhat different from selection pressures on central populations. For example, individuals in the range centers will be living in an environment to which they are well adapted and will, for the large part, face their greatest challenges in the problems of social tolerance. Fitness in these areas, by the argument given above, is held to roughly one by virtue of high population densities. Thus marginal populations are more likely to be responding primarily to stresses originating with the physical and biological environments while selective pressures in central populations are more apt to involve intraspecific social competition. Central populations may evolve a greater ability to intimidate sexual competitors or attract mates but be less resistant to certain types of (say) temperature stress.

As physical conditions deteriorate toward range periphery, more energy must be used by individuals to maintain normal activities. Time available for such activities may decline due to necessities of survival not encountered at range center. As a result of tighter time and energy budgeting there will be less reserve time and energy for meeting new contingencies, so that the fitness set with respect to almost any set of conditions (other than social) occurring over time becomes progressively more concave. Thus individuals in marginal populations at any one point in space and time are apt to be adapted to a narrower range of living conditions (specialized, or *stenotopic*) than their center-range relatives. Another consideration leads to the following conclusion: If individuals choose their habitat patches in order of their suitability for survival, then marginal populations, their numbers being low, will more nearly be able to fit into one patch type than central populations in which high population densities necessitate spillover into less-desirable patch types. For either or both of these reasons, "Bird species, as is well known, are stenotopic at the margins of their distribution areas, whereas they are much more eurytopic (generalized) in the center of their breeding range" (Hilden, 1965). The above arguments are, of course, highly simplified, the real world being a vast collection of patch types and gradients. The conclusions reached, therefore, will not always apply although they should have significance as rough predictors.

## 4.4.2 Genetic Isolation

One characteristic often found in peripheral populations is asexuality or self-fertilization. This is particularly true in plant populations where, for example, the cobwebby gilia of California shows outbreeding over part of its range but becomes self-fertilizing and develops incompatibility barriers in other areas (Wallace, 1957; Grant, 1966). Wallace believes the areas of selfing in the cobwebby gilia to be marginal. The same is true of the grasses *Agrostis tenuis* and *Anthoxanthum odoratum*, in which marginal populations in areas adjacent to

mines show heavy metal tolerance and self-fertility while nontolerant plants a short distance away are strict outbreeders (Antonovics, 1968). In animals the situation may be more unusual but nevertheless occurs. Darevsky (1966) notes a correlation between natural parthenogenesis and so-called extreme environmental conditions in some invertebrates (see also Jahn, 1941; Suomalainen, 1947, 1953) and in the Caucasian rock lizard, *Lacerta saxicola*. There are, in this latter "species," 19 distinct geographic morphs. Four of these, those which occur at the species' range periphery, in mountains, are parthenogenic. These individuals occasionally, where sympatry occurs, will interbreed with the normal, bisexual form producing sterile, triploid offspring. In addition, Wright and Lowe (1968) point out that the habitats occupied by parthenogenetic species of lizards (*Cnemidophorus sp.*) may be characterized by such terms as marginal, edge, extreme, etc.

A number of possible explanations exist for this tendency. First, and perhaps most important, peripheral populations are generally of low population density so that unless well-developed social systems have evolved there is a smaller chance of the meeting of opposite sexes at any given time. Among plants and sessile animals which cannot move about, this factor may be very important. Self-fertilization or other asexual means of reproduction allow for rapid colonization in new areas. In the North American plant genus *Leavenworthia*, there are three selfing species with large ranges and three outbreeders with smaller ranges (Rollins, 1967). In the lizard *Lacerta*, these explanations are less likely to be true and one is tempted to search for other possible alternatives. Suppose, in a peripheral population, that a gene or gene combination beneficial to survival under the existing submarginal conditions appears. Immediately adjacent to the range limits, in extreme marginal areas where fitness lies below one, there must be a net influx of individuals from more central parts of the range. Gene flow from the central population, where the trait is more than likely not advantageous, will tend to dilute the frequency of this new innovation, retarding or even blocking its incorporation into the marginal population. The process of adaptation to the conditions at the periphery will essentially cease.

There is an escape from this dilemma. Any genetic change that will effectively isolate peripheral individuals from immigrants, genetically, will be indirectly beneficial since it will subsequently allow adaptation to occur. Isolating mechanisms such as polyploidy, apomixis, parthenogenesis, or self-fertilization will be selected for in spite of the various drawbacks in a normal (central) population. It is Antonovics' (1968) contention that self-fertility in the grasses he studied has evolved as a pollen flow blocking mechanism in the peripheral mine populations. It seems possible that parthenogenesis in the peripheral lizard populations studied by Darevsky has evolved for the same reason.

# Discussion I

## I.1 MULTIPLE ALLELES AND EPISTASIS

It would be unfortunate to give the impression that all or even most gene loci are characterized by only two allelomorphs. Most are characterized by several. Furthermore it should be noted that expressions for the rate of gene frequency change are likely to be complicated by epistasis. Fortunately the appropriate expressions are similar to those derived for the two-allele case. For example, the reader should, by now, easily be able to show that, for multiple alleles

$$p_k' = \frac{2(p_k^2 W_{kk}) + 1(2\sum_{j \neq k} p_k p_j W_{kj})}{2\overline{W}},$$

where $p_k$ is the frequency of the $k$th allele and $W_{ij}$ is the fitness of the $ij$ genotype. It follows that

$$\Delta p_k = \frac{2(p_k^2 W_{kk} + \sum_{j \neq k} p_k p_j W_{kj} - p_k \overline{W})}{2\overline{W}} = \frac{p_k(\sum_i p_i W_{ki} - \overline{W})}{\overline{W}}.$$

Now since

$$\sum_{i \neq k} p_i + p_k = 1,$$

it follows that

$$\sum_{i \neq k} \Delta p_i + \Delta p_k = 0,$$

so that the frequency of all other alleles must drop by $\sum_{i \neq k} \Delta p_i$ with a change, $\Delta p_k$. If the change in $p_k$ does not alter the relative proportions of the other alleles, then $\Delta p_j / \Delta p_k$ is

$$-\frac{p_j}{\sum_{i \neq k} p_i} = -\frac{p_j}{1 - p_k}.$$

We can thus write

$$\sum_{i,j} W_{ij} \frac{\partial(p_i p_j)}{\partial p_k}$$

$$= \sum_{i,j} W_{ij} \left( p_i \frac{\partial p_j}{\partial p_k} + p_j \frac{\partial p_i}{\partial p_k} \right)$$

$$= \sum_{i,j \neq k} W_{ij} \left[ p_i \left( -\frac{p_j}{1-p_k} \right) + p_j \left( -\frac{p_i}{1-p_k} \right) \right] + \sum_{i \neq k} W_{ik} \frac{\partial(p_i p_k)}{\partial p_k}$$

$$+ \sum_{j \neq k} W_{kj} \frac{\partial(p_k p_j)}{\partial p_k} + W_{kk} \frac{\partial p_k^2}{\partial p_k}$$

$$= -2 \frac{\sum_{i,j \neq k} W_{ij} p_i p_j}{1-p_k} + 2 \sum_{j \neq k} W_{kj} \left( p_j - \frac{p_j p_k}{1-p_k} \right) + 2 W_{kk} p_k$$

$$= 2 \left( \frac{-\sum_{i,j \neq k} p_i p_j W_{ij} - \sum_{j \neq k} p_j p_k W_{kj}}{1-p_k} + \sum_{j \neq k} (p_j W_{kj}) + W_{kk} p_k \right)$$

$$= 2 \left[ \left( \frac{-\sum_{i,j \neq k} p_i p_j W_{ij} - 2\sum_{j \neq k} p_j p_k W_{kj} - p_k^2 W_{kk}}{1-p_k} \right. \right.$$

$$\left. + \frac{\sum_{j \neq k} p_j p_k W_{kj} + p_k^2 W_{kk}}{1-p_k} \right) + \left( \sum_{j \neq k} p_j W_{kj} + p_k W_{kk} \right) \Bigg]$$

$$= 2 \left[ \left( \frac{-\sum_{i,j \neq k} p_i p_j W_{ij} - 2\sum_{j \neq k} p_j p_k W_{kj} - p_k^2 W_{kk}}{1-p_k} + \frac{\sum_{j \neq k} p_j p_k W_{kj} + p_k^2 W_{kk}}{1-p_k} \right) \right.$$

$$\left. + \left( \frac{\sum_{j \neq k} p_j W_{kj} + p_k W_{kk} - \sum_{j \neq k} p_j p_k W_{kj} - p_k^2 W_{kk}}{1-p_k} \right) \right]$$

$$= 2 \left[ -\frac{\sum_{i,j} p_i p_j W_{ij}}{1-p_k} + \left( \sum_{j \neq k} p_j p_k W_{kj} + p_k^2 W_{kk} + \sum_{j \neq k} p_j W_{kj} - \sum_{j \neq k} p_j p_k W_{kj} \right. \right.$$

$$\left. \left. + p_k W_{kk} - p_k^2 W_{kk} \right) \frac{1}{1-p_k} \right]$$

$$= 2 \left\{ -\frac{\overline{W}}{1-p_k} + \left[ \left( \sum_{j \neq k} p_j p_k W_{kj} - \sum_{j \neq k} p_j p_k W_{kj} \right) + (p_k^2 W_{kk} - p_k^2 W_{kk}) \right. \right.$$

$$\left. \left. + \left( \sum_{j \neq k} p_j W_{kj} + p_k W_{kk} \right) \right] \frac{1}{1-p_k} \right\}$$

$$= 2 \left[ \frac{-\overline{W} + (0) + (0) + \sum_j p_j W_{kj}}{1-p_k} \right] = 2 \frac{\sum_j p_j W_{kj} - \overline{W}}{1-p_k} .$$

Thus

$$\Delta p_k = \frac{p_k(1-p_k)}{2\overline{W}} \sum_{i,j} W_{ij} \frac{\partial(p_i p_j)}{\partial p_k} .$$

But

$$\overline{W} = \sum_{i,j} p_i p_j W_{ij}, \qquad \text{so} \qquad \frac{\partial \overline{W}}{\partial p_k} = \sum_{i,j} p_i p_k \frac{\partial W_{ij}}{\partial p_k} + \sum_{i,j} W_{ij} \frac{\partial (p_i p_j)}{\partial p_k}.$$

Hence

$$\sum_{i,j} W_{ij} \frac{\partial (p_i p_j)}{\partial p_k} = \frac{\partial \overline{W}}{\partial p_k} - \left( \overline{\frac{\partial W}{\partial p_k}} \right),$$

and

$$\Delta p_k = \frac{p_k(1 - p_k)}{2\overline{W}} \left( \frac{\partial \overline{W}}{\partial p_k} - \overline{\frac{\partial W}{\partial p_k}} \right). \tag{I.1}$$

In the absence of epistasis $\overline{\partial W/\partial p_k} = 0$, and the similarity of (I.1) and (1.12) is immediately apparent. A similar sort of correction can be made for assortative mating. Note that since $\overline{\partial W/\partial p_k}$ may exceed $\partial \overline{W}/\partial p_k$, selection may act to decrease $\overline{W}$.

When applying the laws of natural selection, you will usually find it much simpler to ignore the epistatic term (the discussion of fitness sets and adaptive functions, 4.2, for example), but this practice may be misleading; since the genetic value of an allele will depend on its *genetic background* at other loci, $\overline{\partial W/\partial p_k}$ is probably never zero. When fitness is a function of the relative abundances of phenotypes, and hence $p_k$, the epistatic factor will not be negligible. What is increased by selection, then, is not really fitness (or, actually, $\ln \overline{W^{1/T}}$), but rather the quantity

$$\int \left( \frac{\partial}{\partial p_k} (\overline{W^{1/T}}) - \frac{\overline{\partial (W^{1/T})}}{\partial p_k} \right) dp_k.$$

This fact should always be borne in mind.

## I.2 THE CONTINUOUS CASE

Suppose that individuals in a population reproduce often and at staggered intervals so that change in population size can be approximated with a continuous function,

$$dn/dt = rn,$$

where $n$ is population size and $r$ is population growth rate. Then it may be more convenient to write Eq. (1.12) as follows:

Let $r_{ij}$ be the growth rate of the $ij$ genotype, and using the same notation as in Chapter 1, note that over a time interval, $dt$, the number of $A_1$ genes ($2np$) increases proportionately by an amount $(1 + r_1 dt)$, where $r_1 = pr_{11} + qr_{12}$. Since the total number of genes increases by $(1 + \bar{r} dt)$, the new frequency of $A_1$ must be

$$\frac{1 + r_1 dt}{1 + \bar{r} dt} p,$$

so that

$$dp = \frac{1 + r_1 dt}{1 + \bar{r}dt} p - p = p \frac{r_1 - \bar{r}}{1 + \bar{r}dt} dt = p(r_1 - \bar{r})\, dt.$$

Thus

$$dp/dt = p(r_1 - \bar{r}) = p[(pr_{11} + qr_{12}) - (p^2 r_{11} + 2pqr_{12} + q^2 r_{22})]$$

$$= pq[p(r_{11} - r_{12}) + q(r_{12} - r_{22})] = \frac{pq}{2} \frac{\partial \bar{r}}{\partial p}.$$

## I.3 THE COST OF SELECTION

Suppose that selection acts simultaneously on a thousand loci each of whose most fit genotype has a one percent selective advantage. Then the most fit genotype imaginable in the population has a fitness $(1 + 0.01)^{1000} = 20{,}959$ times that of the hypothetical average genotype. The number of genetic deaths which occur because the average genotype has a fitness of only $\bar{W} < W_{max}$, is $W_{max} - \bar{W}$, and the fraction by which fitness of the population is decreased by virtue of the existence of less than maximally fit individuals is

$$L = \frac{W_{max} - \bar{W}}{W_{max}}.$$

This quantity is known as the *cost of selection*, or the *genetic load*.

In species which produce tremendous numbers of offspring, most of which die, it is not disturbing that there should be such great discrepancies in fitness as in the above example. While the average mosquito merely replaces itself each generation ($\bar{W} = 1$), some mosquitos will fail to reproduce and others may give rise to thousands of reproductively successful offspring. In such less fecund species as most vertebrates, however, simultaneous selection for many traits presents us with a problem. The human female is probably incapable of producing more than twenty children. Hence $W_{max}$ is on the order of $20/2 = 10$. Thus if 1000 loci are experiencing selection via persons after their birth, $\bar{W}$ must be about $10/20{,}959 = 0.0005$, and the human race must be dying out very rapidly. This, clearly, is not the case. Are we, then, to conclude that selection acts on only a few loci at any time, or that most selective advantages are extremely small (or both), or are there other ways out of the dilemma?

There may be several escapes, two of which are presented here. Wallace (1970) claims that where genetic deaths increase with population size (the population increases until it runs short of some resource and individuals either die or fail to reproduce as a result), the cost of selection may be alleviated.

If the load is due to the selection of new genes (*substitutional load*) this claim seems untrue. Load due to the maintenance of deleterious alleles (*segregational load*) is alleviated and may be eliminated by such density dependency (see Discussion III, Frequency and Density-Dependent Selection).

Frequency-dependent selection may alleviate and even eliminate segregational genetic load. Suppose a population is polymorphic at 1000 loci. This
would appear to reflect a large number of heterotic loci and thus a great
discrepancy in fitness between the average individual and one heterozygous at
all loci. But perhaps a large number of these loci are characterized by genotypes
whose fitness varies inversely with their frequency of occurrence. In the
simplest case, involving complete dominance, equilibrium occurs when
$W_1 = (W_2 = W_3)$. If, as an example, $W_1 = 1.1 - p$, $W_2 = W_3 = 0.75 + p$, then

$$1.1 - p = 0.75 + p, \quad \text{so} \quad p = 0.35/2 = 0.175.$$

Note that $W_{max}$ (with respect to this one locus) $= \overline{W}$, so that there is no genetic
load.

## I.4 EFFECTIVE POPULATION SIZE

It may be misleading, on occasion, to apply the principles of natural selection
to populations without regard to their size. Very small populations clearly will
be influenced by selection, but the overriding changes, as well as the most obvious
differences between such populations, may be statistical in origin, resulting from
random drift or the founder effect. When we wish to assess the effects of drift
within a population we must know the size of the population with which we are
working. But is the "size" of a population merely the number of individuals in it?
Clearly not, for where breeding is not random the population may more closely
resemble a collection of partly isolated subpopulations. We thus speak of *effective
population size*, a statistical concept, the value of which may be calculated from
knowledge of the mating system, sex ratio, true population size, and, possibly,
other parameters. The effective size of a population is that number of individuals
which theoretically would demonstrate the amount of genetic drift observed were
there random breeding. The effective population size is almost always less than
the actual number of individuals in a population.

## I.5 ADAPTEDNESS

The concept of *adaptation* seems, at first, a simple one. Natural selection is an
adaptive process through which adaptations are gained or perfected, rendering a
species adapted to its environment. This process can act in a wide variety of ways
on a wide range of characters. The frequency of genes and gene complexes
affecting various traits are modified, the coevolution and close linkage or separation
of loci are attained, and modifiers are selected which alter the exact expression
or degree of expression (penetrance) of genes. Because of mutation, recombination, migration, and assortative mating, variance is maintained and slowly
released so that selection virtually never ceases. Where highly specific responses
to stimuli are advantageous, selection can act to closely interlock genetic systems
in such a way that phenotypic expression is well buffered against minor genetic

or environmental variations (*canalization*).    Where, on the other hand, great plasticity of response is beneficial, buffering systems will not evolve, or if already present can be expected to degenerate, resulting in low heritability values.    It seems likely, for example, that the increased neural development in higher vertebrates has made learning an efficient enough tool in coping with environmental changes that stereotyped responses and their corresponding genetic buffering systems have degenerated.    This allows more plasticity in behavior and thus increases still further the value of learning.

Suppose, however, that we ask the question: What is "adaptedness"?    Or suppose we wished to be able to measure the degree to which a species (or a population) was adapted to its environment, or predict the nature of adaptations in an as yet unstudied species.    A number of difficulties immediately present themselves.    For the moment, assume that there is in fact an optimal pheno-type for any given environmental situation.    Suppose, for example, that as environmental fluctuations occur, tails in mice inhabiting the environment take on varying optimal lengths.    Since the tail lengths of mice in one generation are the results of natural selection in preceding generations it is clearly impossible for tail length always to be optimal.    There must always be a lag between ideal and realized characters.    A hypothetical situation is pictured below in Fig. I.1.    Even if we assume that an optimal value exists, the fact that environmental changes are not perfectly predictable makes quantitative measure of the value of optimal characteristics and short-term prediction of future adaptive trends impossible.

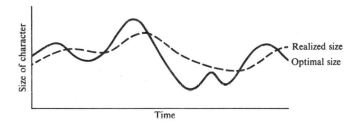

Fig. I.1. Optimal and realized phenotype over time: the evolutionary lag.

There is another problem surrounding the prediction of adaptations. Lewontin and White (1960) measured fitness in a large number of populations of a grasshopper, *Moraba scurra*, each displaying a different frequency combination of two genes.    Using frequency of the first and frequency of the second gene as axes, the data were plotted as isoclines of fitness.    The result resembles a contour map on which the high points, or *adaptive peaks*, represent high fitness values and the low points, or *adaptive valleys*, represent low fitness values.    The authors point out (see also Lewontin, 1965, Wright, 1956a) that natural selection modifies gene frequencies in the direction of the adaptive peaks.    Which peak is climbed, however, may depend on a number of factors, and the peak climbed may not be the highest one (see fitness sets, Chapter 4). Bateson (1963) points out that before an animal can use a supposedly beneficial trait it must be able to cope with that trait's side effects. A long neck would be of no advantage to a giraffe who had a heart too small or weak to accomodate the increase in required work load. Which potentially useful adaptation is evolved (which adaptive peak is climbed) depends on the preadaptive nature of large numbers of other characteristics (see also discussion of the founder principle, Chapter 1). Such considerations may render even the most beautifully reasoned predictions of adaptation useless unless backed by considerable amounts of supporting data. The theory of optimal genetic response represents a useful point of departure for ecological model building but has severely limited usefulness and must be considered in this light.

Finally there are both conceptual and semantic difficulties surrounding the word adaptedness. Those who wish to equate adaptedness with fitness should note that fitness is the measure of the success with which one genotype contributes to the next generation; average fitness is defined as the number of genes (or individuals) of one generation divided by the similar measure in the parental generation in the absence of immigration and emigration. It is important to note that generally it can have no meaning when applied to individuals, since one cannot trace reproductive success of a sexually reproducing animal independently of its mate and its offspring, all of whom will generally possess different genotypes. Ecologically, a fitness of greater than one implies a growing population and herein lies the basis for much of the semantic confusion surrounding the joining of genetics and ecology. If selection acts to increase fitness, then, by definition it would seem, selection acts to increase rate of population increase.

But populations do not forever grow, nor do they, for not having $\overline{W} > 1$, necessarily go extinct. Selection may actually *decrease* population size (and fitness may also be lowered through inbreeding or non-random mating). The apparent paradox is really very simply discarded when one realizes that selection increases fitness only for genetic change, that is, assuming an unchanging environment. But as selection changes the nature of a population, or its size, the environment is, by definition, also changed. If the change is beneficial to the population, $\overline{W}$ will, in fact, rise. However, there is no *a priori* reason to believe that the particular environmental change brought about will be beneficial and, in view of the already fine-tuned nature of the population experiencing it, it is reasonably certain that the change will be deleterious. Thus a selected genetic change will act in such a way that, if the environment were to remain unchanged, fitness would increase and the population grow more rapidly. But the resulting alterations in the environment may render the genetic change disadvantageous so that, actually, fitness falls. Professor Warburton (personal communication) relates the hypothetical case of a horse population which is capable of cropping the grass to a height of two inches and in doing so gains sufficient energy to maintain a stable population. Suppose that a new genetic innovation suddenly appears which enables its owners to crop the grass to one inch. Momentarily, the energy available to the population increases, until enough forage is eaten to one inch that the old genotypes are incapable of surviving. A new equilibrium population density will be reached which may or may not be higher than the original one. As adaptations of this sort continue to appear, there will eventually be a point reached at which the grass can no longer replace itself so efficiently and its productivity goes down. At this point less food is available to the population which must consequently decline. Finally, when the shoots of grass are cropped short enough, no grass reproduction occurs, the food supply disappears, and the horses go extinct. Selection responds to the *relative* fitness values of different genotypes and a genotype may increase in relative fitness by dragging down the fitnesses of other genotypes. Clearly, adaptedness and fitness cannot be the same.

What, then, is adaptedness? A number of suggestions have been made. One might consider a species which maintains high population densities to be well adapted. This makes good sense in that central populations which are well adapted to their local circumstances are generally more dense than marginal populations. This sort of reasoning, however, is quite circular and the above hypothetical example of the horses should make it clear that selection, the process of adaptation, does not necessarily result in increased population size. Population stability in the sense of amplitudes of fluctuation or sensitivity to environmental perturbation is another possible measure of adaptedness. This measure, however, would classify a tremendous number of species such as tent caterpillars, japanese beetles, and locusts as very poorly adapted because of their periodic dramatic changes in population. Thoday (1953, 1958a) and Slobodkin (1968) have suggested that long-term persistence of a species is a good measure of adaptedness. This concept is an admirable one and probably the best to date. Its only weakness, unfortunately a serious one, is that adaptedness then becomes a quantity measurable only upon a species' final demise. It might be modified to deal with local populations on a statistical basis, however, in which case it becomes slightly more practical. Nevertheless, problems of definition of local populations and local extinction make an application difficult and probably, as a meaningful contribution to ecological knowledge, not worth the effort.

In conclusion it might be most practical to say that through natural selection species gain adaptations, thereby adapting to their environments, but "adaptedness" is probably a useless term.

## BIBLIOGRAPHY, SECTION I

Allison, A. C., 1955, Aspects of polymorphism in man, *Cold Spr. Harbor Symp. Quant. Biol.* **20:** 239–255.

Anderson, P. K., 1964, Lethal alleles in *Mus musculus,* local distribution and evidence for isolation of demes, *Science* **145:** 177–178.

Antonovics, J., 1968, Evolution in closely adjacent plant populations, V, Evolution of self-fertility, *Heredity* **23:** 219–238.

Arnheim, N., and C. E. Taylor, 1969, Non-Darwinian evolution: consequences for neutral allelic variation, *Nature* **223:** 900–903.

Bateson, G., 1963, The role of somatic change in evolution, *Evol.* **17:** 529–539.

Beach, F. A., and B. J. LeBoef, 1967, Coital behavior in dogs, I, Preferential mating in the bitch, *Anim. Beh.* **15**: 546–558.

Blair, F. W., 1947, Variation in shades of pelage of local populations of the cactus mouse (*Peromyscus eremicus*), in the Tularosa Basin and adjacent areas of southern New Mexico, *Contr. Lab. Vert. Biol.* **37**: 1–7.

Brown, J. L., 1970, Cooperative breeding and altruistic behavior in the Mexican jay, *Aphelocoma ultramarina, Anim. Beh.* **18**: 366–378.

Bullock, T. H., 1955, Compensation for temperature in the metabolism and activity of poikilotherms, *Biol. Rev.* **30**: 311–342.

Cain, A. J., J. M. B. King, and P. M. Sheppard, 1960, New data on the genetics of polymorphism in the snail *Cepaea nemoralis* L., *Genetics* **45**: 393–411.

Cain, A. J., and P. M. Sheppard, 1954, Natural selection in *Cepaea, Genetics,* **39**: 89–116.

Carrick, R., 1963, Ecological significance of territoriality in the Australian magpie, *Gymnorhina tibicen, Proc. XIII Intern. Ornith. Congr.* pp. 740–753.

Chabora, A. J., 1968, Disruptive selection for sternopleura chaeta number in *Drosophila melanogaster, Amer. Natur.* **102**: 525–532.

Clarke, B., 1966, The evolution of morph-ratio clines, *Amer. Natur.* **100**: 389–402.

Clarke, C. A., and Sheppard, P. M., 1966, A local survey of the distribution of the industrial melanic forms in the moth *Biston betularia* and estimates of the selective values of these in the industrial environment, *Proc. Roy. Soc.* B **165**: 424–439.

Conoway, C., and C. Koford, 1963, Estrous cycles and mating behavior in a free-ranging band of rhesus monkeys, *J. Mammal.* **45**: 577–588.

Cooch, F. G., 1961, Ecological aspects of the blue/snow goose complex, *Auk* **78**: 72–89.

Cooch, F. G., 1963, Recent changes in distribution of color phases of *Chen c. caerulescens, Proc. XIII Intern. Ornith. Congr.* pp. 1182–1194.

Cooke, F., and F. G. Cooch, 1968, The genetics of polymorphism in the goose *Anser caerulescens, Evol.* **22**: 289–300.

Cotterman, C. W., 1940, A calculus for statistico-genetics, Unpublished thesis, Ohio State University, Columbus, Ohio.

Creed, E. R., 1971, Industrial melanism in the two-spot ladybird and smoke abatement, *Evol.* **25**: 290–293.

Creed, E. R., W. G. Dowdeswell, E. B. Ford, and K. H. McWhirter, 1959, Evolutionary studies on *Maniola jurtina:* the English mainland, 1956, 1957, *Heredity* **13**: 363–391.

Creed, E. R., W. G. Dowdeswell, E. B. Ford, and K. H. McWhirter, 1963, Evolutionary studies on *Maniola jurtina:* the English mainland, 1958–1960, *Heredity* **17**: 237–257.

Crow, J. F., 1955, General theory of population genetics, synthesis, *Cold Spr. Harbor Symp. Quant. Biol.* **20**: 54–59.

Crow, J. F., and M. Kimura, 1965, Evolution in sexual and asexual populations, *Amer. Natur.* **99**: 439–450.

Crow, J. F., and M. Kimura, 1969, Evolution in sexual and asexual populations: reply, *Amer. Natur.* **103**: 89–91.

Crow, J. F. and M. Kimura, 1970, *An Introduction to Population Genetics Theory,* Harper and Row, New York.

Crozier, R. H., 1970, Coefficients of relationship and the identity of genes by descent in the Hymenoptera, *Amer. Natur.* **104**: 216–217.

Darevsky, I. S., 1966, Natural parthenogenesis in a polymorphic group of Caucasian rock lizards related to *Lacerta saxicola, J. Ohio Herp. Soc.* **5**: 115–152.

Darnell, R. M., P. Abramoff, and E. Lamb, 1965, Matroclinous inheritance and clona structure of a population of Mexican poeciliid fish, *Amer. Zool.* **5**: 204.

Darwin, C., 1898, *The Descent of Man and Selection in Relation to Sex,* Appleton, N.Y.

DeVore, I., and S. L. Washburn, 1963, Baboon ecology and human behavior, in Howell, F. C., and F. Bourliere (Eds.), *African Ecology and Human Evolution,* Wenner–Gren Foundation, Aldine, Chicago.

Dobzhansky, T., and P. C. Koller, 1939, Sexual isolation between two species of *Drosophila:* A study on the origin of an isolating mechanism, *Genetics* **24:** 97–98.

Dobzhansky, T., and N. P. Spassky, 1962, Genetic drift and natural selection in experimental populations of *Drosophila pseudoobscura, Proc. Nat. Acad. Sci.* **48:** 148–156.

Dowdeswell, W. G., and E. B. Ford, 1953, The influence of isolation on variability in the butterfly *Maniola jurtina* (L), *Symp. Soc. Exper. Biol.* **7:** 254–273.

Dunbar, M. J., 1960, The evolution of stability in marine environments: Natural selection at the level of the ecosystem, *Amer. Natur.* **94:** 129–136.

East, E. M., 1936, Heterosis, *Genetics* **21:** 375–397.

Ehrman, L., 1965, Direct observations of sexual isolation between allopatric and between sympatric strains of the different *Drosophila paulistorum* races, *Evol.* **19:** 459–464.

Ehrman, L., 1966, Mating success and genotype frequency in *Drosophila, Anim. Beh.* **14:** 332–339.

Ehrman, L., 1967, Further studies on genotype frequency and mating success in *Drosophila, Amer. Natur.* **101:** 415–424.

Ehrman, L., 1970, The mating advantage of rare males in *Drosophila, Proc. Nat. Acad. Sci.* **65:** 345–348.

Emlen, J. T., 1954, Territory, nest-building, and pair-formation in the cliff swallow, *Auk* **71:** 16–35.

Falconer, D. S., 1960, *Introduction to Quantitative Genetics,* Ronald Press, New York.

Felsenstein, J., 1965, The effect of linkage on directional selection, *Genetics* **52:** 349–363.

Fisher, R. A., 1928, The possible modification of the response of the wild type to recurrent mutations, *Amer. Natur.* **62:** 115–126.

Fisher, R. A., 1958, *The Genetical Theory of Natural Selection* (second edition), Dover, New York.

Ford, E. B., 1964, *Ecological Genetics,* Broadwater Press, Welwyn Garden City.

Grant, V., 1966, The selective origin of incompatibility barriers in the plant genus *Gilia, Amer. Natur.* **100:** 99–118.

Guhl, A. M., and D. C. Warren, 1946, Number of offspring sired by cockerels related to social dominance in chickens, *Poultry Sci.* **25:** 460–472.

Haldane, J. B. S., 1932, *The Causes of Evolution,* Longmans, Green, London and N.Y.

Haldane, J. B. S., 1939, The theory of the evolution of dominance, *J. Genet.* **37:** 369–374.

Haldane, J. B. S., 1955, Population genetics, *New Biology,* **18:** 34–51.

Hamilton, T. H., and R. H. Barth, 1962, The biological significance of seasonal change in male plumage appearance in some new world migratory bird species, *Amer. Natur.* **96:** 129–144.

Hamilton, W. D., 1963, The evolution of altruistic behavior, *Amer. Natur.* **97:** 354–356.

Hamilton, W. D., 1964a, The genetical evolution of social behavior, I, *J. Theoret. Biol.* **7:** 1–16.

Hamilton, W. D., 1964b, The genetical evolution of social behavior, II, *J. Theoret. Biol.* **7:** 17–52.

Haskins, C. P., E. F. Haskins, and R. E. Hewitt, 1960, Pseudogamy as an evolutionary factor in the poeciliid fish *Molynesia formosa, Evol.* **14:** 473–483.

Herbert, J., 1968, Sexual preference in the rhesus monkey (*Macaca mulatta*) in the laboratory, *Anim. Beh.* **16:** 120–128.

Hershkowitz, P., 1962, The evolution of neotropical cricetine rodents, *Fieldiana, Zool.* No. 46.

Herter, K., 1924, Temperature sense, *Z. vergl. physiol.* **1:** 221–388.

Herter, K., 1925, Temperature sense, *Z. vergl. physiol,* **2:** 226–232.

Hilden, O., 1965, Habitat selection in birds, *Ann. Zool. Fenn.* **2:** 53–75.

Hovanitz, W., 1953, Polymorphism and evolution, *Symp. Soc. Exp. Biol.* **7:** 238–253.

Hubbs, C. L., and L. C. Hubbs, 1932, Apparent parthenogenesis in a form of fish of hybrid origin, *Science* **76:** 628–630.

Jahn, E., 1941, Uber parthenogenese bei forstshadlichen Otiorrhynchus arten in den Wahrend der Eiszeit vergletscherten Gebieten det Ostalpen, *Zeitschr. Angew. Entom.* **28**: 366–372.

Jain, S. K., and A. D. Bradshaw, 1966, Evolutionary divergence among adjacent plant populations, *Hered.* **21**: 407–441.

Johnston, R. F., 1954, Variation in breeding season and clutch size in song sparrows of the Pacific coast, *Condor* **56**: 268–273.

Jones, D. F., 1917, Dominance of linked factors as a means of accounting for heterosis, *Genetics* **2**: 466–479.

Kalmus, H., and S. Maynard Smith, 1967, Some evolutionary consequences of pegmatypic mating systems, *Amer. Natur.* **100**: 619–633.

Karn, M. N., and S. Penrose, 1951, Birth weight and gestation time in relation to maternal age, parity, and infant survival, *Ann. Eugen.* **16**: 147–164.

Kettlewell, H. B. D., 1955, Selection experiments on industrial melanism in the Lepidoptera, *Hered.* **9**: 323–342.

Kettlewell, H. B. D., 1956, Further selection experiments on industrial melanism in the Lepidoptera, *Hered.* **10**: 287–301.

Kettlewell, H. B. D., 1958, A survey of the frequencies of *Biston betularia* (L.) (LEP.) and its melanic forms in Great Britain, *Hered.* **12**: 51–72.

Kettlewell, H. B. D., 1961, The phenomenon of industrial melanism in Lepidoptera, *Ann. Rev. Ent.* **6**: 245–262.

Kettlewell, H. B. D., and R. J. Berry, 1969, Gene flow in a cline, *Amathes glariosa* Esp., and its melanic form, *Hered.* **24**: 1–14.

Kimura, M., 1958, On the change in population fitness by natural selection, *Hered.* **12**: 145–167.

Kimura, M., 1965, Attainment of quasi-linkage equilibrium when gene frequencies are changing by natural selection, *Genetics* **52**: 875–890.

King, J. L., and T. H. Jukes, 1969, Non-Darwinian evolution, *Science* **164**: 788–798.

King, J. W. B., 1950, Pygmy, a dwarfing gene in the house mouse, *J. Hered.* **41**: 249–252

King, J. W. B., 1955, Observations on the mutant "pygmy" in the house mouse, *J. Genet.* **53**: 487–497.

Kluge, A. G., and M. J. Eckardt, 1969, *Hemidactylus garnotii* (Dumeril and Biron), a triploid all-female species of a gekkonid lizard, *Copeia* 651–664.

Lerner, I. M., and E. R. Dempster, 1951, Attenuation of genetic progress under continued selection in poultry, *Hered.* **5**: 75–84.

Levins, R., 1964, The theory of fitness in a heterogeneous environment, IV, The adaptive significance of gene flow, *Evol.* **18**: 635–638.

Levins, R., 1968, *Evolution in Changing Environments,* Princeton Univ. Press, Princeton, N.J.

Lewontin, R. C., 1965, Selection in and of populations, in Moore, J.A. (Ed.), *Ideas in Modern Biology,* XVI Intern. Congr. Zool.

Lewontin, R. C., 1967, Population Genetics, *Ann. Rev. Genetics* **1**: 37–70.

Lewontin, R.C., and J. L. Hubby, 1966, A molecular approach to the study of genic heterozygosity in natural populations, II, Amount of variation and degree of heterozygosity in natural populations of *Drosophila pseudoobscura, Genet.* **54**: 595–609.

Lewontin, R. C., and M. J. D. White, 1960, The interaction between inversion polymorphism of two chromosome pairs in the grasshopper *Moraba scurra, Evol.* **14**: 116–129.

Lowe, C. H., and J. W. Wright, 1966, Chromosomes and karyotypes of cnemidophorine teiid lizards, *Mamm. Chrom. Newsletter* **22**: 199–200.

Mainardi, D., F. M. Scudo, and D. Barbieri, 1965, Assortative mating based on early learning: population genetics, *L'Ateneo Parmense* **36**: 583–605.

Malecot, G. M., 1948, *Les Mathématiques de l'hérédité,* Masson, Paris.

Marler, P., 1956, Behavior of the chaffinch, *Fringilla coelebs, Beh. Suppl.* **5**: 1–184.

Maslin, T. P., 1962, All-female species of the lizard genus *Cnemidophorus* (Teiidae), *Science* **135**: 212–213.

Maslin, T. P., 1971, Conclusive evidence of parthenogenesis in three species of *Cnemidophorus* (Teiidae), *Copeia* 156–158.

Mather, K., 1953, Genetic control of stability in development, *Hered.* **5**: 75–84.

Maynard Smith, J., 1962, Disruptive selection, polymorphism, and sympatric speciation, *Nature* **195**: 60–62.

Maynard Smith, J., 1964, Kin selection and group selection, *Nature* **201**: 1145–1147.

Maynard Smith, J., 1965, The evolution of alarm calls, *Amer. Natur.* **99**: 59–63.

Maynard Smith, J., 1968, Evolution in sexual and asexual populations, *Amer. Natur.* **102**: 469–473.

Merrill, D. J., 1960, Mating preferences in *Drosophila, Evol.* **25**: 525–526.

Morris, D., 1954, The reproductive behavior of the zebra finch (*Poephila guttata*), *Behavior* **6**: 271–322.

Mountfort, G., 1956, The territorial behavior of the hawfinch (*Coccothraustes coccothraustes*), *Ibis* **98**: 490–495.

Muller, H. J., 1932, Some genetic aspects of sex, *Amer. Natur.* **68**: 118–138.

Neill, W. T., 1964, Viviparity in snakes: some ecological and zoogeographic considerations, *Amer. Natur.* **98**: 35–55.

Newman, M. T., 1966, Adaptation of man to cold climates, in J. Bresler (Ed.), *Readings in Human Ecology,* Addison-Wesley, Reading, Mass.

Noble, G. K., and H. T. Bradley, 1933, The mating behavior of lizards; its bearing on the theory of sexual selection, *Ann. N.Y. Acad. Sci.* **35**: 25–100.

Oldham, R. S., 1967, Orienting mechanisms of the green frog, *Rana clamitans, Ecol.* **48**: 477–491.

Orians, G. H., and G. M. Christman, 1968, A comparative study of the behavior of redwing, tricolor, and yellow-headed blackbirds, *Univ. Calif. Publ. Zool.* **84.**

Peterson, R. T., 1941, *A Field Guide to the Western Birds,* Houghton Mifflin, Boston.

Robel, R. J., 1967, Booming territory size and mating success of the greater prairie chicken (*Tympanuchus cupidopinnautus*) *Anim. Beh.* **14**: 328–331.

Rollins, R. C., 1967, The evolutionary fate of inbreeders and nonsexuals, *Amer. Natur.* **101**: 343–351.

Sadoglu, P., 1967, The selective value of eye and pigment loss in Mexican cave fish, *Evol.* **21**: 541–549.

Satchell, J. F., 1967, Colour dimorphism in *Allolobophora chorotica* Sav. (Lumbricidea), *J. Anim. Ecol.* **36**: 623–630.

Scharloo, W., 1964, The effect of disruptive and stabilizing selection on a cubitus interruptis mutant in *Drosophila melanogaster, Genetics* **50**: 553–562.

Scharloo, W., 1971, Reproductive isolation by disruptive selection: Did it occur? *Amer. Natur.* **105**: 83–86.

Scharloo, W., M. S. Hoogmoed, and A. Ter Kuile, 1967, Stabilizing and disruptive selection on a mutant character in *Drosophila, I,* The phenotypic variance and its components, *Genetics* **56**: 709–726.

Scholander, P. F., 1955, Evolution of climatic adaptations in homeotherms, *Evol.* **9**: 15–26.

Selander, R. K., 1965, On mating systems and sexual selection, *Amer. Natur.* **99**: 129–141.

Sheppard, P. M., 1951, Fluctuations in the selective value of certain phenotypes in the polymorphic land snail, *Cepaea nemoralis, Hered.* **5**: 124–134.

Sheppard, P. M., 1960, *Natural Selection and Heredity,* Harper and Row, New York.

Sibley, C. G., 1957, The evolutionary and taxonomic significance of sexual dimorphism and hybridization in birds, *Condor* **59**: 166–191.

Sibley, C. G., 1961, Hybridization and isolating mechanisms, in W. F. Blair (Ed.), *Vertebrate Speciation,* University of Texas Press, Austin.

Siegel, R. W., 1955, Mating types in *Oxytrichia* and the significance of mating type systems in ciliates, *Biol. Bull.* **110**: 352–357.

Skutch, A. F., 1940, Some aspects of Central American bird life, *Sci. Monthly* **51**: 409–418.

Skutch, A. F., 1961, Helpers among birds, *Condor* **63**: 198–226.

Slobodkin, L. B., 1953, An algebra of population growth, *Ecol.* **34**: 513–519.

Slobodkin, L. B., 1968, Toward a predictive theory of evolution, in R. C. Lewontin, (Ed.), *Population Biology and Evolution*, Syracuse University Press, Syracuse, N.Y.

Smith, D. G., The role of the male's epaulet in the redwing blackbird (*Agelaius phoeniceus*) social system, *Behavior,* in press.

Sorenson, M. W., and C. H. Conoway, 1966, Observations on the social behavior of tree shrews in captivity, *Folia Primat.* **4**: 124–145.

Stalker, H., 1956, On the evolution of parthenogenesis in Lonchoptera (Diptera), *Evol.* **10**: 345–389.

Struhsaker, J. W., 1968, Selective mechanisms associated with intraspecific shell variation in *Littorina picta* (Prosobranchia: Mesogastropoda), *Evol.* **22**: 459–480.

Sturtevant, A. H., and K. Mather, 1938, The interrelations of inversions, heterosis, and recombination, *Amer. Natur.* **72**: 447–452.

Sudd, J. H., 1967, *An Introduction to the Behavior of Ants,* St. Martins, New York.

Suomalainen, B., 1947, Parthenogenese und Polyploidie bei Rüsselkäfern (Curculionidae), *Hereditas* **33**: 425–456.

Suomalainen, B., 1953, Die Polyploidie bei den parthenogenetischen Rüsselkäfern, *Zool. Anz. Suppl.* **17**: 280–289.

Thoday, J. M., 1953, Components of fitness, *Symposium Society Experimental Biology* **7**: 96–113.

Thoday, J. M., 1958a, Natural selection and biological progress, in S. A. Barnett (Ed.), *A Century of Darwin*, Heinemann, London.

Thoday, J. M., 1958b, Effects of disruptive selection: experimental production of a polymorphic population, *Nature* **181**: 1124–1125.

Thoday, J. M., and J. B. Gibson, 1962, Isolation by disruptive selection, *Nature* **193**: 1164–1166.

Thoday, J. M., and J. B. Gibson, 1970, The probability of isolation by disruptive selection, *Amer. Natur.* **104**: 219–230.

Thoday, J. M., and J. B. Gibson, 1971, Reply to Scharloo, *Amer. Natur.* **105**: 86–88.

Trivers, R. L., 1971, The evolution of reciprocal altruism, *Quart. Rev. Biol.* **46**: 35–57.

Verner, J., 1963a, Aspects of the ecology and evolution of the long-billed marsh wren, Ph.D. thesis, University of Washington, Seattle.

Verner, J., 1963b, Song rates and polygamy in the long-billed marsh wren, *Proc. XIII Intern. Ornith. Congr.* pp. 299–307.

Wallace, B., 1957, Influence of genetic systems on geographic distribution, *Cold Spr. Harbor Symp. Quant. Biol.* **24**: 193–204.

Wallace, B., 1970, *Genetic Load, Its Biological and Conceptual Aspects*, Concepts in Modern Biology series, Prentice Hall, Englewood Cliffs.

Warburton, F. E., 1967, Increase in the variance of fitness due to selection, *Evol.* **21**: 197–198.

Warriner, C. C., W. B. Lemmon, and T. S. Ray, 1963, Early experience as a variable in mate selection, *Anim. Beh.* **11**: 221–224.

Washburn, S. L., P. C. Jay, and J. B. Lancaster, 1965, Field studies of Old World monkeys and apes, *Science* **150**: 1541–1547.

Watt, W. B., 1968, Adaptive significance of pigment polymorphisms in *Colias* butterflies, I, Variation of melanin pigment in relation to thermoregulation, *Evol.* **22**: 437–458.

Wichterman, R., 1967, Mating types, breeding systems, conjugation, and nuclear phenomena in the marine ciliate *Euplotes cristatus* Kahl from the gulf of Naples, *J. Protoz.* **14:** 49–58.

Wright, J. W., and C. H. Lowe, 1968, Weeds, polyploids, parthenogenesis, and the geographical and ecological distribution of all-female species of *Cnemidophorus, Copeia* 128–138.

Wright, S., 1934, Physiological and evolutionary theories of dominance, *Amer. Natur.* **68:** 25–53.

Wright, S., 1937, The distribution of gene differences in populations, *Proc. Nat. Acad. Sci.* **23:** 307–320.

Wright, S., 1945, Tempo and mode in evolution: a critical review, *Ecol.* **26:** 415–419.

Wright, S., 1948, On the roles of directed and random changes in gene frequency in the genetics of populations, *Evol.* **2:** 279–294.

Wright, S., 1956a, Modes of selection, *Amer. Natur.* **90:** 5–24.

Wright, S., 1956b, Classification of the factors of evolution, *Cold Spr. Harbor Symp. Quant. Biol.* **20:** 16–24.

Wright, T. D., and A. D. Hasler, 1967, An electrophoretic analysis of the effects of isolation and homing behavior upon the serum proteins of the white bass (*Roccus chrysops*) in Wisconsin, *Amer. Natur.* **101:** 401–414.

Wynne-Edwards, V. C., 1962, *Animal Dispersion in Relation to Social Behaviour,* Oliver and Boyd, Edinburgh.

Young, H., 1963, Age-specific mortality in the eggs and nestlings of blackbirds, *Auk* **80:** 145–155.

# Section II
# The Ecology of
# Individuals

# 5
# Defense against Predation

In Section I we noted that natural selection acts to increase the value of a quantity roughly described by $\ln(\overline{W^{1/T}})$, within any given set of environmental conditions. The biological variables which determine this quantity are survival, mating success (and age), and fecundity. Thus a gene combination which predilects its owner to some morphology, behavior, etc., increasing survival, mating success, or fecundity such that $\ln(\overline{W^{1/T}})$ rises, eventually increases in frequency in the population. If we work under this assumption, and remember the basic possibilities and limitations of genetic mechanisms, it should be possible to construct a theoretical framework predicting patterns of individual adaptations. We use this approach in the next four chapters on the ecology of the individual.

Before we proceed, a number of difficulties besetting an evolutionary approach to individual ecology must be mentioned. First, there are a great number of environmental parameters which may affect the direction of natural selection. The evolution of one trait may depend on the existence of some other, or the chance coevolution of still another. It may be possible to work out all the details of one situation (say, age of nesting) for one species, considering all impinging variables; but the resulting hypotheses, while they may be accurate, must be restricted in scope. More significantly, basic biological relations may be lost in the mire of details. On the other hand, more general hypotheses (models) depend on a large number of simplifying assumptions or simply the ignoring of many parameters which, we hope, are of relatively minor significance. This simple model approach, which we follow in this book, clearly cannot yield quantitative theories, but merely rules of a very rough nature. Thus we may predict a positive correlation between avian clutch size and latitude; we expect a correlation only and are not surprised at (perhaps numerous) exceptions. We sacrifice specificity and great reliability in the interests of generality and a glimpse at basic biological relations.

Such rule-of-thumb hypotheses have the unpleasant property of being unfalsifiable. They can be "tested" only by the accumulation of great amounts of natural history information. Such information accumulates slowly. To make matters worse, the clever evolutionist can guess "reasons" for almost any character observed in nature, thereby adding vast numbers of often nonsensical hypotheses to the already overflowing collection of inadequately tested ideas.

These facts make the evolutionary study of individual ecology dangerously susceptible to circularity and bias.

In this book, limited space, author's patience, and readers' span of attention preclude inclusion of adequate supporting data for most concepts. The student should supplement his reading to convince himself that most of the ideas presented here have considerable (although seldom overwhelming) support, but also remain aware of the dangers of speculation in this area.

A final point must be made. There is, at least in most cases, no way to reconstruct the evolution of a trait. The factors which led to the evolution of a trait may have been entirely different from those which maintain it today. When we assess the adaptive significance of protective coloration, flock feeding, or territoriality, for example, we are looking at the reason for these traits' continued existence and *not* their origin.

Clearly, one result of selection for reduced mortality is the evolution of various mechanisms of defense against predators. In this chapter we divide such mechanisms into four categories. The first involves chemical and anatomical adaptations, the second morphological adaptations, the third behavioral, and the fourth patterns of codispersion with other organisms. This breakdown is convenient for purposes of discussion, although the categories are in no way exclusive.

## 5.1 CHEMICAL AND ANATOMICAL MECHANISMS

A common form of chemical defense is the production or incorporation of noxious substances and toxins. Many plants store alkaloids or other chemical products in their tissues (for a recent review see Levin, 1971). This mechanism is also widespread in amphibians where, for example, the parotid glands of toads exude a bitter-tasting liquid. Anyone who has observed the behavior of a dog unfortunate enough to have mouthed a toad will have some feeling for the noxious taste of this substance. The dangers to a predator of devouring tropical frogs of the family *Dendrobatidae* can be attested to by those South American Indians who use a preparation of the skins of these frogs to poison their arrows. Among other groups of animals, certain mollusks are apparently noxious-tasting to potential predators (Paine, 1963), and both unpalatability and gastronomic dangers lurk very commonly in insects (L. P. Brower and J. V. Z. Brower, 1964; L. P. Brower, J. V. Z. Brower, and J. M. Corvino, 1967; L. P. Brower *et al.*, 1968). Brower and Brower (1964) have shown that some, and probably many, insects have incorporated plant poisons, originally evolved by the plant to repel insect attack, for use in their own defense. A classic case is that of the monarch butterfly, *Danaus plexippus*, which incorporates cardiac glycosides from juices of its food plant, the milkweed, *Asclepias sp.* Naive bluejays, *Cyanocitta cristata*, tested on these butterflies respond to their first meal with vomiting and decline to accept further individuals (L. P. Brower *et al.*, 1967; L. P. Brower *et al.*, 1968; L. P. Brower, 1969). Experiments in which monarchs and other insects were raised on their usual, toxic food plants and on non-toxic plants, and subsequently offered to naive bluejays indicated predator avoidance of the first but not the second set of insects (L. P. Brower *et al.*, 1968, L. P. Brower, 1969).

Among the nudibranchs, a group of gastropod mollusks, some species have evolved the ability to incorporate into their own skin the stinging cells (cnidoblasts) of hydrozoans. These cells, when triggered, eject small barbs into the source of disturbance and are normally used by hydrozoans in prey capture and defense.

A defense mechanism which is now known to be quite common is the active spraying of toxic substances by insects. Brower and Brower (1964) report that the pentastomid bug *Carpocoris purpureipennis* sprays an aldehyde, prepared in special glands, which is capable of paralyzing ants and other insects at a distance. The bombardier beetle, *Brachinus sp.*, possesses two glands at the tip of its abdomen, each of which consists of two chambers. When startled, the beetle squeezes the contents of one (hydroquinones and hydrogen peroxide) into the other (containing catalases and peroxidases) with literally explosive results. The release of free oxygen blasts quinones out of the glands at about 100°C. The beetle is capable of aiming the noxious spray at attackers (Aneshansley and Eisner, 1969).

Many species possess mechanisms whereby dying individuals emit chemical substances which act as warning signals to others. This, then, is a defense mechanism, although of a somewhat different sort from those mentioned above. The freshwater snail, *Heliosoma nigra*, possesses a chemical warning system and responds to the crushing of a nearby compatriot by dropping off the vegetation on which it finds itself and burrowing rapidly into the mud (Kohn, 1961, citing Kempendorff).

The use of chemical communicating substances (*pheromones*) for warning purposes is widespread and particularly well developed among ants. Here, the warning, or "alarm" pheromone, if in sufficient quantity, triggers rapid dispersal of nearby individuals. It is highly volatile and fades rapidly unless reinforced, thereby minimizing overreaction (Wilson, 1963; Butler, 1967).

Two interesting questions arise with respect to these chemical defense mechanisms. First, if predators learn or evolve the ability to exploit food poisons or warning pheromones to their own advantage, these mechanisms lose effectiveness and may actually become disadvantageous. Thus we must view the relationship between predators and their prey as a continuous evolutionary race. Ehrlich and Raven (1965) have explored this view as an explanation for the tremendous diversity observed in insect and angiosperm species. When an angiosperm plant evolves an effective toxin protecting it from insect attack, it competes better with other plant species over a wider range of habitats. As such, many ecological niches once closed are now open to it for colonization. Geographic spread and increased variability and, later, isolation result in an adaptive radiation of new species.  Eventually, however, some insect species evolves the ability to resist the toxicity of the plant poison, thereby providing itself with a great diversity of food species on which to specialize.  An adaptive radiation of insect species follows. The process may have been repeated over and over again.

The second question relates to the means by which unpalatability and toxicity evolved.  Experiences with noxious prey may teach a predator to avoid similar-appearing food in the future (see 5.2.3).  A bad-tasting organism thus benefits other individuals in its neighborhood although it may, itself, die in the process of its predator's education.  A toad which has been chewed and rejected by a raccoon may survive as a result of its parotid exudate, but a poisonous butterfly which has had its wings removed as a prelude to ingestion gains little by its predator's subsequent rejection of its crushed body.  Toxins, in the latter case, do not convey an adaptive advantage on their bearer, and because they require energy in their production (or incorporation) they may, in fact, be disadvantageous.  In such instances, unpalatability and toxicity must have evolved through kin selection (or perhaps group selection—see 3.1 and 3.2).  The avoidance pattern learned by a predator at one (or more) individual's expense will benefit adjacent, similar-appearing individuals which are genetically related to the victim(s).  Placement of eggs in one or a few concentrated masses is the only likely means of achieving genetic relatedness of adjacent butterfly individuals, and may be a behavior pattern necessary to the evolution of unpalatability and toxicity.

One often thinks of teeth and claws as defensive weapons.  This, however, may be inaccurate.  It is difficult to imagine rodent incisors as effective protection against the attacks of weasels, hawks, or owls, or bovid hoofs as protection against lions.  However, in some cases where prey are larger than their predators, as in the case of moose hunted by wolves, a prey animal which holds its ground is generally immune from attack.

Anatomical defense mechanisms are largely devices for escape.  They take three basic forms.  First, speed and maneuverability are often effective escape mechanisms; predators generally strike the weak, sick, and slow.  Thus the nature of ruminant limbs and digitograde gait (walking on the tips of the digits), both serving to enhance speed, may have evolved at least in part as defense mechanisms. Second, and perhaps more important, defense takes the form of ability to break away.  The "hair" of hairy caterpillars pulls away, and some salamanders and lizards have evolved fracture planes which serve to break off (*autotomize*) portions of the tail which may be in the grasps of a predator, and seal off blood flow to the sloughed member.  In some cases the severed tail wriggles violently, serving to distract the predator as the prey escapes.  Autotomy is more developed in other species such as the nereid worms in which varying amounts of the tail end break off and, freed from the inhibitory effects of the brain, wriggle violently.  Often the head end, which burrows into the sand or mud in escape, regenerates a new tail. Sea cucumbers may spit out their viscera and later regenerate them.  The leathery shell left behind is poor food for the predator which attacks the disposable insides.

## 5.2 MORPHOLOGICAL MECHANISMS

### 5.2.1 Crypsis

Everyone knows examples of protective coloration. Birds nesting in grass lay light, splotchy eggs, grouse and goatsuckers such as nighthawks or whip-poor-wills blend into their background habitat. Arctic hares are white in snow, brown in other areas, and in some places with seasonal differences they change color over the year. Buxton (1923) notes that essentially all animals of the desert are sandy-buff colored—except the tenebrionid beetles which are black. These beetles are poisonous and may be advertising that fact—see 5.2.2.

Protective coloration has been well studied in mice where Blair (1947), for example, examined the genetics of crypsis in *Peromyscus maniculatus* in the Tularosa Basin of New Mexico. Here, buff and gray color, with the former dominant, are determined by a two-allele system at a single locus. Blair examined mice (and gene frequencies, under the assumption of random breeding) at five areas. The results, given in Table 5.1, show fair agreement between mouse and background color.

**Table 5.1.** Adaptive coloration in *Peromyscus maniculatus* in New Mexico

| Number of mice collected | | Frequency of buff gene | Soil color | Area |
|---|---|---|---|---|
| Buff | Gray | | | |
| 88 | 20 | 56.76 | Dark red | Tularosa |
| 52 | 13 | 54.51 | Dark pinkish gray | Salinas |
| 36 | 21 | 36.46 | Pale pinkish gray | Lone Butte |
| 79 | 97 | 24.87 | Pale pinkish gray | Alamagordo |
| 6 | 7 | 24.13 | Creamy white | White Sands |

Crypsis involves not only color, but also color pattern and color distribution. Fish are very often dark dorsally and light colored ventrally, the classic idea being that predators looking down on the fish see a dark animal against a dark background while predators looking up see a light animal against a light background. Birds also often show this pattern, as do mammals. In the latter cases this *counter shading* may serve in a slightly different fashion. Sunlight or moonlight shining on a dark back makes it appear lighter, while shadow darkens the light belly. Counter shading thereby acts to make the animal appear more uniformly colored and, possibly, to aid in crypsis. A similar situation exists in insects. The caterpillar of *Papilio podalirium* is light colored ventrally, with a slightly darker colored back. It crawls along the tops of branches, the sun hitting its dark back more strongly than its lighter belly, eliminating the contrast in color. This presumably helps make the creature more cryptic since the branches on which it crawls are very dull and nonreflective and thus, themselves, show little or no contrast between top and bottom faces. The caterpillar of the hawk moth,

*Smerinthus ocellatus*, crawls along the *bottom* of branches and is light on its dorsum, dark on its ventrum (Portmann, 1959).

Many species have the ability to change their coloration to match the background of the moment. The most familiar case, perhaps, is the African chameleon which possesses black, yellow, and red chromatophores, and a structural blue layer of skin. The changes in color are not strictly protective, though, being also affected by "emotion." Somewhat less accurate in their ability to match colors are the American *Anolis* lizards which may turn various shades of brown, green, or patterned brown and green. More subtle color changes occur in frogs which become darker on dark, wet substrates than on light, dry substrates. Flounders and various other fish turn dark when their heads are held against a dark-colored background (Nikolsky, 1963). Color change is also well developed among many invertebrates. The prawn *Hippolyte*, for example, is green when on green algae, brown or reddish when on brown or red algae, and violet when resting on the violet corallines. An interesting sidelight on the prawn is that its ability to change its protective coloration degenerates with age. The young animals require only a few seconds to effect a change, older individuals may take up to 24 hours, and the very old can change color only with a molt (Portmann, 1959) (see discussion of selection for age-specific mortality in 9.4).

All of the above examples seem to be cases of protective coloration. Perhaps it is because this appears so obvious that only a few experiments have been performed to test whether protective coloration really protects. Three examples of such experiments are given below. The mosquito fish, *Gambusia patruelis*, adjusts rapidly to light–dark background changes. A fish kept on a white substance and switched to a black one changes its color from off-white to dark gray in a matter of minutes. Still greater adjustment, to black coloration, is effected by a secondary, long-term physiological mechanism. Sumner (1934, 1935) took long-term black-adjusted and short-term black-adjusted (gray) fish and against a black background subjected them to predation by blue-green sunfish, *Apomotis cyanellus*, or Galapagos penguins, *Spheniscus mendiculus*. The same experiment was done with long- and short-term white-adapted fish against a white background. In both cases, both predators took significantly more of the less cryptic (short-term adapted) individuals.

Popham (1941) ran similar experiments with the aquatic beetle, *Arctocorisa distincta*. Nine ponds with different water coloration were examined, and in eight of the nine the color of the insect larvae matched that of the water. Popham also found some indications that, in the laboratory, the larvae tended to choose the appropriate background color when given a choice, trying to escape if placed into water of the wrong color. Also, eggs laid in an aquarium with intermediate-colored background and then transferred to an aquarium with a light or dark background, hatched into light- or dark-colored larvae. Nymphs in the last or next to last instar, if placed on a different-colored background at least three hours before molting, emerged with the color appropriate to their latest background.

Popham let fish, *Leuciscus erythrophthalamus*, prey upon larvae and found significantly more mortality among larvae poorly color matched with their background.

Dice (1947) used barn owls, *Tyto alba*, as predators and found that matching of coat color to substrate color was important for the survival of the deer mouse, *Peromyscus*, even at night under conditions of very dim light.

Crypsis involves shape as well as color and color pattern. (It undoubtedly also occurs in the form of scent matching but has not been studied in this sense owing to our human lack of efficient sensory apparatus.) Forest chameleons in Africa live in dark, multicolored green and brown environments where color change would be less useful than in savanna country. In these species the ability to change color is less developed, but the animals possess marvelously grotesque bodies resembling old broken branches. Shape crypsis is especially common among insects where we find lepidopterans shaped (and colored) remarkably like leaves, orthopterans looking like leaves or twigs, and others resembling leaves, twigs, or, very commonly, thorns. Walking sticks have evolved a swaying behavior which may supplement their cryptic shape by simulating the effects of wind-caused motion.

If it is advantageous to merge into the background, then clearly it is advantageous to throw no tell-tale shadows. Shadows may be eliminated in at least three ways. Natural selection may favor dorso-ventral flattening. An example of this strategy is found in the horned toads, *Phrynosoma*, of the American southwest. Many bugs are also flattened and a number of insect species have evolved lateral flaps which serve the same purpose of minimizing shadow. A second way to eliminate obvious shadows is to crouch low, a behavior found universally in young, cryptic, precocial birds. A corollary of this strategy is the preferential distribution of individuals into depressions in the substrate, a practice commonly found, for example, in crabs. Third, shadows may be minimized by proper alignment of the body. Thus (it is supposed) cryptic butterflies face towards or away from the sun and lean at appropriate angles on their perch so as to align their wings in the way that throws the least shadow. Unfortunately, butterflies control their body heat by soaking up or avoiding sunlight, a fact which considerably complicates the matter of interpreting wing alignment.

### 5.2.2 Aposomatic Coloring

Protective coloration is often advantageous, as is attested to by its commonness. On the other hand, bright, contrasting (*aposomatic*) coloration is also quite common. The reason is simply that in certain cases, specifically involving noxious-tasting, poisonous, or dangerous animals, it pays to advertise. If individuals possess stings, for example, it is clearly advantageous to convey this information to predators before they attack. By standing out brightly the prey species not only makes clear its difference from cryptic, palatable prey, but may jolt the predator out of any nondiscriminating "state of mind." It is unimportant that the prey is more readily found since they are avoided as a result.

Aposomatic or advertising characters may be olfactory (our limited sensory capacities again hinder us here) or auditory as well as visual. The rattle of the rattlesnake seems to be aposomatic (although some have suggested that it scares off potential predators by startling them) and Sudd (1967) has suggested the same function for the whistling noise of the ant *Megaponera foetans* and the stridulation of the ants *Messor sp.* and *Aphaenogaster testaceus*. The latter two stridulate when disturbed. The clicking of certain arctiid and ctenuchid moths— which seem to discourage bats (Dunning, 1968)—may also be an example of a nonvisual aposomatic character.

Aposomatic coloration is particularly common in insects which possess stings or those which, as mentioned previously, synthesize repellents or incorporate plant poisons to their own advantage. Most hymenopterans, for example, are brightly colored. Grasshoppers and locusts, which spit out noxious substances, are often characterized by yellow and black or red and black wings. Milkweed bugs, *Oncopeltus*, which become bad tasting as a result of their food, are orange and black. The iridescence of some beetles also may be aposomatic. Noxious caterpillars are often brightly colored. The bright colors of some frogs (red, yellow, black in various combinations) in Central and South America are clearly aposomatic. The skins of the atelopodid and dendrobatid frogs are used for poisoning arrow tips (Judy, 1969) and the atelopidtoxin of the bright yellow, orange, and black *Atelopus zeteki* has an LD–50 of $16\mu g$ per kg in mice.

Cott (1940) has noted that, at least in insects, aposomatic coloration follows definite patterns. By far the most common combinations are red, black, and white; yellow, black, and white; red and black; orange and black; yellow and black; and white and black. This suggests one or both of two things. A limited number of patterns limits the amount of information a predator must learn in order to minimize errors in feeding judgment. As such it enhances the effectiveness of aposomatic coloration and thereby makes convergence in aposomatic pattern selectively advantageous. Also, limited numbers of patterns make inborn tendencies of avoidance possible. Genetically induced predilections to avoid or rapidly learn avoidance of certain patterns, when possible, is clearly advantageous and should, we suspect, have evolved.

Inborn tendencies to avoid certain stimuli do seem to occur. Frings and Frings (1964) have suggested that the negative reactions of naive humans and other animals to snakes and spiders is genetic. Joslin *et al.* (1964) found a strong fear of snakes in wild-raised rhesus monkeys, but a mild fear even in totally naive, laboratory-raised individuals.

Not all brightly colored animals are unpalatable, nor are all bad-tasting creatures aposomatic. The monarch butterfly, although possessing an orange and black color pattern, is far from flashy even though it is usually highly distasteful as a result of feeding on milkweed plants. Some animals are brightly colored by virtue of mimicking aposomatic species (see 5.2.3, below). Nevertheless, there is often a good correlation between brilliance and acceptability as food. Prop (1960) studied the protective mechanisms in larvae of the sawfly genus, *Diprion*, and found *D. pini* and *D. sertifer* to be palatable to birds and inconspicuous. *D. virens* and *D. frutetorum* appear to be noxious tasting and very conspicuous. Intermediate both in taste and appearance are *D. nemoralis* and *D. simile*.

The intermediate case above raises an important question: The selective advantages of either extreme case are clear, but how might an intermediate case evolve? A likely explanation lies in environmental heterogeneity, fluctuations from situations of high population density (plenty of food for predators) to low. In years of plenty, predators avoid bad-tasting species so that noxious prey will do well to be aposomatic. In years of famine, no prey can be bypassed, even relatively bad tasting ones (see 7.2.1), and crypsis is advantageous. If the prey species in question possesses the behavioral requisites for minimizing the disadvantages of inappropriate coloration—by preferentially choosing its background, perhaps,—then its fitness set with respect to surviving predation may be convex (see 4.2.2). If so, the optimal phenotype is intermediate. Since a morph intermediate between cryptic and aposomatic appearance will experience less selection pressure for bad taste than a perennially obvious morph and more selection pressure for bad taste than a perennially obscure-looking morph, its palatability should also be intermediate. It would be interesting to know whether *D. nemoralis* and *D. simile* are more plastic in their behavior or more accurately choose their substrate than the four species occupying one specialized niche or the other.

Bright coloration may also act as a defense mechanism for a somewhat different reason. Coppinger (1969, 1970) has shown that naive bluejays, redwings, *Agelaius phoeniceus*, and grackles, *Quiscalus quiscula*, avoid novel food items. It is a well-known behavioral phenomenon that almost any stimulus, in moderation, evokes a positive response while the same stimulus at higher intensities elicits negative reactions. Coppinger suggests that if an insect is sufficiently different, perhaps through being brightly colored, it will register in a predator's mind as a strong novelty stimulus and thus evoke an avoidance reaction.

A defense mechanism related to aposomatic appearance is the startle effect. Here, some sound, sight, or, possibly, smell frightens the predator into dropping the prey—the hiss of a snake, perhaps; the sudden spread of a cobra; or the clicks of click beetles. Game birds, when come upon, explode from the ground, pistol shrimps make loud cracking sounds, and many small vertebrates scream piercingly. The Carolina locust, *Dissosteira carolina*, combines its yellow and black aposomatic coloring in flight with a loud crackling sound which may have startle value (Wynne-Edwards, 1962).

The classical examples of startle mechanisms are the eye spots found in many species, especially commonly on the wings of lepidopterans. The eye spot is said to mimic a vertebrate eye, giving a predator the impression, when it is suddenly flashed, that he himself is being eyed as a meal. It is possible, though, that eye spots serve a defensive function for other reasons. The grayling butterfly,

*Eumenis semele*, has a bright eye spot on its wing tip which is flashed briefly when the creature alights. Sheppard (1960) feels the eye spot to be an attention getter which will be grabbed at by any predator in the immediate vicinity. If the spot is attacked, the butterfly immediately leaves having lost only a wing tip; if not, the wings are folded and the eye spot disappears, revealing a very cryptic underwing.

In the real world there may be many kinds of predators and many kinds of background into which a prey species may blend. It may, perhaps, decrease attention from one predator if a prey is cryptic, from another if the prey is aposomatic. An educated estimate of the strategy followed by a species under such complex circumstances can be made with the use of the visual aid, the fitness set (see 4.2.2). Suppose $P_1$ is the probability of escaping a lifetime of attack by predator 1, and let $P_2$ be the corresponding probability with respect to predator 2. In one case (Fig. 5.1a) behavioral plasticity may produce a convex fitness set and genetically intermediate individuals. If enough predators and backgrounds are present, however, so that the dimensionality of the fitness set is high, then the projection of the set on at least some two-space is bound to be concave as in Fig. 5.1(b). In this case disruptive selection over time or space may lead to polymorphism. The expected number of morphs should increase with the complexity of the environment and the number of predators.

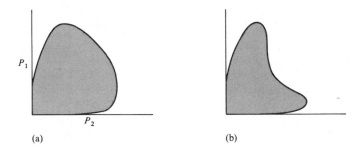

(a)                                        (b)

**Fig. 5.1.** Fitness sets for prey in a multi-predator environment.

### 5.2.3 Batesian Mimicry

Prey species not only mimic twigs, leaves, and thorns, but they also mimic each other. When a palatable species evolves an appearance similar to that of an unpalatable one it is known as a *Batesian mimic*. The mimicked species is the *model*. The rationale is simple enough: By resembling noxious creatures, palatable individuals may be confused for their model by predators, thereby gaining protection. As with aposomatic coloring, mimicry may be olfactory or auditory, but visual mimicry is by far the best known.

A classic case of Batesian mimicry involves the unpalatable monarch butterfly (the model) and the viceroy, *Limenitus archippus* (the mimic). In this case the mimic is amazingly similar in appearance to its model except for its possession of an extra black wing bar. The mimic appears to have improved on its similarity to a monarch by being smaller than its model. In situations of choice, the predator is faced with two prey, virtually identical except for a slight size differential. Since large prey are generally preferred to small (see 7.1.3), the viceroy should generally escape even if the predator attacks.

Batesian mimicry is quite common, particularly in butterflies, the papilionids (swallowtails) being exceptionally well known as models. The dronefly is usually considered to be a mimic of the honeybee in appearance as well as sound, although the resemblance to stinging worker bees is nowhere near so good as the resemblances to stingless drones. This observation would ordinarily raise questions as to the validity of the mimetic relationship except that the Browers (1966) have shown that the rather poor mimicry of workers nevertheless seems to be effective (see below).

If one wishes to verify the existence of intuitively obvious cases of Batesian mimicry, the first question that must be settled is whether predators really confuse look-alike prey and consequently eat fewer of the palatable species. Both L. P. Brower and J. V. Z. Brower have performed numerous experiments on this subject, all of them with positive results, one of which (Brower and Brower, 1966) is reported briefly below. Experimental toads (predators) were offered first models (stinging honeybees) and then mimics (droneflies). Control toads were offered no bees. Both groups were allowed an alternative food of mealworms. The results (Table 5.2) show a significantly higher ($p < 0.001$) rate of rejection of droneflies by the experimental toads than by the controls. Similar results were found using jays or starlings, *Sternus vulgaris*, as predators and mealworms painted with various color patterns and dipped in quinine dihydrochloride (models).

**Table 5.2.** The results of mimicry experiments by Brower and Brower (1966)

|  | Model (honeybees) | | Mimic (droneflies) | | Alternative food (mealworms) | |
|---|---|---|---|---|---|---|
| Number | eaten | rejected | eaten | rejected | eaten | rejected |
| Experimental toads | 36 | 44 | 26 | 34 | 140 | 0 |
| Control toads | — | — | 110 | 30 | 140 | 0 |

Of course the next, and essentially clinching, argument would be that mimics are preyed upon less than otherwise identical, nonmimetic prey in natural situations. To test this Brower *et al.* (1964) released artifical (painted) mimetic and control male butterflies in Trinidad. Virgin females of the same species were then captured and held. In this as well as many other species, virgin females exude a sex pheromone to which the males are extremely sensitive. The response of the male is to fly upwind and then upgradient to the female. Capitalizing on this fact, it was an easy matter to subsequently recapture surviving mimic and control males. For models, *Parides neophilus* and *Heliconius errato* were chosen; both were known to be unpalatable to local birds. *Hyalophora promethea* males served as mimics and controls. No difference was found in the percent of mimic and control butterfly returns over several days. The experiment was repeated (Brower, Cook, and Croze, 1967), this time using *Parides anchises* as the model. In these experiments, an apparent trend in the earlier work was confirmed: More mimics than controls returned during the first few days (suggesting an advantage to mimicry), but fewer returned later (suggesting the opposite). No overall advantage was clear when the results were averaged over several days. The interpretation of these data was that too many mimics had been released in a local area and that while mimicry was at first advantageous, the predators soon learned to recognize the difference between model and mimic, rendering mimicry useless. Finally, once this occurred, the brighter coloration of the mimics (the models, like most noxious-tasting insects, are aposomatic) made them more conspicuous and hence put them at a net disadvantage. The same sort of results were reported from still later work in the same area (Cook *et al.*, 1969). The results are not terribly persuasive but are suggestive and, along with the overwhelming amount of circumstantial evidence in the form of apparent, existing model–mimic complexes, make a fairly convincing case against nonbelievers of Batesian mimicry.

The evolution and genetics of mimicry have been studied in great detail in a number of cases, and although the details need not concern us here, it is worth looking at some of the evolutionary questions that have been raised. There was much early concern over whether mimicry could really evolve, since it seemed unlikely that the mimic should, in one pre-adaptive step, come to closely resemble an appropriate model. We now know that such a step is unnecessary. Sexton (1960) settled this with a proof that (at least in his one case) only a partial resemblance will benefit the mimic. He used *Anolis* lizards as predators on *Tenebrio* larvae which were made into partial mimics by gluing onto them the prothorax or elytra of a lightning bug, *Photinus*, a noxious-tasting, aposomatic insect. If only the prothorax was used the lizards showed no tendency to avoid the larvae. Some avoidance occurred when only the elytra was used, however. With both prothorax and elytra glued in place, avoidance was marked. Sexton also points out that a model need not be extremely noxious, but only less preferred—that is, a predator takes encountered models less often than potential mimics. This latter point, and/or the matter of advantage of partial mimicry, has also been made by Sheppard (1959), Ford (1960), Duncan and Sheppard (1965), and Holling (1965). Sheppard (and others) have shown, as we might intuitively expect, that mimicry may often involve a switch gene (4.2.3) acting on a gene complex. Improved mimetic resemblance is evolved through selection on the gene complex while frequency of mimics reflects the frequency of the switch gene.

If we accept the existence of mimicry we must resolve another apparent difficulty in its evolution. One of the most noticeable aspects of the Browers' various sets of data is the inconsistency with which models are rejected; the toads and starlings which act as predators seemed to forget the nastiness of models very quickly. This problem can be surmounted in any of three ways:

**1.** We assume that model–mimic complexes are important in the diets of predators in nature so that model–predator contacts are common and little forgetting occurs.

**2.** We note that toads and starlings are not likely predators of model–mimic complexes in nature—at least of complexes involving bees or painted, quinine-dipped mealworms—and assume that the forgetting of palatability in this case is an artifact of the laboratory environment.

**3.** We assume that there is a strong genetic predilection to learn to avoid certain patterns but that the patterns characterizing bees and painted mealworms—these species not being normal food items—are not among them.

The first possibility is a weak one. Butterflies (one of the most common groups in which mimicry occurs) may not be important in the diets of birds, their prime predators. Urquhart (1957) believed them to be unimportant and used this assertion to try to discredit the idea of mimicry in general. The importance of butterflies in avian diets is contested.

The second possibility is still weak. It does not solve the learning problem, but only postpones it until data are gathered showing errors being made under more natural situations.

With respect to the third possibility, suppose that rigid avoidance of certain patterns (for example that exhibited by the monarch butterfly) is inborn. This would clearly be disadvantageous, since creatures bearing these patterns may not be distasteful everywhere. Petersen (1964) relates experiments in which monarchs were eaten regularly by brown thrashers, cardinals, grackles, robins, and English sparrows in Iowa (with alternative food available in abundance), and by scrub jays in Colorado. The experiments were not well controlled but little doubt remains as to their interpretation. The results show that monarchs are not everywhere distasteful (see 5.1). Furthermore, an automatic rejection of certain patterns immediately eliminates mimics as food items. Clearly it would be advantageous to first evolve not an automatic rejection, but a strong genetic predilection to learn and retain rigidly a pattern of avoidance, coupled with a decision mechanism allowing some chance of correctly distinguishing model from mimic. This is the third, and most likely, way around the learning and forgetting problem suggested by the Browers' data. It does not require that the model–mimic complex be important in the diet of some predator.

The importance of mathematical models in ecology is that they suggest new ideas and generate testable hypotheses. With this in mind we explore mimicry further by presenting a simple model worked out jointly by J. M. Emlen and J. S. Farris (unpublished). We assume at least a limited ability on the part of the predator to recognize variety in the trait (or traits) responsible for the similarity in appearance of model and mimic. In the simplest case, presented here, we examine the predators' response to one character (possibly color) taking on continuous values $\{Q\}$. We let the distribution of $Q$ for model and mimic be, respectively, $f(Q|S_i)$, $f(Q|S_j)$ where $S_i$, $S_j$ denote species $i$ (model) and species $j$ (mimic). We read $f(Q|S_i)$ as "The probability density of $Q$ given that the set of animals examined are $S_i$."

$$\int_B^A f(Q|S_i) \, dQ$$

is the probability that, given a random encounter with $S_i$, the observed value of $Q$ lies between $B$ and $A$. Suppose now that as observed $Q$ increases, the probability that the creature observed is a mimic steadily increases. In other words, $f(Q|S_j)/f(Q|S_i)$ always increases with $Q$. Then a simple strategy on the part of the predator is to accept an encountered item as mimic rather than model if the observed value of $Q$ is larger than some critical value, $\hat{Q}$. To see the meaning of $\hat{Q}$, note that $f(Q) = f(Q|S_j)\mu_j + f(Q|S_i)\mu_i$, where $\mu_i$, $\mu_j$ are the relative frequencies of encounter with model ($S_i$) and mimic ($S_j$), respectively. Also

$$f(S_j|Q) \, f(Q) = f(S_j \wedge Q) = f(Q \wedge S_j) = f(Q|S_j)\mu_j.$$

Thus

$$f(S_j|Q) = \frac{f(Q|S_j)\mu_j}{f(Q)} = \frac{f(Q|S_j)\mu_j}{f(Q|S_j)\mu_j + f(Q|S_i)\mu_i}.$$

This gives the probability that a creature possessing the value $Q$ is, in fact, a mimic. We say that when this probability reaches some critical value—depending on hunger level and the food "value" of the model–mimic complex as a whole, in the presence of alternative foods—then the predator will take the risk of deciding to treat his potential quarry as a mimic. That is, the prey is treated as if a mimic when $Q$ is such that $f(S_j|Q) \geqslant k = $ some constant. $\hat{Q}$, then, is given by

$$\frac{f(\hat{Q}|S_j)\mu_j}{f(\hat{Q}|S_j)\mu_j + f(\hat{Q}|S_i)\mu_i} = k. \tag{5.1}$$

The probability that, if the prey is a mimic, it is treated as such, will be written $q_j$. Its value is given by

$$q_j = \int_{\hat{Q}}^{Q_{max}} f(Q|S_j) \, dQ.$$

Similarly

$$q_i = \int_{Q_{min}}^{\hat{Q}} f(Q|S_i) \, dQ. \tag{5.2}$$

If prey believed to be mimics are eaten a fraction, $p_{js}$, of the time, $p_{is}$ being the appropriate measure for models, then

$$p_{is}^* = P[q_i p_{is} + (1 - q_i)p_{js}]$$

$$p_{js}^* = P[q_j p_{js} + (1 - q_j)p_{is}],$$

(5.3)

where $p_{is}^*$, $p_{js}^*$ are the fraction of models and mimics actually eaten, and $P$ is the probability that the predator will not ignore the complex completely (because of the inherent risk in doing so). We note now that in the course of becoming a mimic (since most models are aposomatic) the probability of any one mimic being noticed increases. Thus we can write $\beta p_{js}$ $(0 < \beta < 1)$ for the fraction of protomimics or nonmimetic morphs eaten.

Clearly the mimetic morph will spread if, and only if, the chances of any one individual being eaten are decreased by becoming mimetic: that is, mimicry will spread only if $p_{js}^* < \beta p_{js}$. Written out fully, this becomes

$$P[q_j p_{js} + (1 - q_j)p_{is}] < \beta p_{js}$$

or

$$\beta > \frac{P[q_j(p_{js} - p_{is}) + p_{is}]}{p_{js}} \quad \text{(conditions for spread of mimicry).}$$

(5.4)

$$(\beta p_{js} - p_{js}^* > 0).$$

Let us now look at some consequences of the model.

**1.** As $\mu_i$ (the relative abundance of mimics) increases, $\hat{Q}$ decreases for any given value of $k$ (Eq. 5.1), so that $q_j$ goes up, $q_i$ down. Furthermore, since this means more, palatable food items, $P$ should rise. Since predators with more good food can afford to be more particular in their food choice, $k$ may also rise, buffering but probably not negating the effects of increased $P$, $q_j/q_i$. Thus an increase in $\mu_i$ makes the conditions in both Eqs. (5.4) less likely to be met. This leads to (or is consistent with) two consequences:

a) The effectiveness of mimicry increases with an increase in the relative abundance of models. This has been pointed out on more intuitive grounds by Sheppard (1959) and Brower (1960).

b) If the mimetic form becomes too numerous, the conditions for its increase will no longer be met and selection will favor the nonmimetic morph. Dimorphism is commonly found among mimics but its causes are not always clear. For example *Papilio glaucus* is characterized by nonmimetic males and variable frequencies of a black female mimetic form. This case fits our expectations to some extent, though, the black mimetic form being more common where the model is more common (Brower and Brower, 1962). Mimetic species also may be polymorphic by virtue of their mimicking several models.

**2.** Note that the quality of mimicry depends on a continuing evolutionary race between the mimic (tending to converge on the model), and the model (tending to diverge from the mimic). It might be reasonable to assume that the quality of mimicry is thus related to the relative amounts of additive genetic variance in the traits contested between mimic and model. To the extent that genes affecting morphology are pleiotropic, the quality of mimicry should be greatest when mimics are eurytopic (generalists) and their models stenotopic (specialists).

3.  As the abundance or relative importance to predators of the model–mimic complex declines, $k$ should increase. The predator, faced with less need for the complex, has less need to take risks and can thus be more cautious. Thus $\hat{Q}$ goes up, and with it $q_i$. The value of $q_j$ declines. Also $p_{is}$ and $p_{js}$ decrease. As a result, the conditions in Eqs. (5.4) are more easily met. Mimicry is more likely to evolve when the potential mimic and its model(s) are not important food items (see above discussion of birds and butterflies). This conclusion is, of course, invalidated if predators are slow to learn and quick to forget aposomatic patterns.

4.  An obvious conclusion is that as $\beta$ increases (i.e. the difference in noticeability to the predator of nonmimic and mimic declines), the conditions for the evolution of mimicry are more easily met. There should be more mimics of relatively dull than bright models.

The conditions behind the above model are great simplifications of nature. Thus the above predictions will be strengthened if they can also be shown to arise from a model making somewhat different assumptions. Suppose the predator does not necessarily look at the values of $f(Q|S_i)$, $f(Q|S_j)$, or even $\mu_i$, $\mu_j$, but instead merely asks a number of yes–no type questions as to the appearance of the creature encountered. We note, from information theory, that the number of questions needed to correctly identify a series of $k$ ($k$ very large) encountered food items, is given by $kH = -k(\mu_1 \log_2 \mu_1 + \mu_2 \log_2 \mu_2)$. But mimicry is characterized by confusion, so that the probability of wrongly assessing the answer to any given question is $m$ ($0 \leqslant m \leqslant \frac{1}{2}$). The number of questions required to correct these "lies" is $-kn[m \log_2 m + (1-m) \log_2 (1-m)] = nkH_m$, where $n$ is the total number of questions asked. Thus $nk = kH + nkH_m$, and $nk = kH/(1-H_m) = $ the number of questions required for correct identification of all $k$ individuals encountered. Suppose the number of questions actually asked is $Nk < nk$. Then for each question the predator is short of $nk$, the chances of correctly assessing all $k$ encounters is cut by one-half. The probability of no errors is thus

$$2^{kN-kn} = 2^{k[N-H/(1-H_m)]}.$$

But this is the probability that encountered individuals are correctly assessed $k$ times in a row, so that the expected probability of a correct assessment at any given encounter is approximately

$$\{2^{k[N-H/(1-H_m)]}\}^{1/k} = 2^{N-H/(1-H_m)} \qquad \text{(see Emlen, 1968b)}.$$

Noting that $H$ is maximized when $\mu_i = \mu_j$ and minimized when $\mu_i$ or $\mu_j$

approach zero, and that the probability of a correct assessment ($q_i$ or $q_j$ in the previous model) $= 2^{N-H/(1-H_m)}$, the reader should be able to verify that this model leads to the same predictions as the previous one. A decline in the importance of the complex should decrease $N$.

Both of the above models indicate an upper limit in the ratio of mimic to model above which mimicry is selected against. Rather than the alternative morph being nonmimetic, however, it may mimic a next-best model. Eventually an equilibrium should be reached in which the proportions of the various morphs vary with the relative frequencies of their respective models. Sheppard (1959) gives an example in which the African butterfly species *Pseudacrea eurytus* mimics various species of *Bematistes*. Sampling was done over several areas so many points were obtained. The results are given in Fig. 5.2.

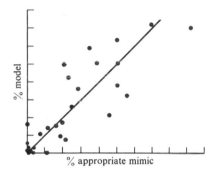

Fig. 5.2. Variation in the number of mimics in relation to the number of models.

### 5.2.4 Müllerian Mimicry

Whether recognition of aposomatic patterns is inborn, learned, or both, there is always the problem that predators may be confused and make errors if the number of such patterns is large. Consequently it may be advantageous for aposomatic species to look alike. Mutual mimicry of this sort is known as Müllerian mimicry. There are several cases of convergence of unpalatable, brightly colored prey species. Many hymenopterans, for example, appear to have converged. The so-called tiger stripe butterfly complex of the neotropics, including species of *Lycoreini*, *Ithomiinae*, and *Heliconiinae*, is a Müllerian complex. It is often stated that while Batesian mimics are generally polymorphic, polymorphism runs contrary to the advantages of Müllerian mimicry. To say that Müllerian mimics are therefore monomorphic would be an oversimplification, however, for what constitutes a Müllerian as opposed to a Batesian complex is not entirely clear. In the tiger stripe complex, *Lycorea* is considerably less unpalatable than its fellow mimics. Is *Lycorea* then Müllerian or Batesian? Was it once a Müllerian mimic on which selection for bad taste was relaxed, or a Batesian mimic which subsequently evolved a bad taste? This continuity of which the two types of mimicry described constitute the extremes was noted also by Brower and Brower (1964).

## 5.2.5 Polymorphism

We have already noted that polymorphism may arise as a result of disruptive selection for aposomatic and cryptic appearance, and as a consequence of high population densities of Batesian mimics. But there are other defense mechanisms leading to polymorphism in prey species. De Ruiter (1952), L. Tinbergen (1960), and others, have noted a tendency among birds (and it may exist in other groups) to behave as if they gained a mental picture of one type of prey and subsequently noticed creatures fitting this *search image* more readily than dissimilar-appearing creatures. (This matter of search image is more fully treated in Chapter 7.) Thus, as the frequency of one morph in a polymorphic species rises, it is eaten disproportionately often, its mortality rate rises, and its fitness falls. The existence of search images leads to what is called *apostatic selection* by predators, in which the relative rate of predation rises more rapidly than the relative rate of encounter. Frequency-dependent selection of this sort may result in the maintenance of several morphs. As a simple example, suppose that the genotypes $A_1A_1$ and $A_1A_2$ imbue their bearers with one morphology with fitness

$$W_1 = c_1\left(1 - \frac{\rho_1}{\rho_1 + \rho_2}\right)$$

where $c_1$ is a measure of the noticeability of the morph by its predators and $\rho_1$, $\rho_2$ are the population densities, respectively, of this morph and the alternative morph shown by $A_2A_2$ individuals. $W_2$ may be written $c_2(1 - [\rho_2/\rho_1 + \rho_2])$. In this simple case, selection increases the frequency of the most fit morph until its fitness falls below that of the other. Equilibrium must occur when $W_1 = W_2$, or $\rho_1/\rho_2 = c_1/c_2$, so that, where $c_1 = c_2$, and $p$ is the frequency of the $A_1$ allele, $p^2 + 2pq = q^2$, or $p = 0.29$.

**Table 5.3.** The maintenance of polymorphism by apostatic selection

| Genotype | $A_1A_1$ | $A_1A_2$ | $A_2A_2$ |
|---|---|---|---|
| Fitness | $c_{11}\left(1 - \dfrac{\rho_{11}}{\rho}\right)$ $= c_{11}(1 - p^2)$ | $c_{12}\left(1 - \dfrac{\rho_{12}}{\rho}\right)$ $= c_{12}(1 - 2pq)$ | $c_{22}\left(1 - \dfrac{\rho_{22}}{\rho}\right)$ $= c_{22}(1 - q^2)$ |
| Conditions for heterozygote superiority | | $c_{12} > c_{11}\left(\dfrac{1 - p^2}{1 - 2pq}\right),$ | $c_{12} > c_{22}\left(\dfrac{1 - q^2}{1 - 2pq}\right)$ |

($p$ = frequency of the $A_1$ allele)
($\rho$ = total density of all genotypes)

Note that, if the likelihood of detection of various morphs, and their frequencies, take on a certain range of values (see Table 5.3), heterozygosity is advantageous and thus heterosis may be favored by natural selection. In this case also, polymorphism will be retained.

An extreme case of apostatic selection may occur when (and if) the existence of many morphs confuses the predator possessing tendencies toward search image formation; this may inhibit predation. Moment (1962) calls natural selection favoring multiple morphs *reflexive selection*. He cites as possible examples of species undergoing such selection the clam *Donax variabilis* and the brittlestar *Ophiopholis*. Both show tremendous morphological variability for which no other explanation seems possible.

## 5.3 BEHAVIORAL MECHANISMS

### 5.3.1 Crypsis

Neither crypsis nor aposomatic coloring are entirely morphological in mechanism. Camouflaged feather pattern would be useless in ducklings or gull chicks which hopped about in view of a predator. Precocial birds universally crouch in response to a parental warning call. Spiders may respond to danger by changing color slightly and becoming still (Kaston, 1965). A particularly good example of the correlation between behavior and morphology in defense is shown by Heatwole (1968) who studied the lizards *Anolis cristellatus* and *A. stratulus* in Puerto Rico. Heatwole would spot a lizard on a tree trunk, walk briskly toward the animal, and note his distance from it when it ran. As a measure of crypsis, he had two people at a distance of 5 meters count the number of different lizards visible on trees at less than 2m height; they then moved in to count the actual number. Those patterns most often missed were ranked most cryptic. It was assumed that lizard predators (birds, snakes, other lizards, mongooses) ranked crypsis in the same order as the human experimenters. Table 5.4 shows the results. As might

**Table 5.4.** Crypsis and escape behavior in Puerto Rican lizards (see text)

| Tree color | Brown | | | Gray | | | Mottled: black, green and white | | |
|---|---|---|---|---|---|---|---|---|---|
| Lizard type | 1 | 2 | 3 | 1 | 2 | 3 | 1 | 2 | 3 |
| Percent unobserved at 5m (measure of crypsis) | 0 | 44 | 52 | 40 | 30 | 33 | 20 | 18 | 16 |
| Approach distance in cm | about 50 | | | about 120 | | | about 150 | | |

($1 = A.$ *stratulus*, $2 = A.$ *cristellatus* females, juveniles, $3 = A.$ *cristellatus* males)

be expected, the lizards behaved as if they knew the extent of their protective coloration and thus assumed that they had been spotted (and ran) much sooner when they blended poorly with their background.

Other examples of behavioral crypsis include the decorator crabs which hold algae and sponges, cut and trimmed to size, over their carapace until they take hold and grow.   This practice is common in *Hyas*, *Pisa*, *Pugettia*, and *Mimulus*. Spider crabs (a general, descriptive name for these creatures) are long and narrow as opposed to other crabs which are square or wide-ovoid shaped.   This body shape permits the crabs to use their legs to reach all parts of their body.   The sponge crab, *Dromia vulgaris*, possesses a last pair of legs specially modified for the task of holding bits of cut sponge over its back (Portmann, 1959).

Mimicry of a sort also exists as a behavioral mechanism.   For example the fish *Chaetodon capistratus* possesses an eye spot near its tail, and completes the illusion of a head at the wrong end by swimming slowly backward.   When disturbed by a predator it suddenly shoots off forward, presumably aiding its escape by confusing its attacker (Cott, 1940).   Some fish (for example the jewel fish) swim in schools which "blow up" and then contract (Noble and Curtis, 1939), a mechanism suggestive of the startle effect mentioned earlier (Baerends and Baerends van Roon, 1950).

Cephalopods utilize a combination of crypsis, startle, and escape.   The squid, whose chromatophores are normally somewhat dark, darkens still more if disturbed, ejects smoke of a similar color, then blanches (making it difficult to see in water) and shoots off at an angle.

### 5.3.2 Escape

Escape is a mechanism which involves much more of interest than just running away.   The moth *Caenurgina erechtea*, for example, has evolved the ability to hear the sounds of bat echolocation.   If these sounds of its predator are weak, the moth responds by moving away.   If the sounds are strong, the moth responds with erratic diving and looping.   *Caenurgina* can't outfly bats but at close range it can outmaneuver them (Roeder, 1962, 1966).

A number of interesting escape mechanisms occur among marine species.   The sea anemone *Stomphya*, when touched by the predatory starfish *Dermasterias*, contracts, releases its foothold, and swims off by violent jerking and bending.   In mollusks, *Physa*, *Limnaea*, and *Heliosoma* make violent escape lunges from leeches (Herter, 1929), and *Tegula funebralis*, the West Coast turban snail, when its head tentacles touch a *Pisaster*, turns 90 degrees and moves off.   If the epipodial tentacles touch this starfish, a greater angle of turn results (Feder, 1963). *Nassarius*, the mud snail, catapults its shell and rolls over forward and to the right up to ten times if contracted by a predatory starfish.   No such reaction is shown to herbivorous starfish such as *Patiria*, (Weber 1924).   The snail *Strombus* "leaps" (quick lunges) away from its predators, *Fasciolaria* and *Conus* (Kohn and Waters, 1966).   The lunge apparently acts to increase the distance from the predator as fast as possible immediately after contact.

Escape may be expedited by group behavior. If, as with many predators, only one or a few individuals can be handled at one time it is advantageous, where clumping occurs (see below) to try to seek cover behind others. Springer (1957) notes that fish often appear "uneasy" on the periphery of schools and try to swim into the inner part, in spite of depression of speed and well-being there due to depressed oxygen pressure (McFarland and Moss, 1967). It is reasonable to think that confusion of a predator might result from coordinated, erratic movements of a number of closely spaced individuals. Many predators must fixate on one prey animal if an attack is to be successful and such behavior by the prey should make fixation difficult. Perhaps for this reason, starlings and limicoline birds close ranks in a tight flock and fly erratically when attacked by a hawk.

In some fish, loose schools may suddenly close in the presence of a predator. (This may be due to frantic efforts to move to the inner part of the school.) A variation on this theme is found in the bobwhite quail, *Colinus virginianus*, which forms characteristic "huddles," several birds standing close together, feathers interlaced, tails in, and heads outward. Gerstell (1939) finds these huddles an aid in temperature control during cold weather, but notes that the individuals are also positioned for the fastest getaway possible. In addition to fast escape, the huddle essentially blows up, birds exploding in all directions, perhaps causing surprise and confusion in a potential predator.

Flight takes on another interesting aspect when it changes from orderly to random. Unpredictable, or random, flight, (panic) might seem a poor strategy, but this may be far from the truth. Fisher (1958b) suggests, in fact, that random behavior may be advantageous. An orderly retreat may be optimal when a prey is not cornered, but as a last resort, a predator may be effectively stymied if the escape patterns are unpredictable rather than orderly.

### 5.3.3 Dispersion

One factor in habitat selection by animals may be the discouragement of predation. Whether it evolved to that purpose or not, the cliff-nesting habit of many colonial sea birds alleviates predation pressure. We might expect hole nesters to be safer from ground predators—and probably no more vulnerable to most aerial or tree predators—than ground nesters. Crook (1965) suggests that some birds, including representatives of the ploceids and tyrannids, minimize nest predation by building nests in thorn bushes, near stinging insect colonies, in well shafts, over water, or near human habitations.

In species which are large enough, grouping may be useful in defense (large bovids or cervids). This should be particularly true in open country. In forests, trees separate the herd, make shoulder-to-shoulder defense difficult, and break up an orderly (or erratic but coordinated) flight pattern. Also a forest provides little opportunity for sighting a stalking intruder—a predator can more easily surprise a group of prey before it closes ranks. Group defense is thus less useful in forests than in open country and large forest herbivores are usually solitary or travel in only small groups. The white-tailed deer, *Odocoileus virginiana*, of North America, the roe deer, *Capreolus capreolus*, of Europe, and the forest duikers of Africa are essentially solitary, forest species while the open country bison *B. bison*, the red deer, *Cervus elaphus*, and most savanna antelope in Africa, are group animals.    Pronghorn, *Antilocapra americana*, are intermediate in behavior, although strictly plains species. Local groups of 12 to 15 or more present a compact group to attacking wolves or coyotes, and the bucks may successfully repel the predators. Smaller groups make no attempt to stand, but rather stampede.

There are other advantages to grouping in open-country animals.   If the chance of one individual spotting a predator before it approaches within some distance is $\alpha$, the chance of $n$ individuals spotting it is roughly $1 - (1 - \alpha)^n > \alpha$. Prairie dogs, *Cynomys sp.*, certainly seem to benefit by a social order favoring several lookouts.  Darling (1937) reports that red deer, when alone, lie with their backs to the wind so as to see in front and smell from behind.  When in groups, the usual situation, individuals face all directions, covering all avenues of predator approach.  This division of labor must certainly work to the advantage of all members of the group.

A different kind of group effect may also discourage predators.  Mobbing, the harassment of a predator by a number of maneuverable prey individuals, is common among bird species, especially those nesting colonially.  It is a common sight to see blackbirds swarming about a flying crow, or crows clustering about and swooping at an owl. There is some question as to whether mobbing is a social, coordinated activity or not, but this is really unimportant to its final effect.  Horn (1966, 1968) suggests that birds in a colony each respond to a predator within some critical radius of their nests.  The result is that over a nesting colony, many birds may simultaneously be attacking the predator.  So long as nests are closely enough spaced so that a sufficient number of individuals are mobbing, the practice is safe and, perhaps, effective at driving the predator off.  Horn (1968) tested the effectiveness of mobbing as a defense mechanism in the brewers blackbird, *Euphagus cyanocephalus*.  Reasoning that where nests are more closely spaced more mobbing individuals will rise, he concluded that mobbing should be most effective in the densest parts of colonies.  Accordingly, he divided all nests into two groups according to whether the distance to the nearest neighbor nest was greater (far) or less (close) than the mean nearest neighbor distance.  He then checked reproductive success in the two groups.   The results (Table 5.5) show the advantage of close nesting and thus, presumably, the advantage of mobbing during the egg-laying and incubation stage (open colonies). The term "open colonies" refers to a two-dimensional pattern of nests. There is one clear disadvantage to mobbing, namely that the rising of birds may give away the position of their nests. In an open colony this disadvantage is minimized, and the net advantage of mobbing appears to be positive. Where nests are strung out along a road or stream bank, on the other hand (linear colonies), the disadvantage should be maximized and we might expect a net disadvantage to mobbing. Table 5.5 shows this to be the case during the egg-laying and incubation period.

During the nestling stage there appears to be no net advantage or disadvantage in dense nesting in either type of colony. This is expected, since at this time little or no mobbing occurs. The best test as to the effectiveness of mobbing is thus to compare the ratios 35 : 16 and 20 : 21 (see Table 5.5) in the open colonies. The ratios are different, by a $2 \times 2$ contingency table test, with $p < 0.05$.

**Table 5.5.** The advantages and disadvantages of colonial nesting (number of predated nests) in the brewers blackbird

|  | Egg-laying and incubation period | | Nestling period | | Total | |
| --- | --- | --- | --- | --- | --- | --- |
|  | far | close | far | close | far | close |
| Open colonies | 35 | 16 | 20 | 21 | 55 | 37 |
|  | $\chi^2 = 6.1$, p 0.05 | | | | | |
| Linear colonies | 9 | 20 | 16 | 10 | 25 | 30 |
|  | $\chi^2 = 4.1$, p 0.05 | | | | | |

The effectiveness of mobbing may vary not only with respect to the nature of the colony (open versus linear) but with other factors. For example, in species with cryptic eggs the disadvantageous aspect of giving away nest position will be less than in species with easily seen eggs. Thus the Arctic and common terns with cryptic eggs, *Sterna paradisaea* and *S. hirundo*, mob predators while sooty terns, *S. fuscata*, without cryptic eggs, do not (Ashmole, personal communication).

Grouping may, in some cases, help in avoiding detection by predators. If prey are individually dispersed a predator will detect, on average, some given number of prey individuals per unit of time. If all prey individuals clump together in groups of $n$, and in the most simplified case, occupy no more space than a single individual, a predator will encounter such groups only one $n$th as often as it will encounter individually dispersed animals. Thus if the predator cannot eat all $n$ at one encounter, fewer clumped than individually dispersed prey are eaten per unit time. The problem can be formulated more quantitatively in the following way. Let

$N$ = total number of individuals over a given area

$n$ = number of individuals per clump

$m$ = number of individuals eaten per encounter of a clump and a predator

$\rho$ = density of individuals in the clump

$r$ = the distance at which a predator notices a clump.

Since one individual occupies essentially a point in space, the area occupied by $n$

individuals may be approximated by $(n - 1)/\rho$. If the clumps are roughly circular, then, they possess a radius of

$$\sqrt{\frac{\text{Area}}{\pi}} = \sqrt{\frac{n-1}{\pi}}.$$

Since the predator can detect these clumps at a distance $r$, we can picture the clumps as shown in Fig. 5.3. The inner circles represent the clumps, the outer circles the limits at which the clumps are detected by predators. Picture the predator, represented by the dashed line, as moving across the area. The expected number of clumps it encounters per unit time is clearly proportional

to the diameter of the outer circles, $2\left(r + \sqrt{\frac{n-1}{\pi}}\right)$, and the number of clumps

per unit space (which is proportional to $N/n$). Thus the number of clumps encountered per time can be written

$$\propto \frac{N}{n}\left(r + \sqrt{\frac{n-1}{\pi}}\right),$$

and the expected number of individuals eaten per time, given that $m < n$,

$$\propto \frac{Nm}{n}\left(r + \sqrt{\frac{n-1}{\pi}}\right).$$

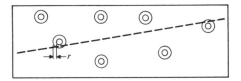

**Fig. 5.3.** (See text for explanation).

The probability that any given prey individual will be eaten over a given period of time is thus

$$\propto \frac{m}{n}\left(r + \sqrt{\frac{n-1}{\pi}}\right). \tag{5.5}$$

When no clumping occurs, $n = 1$, so that $m = 1$, and the expression becomes

$$\propto mr = r. \tag{5.6}$$

To find the conditions under which clumping is individually advantageous and, therefore, selected for, we simply set the value of the expression (5.5) less than that of expression (5.6.)

$$\frac{m}{n}\left(r + \sqrt{\frac{n-1}{\pi}}\right) < r, \quad \text{or} \quad r > \frac{m}{n-m}\sqrt{\frac{n-1}{\pi}}: \quad m < n. \tag{5.7}$$

Where the clump is stationary so that predators can return again and again, no search is required for the second and subsequent encounters so that, effectively, $m$ becomes very large and may equal $n$. In this case no advantage to clumping exists. In fact clumping may be disadvantageous.

Given that the clumps move about and that $m < n$, however, it can be seen that the conditions for the evolution of clumping are more often met if $\rho$ is high or $r$ is large. We might expect the density of clumps of prey animals to reflect a balance of advantages arising from this kind of defense and various disadvantages associated with close contact of individuals. The variable, $r$, is really the most interesting. In woodland, for example, or among very cryptic or nocturnal species, $r$ will be small. Thus it is likely that clumping will be an effective defense mechanism less often in forest species, cryptic species, or nocturnal species than in open-country, less cryptic, or diurnal species. As we have noted before, woodland species are more often solitary than plains species. It is also true that nocturnal animals are usually more likely to be solitary than diurnal animals. When $r$ is *very* large, however, the predator may never lose sight of the prey, and the advantage of clumping again disappears. Thus clumps of ungulates on short grass prairies cannot be avoiding predator notice, unless populations are very sparse or land features exist which will prevent sighting of prey over long distances.

Brock and Riffenburg (1963) have derived the same model as above, only for the three-dimensional case. The general conclusions are the same. In applying the model to fish schools they note first that, in general, prey species school but predatory species do not. This suggests that schooling is in many cases a defense mechanism, probably for avoiding detection by predators. They then note that when $r$ is small, in murky water and at night, fish schooling tends to be absent.

Aposomatic species, like cryptic species, are safer if not detected, but once seen, do well to emphasize their brilliance. For this reason, the advantages of clumping may be greater in bright, as opposed to dull, prey species. Furthermore, bright species can be seen at greater distances ($r$ is larger) so that, according to the above model, the conditions for clumping are more likely to be met. Aposomatic insects, then, should be more clumped than cryptic insects, all else (other reasons for clumping or dispersing) being equal. De Ruiter (1952) has noted that the aposomatic species *Phalera bucephala* and *Vanessa io* are highly clumped in their local distributions, while stick caterpillars are not.

## 5.4 CODISTRIBUTION

There are a number of close relationships (mutualism) between animals and plants, or between different animals which serve to protect one or both species from attack. These will be discussed in 12.4. For the moment, we consider the advantages of more loosely defined patterns of dispersion, patterns which may ultimately lead to the evolution of mutualism or mutual avoidance.

Picture a situation among two species of mice or insects (say) inhabiting a field, in which one enjoys the advantage, in the face of the various predators present, of being the less-favored food. If individuals of this species can be in the company of individuals of the preferred food species when an attack comes, they will be least likely to be eaten, since a choice is available to the predator. Individuals of the preferred food species, on the other hand, will suffer less predation if the predator is given fewer clear-cut food choices—that is, when the preferred prey avoids the company of the poorer food species. Thus the poor . food species should attempt to associate closely with the better food species, and the latter should try to avoid the company of the former. Two conclusions follow: The poor food species should be docile, appeasing the aggressive tendencies of the other species in order to more easily mingle, while the preferred food species is more likely to be aggressive in its behavior; and patterns should include mixed populations and populations of the preferred food species who have either successfully driven out the other species or colonized new areas one step ahead of their persistent followers. The reader might keep these suggestions in mind when considering active competition, interference, and habitat displacement (Chapter 12).

# 6
# Reproductive
# Ecology

Natural selection favors any trait which increases rate of gene spread. As such, it acts to favor the choice of appropriate breeding areas and breeding times as well as the concentration of reproductive efforts at certain ages. The interesting questions in the study of reproductive ecology stem from the fact that opposing selection pressures operate with respect to all these factors. The area which is safest from attack by predators may be poor with respect to the gathering of food by or for the young. The time of year when food is most plentiful is often the time when predators are most active. The advantage of short generation time is balanced by the advantages of increased reproductive success with age due to accumulated experience. All reproductive strategies may be thought of as compromises reflecting balances of selection pressure on parental survival, survival of offspring, fecundity, and other factors.

## 6.1 REPRODUCTIVE SITE

Two sorts of reproductive site exist: The first is the area in which mating occurs; the second is the area of production of young. In some cases, the first may be subdivided into one area for pair formation and another area for copulation. The nature of a mating site may or may not be predictable. The usual function is merely to bring the sexes together so that almost any well-defined area may do, regardless of its value as a habitat for feeding, sleeping, or other activity. Ants, although they may copulate in the air, usually do so at special nesting stations. *Myrmeco ruginoides* uses for its nesting station a bare, flat surface near some vertical feature. The tops of hills often serve as mating areas (Sudd, 1967). The intertidal snail *Thais lamellosa* spawns in huge clumps (depending on the size of the local boulder on which they congregate), often within quite specific tidal levels. Since the young are found at quite different (higher) tidal levels and there is no evidence for different egg mortality with tidal level, the spawning grounds appear to be chosen for the benefit of the mating adults (Emlen, 1966b). Marine turtles, *Chelonia mydas*, which are not likely to see each other often, spawn colonially. The females lay their eggs along certain beaches on certain days, and on their return to sea are inseminated by males waiting offshore. The sperm are retained and used to fertilize the next year's egg crop (Carr and Giovanelli, 1957). In the case of some terrestrial turtles such as the American box turtle, *Terrapene sp.*, no special mating site is used. Individuals of different sexes simply mate when they meet. The females are capable of storing the sperm in viable condition for up to four years (Carr, 1952; Oliver, 1955). Certain ponds act as breeding aggregation sites in many frogs and toads and special areas of lakes are used by salamanders. Hibernating sites double as mating sites for many snakes (Oliver, 1955). Birds may require a high vantage point or some specific sort of perch from which they can sing if an area is to serve for breeding purposes.

Often the mating site also serves as the spawning site. Here the requirements are often less obscure and the nature of the site more predictable. Spawning sites are, presumably, chosen in terms of their value to offspring (as well, perhaps, as simply an aggregating site for adults).

For physical protection, ants living in the tundra nest in cushion plants. This practice is useful for escaping the wind. Nesting under rocks would also be a means of avoiding wind but would necessitate losing the warmth of the sun. Ants do not nest under rocks in the tundra (Sudd, 1967). Termites, *Constrictotermes cavifrons* and *Procubitermes niapuensis*, avoid rain damage to their tree-trunk nests by constructing chevron ridges on the trunks immediately above. The inverted V's act to drain water around the sides of the nest (Allee *et al.*, 1949). Goldfinches, *Carduelis c.*, in northern Europe nest on the north sides of trees between 12 and 30 feet up for maximum protection from both sun and the prevailing southwesterly winds (Conder, 1948). One of the most interesting responses to local weather conditions by nesting birds is reported by Horvath (1964) for the rufous hummingbird, *Selasphorus rufus*. Birds of this species appear to achieve a (presumably optimal) nest temperature in the spring by placing their nests in the lower branches of conifers. At this time of year there is little ground radiation of heat and the thick branches of the conifers prevent too much sunlight from hitting the nest. Later in the summer, however, the ground radiates heat and conifer branches above the nest would act to hold this heat. Overheating is best avoided by moving up from the lower branches of conifers into the tops of deciduous trees and taking advantage of the cooling effect of transpiration through the leaves. Of course, there may be other reasons for the birds' behavior. Orians (personal communication) suggests they may be responding to availability of food which shifts from low to high tree levels during this period.

Food available should clearly be an important consideration in selection of a spawning site. In the case of species with parasitic or scavenger larvae such as botflies or house flies, food, in the form of animal tissues or decaying or fecal material, is of primary importance. Marsh-breeding birds feed their young on insects emerging primarily in the marsh and tend to nest in those marsh areas with the highest food density. Nest placement for a species disposed toward coloniality as opposed to territoriality may depend greatly on food distribution, as we shall see in Chapter 8.

Protection from predation is another consideration in nest placement. We saw in Chapter 5 that some bird species place their nests in "unpleasant" or "inaccessible" places, such as among stinging insects, in well shafts, or over water. Along coast lines, colonial sea birds often seem to "prefer" islands rather than the mainland as colony sites. This is undoubtedly due, in part, to the fact that island areas are insufficient to support predator populations.

In many cases, a quick glance at an animal's general ecology vis-à-vis climatic, defensive, and feeding requirements allows for reasonably accurate predictions as to the nature of that animal's spawning sites. We would expect marsh-feeding birds to nest on marshes, tundra ants to nest in cushion plants, and colonial sea birds to prefer island colonies. In other cases, the adaptive values of spawning site characteristics are not at all obvious without detailed study. Meanley and Webb (1963). for example, note that the nesting success (number of eggs successfully fledged) in redwing blackbirds, *Agelaius phoeniceus*, rises from 45% for pairs nesting at heights below 2 feet to 62% for those nesting above 4 feet. Holcomb and Twiest (1968) found nesting sucess values of 17.2%, 22%, and 34.8% respectively for redwings nesting lower than 2 feet, between 2 and 4 feet, and above 4 feet. Why these differences in mortality with nest height occur is not known; they may be related to temperature balance, accessibility to predators, or other factors. A serious question is that, given adequate, high nesting sites, why do redwings nest below 2 feet?

## 6.2 TIMING

A number of species are aseasonal breeders—they show no preferred breeding season. These include deep water marine animals (although some seem to be seasonally rhythmic—A. Schoener, 1968) and a few tropical animals. Bornean rain forest frogs, for example, breed all year with no sign of seasonal preference (Inger and Bacon, 1968). Most species, however, even tropical ones, show distinct seasonal breeding peaks.

Timing of production of young is usually—although not always—influenced by the time of mating. The latter, or both, may be under the influence of light: dark cycles. In many fish, increasing ratios of light to dark intervals in the spring promote faster gonadal development. The opposite is true of those species breeding in the fall. Occasionally, reproduction is triggered by some critical amount of sunlight. The cichlid fish, *Tilapia macrocephala*, is sexually active on sunny days, but not in cloudy weather. Also, increasing temperatures act to induce gonadal maturing in this species (Aronson, 1965). Iersel (1953) has noted that sudden temperature rises affect reproductive behavior in sticklebacks (*Gasterosteus*) and some other fish. Cyprinid fish respond sexually to rain.

Rainfall often stimulates migrations to breeding ponds, calling by males, and copulation in many frogs and toads (Aronson, 1965). Biological and social factors also play a role. In some birds, once the gonads start to mature (under the influence of light: dark ratios), weather, food supply, and presence or absence of suitable nest sites may affect further gonadal development. Cold or rainy weather, hunger, or lack of nesting material and nest sites inhibit sexual activity, especially in females (Marshall, 1959), and it is likely that in males, gonadal maturity and territorial defense work in a positive feedback relation.

The type of reproductive cue utilized, of course, must be a function of the size, physiology, and other features of the animal. Small and poikilothermic species, with little homeostatic ability in bad weather, might be expected to respond primarily to temperature or moisture changes, while large homeotherms, better buffered against inclement weather but in constant need of nourishment, might respond more often to changes in food supply. In temperate regions where temperature and photoperiod vary markedly with season, response to these variables is more likely to occur than in the tropics where rainfall changes are proportionately much more pronounced.

Proximate factors are merely mechanisms which bring about adaptive response, and they need not be directly related to the ultimate factors. Thus we do not generally make love out of a desire to procreate, but rather as a response to more immediate stimuli. A bird which nests in the spring does so not because it "knows" that this is the optimal time to nest, but because it is somehow driven by stimuli correlated with the onset of spring. There may be many cues appropriate as stimuli to an act. Which of them is siezed upon during the course of evolution is generally unpredictable. We now look at ultimate factors affecting reproductive timing.

## 6.2.1 Bring the Sexes Together

The timing of reproduction into peaks serves all or any of three purposes: It brings the sexes together, assures that the greatest number of offspring survive to maturity, and it maximizes the parents' chances (in species breeding more than once) of surviving to the next breeding period.

In many species, there are either common social contacts between the sexes (most higher vertebrates, social insects, for example) or dense aggregations during the breeding season (many intertidal species such as *Thais, Uca*). In these animals, the bringing together of the sexes has been solved by other means. In many species, however, timing may be critical to the uniting of egg and sperm. A spectacular example of synchronized breeding activity is found in the palalo worm, *Eunice viridis*, in the South Pacific. These annelids swarm over certain coral reefs primarily on one day in the year—some time during the last quarter of October or November in Samoa, November or December in Fiji. Because of the sacrifice on the part of a breeding worm, timing is vital. *Eunice*, at the time of swarming, loses its head. The headless, or *epitokous*, part of the body, freed from neural inhibition by the hind brain, writhes violently in a mass of other bodies and splits open, releasing its gametes. Improper timing means death without gene passage. In ants, winged sexuals perform their "nuptual flights" not only in certain areas or types of habitat, but often at very specific times of day, only on certain days (Wilson, 1963). The grunion, *Leuresthes tenuis* (a fish), spawns three or four days after the new or the full moon, corresponding to the highest tides of the month. One to three hours after the tides begin to recede, females, hotly pursued by males, wiggle up onto the beach with the highest waves and, by properly moving the tail, half bury themselves in the sand and there deposit their eggs. The males, forming a thrashing circle about the female, release sperm and fertilize the eggs. The next wave is caught and ridden back to sea. In this manner, the eggs are protected from cannibalism and many other marine dangers until the next spring tides bring waves up the beach to them—at which time they hatch and the young ride the waves out.

## 6.2.2 Parental Survival and The Successful Raising of Young

From the standpoint of parental care, reproductive timing may be considered as a compromise between time, energy, and risk factors, optimal for the production of the largest possible number of surviving offspring.  Since parental care, as well as fecundity, depends on the vigor of the parents, and since fitness may often depend on parental survival to subsequent breeding seasons, reproductive timing must also consider parental well-being.  Of course, when food is plentiful for the young, it is generally also plentiful for the adults, so that these two considerations usually lead to the same timing strategy.  Considerations of parental well-being will be specifically discussed later.

It is obvious that, from the standpoint of survival of both parents and offspring, young should appear when food and water supply are adequate and risk of predation is minimal.  Yellow-headed blackbirds, *Xanthocephalus xanthocephalus*, time their reproduction so that the young hatch just as the local damselfly population emerges.  The tenerals, clinging to the cattails or bullrush stems, make easy targets for foraging birds (Orians, 1966; Willson, 1966).  The great horned owl, *Bubo virginianus*, lays eggs in mid-December in Florida and in mid-February in Pennsylvania and Iowa.  This timing results in the hatching of young owls shortly before the production of young rodents, the prime food source, but at the time of poorest cover for the prey.

In moist, temperate areas, herbivores typically avoid bearing young in the winter and wait until the plants leaf out in the spring.  Predators, which rely on herbivores for food, generally do the same.  In dry areas where productivity depends on infrequent rains, the usual strategy is to produce when the rains come.  In the case of fish, rains produce runoff, and thereby the accumulation of new nutrients and thus increased productivity in their habitat.  The rains also signal the end of the season of receding water, meaning eggs can be laid with less danger of desiccation.  In Texas, the frog *Rana pipiens* has two yearly reproductive peaks corresponding to the wettest parts of the year.  The tree frog, *Microhyla olivacea*, reproduces anytime between April and September in direct response to rainstorms. *Acris crepitans*, another frog, on the other hand, is more strictly limited to an aquatic environment and although rainfall may be important, as in the case of the fish mentioned above, temperature seems to be the most important factor determining breeding and spawning (Blair, 1951).  In the American southwest, lizards lay their eggs during the wet season(s).  Where rainfall and temperature are nearly invariant and seldom or never too high, as in the Bornean rain forest, lizards breed all year (Inger and Greenberg, 1966).  Iguanas, *Iguana sp.*, breed in the fall dry period in the western United States so as to be able to lay their eggs at the beginning of the rainy season (Rand, 1968).

In Africa, water and swamp birds breed mostly in the rainy season (or early dry season) when insects and cover are most abundant. Grassland birds follow the same strategy. Strictly insectivorous birds compromise their need for insects less with their need for cover, and usually breed earlier in the rainy season. On the other hand, large ground birds often, and raptors generally, nest during the dry season. The reason is not obvious for large ground birds, but the raptors may be taking advantage of the number of newly fledged young grassland birds and the lack of protective cover that follows the drying up of the land (Moreau, 1950; Benson, 1963).

Among Central American birds, the nectar drinkers (such as hummingbirds) nest in the dry season when the most flowers are in bloom. Grass seed feeders nest at the time when grass seeds are most abundant, early in the rainy season. Raptors, as in Africa, nest in drier parts of the year, except for the kites which catch insects in the air and nest early in the rainy season, the time when insects are most abundant. Insect feeders of other sorts also nest at this time (Skutch, 1950).

Among small, desert mammals, water shortage reflects importantly on milk production by females. Thus during dry or hot times of the year, when water is particularly scarce or being used in temperature regulation, most small mammals don't reproduce. Most births occur in the wet seasons. Those animals, such as *Dipodomys* or *Perognathus*, however, which produce highly concentrated urine and efficiently utilize metabolic water, do not face the same problem and breed all year (Schmidt-Nielson and Schmidt-Nielson, 1952). Large mammals are somewhat more buffered from physical environmental changes than small mammals, but nevertheless require sufficient water to produce milk. Thus in a climate with a rainy season from July to September, peccaries breed all year, but with an obvious peak in July (Sowles, 1966).

To escape from predation may require a compromise with the time of the best temperature or food supply. Both young and parents may be particularly vulnerable to attack up until the time of spawning (or nesting). The eider duck, *Somateria mollissima*, where it nests on Arctic islands, waits until well after food is available to begin reproduction. This strategy is quite different from that followed by most Arctic birds, which breed earlier in "anticipation" of an emerging food supply. The late breeding in eider ducks appears to be correlated with the presence of foxes: late nesting allows the ice linking the islands and mainland to melt, eliminating a land bridge over which foxes might cross to prey on duck eggs (Lack, 1954).

The situation is not always so simple, however. Although prey species may shift nesting seasons so as to avoid danger from nesting predators, the predators shift their reproduction to coincide with peak prey production. Since, in many cases, temperature and vegetation emergence set rather severe time limits on prey species reproduction, the sort of strategy applied by the eider duck may seldom be possible. Kruuk (1964) has suggested that prey might escape some predation by nesting quickly and synchronously. This would not only make it difficult for predators to optimally time their own reproduction but would glut the food market very temporarily for predators whose populations may be held down by food shortages at other times of year. Predators, therefore, would not be allowed the advantage of the drawn-out food supply necessary for their own population expansion. An excellent example of this strategy is found in the periodic cicadas which emerge from the ground, metamorphose, and lay their eggs (back in the ground) in huge numbers over only a few days, once every 13 or 17 years (Lloyd and Dybas, 1966).

It is possible that, with reduced predation, bird nestlings could afford to develop more slowly. Slow development would mean less food demand and, to the extent that the number of nestlings is correlated with the ability of the parents to provide adequate food (see 6.3 below), allow larger clutches. Evidence on this point shows an average development period of 17 to 19 days for hole-nesting (low predation rates) and 11 to 13 or more days for open-nesting (high predation rates) altricial birds (Lack, 1948b; Nice, 1957). Clutch-size data also fit the hypothesis and are discussed later. Good year-round weather conditions in the tropics might suggest that, here also, birds prolong nesting in the interests of large clutches. Predation is high in the tropics, though, and it is better to risk destruction in several small doses than to have a whole year's reproductive efforts wiped out. In the tropics, bird clutches are usually small. The prolonged season is efficiently made use of through the production of several, spaced clutches.

Parental well-being may affect reproductive timing profoundly. First, in species with parental care, the parents must be sufficiently vigorous to gather food and protect their young. This is true even though they may already have spent considerable time and energy toward the production of eggs. In all species which breed more than once in their lifetimes, the parents must compromise the advantage of lavishing great time and energy on their present young with the advantage of living to reproduce again either later in the same season or in future years. If the energy drain has been too great, lowered resistance to disease, predation, or exposure may prohibit future gene passing. In his 1954 book, Lack hypothesized that reproductive timing in birds was directly related to the availability of food for the young. In 1966, he modified his views somewhat, noting that the egg-laying process was very energy draining, and that reproduction must occur at such a time that the female has adequate food before egg laying. Two pieces of evidence for his later view are given here (both cited in Lack, 1966). First, egg laying is, as he suggests, very taxing. The great tit, *Parus major*, for example, lays 9 or 10 eggs, each weighing about 1.75 g, over a 9 or 10 day period. This is a production of 17.5 g of egg by a female of average weight 17.3 g, over a period only slightly greater than a week. Second, the timing of nesting may be delayed for no obvious reason other than the need of the female for food prior to egg laying. For example, the pied flycatcher, *Muscicapa hypoleuca*, lays its eggs at such a time that the young hatch after the peak in the spring caterpillar population. The young thus grow up during a decline in the food population. One would expect earlier breeding to be advantageous with respect to survival of the young and, in fact, the success with which young survive to maturity is greater among those young hatched somewhat earlier than the average (Lack, 1966;

Perrins, 1970). Other bird species are similar in this respect. Furthermore, one might expect species laying more than one clutch to spread their breeding activities over that part of the breeding season best for survival of the young. In fact, even the first clutch is usually laid after the food supply peak. Subsequent clutches have much lower survival, as might be expected. Strictly in terms of producing healthy offspring, clutches laid earlier than average are more successful. The usual practice of nesting "too late" appears to be related to the female's need for energy in the egg-laying process.

Parents, as well as young, may have their energy reserves further depleted by the need to molt. Since feathers make up close to 10% of some birds' weights, there is no doubt that a turnover of feathers is an energy drain. Whereas, in the tropics, energy is constantly available and birds may molt slowly throughout the year, in the temperate and Arctic zones, many birds must lay eggs, feed young, molt, and possibly build up energy reserves for the southward migration all within the period of productivity. In the Arctic, particularly, the problem may be critical, with severe limiting effects both on the dates and duration of nesting. Fast molting, in addition, may leave birds incapable of efficient flight and thus extremely vulnerable to predation. In the Cassin's auklet, *Ptychoramphus aleutica*, the feathers are lost rapidly, but the primaries only one at a time so that flight is always possible. Molting occurs during the second half of the very short nesting season and is related in its rate to nesting success. Females feeding young molt 40% more slowly than females without young (Payne, 1965b). The budgeting of time and energy in the Cassin's auklet thus represents a compromise between nesting needs and molting (and other) needs. If nesting activities are successful, the budget is biased in favor of reproduction, if not, in favor of a successful retreat until the following year.

The whistling swan, *Olor columbianus*, arrives in the Arctic and lays its eggs before the thawing of the ground so that the eggs hatch at the same time the first food becomes available. The young fledge just in time to molt before the food is gone. In the Antarctic, the emperor penguin, *Aptenodytes forsteri*, lays one egg in late June or July (mid-winter) which hatches in August or early September. This is still very early for anything but a very meager food supply, but is necessary if the young is to reach independence before the next winter sets in (Welty, 1962).

Finally, it is important to note that the time of breeding may vary in a definite fashion within a population. For example, it is common, at least among colonial sea birds, for the older individuals to breed before the younger ones. Coulson and White (1956) discuss reproduction in the kittiwake gull, *Rissa tridactyla*, and conclude that age is highly important in determining the timing of reproduction. Young birds, when stimulated by the activity of older individuals mixed with them, tend to converge in breeding time with their elders, but they nevertheless consistently breed later. Older birds and pairs with more prior experience generally breed earlier, establish themselves in the "best", central positions in the colony (see Chapter 8), and have significantly greater success in fledging young (Coulson and White, 1958; Coulson, 1966).

### 6.2.3 The Balance of Advantages

There are selection pressures affecting both the time of mate selection and copulation and the time of egg or young production. Development time binds these acts together and is not necessarily flexible enough to permit optimal timing for both. However, a number of groups have evolved the means to delay fertilization or implantation in order to gain the best of both worlds. Many species, such as the marine and terrestrial turtles mentioned above, store sperm and fertilize eggs later, consonant with the best laying or hatching time. In mammals there is no sperm storage, except in bats and possibly bears (Asdell, 1946), but many species can delay the implantation and onset of development of their young. Delayed implantation occurs in bats, *Chiroptera*. The fruit bat, *Eidolon helvum*, of Uganda, for example, mates in the spring (why this is advantageous is not known), but no pregnancies are observed until the following October. The young are born in February and March, just before one of two rainfall peaks in April and May (Mutere, 1967). Delayed implantation also occurs in the Texas armadillo, *Dasypus novemcinctus*, where at least a three-week lapse occurs between copulation and implantation (Patterson, 1913, in Asdell, 1946), the roe deer, *Capreolus capreolus*, which mates in July and August and becomes pregnant in December or January (Short and Hay, 1966), the Alaskan fur seal, *Callorhinus ursinus*, and many carnivores. The badger, *Meles meles*, copulates in July or August in southern England and implantation occurs in December or January (Bourliere, 1956). Martens, *Martes*, and weasels, *Mustela*, are well-known examples of mammals with delayed implantation (see Asdell, 1946).

There may exist one optimal period during which success in raising or producing young balances parental well-being for maximum fitness. But many animals do not time their reproduction quite so perfectly. There are several reasons for this. If an unexpected disaster or turn of weather destroys a nest, all would be lost to the species nesting at this time. Accordingly, it may be advantageous (over a long time period — group selection) to begin nesting a little earlier than otherwise expected. The animals can then spread two or more clutches over a longer, slightly less optimal period (on average), rather than throw everything into one big effort over a better, shorter period of time. Also, eggs may be laid or young produced in several batches simply because it is physically possible to produce only so many young at once. After one spawning bout a mother may retreat from her reproductive activities, feed, and develop a new set of oocytes.

The general picture, then, is this. In species with parental care, reproduction should be well timed and one-shot where the potential breeding season is short, and less well timed, with multiple clutches, when the season is long. Individuals which can lay only a few eggs at a time, or can care for only a few young at a time, and for whom the potential breeding season is long, should also spread their spawning activities. Species which lay or are capable of laying large egg masses may spawn once or several times during the year. The latter strategy should be most often found, and most pronounced, in species inhabiting unpredictable environments in which accident or disaster are high possibilities. In this connection it is interesting to note that the snails *Thais lamellosa* of the North American west coast and *T. lappillus* of Britain and Europe, lay their eggs during a single spawning period low in the intertidal. The low intertidal is less subject to predation by shore birds and mammals, experiences less exposure, and runs fewer risks of catastrophic destruction by floating debris and shifting sands than the mid-intertidal. In the latter, more unstable area, the related snail *Thais emarginata* lays several clumps of eggs throughout the year, with two peaks of spawning activity just preceding the two peaks of settling barnacle spat, the primary food of small individuals (Emlen, 1966b).

## 6.3 THE NUMBER OF YOUNG

It is obvious that, if a female has only so much energy to put into egg production during one spawning period, she must make a compromise between egg number and energy content per egg. It is immaterial whether the energy is in the form of yolk, as in most insect eggs, *nurse eggs* upon which developing young feed, as in some snails such as *Thais* (see Thorson, 1950), or parental care. (Of course where the energy spent is largely in the form of parental care slower development will permit larger numbers of young.) It is an interesting comment on the growth of ecological thought that Svärdson, as late as 1948, felt it necessary to spell out this fact.

The necessity of compromise on reproductive energy expenditure, although very simple in concept, opens up a number of avenues for exploring correlations between environmental factors and family size. For example, in cold, fresh water species, where food production is low, a young animal must be provided with a good start in life so as to be able to withstand the food shortage characteristic of such habitats. Without parental care, the best strategy is to provide the young with a food supply of yolk. Cold water species generally lay fewer, yolkier eggs than warm-water relatives. This is true, for example, in the frog *Rana pipiens* (Moore, 1949) and frogs in general (Moore, 1942).

Barnacles lay very large, yolky eggs in the far north and in cold areas, and smaller eggs at tropical latitudes and in other warm places. In areas where winters are cold and summers cool, the barnacle *Balanus balanoides* lays larger eggs than in warm areas, regardless of latitude (Barnes and Barnes, 1964). There are also other reasons for expecting fewer, larger eggs per clutch in cold-water species. Boreo-Arctic, marine, planktonic feeders must deal with larger food particles than their colleagues to the south (for the reason given above with respect to eggs— among others). To handle the larger food particles, young animals must be larger than in warm climates, and larger offspring are produced by larger eggs.

Clearly, there are factors other than temperature which affect the allocation of energy in egg production. In such groups as fish, size of fry may determine success in social competitive encounters and thus, indirectly, feeding success, avoidance of injury, or mating success. Larger, yolkier eggs give rise to larger young (Svärdson, 1948). It is a fact, at high altitudes, where physically marginal conditions are believed to result in lower population densities with consequent low levels of competition (see 15.3.3), that trout lay larger numbers of smaller eggs than fish at lower altitudes (Svärdson, 1948). (This is not to imply that other reasons do not exist for egg number differences in high and low altitude trout.) Among seed plants, species inhabiting woodland undergo severe competition for light, those individuals getting a head start in growth generally being the most successful. A rich food supply—and thus fewer, larger, seeds—should be a more likely strategy here than in open country. In fact, woodland plants usually produce larger seeds than field plants (Salisbury, 1942).

Predation also affects reproductive strategy: Where predation is high and the size of large eggs and or young is, in itself, not sufficient deterrent to being eaten, it may be good strategy to produce many small eggs. Among marine invertebrates, those species with the longest—and thus most predation-prone—planktonic larval stages are generally the ones which produce the largest numbers of eggs (Thorson, 1950). Of 23 species of Central American legumes (plants of the pea family) attacked regularly by bruchid beetles, average seed size is 0.2623 g and average number of seeds produced per cubic metre of canopy is 1020.00. Of 13 legume species not attacked, the corresponding numbers are 3.0065 and 13.89 (Janzen, 1969).

Species which produce large numbers of young with the result that some disperse to beneficent areas and survive, are known as *opportunistic*. Opportunism plays an important role in ecological relationships and will be referred to often throughout the rest of this book.

The strategy of energy allocation may be complicated by still another consideration: the laying of eggs in one location to avoid predation in another. Marine turtles, for example, greatly alleviate predation pressure on their eggs by burying them in sand on the beach rather than in open water. However, to minimize predation between hatching and the acquisition of safety through size, as well as to increase the capability of crawling to the water after hatching, these turtles must provide their young with the necessary food to grow quite large in the egg. Marine turtles clearly have no choice but to lay large, yolky eggs so long as they insist on depositing their eggs on beaches.

Where the environment is coarse-grained patchy (see 4.2.1), or changing in time, there may be an advantage in producing large numbers of small eggs—that is, to be opportunistic. In the former instance, there may be selection for great dispersal powers so that the young escape competition with their parents and yet settle in the patch type to which they are specialized (see Chapter 4). In most animals of small size—for whom the environment is more coarse grained—adequate dispersal carries a high risk and a 'Gatling gun' approach is the best strategy. Where the environment changes in time, the same arguments hold. Small species possess short life spans and, as such, view temporal changes more often as being coarse grained. They must, more regularly, disperse as a means of escape from change and so evolve high fecundities. It is interesting, during early secondary succession (see 13.5), when changes occur rapidly in time, that plant seeds are usually smaller and more numerous, with more developed mechanisms for dispersal, than when change is slower, in late successional stages. In the amniotes, particulary birds and mammals, behavioral plasticity and sophisticated habitat selection abilities allow for high survival of young, and it is thus an advantage to have few, large young.

Cody (1966) adds a comment on the relation between number of young and heterogeneity of the environment. He notes that, as the environment becomes less stable, populations are held at lower levels and thus are less often faced with problems of competition for resources or social tolerance. Under these circumstances, selection favors the ability to seize upon good opportunities and multiply. That is, selection favors increased fecundity rather than survivorship (see also Discussion, Section III). If this is true, then the clutch size of birds should increase with latitude (temperate regions are considered to be less stable than tropical regions), which they do (Moreau, 1944), and with distance from stable, coastal areas. Johnston (1960) shows this to be the case in the song sparrow, *Melospiza melodia*, whose clutch size is higher in Ohio than on the west coast.

### 6.3.1 Energy Supply

For any given energy allocation strategy—that is, for any species which expends a certain amount of energy per egg—increased energy intake should reflect itself in increased clutch size. This hypothesis is supported by Hoddenbach and Turner (1968), who show a correlation between the size of the first clutch of the season in American lizards and the productivity of the previous winter, which is when the lizards are gathering and storing energy in fat bodies for use in reproduction. Hahn and Tinkle (1965) had noted the same relationship three years previously.

With respect to species with parental care, the situation is far more thoroughly studied. Such a species can choose between two extremes; putting all reproduction energy into may eggs and exercising only token care, or producing very few eggs but subsequently lavishing care on them.

Note here that parental care is very widespread and varied in expression. Parental care is found not only among birds and mammals but among wolf spiders, *Lycosidae*, which carry their young about on their backs. Pisaurid spiders construct special nurseries and watch over their offspring. In the spider *Coelotes* the female shares food with her young and may feed them with regurgitated food. *Theridion saxatile* moves its egg cases about to keep them at the best temperature for development and *Lycosa pirata*, which needs moisture, burrows in sphagnum and, during the heat of the day, stands on its head to keep cool while holding its egg case out in the air for warmth (Kaston, 1965). Many crustaceans carry eggs and often young about, and octopi place their eggs in crannies and fan them to keep them aerated. Many fish, such as the sticklebacks, also fan their eggs. Male seahorses carry eggs around in a pouch. Sticklebacks guard their young after hatching, and some cichlid fishes carry their young in their mouths. Among the amniotes, powerful and/or poisonous snakes, such as the boas and vipers, guard their eggs (Niell, 1964).

Red deer females can nurse more young more easily in good food years.

Darling (1937) found 92% of the females examined to be pregnant in good food years, 33% with one young, 60% with two, and 7% carrying triplets. By contrast in bad food years only 78% were pregnant, 81% of these with one young, and only 18% and 1% with two and three. Pigs and rabbits, when domesticated and well fed, have larger litters than their wild relatives who must seek out their nourishment. Lions have larger litters in good food years (up to four or five) than in bad years (two or three) (Stevenson-Hamilton, 1937, cited in Lack, 1954). Also in non-viviparous species with parental care, the number of young is directly related to food supply.

In the past it was usually thought that birds laid the maximum number of eggs they were capable of producing. However, it is clear that this is not so in all species. Many birds lay a definite number of eggs and if one is removed during the laying period, will replace it. Taking advantage of this fact—and also of an unfortunate experimental subject—one worker methodically removed the eggs of a yellow-shafted flicker, *Colaptes auratus*, and induced the bird to lay 71 eggs in 73 days (Phillips, 1887, cited in Bent, 1939). The same sort of indignities have been visited upon the willow ptarmigan, *Lagopus lagopus*, and the ostrich, *Struthio camelus*, with similar results (Welty, 1962). The case of the domestic chicken is, of course, well known. Clearly, birds are capable of laying more eggs than they do. It seems reasonable to rephrase the old hypothesis and note that birds, and for that matter all animals displaying parental care, produce the maximum number of eggs they are capable of raising. Many species do not replace missing young so the above experiments cannot be performed on them. Nevertheless, it seems reasonable that in these species, too, the number of eggs laid is related to the energy available for egg production.

The gist of the above argument is that animals produce that number of eggs which result in the maximum number of young surviving to reproduce. This hypothesis is almost tautological to an evolutionist and has been championed for a number of years by Lack (1954, 1966, and earlier work). We refer to it as Lack's hypothesis. The argument can best be presented through the use of a simple graph (Fig. 6.1). As the number of eggs increases, the number of hatchlings increases. However, more young means more mouths to feed and results in lower food levels for each offspring. Coulson and White (1958) have shown, for example, that in kittiwakes, young alone in the nest were fed enough to gain 15.81 g per day, while young sharing their nests with another offspring gained only 14.75 g (parents breeding for the first time). Experienced parents were able to raise these figures to 16.15 and 15.57. Because food brought to the young drops on a *per capita* basis as egg number increases, the probability that a young bird will live to fledge also drops. Furthermore, less food per individual results in lower weight at fledging which has been shown to be detrimental to later survival (see, for example, the work of Perrins, 1965, on the great tit, *Parus major*). Thus the probability of survival falls as the number of eggs laid increases. Fitness, the ratio of offspring surviving to reproduce to parents, is simply the number of hatchlings times their probability of survival, divided by two (parents), and is plotted against egg number in Fig. 6.1(b).  In this hypothetical case, fitness is increased by any change in clutch size toward 3.

There are many data to support Lack's hypothesis, both direct and indirect. Indirect evidence will be presented first.  If food supply increases from one year to the next, more food can be gathered with the same time and energy expenditure, so that larger clutches should be favored.  If clutch size is under genetic control and individuals tending to lay large clutches are selected for in such years, larger clutch sizes should lag good food years.  If selection has favored behavioral plasticity rather than genetic control of clutch size, large clutch years should correspond to good food years.  There is abundant evidence for the latter.  MacArthur (1958) noted an increase in the average clutch size of eastern United States wood warblers from about 5.1 (normal years) to about 5.8 during an outbreak of budworms.  Hawks and owls have larger clutches in years of high vole and mouse populations.  The swift, *Apus apus*, lays three eggs in good food years.  In poor years, only two may be laid or, as the weather gets cold and rainy (a harbinger of insect scarcity), one of the three eggs may be removed from the nest (Koskimies, 1950).

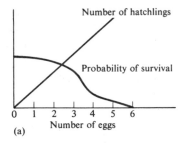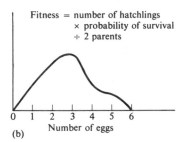

Figure 6.1

Indirect evidence for Lack's hypothesis comes also in the form of latitudinal gradients (although other factors may also be involved here).  In temperate regions, nesting takes place during that part of the year when days are longest, offering the greatest amount of time for foraging.  Furthermore, productivity is concentrated at this time of year.  In the tropics, days are shorter in the late spring and summer and productivity, although higher annually, is spread over time. As a result, birds may be expected to gather more food during a day in temperate areas, and this leads us to expect larger clutches than in the tropics.  This expectation is increased by the argument that in temperate regions birds die during the winter or during migration, so that the population reaching the breeding grounds may be less than what the food supply will support.  This makes food easier to get and suggests larger clutch sizes (Ashmole, 1961; Lack and Moreau, 1965). In passerines this seems to be true.  Moreau (1944) found that of 15 species common to south Africa (south of 25° latitude) and Europe (50° to 60° north latitude), 4 exhibited the same clutch size in both places, 10 had larger clutches in Europe, and only 1 laid more eggs in Africa. Comparing south Africa and equatorial Africa (within 5° of the equator), the corresponding figures are 83, 54, and 4 for passerines, and 56, 17, and 3 for nonpasserines. Lack (1948a) states that tropical nesters nearly always have smaller clutches than their temperate counterparts in the same family. In the *Caprimulgidae* and *Burhinidae* the ratios of tropical to temperate clutch sizes (average) are 0.95 and 1.00, but birds in these families are twilight feeders and so no difference is expected. The fact that some precocial birds (with no parental feeding, giving us no reason to expect latitudinal differences in clutch size) show the same latitudinal trend indicates that there is more to the argument than presented above, but does not necessarily negate the evidence for the relation between food supply and clutch size.

Direct evidence in support of his hypothesis has been gathered by Lack. Data for Swiss starlings (early broods) are given in Table 6.1 (after Lack, 1948b). Data for great tits (Lack, 1966, and others) do not show such a good fit, but when mean clutch size over the years 1947 to 1963 is compared with mean clutch size weighted according to number of young per nest that survive for three months (a measure of optimal mean clutch size), the agreement is striking (Table 6.2). In 12 out of 17 cases, an increase (or decrease) in mean optimal clutch was paralleled by an increase (or decrease) in mean clutch size ($p = 0.07$). A corner test (Fisher exact probability) based on a scatter diagram, plotting mean optimal clutch against mean clutch size, yields $p = 0.036$, indicating a significant correlation between the two.

Table 6.1. Optimal and observed brood sizes in Swiss starlings (from Lack, 1948b)

| Brood size | Number of broods observed | Number of young per nest recovered three months after fledging |
|:---:|:---:|:---:|
| 1 | 65 | 0 |
| 2 | 164 | 0.04 |
| 3 | 426 | 0.06 |
| 4 | 989 | 0.08 |
| 5 | 1235 | 0.10+ |
| 6 | 526 | 0.10 |
| 7 | 93 ⎫ | 0.10+ |
| 8 | 15 ⎭ | |
| 9 | 2 | — |
| 10 | 1 | — |

Closely related to the strategy of varying clutch size with food level is another strategy followed commonly by raptors and some other birds. Whereas most birds do not incubate until all eggs are laid—or, at least, do only very little incubating—

raptors generally begin incubating as soon as the first egg is laid. Thus while the young of most birds develop synchronously, raptor young in the same nest may be at different stages of development. The result is that the older and, therefore, larger young, who hold their gaping beaks the highest, get fed first. Only when the largest bird is satiated and stops begging does the next largest bird get fed— and so on down the line. In this way the number of birds fed is related to the amount of food available, very little food is wasted on young that won't survive and the maximum possible number of healthy young are raised.

**Table 6.2.** Optimal and observed brood sizes in great tits

| Year | Mean clutch size | Mean optimal clutch |
|------|------------------|---------------------|
| 1947 | 9.67 | 10.99 |
| 1948 | 11.40 | 11.36 |
| 1949 | 8.96 | 7.62 |
| 1950 | 8.42 | 9.19 |
| 1951 | 6.57 | 5.40 |
| 1952 | 8.89 | 9.09 |
| 1953 | 9.30 | 9.95 |
| 1954 | 8.61 | 8.36 |
| 1955 | 8.79 | 8.95 |
| 1956 | 9.05 | 8.47 |
| 1957 | 8.22 | 8.36 |
| 1958 | 9.15 | 9.61 |
| 1959 | 8.02 | 8.66 |
| 1960 | 7.29 | 10.14 |
| 1961 (one area) | 7.07 | 9.26 |
| 1961 (second area) | 7.84 | 8.36 |
| 1962 | 9.35 | 8.77 |
| 1963 | 9.24 | 11.34 |

The relative merits of the two strategies for maximizing fitness are not entirely clear. One may speculate that where food supply is unstable the raptor strategy is best. Birds with long incubation and nestling periods have a greater chance of encountering a food crash and thus experience effectively less stable food conditions than species with short incubation periods. Species which aggravate the results of poor food or weather conditions by being food specialists or possessing very high metabolic rates which must be maintained are in the same category. This, perhaps, explains why the raptor strategy is followed not only by predators and other large species (large birds generally have long incubation periods) such as pelicans, herons, storks, cranes, boobies, and penguins, but also by such specialists and/or tiny (high metabolism rate) species as swifts and hummingbirds. This hypothesis is very incomplete and valid only in a very loose way, but does indicate that passerines should follow the first strategy of varying egg number, which they do.

It should also be pointed out that the two strategies are not exclusive. It is possible that a female varies the number of eggs she lays according to food supply and also incubates with the appearance of the first egg. The swifts, for example, as mentioned above, may reduce their normal clutch of three to two in bad weather or voluntarily reduce the clutch by removing an egg. They also incubate with the first egg (Koskimies, 1950). Also, we find that many passerines, following scheme one with their first clutch, follow the second strategy in subsequent clutches (Lack, 1966). Finally, neither scheme explains the limitation of egg number by precocial birds. In this case, clutch size may be determined by the energy available to laying females, the number of eggs an incubating individual can cover, or the degree to which large numbers of young might be more noticeable to predators, but the situation is not well understood.

## 6.3.2 Predation

If egg size does not vary, it seems clear that egg number must vary strictly as a function of the energy available for egg production and parental care. The situation, though, is not so rigid as it may sound. If, for example, an animal displaying parental care is able to locate its nest in such a place that it is safe from predation, it can then afford to prolong the nestling period without fear of nest destruction. Longer nestling periods, if the young undergo equal amounts of growth, mean a slower growth rate and thus lower daily food requirements for each individual offspring. Because each young needs less per day, it is possible to raise larger numbers of young. Thus, where parental care is involved, unlike the situation where parental care is absent, high predation may lead to smaller clutches. This hypothesis has been examined by Lack (1948a) and data have also been gathered by Nice (1957). Nice showed that among birds nesting in holes, the percent of young hatching (78.3%) and fledging (66.8%) were higher than the similar values for open-nesting birds (64.7% and 46.9%). Thus hole nesters suffer lower mortality. Lack gives the results shown in Table 6.3 for European passerines, and Nice (1957) presents the results shown in Table 6.4 for North American passerines. Ricklefs (1968) shows that birds have slower growth rates in the tropics in spite

**Table 6.3.** The relation between nest location, clutch size, and nestling period in European birds (after Lack, 1948a)

| Predation | Nest location | No. species examined | Average clutch size | Nestling period |
|---|---|---|---|---|
| Lightest | Hole | 18 | 6.9 | 17.3 |
| ↓ | Roofed nest | 7 | 6.8 | 15.9 |
| | In a niche | 5 | 5.5 | 14.9 |
| Heaviest | Open | 54 | 5.1 | 13.2 |

of heavier predation there (Skutch, 1949, 1954, 1960). This would suggest larger clutches—which do not generally occur. It may be that slow development and consequent lowered food needs is overbalanced by the shorter hours for foraging—that is, clutch sizes are smaller in the tropics because of the food supply argument presented in 6.3.1. This, however, leaves us with the nagging problem of slow development rates in the face of heavy predation. Perhaps the best explanation is offered by Skutch (see references above) who suggests that, because of the heavier predation, parents make fewer trips to the nest and thereby avoid attracting predators to their young. Also the young of tropical birds are silent compared with those of temperate species, lending support to Skutch's hypothesis. This explanation would account both for the smaller clutch sizes and the slower rates of development in the tropics. Clearly, a number of factors are interrelated and we can only presume that what we observe represents a balance of opposing selection pressures.

Table 6.4. The relation between nest location, clutch size, and nestling period in North American birds (after Nice, 1957)

| Nest location | Average clutch size | Nestling period |
|---|---|---|
| Hole | 5.4 | 18.8 |
| Open | 4.0 | 11.0 |

## 6.4 AGE OF REPRODUCTION

A population which increases its number geometrically by an amount $\overline{W}$ every generation, increases by a factor of $\overline{W}^{1/T}$ every time unit, where $T$ is generation time. Thus when $\overline{W} > 1$, the rate of increase is enhanced if $T$ declines. That is, in growing populations, natural selection favors short generation spans (early reproduction). Lamont Cole (1954) reached this same conclusion by means of slightly different, population ecological, considerations; since populations grow in a manner analogous to principal accruing compound interest, and since early age of reproduction is analogous to frequent collecting of interest, young born early contribute more to population growth than young born late. It can be argued that in stable populations ($\overline{W} = 1$), neither early nor late reproduction is favored, while in declining populations ($\overline{W} < 1$), selection favors delayed maturity. The facts are not quite so simple, though. In any population, unless really drastic decline is taking place (in which case the population must rapidly stabilize or go extinct), there will be some genotypes whose numbers are increasing. These genotypes experience selection for early reproduction and, by virtue of their growth in the face of stability or decline of other genotypes, come to dominate the population. Thus, in almost all cases, selection should favor early maturity (see Chapter

9 for more details).   Of course, there may be opposing selection pressures or mechanical difficulties which make early reproduction impossible and we must view observed age-specific fecundity values as selective equilibria.   The ecologically interesting questions pertain to the factors opposing selection for early reproduction.

One problem which must be faced by all individuals of species breeding more than once (*iteroparous* species) is that too much energy spent on the current brood or on associated courting and territorial behavior means increased chances of dying before the next production of young.   Reproduction may be an expensive process.   For example, it is estimated that a redwing male loses about 8.7% of its weight during the breeding season, from March to July.   Females lose roughly 7.6% of their weight between May and July when most feeding of young takes place (Brenner, 1967).   Weight loss over the incubation period runs between 10% and 20% in tree sparrows, *Spizella arborea* (Heydweiler, 1935), 9% in the song sparrow, *Melospiza melodia* (Nice, 1937), 7% to 14% in the bullfinch, *Pyrrhula pyrrhula* (Newton, 1966), and 6% to 8% in ring doves, *Streptopelia risoria* (Brisbin, 1969).   Considering this fact, it might be expected that birds who raise the optimal number of offspring in one year are less likely to breed successfully the next, while fitness is maximized by pairs which raise, in any one year, slightly fewer than the "optimal" number.   It is true that where Lack's hypothesis on clutch size is inaccurate, it usually errs in the direction of "too few" eggs being laid.

Loschiavo (1968) has gathered direct data on the relation between the number of eggs laid by groups of the beetle *Trogoderma parabile* and length of life (from emergence), and finds an almost linear, inverse relation (Table 6.5).   Females laying no eggs, found in several of the groups, had an average life span of 20.8 days.   Tinkle (1969) plotted mean annual adult survivorship against total fecundity per season for 13 lizard species and found the same pattern.   These latter data are less conclusive in support of the notion that reproduction is a risky procedure

Table 6.5. The effect of reproductive effort on subsequent life span in the beetle *Trogoderma parabile* (from Loschiavo, 1968)

| Number of eggs laid | Length of life in days |
| --- | --- |
| 21.5 | 16.5 |
| 64.8 | 11.9 |
| 67.8 | 11.9 |
| 67.9 | 11.0 |
| 72.9 | 10.7 |
| 78.2 | 13.0 |
| 82.7 | 9.9 |
| 89.0 | 10.3 |

because they compare differences between rather than within species. They are, however, highly suggestive.

In many species, growth is indeterminate and large individuals produce larger numbers of (and often physically larger) eggs. In the arrow worm, *Sagitta elegans*, the maximum number of eggs a female can produce is proportional to the 2.46th power of her length (McLaren, 1966). In fish, Nikolskii (1963) gives the expression for egg number, $300L - 10,000$, where $L$ is length, in *Salmo salar*. The expressions for *Percina caprodes* and *Etheostoma spectabile* are, respectively, $20L - 1300, 6.25L - 219$ (Hubbs, 1958). In reptiles and amphibians, larger females usually lay larger eggs and, often, larger numbers of eggs (Tinkle *et al.*, 1970). In salamanders, clutch size multiplied by ovum size ("clutch volume") bears a linear relation to body volume (Salthe, 1969). In all such species, regardless of selection for early reproduction, maximum fecundity will occur well after first reproductive capability. Reproductive activities at one age may diminish chances of surviving to reproduce at later ages in these species, and so selection pressure opposes early reproduction. Actual age-distribution of reproductive effort and minimum reproductive age will be determined by the relative strengths of the opposing selection pressures. In general, one would expect reproductive success to be affected more by egg number than sperm number, so that selection for late maturity should be relatively stronger in females than males. In species with indeterminate growth, males usually, but not always, reproduce earlier than females.

In many species, both those with indeterminate growth and those reaching a maximum size somewhere near the age of first reproduction, there is an important courtship phase. As was pointed out earlier (Chapter 3), it is usually the male who most exerts himself in this period, first (perhaps) defending a territory, and then wooing a female. In these activities an experienced male is, as might be expected, more successful. Thus young males have less chance of successfully mating and selection for early reproduction is opposed. With lessened selection for efficiency in reproductive behavior early in life, the male portion of populations will respond to any existing selection pressures opposing mature sexual characteristics, and delayed maturity should result. In birds and mammals particularly, where experience through learning is likely to be very important, we usually find females reaching reproductive age sooner than males. In the vole *Microtus ochrogaster*, females are capable of reproduction at an age of 25 days, males not until an age of 45 days. In the walrus, the figures are four to five years and five to six years (Bourliere, 1956). The same is true in the higher primates. In passerine birds, males may be physiologically capable as early as females, but are often behaviorally and/or morphologically retarded.

Experience is also important during the parental care period where, at least in some birds, both sexes are involved. Thus Coulson and White (1958) have shown that kittiwake gulls nesting for the first time lay, on the average, 1.83 eggs per nest, of which 0.66 fledge. Birds nesting for the second time lay 2.06 eggs and fledge 1.21 young. For birds nesting their third and subsequent times, the corresponding numbers are 2.35 and 1.63.    Delayed maturity of both sexes is most pronounced, as expected, in such species as colonial sea birds.    These must utilize their experience to the fullest, both to win an adequate territory in the size-limited and thus crowded nesting colonies, and to successfully build nests that will hold eggs on the often very narrow cliff ledges.

Species whose chances of survival from one breeding period to the next are high, and who are capable of storing food energy during this time, will gain little by putting their last reserves of energy into one reproductive period and dying. Species of this sort will show greatest fitness by sacrificing reproductive success very slightly in any one year, thereby surviving to reproduce the next.    It is the small species, suffering high mortality and often specialized to life during the breeding season, whose interests are best served by one mass reproductive effort (*semelparity*).    Semelparity is the rule among insects, for example, and where *iteroparity* (breeding throughout life) is found in this group, it is usually associated with long life expectancy.    The queens of social insect colonies, in their stable, safe environment, can expect to live long periods of time, and are reproductively active over much of their lives.    Iteroparity is the rule among the larger animal species, which suffer less uncertainty in their futures.

While uncertainty of survival in adults may lead to semelparity, however, uncertainty of survival in offspring may lead to iteroparity.    If the chance of a whole clutch being destroyed is sufficiently high, the advantage of spreading reproductive efforts over several clutches will, in time, overshadow any advantages of semelparity.    The relative uncertainty of the environment to parents as opposed to offspring would seem the vital factor in the evolution of iteroparity or semelparity.    Unfortunately, data bearing on this point are lacking.

## 6.5 MATING SYSTEMS

The variety of existing mating systems is tremendous, but to simplify discussion we define only three broad categories.    The sexual relationship is said to be *promiscuous* when no pair bond—beyond copulation—is formed.    *Polygamy* is the situation when one individual forms bonds with more than one individual of the other sex.    It is called *polygyny* when one male mates with several females, *polyandry* when one female mates with several males.    Finally a pair of one male and one female constitutes *monogamy* which may be lifelong or lasting over only one breeding season.    Permanent monogamous pair bonding is found in such species as the American cardinal, *Richmondena cardinalis*, magpies, *Pica pica*, ravens, *Corvus corax*, common terns, *Sterna hirundo*, occasionally gulls, *Larus sp.* (Welty, 1962), and, perhaps, some mice, *Peromyscus polionotus* (Blair, 1951).    Most birds are temporarily monogamous and most mammals promiscuous.

In species in which the sexes seldom meet, or the requirements of proper timing make the time and energy expenditures of pair formation disadvantageous, promiscuity without courtship results.    Where very rapid reproduction is not important—nonopportunistic species— then choice of a mate may become important to successful gene passage. The males of promiscuous species of this sort depend for their genetic contribution on attracting as many females as possible and thus experience very strong sexual selection. Such brilliance as is seen in the males of mannikans and birds of paradise is due to sexual selection operating on promiscuous males.

Where pair bonds are formed, it would be interesting to know why some are monogamous, others polygynous (polyandry is rare and will not be treated here). Verner (1964) has made a good start at answering this question; Table 6.6 is paraphrased from his work. Verner and Willson (1966) have suggested that one factor which might swing the net advantage from one system to the other is food supply. Where food is abundant, the male may increase his genetic contribution to future generations by ignoring his young, thereby condemning them to a nestling life of less food and more uncertainty, and spending the time otherwise spent in parental care attracting and inseminating more females. In situations of food plenty, females can more easily afford to forgo the aid of the male. Thus polygyny (and promiscuity) should be more common when food is plentiful than when food is scarce. To support this hypothesis, Verner and Willson note first that the productivity per unit volume (productivity density), what they feel to be an appropriate measure of food abundance, is highest in marshes, less high in fields and savannas, and least in forests (although total productivity may be greatest in forests). They then point out that of 291 passerine bird species in North America only 14 are polygamous or promiscuous; 13 of these are marsh breeders, the 14th a savanna breeder. The hypothesis accurately predicts that, for one species of wren, *Troglodytes troglodytes*, birds nesting on St. Kilda island where little food is available are monogamous, while in English gardens, where there is much food, they tend to be polygamous (Armstrong, 1955). Crook (1964, 1965) has also collected many supporting data. Of 23 field species of weaver birds, *Ploceidae*, most are polygamous or promiscuous, while of 22 forest species, most are monogamous. Data gathered by Lack (1968) show some correlation, world-wide,

Table  6.6

|  | Advantages of polygamy | Disadvantages of polygamy |
|---|---|---|
| To a male | More offspring for reproducing males<br>Greater chance of mating with a "superior" female | More energy needed for courtship<br>More time needed in territorial defense<br>Risk of stolen copulations |
| To a female | Greater chance of mating with a "superior" male<br>Greater chance of nesting on a "good" territory | Less time to care for young<br>Less protection and help in parental care from the male |

between habitat (and thus, presumably, food abundance) and mating system in passerines, although the relation is not significant (Table 6.7). One possible explanation for the poor fit with expectation in Table 6.7 is the following: Some open country bird species nest in dense colonies so that, in spite of high productivity density, they have to travel great distances for food. Because of the consequent need for long absences from the nest and the cost of food acquisition, paternal care is needed (Davis, 1952) and monogamy is a necessity. Both the quelea, *Q. quelea*, and the social weaver fit in this category.

**Table 6.7**

|                | Number of passerine families whose members are | | |
|                | Monogamous | Polygynous | Promiscuous |
|----------------|------------|------------|-------------|
| Forest nesters | 25         | 0          | 2           |
| Savanna nesters| 20         | 2          | 0           |

In spite of the supporting data, however, there are theoretical reasons for doubting the Verner–Willson hypothesis. First, food abundance is a relative, not an absolute factor. To a bird which raises few enough young, a forest may supply more than enough food, while to another bird which raises many, food may be inadequate, even with paternal help, to maintain the nest. The hypothesis may be valid if:

**1.** The environment is more temporally heterogeneous in high productivity density areas—rendering populations low relative to their food supply due to heavy off-season mortality, or

**2.** The factor limiting the number of nests in high productivity density areas is not food, but space so that the breeding—as opposed to the total—population is low relative to the available food supply. The latter situation is likely when breeding areas are highly restricted and not all feeding occurs on the nesting territory. In either case, the critical factor is not food abundance, but heterogeneity. It is convenient to Verner and Willson's hypothesis that forests tend to be homogeneous, savannas and marshes heterogeneous in food distribution and that marshes generally constitute restricted nesting grounds scattered like islands over the surrounding countryside.

Orians (1970), in a well-thought-out review of ideas on mating systems, reiterates the correlation between need for paternal care and mating system but is far more cautious in linking paternal care with food shortage. He points out that, all else being equal, a female will always benefit if she does not need to share her male. Polygyny comes about when all else is not equal—when different males possess strikingly different characteristics or territories. Where enough heterogeneity exists so that one male can provide for two females and their offspring more efficiently than another can provide for only one female and her offspring, polygyny is favored.  In birds, there is a high correlation between polygyny (or promiscuity) and lack of paternal care (Verner and Willson, 1969), and between paternal care and environmental heterogeneity.  There seems to be some correlation between heterogeneity and food abundance.  The correlation between lack of paternal care and food abundance is indirect.

The argument that when paternal care is not necessary, or not very useful, polygyny or promiscuity are likely to evolve, applies to many groups of animals, including mammals.  Here, of course, direct feeding of the young is accomplished, at least in the early stages of the offspring's life, by the mother alone.  Females which don't have to chase and capture prey, but can eat continuously, have no need for male help.  Herbivores fit into this category and are virtually all promiscuous.  Omnivores, such as oppossums and bears, also tend to be promiscuous. Among hunting mammals, on the other hand—with the exception of those which wait and pounce rather than wander about—it may be very difficult for the female with young to feed herself.  Terrestrial mammalian carnivores are generally, though not always, monogamous (Eisenberg, 1966).

The arguments above, on monogamy versus polygyny or promiscuity, apply nicely to most terrestrial vertebrates but are generally far off the mark when applied to most marine or aquatic fish and invertebrates.  There are two basic reasons for this.  First, parental care consists largely of guarding, not feeding, so that there is no great need for the efforts of two parents.  Second, many members of these groups display indeterminate growth.  This being so, a female can contribute far more to the ancestry of future generations if she is able to leave her offspring in the care of the male, build her energy reserves, and thereby grow in size.  The male, not benefiting greatly in terms of sperm production by growing, benefits by filling the parental role of the absent female.

In fish, where parental care exists, the pair bond is very short (we speak of monogamy and polygamy, but this is stretching a point), and usually the female leaves the male behind to guard the young.  Of course, parental care—unless the young are carried about by a parent—is possible only where nest sites are available.  As one might expect, most fish which build nests and display parental care are shallow-water or bottom species.  Deep-water, pelagic species are generally in no position to care for their young.  Pair bonding is of no apparent advantage, and most such fish are indiscriminately promiscuous, with free-swimming or planktonic young.

Theoretical ecology has many predictive hypotheses that are, in general, correct, but suffer many exceptions.  It is, therefore, highly satisfying to be able to end a discussion with a story of unusually great appeal to evolutionists.  Monogamous species may or may not change mates between successive breeding seasons. Since poor reproductive success may, in part, be due to poor choice in mate, we should like, as evolutionists, to believe that a change in mate is more likely to occur following poor rather than good reproductive seasons.  Coulson (1966) in his work on kittiwakes confirms our expectation.  Among 325 females whose eggs successfully produced young, only 28% were found with different mates the next year. Among 57 whose eggs failed to hatch, 36 of them (63%) changed partners. The difference is significant ($\chi^2 = 26.8$, $p < 0.005$).

## 6.6 SEX RATIOS

It is not by any means always clear that sex ratio is a meaningful concept. In many small aquatic forms (water fleas and rotifers, for example) parthenogenesis is the rule, and *gonochorism* (separate sexes) occurs only during certain seasons or times of stress. Some species, while occasionally reproducing sexually, depend largely on vegetative procreation.

Many species, both plant and animal, are hermaphroditic. Hermaphrodites may be simultaneously male and female or may be first male and then female (*protandrous*) or vice versa (*protogynous*). Larvae of the slipper limpet, *Crepidula fornicata*, metamorphose first into males and, as they grow, change into females. This species forages during part of the day and at other times is often seen in stacks, several animals deep, the larger females on the bottom, males on the top, with intermediate forms in between. The common Atlantic sea bass, *Centropristes striatus*, shows the opposite trend, starting its adult life as a female and then, sometime after its second year (usually between age five and seven) becoming male (Lavenda, 1949). The serranids, the family to which *Centropristes* belongs, also includes protandrous and simultaneous hermaphroditic species and gonochorists (Chan, 1970).

*Anthias squamipinnis*, a reef fish studied by Fisbelson (1970) in the gulf of Aqaba and Suez, also changes sex, but with an interesting variation. Aquarium studies show that if a number of females but no males are kept together, one of the females will reverse its sex. If this new male is removed, another female will switch and take the male role, and so on. In its natural habitat this fish is found in schools of hundreds or thousands of individuals consisting mostly of adult and subadult females, with only a few territorial males.

The relative advantages of gonochorism and simultaneous or consecutive hermaphroditism are very poorly understood and will not be discussed here.

In some species, populations are characteristically very sparse and individuals rarely contact one another. Deep sea fish often live in such sparse populations and some have gotten around the consequent difficulty of finding mates by making the male an obligate parasite on the female (for references, see Nikolskii, 1963).

In the above cases it is difficult to assess the ecological significance of sex ratio. In the remainder of this section we examine sex ratio in gonochorists.

For many years it was believed that mating systems reflected sex ratio. Clearly if females outnumbered males, the species would benefit from a polygynous system. While the argument sounds valid, it does not fit the facts. In addition, it only puts off the question of skewed sex ratios. The mating system argument presented earlier in this chapter is far more satisfactory and, as shown below, leads us to an understanding of sex ratios as well.

First, we reiterate the comments in Chapter 3 that in polygynous (and in the case of birds and mammals who choose mates, promiscuous) species, sexual selection often produces males that are bigger and often more mortality prone than females (see also Willson and Pianka, 1963). Crook (1964) reports that 21 of 23 field-nesting weaver birds (polygamous or promiscuous) are sexually dimorphic, while 20 of 22 forest nesters (monogamous) are monomorphic.

Next, we paraphrase an argument by Fisher (1958a). Each offspring receives an equal number of genes from its mother and its father. Thus, in any given breeding season, the total number of genes passed by all reproducing males and females must be equal. If males outnumber females, then each male, on average, must pass slightly fewer genes than the average female. Thus males are "worth" less than females and natural selection will favor the production of females until the sex ratio reaches 50:50. There is a little more to the argument, though, as Fisher himself pointed out. Suppose males outnumber females two to one (so that, on the average, a male is worth only one-half as much as a female), but that three males can be produced for the cost of one female. The value of a clutch of all males is thus proportional to $\frac{1}{2} \times 3 = 1.5$, while the value of a clutch of all females is proportional to $1 \times 1 = 1.0$, and males are favored even though they contribute less to future generations.

The problem can be formulated more rigorously as below: By Fisher's argument, the value of one male reproducing at age $x$, as a contributor to the ancestry of future generations, is $1/M(x)$ where $M(x)$ is the number of reproducing males in the population at the time the male in question is age $x$. If the probability that the average male survives to age $x$ is $l_x$ and the total population of reproducing males does not change with time, the total contribution, over his lifetime, of a male (where $\alpha$ is minimum reproducing age), is

$$\frac{l_\alpha}{M(\alpha)} + \frac{l_{\alpha+1}}{M(\alpha+1)} + \cdots = \frac{\sum_{x=\alpha}^{\infty} l_x}{M}.$$

Where $M_x$ is the number of males of age $x$ in the population the contribution per male can be written

$$\frac{\sum_{x=\alpha}^{\infty} l_x}{M} = \frac{\sum_{x=\alpha}^{\infty} l_x}{M_0 \sum_{x=\alpha}^{\infty} l_x} = \frac{1}{M_0}.$$

If a breeding pair produces $m_0$ males (subscript denotes age 0), its total contribution, by way of males, is

$$C_m = m_0/M_0 = m_x/M_x, \text{ for any } x.$$

For females, the corresponding expression is

$$C_f = f_0/F_0 = f_x/F_x.$$

Total contribution is given by

$$C = C_m + C_f = \frac{m_x}{M_x} + \frac{f_x}{F_x}.$$

We now find isoclines of fitness (contribution) by setting

$$dC = \frac{\partial C}{\partial m_x} dm_x + \frac{\partial C}{\partial f_x} dF_x = 0.$$

When this expression holds, $dm_x/df_x$ is given by

$$-\frac{\partial C/\partial f_x}{\partial C/\partial m_x} = -M_x/F_x.$$

But since selection acts similarly on all breeding pairs, $m_x/f_x \to M_x/F_x$, so that

$$dm_x/df_x \to -m_x/f_x, \quad \text{or} \quad m_x f_x = k. \tag{6.1}$$

The best strategy for a breeding pair to follow is to maximize the value of $m_x f_x$ (see fitness sets and adaptive functions, 4.2.2).

MacArthur (1965) came to the same conclusion and considered the appropriate value of $x$ to be "reproducing age", a consideration which has led to much confusion (J. M. Emlen, 1968c, 1968d; Leigh, 1970). To avoid this confusion we digress a moment and define a simple fitness set. Suppose that the energy spent to maintain one nesting male through a given time of life does not vary with the ratio of male to female offspring. If $E_{m_x}$, $E_{f_x}$ are the amounts of energy required of a parent to sustain one male and one female through the age interval $x$ to $x + 1$, $E$ is the total energy available for parental care, and parental care ends when offspring reach age $p$, then

$$\sum_{x=0}^{p} m_x F_{m_x} + \sum_{x=0}^{n} f_x E_{f_x} = E. \tag{6.2}$$

Where $l_{m_x}$, $l_{f_x}$ are probabilities of males, females surviving to age $x$ (survivorship),

$$\sum_{x=0}^{p} (m_0 l_{m_x} E_{m_x} + f_0 l_{f x} E_{fx}) = m_0 E_m{}^* + f_0 E_f{}^* = E, \text{ and}$$

$$m_0 = \frac{E - f_0 E_f{}^*}{E m^*}. \tag{6.3}$$

This can be rewritten,

$$m_x = \frac{l_{m_x}}{E_m{}^*}\left( E - \frac{f_x E_f{}^*}{l_{f_x}} \right).$$

Substituting into Eq. (6.1) this becomes

$$m_x f_x = \frac{l_{m_x}}{E_m{}^*}\left( E f_x - \frac{E_f{}^* f_x{}^2}{l_{f_x}} \right),$$

which is maximized (differentiating with respect to $f_x$ and setting the derivative equal to zero) when

$$f_x = \frac{E l_{f_x}}{2 E_f{}^*}, \quad \text{so that} \quad f_0 = \frac{E}{2 E_f{}^*}.$$

Similarly

$$m_0 = \frac{E}{2E_m^*} \cdot$$

The energy spent on successfully raising (to age $p$) $m_p$ males and $f_p$ females, then, is

$$m_0 E_m^* = \tfrac{1}{2}E,$$

$$f_0 E_f^* = \tfrac{1}{2}E.$$

A breeding pair will contribute maximally to the ancestry of future generations by equalizing the total amount of energy spent on raising males and on raising females, including those individuals which die before reaching independence. Of course, the boundary of the fitness set need not be linear, as in Eq. (6.2), in which case the above conclusion will be inaccurate. There seems no reason to believe, however, that in most cases the conclusion will not be approximately true.

The values $m_0 E_m^*$, $f_0 E_f^*$, represent the total energy spent on males and females, and not all young on whom energy is expended survive to fledge. Thus if males, due to sexual selection later in life or other reasons, suffer higher nestling mortality than females, we might expect them to outnumber females at hatching but be outnumbered at fledging. During the nestling stage, if their sizes are roughly equal—take equal amounts of energy to mature—males and females should be roughly equal in number. Most nestling birds are very difficult to sex (partly because they are about equal in size), but Selander (1960, 1961, 1965) believes the nestling sex ratio in grackles, *Quiscalus major* and *Q. mexicana*, to be $1:1$. In man, where males are known to suffer greater mortality in childhood, the secondary sex ratio (at birth) invariably favors males ($50.7\%$ in American Negroes, $53.5\%$ in Koreans) (Hunt *et al.*, 1965,) while females outnumber males in maturity. Males of the tropical bird group known as oropendolas are roughly twice the size of females at the time of fledging. Therefore we may presume that the cost of raising males is considerably more than the cost of females in this species. By the energy expenditure argument, fledgling sex ratio should show a significant bias in favor of females. The observed ratio of females to males is, in fact, between $5:1$ and $10:1$ (N. G. Smith, 1968). Where mating systems are polygynous or promiscuous (and involving choice of a mate) sexual selection more strongly affects male survival so that greater inequalities between hatching and fledging ratios should be expected. The adult sex ratio should invariably favor females. Thus the mating system undoubtedly affects sex ratio, but sex ratio need not affect mating system (Willson and Pianka, 1963).

To those who worry about genetic mechanisms for skewed sex ratio, we emphasize the existence of nondisjunction as a genetically controlled process, differential mobility and viability of $x$ and $y$ sperm (undoubtedly under some amount of $x$-linked autosomal control (Morgan, *et al.*, 1925; Gershenson, 1928; Novitsky, 1947)), and meiotic drive (Sandler and Novitsky, 1957).

## 6.7 BROOD PARASITISM

When individuals of one species use the nest or the help of another to raise their young, they are said to be brood parasites.

Brood parasitism is fairly common in insects and birds. In the former group, it may take either of two forms. One form exemplified by the small workerless ant, *Labauchena daguerrei*, of Argentina, involves the entry of four to six queens into the nest of the fire ant, *Solenopsis saevissima*. Here, the workerless ant queens mount the host queen and chew off her head. In this way, they remove her as a competitor and pick up enough of the local scent to be accepted into the colony. The fire ant workers subsequently care for the workerless ant queens and their larvae (Emerson, 1958). The slave-making ant, *Formica sanguinea*, epitomizes the other sort of brood parasitism found in insects. Individuals of this species make periodic forays to other ant colonies where they steal pupae, bringing them back to their own colony. When these pupae mature they care not only for new batches of stolen pupae but for the young of their abductors as well. Slave-making ants may occasionally move into the raided nest and set up a new colony. Talbot and Kennedy (1940) describe in detail a raid which they observed; their report reads like a war news documentary. As the red slave-makers searched for and opened entrances into the raided nest, the occupants were busy plugging entrances. Adults were left alone by the raiders except when they were dragged, bodily, out of the way of the slave-makers. Occasionally, two slave-makers would pull an adult apart in their uncoordinated efforts to remove him. Long streams of raiders coming to the nest and leaving with stolen pupae were present throughout the day. A few of the raiders remained overnight to insure possession, and the plundering continued the next day. Ants from the attacked colony formed passive groups and appeared to be waiting for the raiders to finally leave. The raids are frequent: In one year, a slave-maker colony carried out 16 raids in 42 forays in one area and 10 raids in 36 forays in another.

Brood parasitism in birds takes the form of one species laying its eggs in the nest of another. It occurs in at least five families: *Cuculidae* (cuckoos), *Indicatoridae* (honey guides), *Ploceidae* (weavers), *Icteridae* (blackbirds), and *Anatidae* (ducks) (Weller, 1959).

Brood parasitism may have profound effects on host species, as indicated by the figures on percentage of nests parasitized by cowbirds, *Molothrus ater*, (Table 6.8) taken from data of Wiens (1963) and Ely (1957, cited in Wiens, 1963), in southern Oklahoma.

Avian brood parasites display a number of interesting adaptations to their mode of life. Most are drab colored, presumably to avoid provoking aggression in their hosts. Payne (1967) reports polymorphism in the European cuckoo, *Cuculus canorus*, and suggests that this results from specific adaptations to local host species. Nonparasitic relatives are monomorphic. The European cuckoo is divided into a number of ecotypes, called "congens", which lay different-appearing eggs in different areas. The eggs mimic those of the most often parasitized local species.

It has been mentioned that some birds lay eggs until some definite number is present. If too many eggs are in the nest, one will be removed. The European cuckoo insures that the ejected egg is not its own by removing one host egg itself. As a young cuckoo grows up, he develops a tendency to react to contact with other young by lunging, thus tipping his nestmates out of the nest. Honey guides accomplish the same end in a different manner. Strong mandibular hooks are used to jab, and often kill, other young.

**Table 6.8.** Rate of parasitism by cowbirds

| | Total no. of nests observed | | No. of nests parasitized | | Percent parasitized (combined data) | first set of data from Weins, second from Ely |
|---|---|---|---|---|---|---|
| Bell's vireo | 17 | 14 | 12 | 10 | 71 | |
| Dickcissel | 15 | 14 | 5 | 1 | 21 | |
| Cardinal | 14 | 4 | 1 | 4 | 28 | |
| Lark sparrow | 12 | 4 | 1 | 3 | 25 | |
| Orchard oriole | 17 | 3 | 1 | 3 | 20 | |
| Blue grosbeak | 13 | 5 | 6 | 2 | 44 | |
| Field sparrow | 1 | 4 | 1 | 3 | 80 | |
| Redwing blackbird | 33 | 73 | 0 | 2 | 2 | |
| Painted bunting | 2 | 5 | 1 | 4 | 72 | |

An interesting and puzzling question is that of brood size in bird brood parasites. Since the parents do not feed the young, Lack's hypothesis is not applicable and the parasites might be expected to lay as many eggs as their physiological limitations permit. Cowbirds lay several "clutches", with rest periods between. But as the breeding season progresses, food becomes more common so that more energy is available for egg production (as opposed to energy for future nestlings, which declines); yet the number of eggs laid per brood declines as in other bird species. Unlike other birds, parasite clutches tend to be larger in the tropics (Payne, 1965a). The factor controlling brood size in these birds remains a mystery.

The matter of the distribution of eggs by avian brood parasites has interested a number of people. Preston (1948) looked at the distribution of cowbird eggs—number per nest—and showed that if the first egg were laid in a carefully chosen nest and the others scattered about randomly, then the observed distribution fitted a theoretical (Poisson) distribution rather nicely (Table 6.9). Mayfield (1965), however, has challenged Preston's conclusion. He pointed out that cowbirds are likely to seek out nests with cowbird eggs already in them since the presence of a cowbird egg indicates a reluctance of the hosts to desert. Because of desertion and avoidance many nests otherwise expected to contain one egg will either never exist or never be observed. Figuring a 10% deficit of one-egg nests, Mayfield makes the calculations shown in Table 6.10, which shows the best fit yet. The real motivation of the bird is still in doubt.

**Table 6.9.** Distribution of cowbird eggs

| | No. of eggs | | | | | |
|---|---|---|---|---|---|---|
| 0 | 1 | 2 | 3 | 4 | 5 | |
| 482 | 135 | 42 | 5 | 0 | 0 | Observed no. of nests |
| 466.8 | 164 | 29 | 3.4 | 0.3 | 0 | Poisson (random placement of eggs) |
| — | 136.8 | 39.1 | 5.6 | 0.5 | 0 | Poisson with careful placement of first egg: discount the zero egg class, subtract one egg and recalculate |

**Table 6.10.** Distribution of cowbird eggs

| | No. of eggs | | | | | |
|---|---|---|---|---|---|---|
| 0 | 1 | 2 | 3 | 4 | 5 | |
| 482 | 135 | 42 | 5 | 0 | 0 | Observed no. of nests |
| 482 | 201 | 42 | 5 | 0 | 0 | Observed, with correction (10 percent of total = 66) |
| 484 | 198.9 | 40.9 | 5.6 | 0.6 | 0 | Poisson (random placement of eggs) |

Finally, the evolution of brood parasitism presents a formidable theoretical problem. It has been attacked with some success by Hamilton and Orians (1965) who suggest that it began with the chance dumping of eggs in others' nests, possibly in response to nest destruction just prior to laying. Such dumping is quite common. Subsequent evolution would depend on whether this chance dumping resulted in high reproductive success or not. Based on this approach and other, rather obvious, considerations, Hamilton and Orians set down the following conditions under which brood parasitism is most likely to evolve.

1. Egg dumping is most likely when there is *no territorial resistance*
2. Young from dumped eggs are more likely to be adequately fed if the protoparasite has a *short incubation period*
3. Food is most likely to be of the kind required if *the protoparasite is omnivorous*
4. Normal sexual behavior is likely to occur in the *protoparasite* only if the species *tends not to imprint*

5. Parent-offspring communication is most likely to be adequate if the *host is altricial* (young remain in the nest)

6. *Brood parasites* are least likely to decimate hosts and thus themselves if they *are rare compared to their hosts.* In addition, hosts will evolve defense mechanisms less rapidly if this is the case.

The authors point out that the highest rates of nest predation (read destruction) occur in the tropics so that there is greater opportunity there for the first step to brood parasitism. This should be particularly true in savanna and arid tropics where clear-cut breeding seasons make proper timing easier for the parasite (although this argument could be used to predict higher numbers of parasites in temperate areas). Brood parasitism is very common in the tropics, especially the arid tropics.

# 7
# Feeding Ecology

Survival and reproduction both require energy; without energy intake fitness would fall to zero. For most species food is not so plentiful that individuals are likely to be consistently satiated. Thus selection favors maximum efficiency in feeding. Those species for which hunger is seldom a problem nevertheless are under the same selective pressure, for other activities (courting, copulating, hiding) demand time, and often fitness increases with increased time spent at these activities.

Selection for efficiency in feeding acts along three avenues:

1. increased ability to recognize food needs and to feed or cease feeding in response to some optimal (with respect to fitness) schedule
2. increased ability to choose the best (again with respect to their effects on fitness) foods
3. increased ability to locate, acquire, and ingest foods.

Studies on feeding ecology thus raise and attempt to answer three basic questions: how much is eaten and why, what is eaten and why, and how food is obtained and why.

## 7.1 QUANTITY EATEN

### 7.1.1 Relation to Availability

C. S. Holling (1965) has described three basic feeding curves. Where an animal consumes food at a rate proportional to its rate of encounter with food items, a plot of food eaten per unit time against food available is a straight line (type I curve, Fig. 7.1a). At some point a saturation level is reached and the curve flattens. This kind of curve is characteristic of animals who wait for their food to come to them, such as filter-feeders or animals soaking up dissolved nutrients. Both mechanisms are very common, the first being found regularly in cnidarians, many annelids, bivalves, tunicates, crustaceans, and other groups, and the latter occurring in virtually every phylum (McWhinnie and Johanneck, 1966; Stephens, 1968).

Type II curves (Fig. 7.1b) are characteristic of most organisms which take a certain amount of time to ingest and perhaps capture their food. At least two explanations exist for the response characterized by the type II curve. If we let $N_A$ be the amount of food consumed, $N_o$ the amount offered, $T_t$ the total time the animal has to feed, and $T_H$ the time required to handle one food item, then

$$N_A \propto (T_t - N_A T_H)N_o, \qquad \text{so that} \qquad N_A = \frac{aT_t N_o}{1 + aT_H N_o}, \qquad (7.1)$$

where $a$ is the proportionality constant. A plot of $N_A$ against $N_o$ shows a rising curve which gradually levels off (Fig. 7.1b) (Holling, 1965).

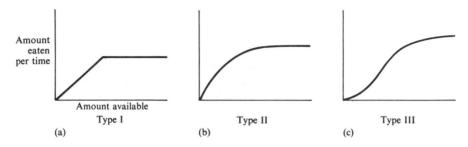

Fig. 7.1. Types of feeding response (see text for discussion).

An alternative explanation for the type II curve is as follows: If $i$ and $j$ denote two food species, and $P_{is}$ and $P_{js}$ are the proportions of encountered items of foods $i$ and $j$ eaten, then

$$\frac{N_{Ai}}{N_{Ai} + N_{Aj}} = \frac{N_{oi}P_{is}}{N_{oi}P_{is} + N_{oj}P_{js}} = \frac{N_{oi}}{N_{oi} + (P_{js}/P_{is})N_{oj}}.$$

Since the total amount of food eaten, $N_{Ai} + N_{Aj}$, should be roughly proportional to the total time feeding takes place, $T_t$, we can write

$$N_{Ai} = \frac{N_{oi}(N_{Ai} + N_{Aj})}{N_{oi} + (P_{js}/P_{is})N_{oj}} = \frac{N_{oi} \,\alpha\, T_t}{N_{oi} + (P_{js}/P_{is})N_{oj}}.$$

Define

$$a = \frac{P_{is}\,\alpha}{N_{oj}P_{js}}, \qquad \text{so that} \qquad N_{oj}\frac{P_{js}}{P_{is}} = \frac{\alpha}{a}.$$

Then

$$N_{Ai} = \frac{N_{oi}\alpha T_t}{N_{oi} + (\alpha/a)} = \frac{aT_t N_{oi}}{1 + (a/\alpha)N_{oi}}.$$

This is equivalent to an expression given by L. Tinbergen (1960), and when $a/\alpha = P_{is}/N_{oj}P_{js}$ is constant, is identical with Holling's (above) equation. The term $a/\alpha$ is, in fact, constant when food $j$ is relatively very common so that $P_{is}/P_{js}$ does not change significantly with a change in the amount of food $i$ offered (see 7.2.1).

Ivlev (1961) argues that if $P$ is the amount of food available, $R$ is the amount of food eaten in the absence of competitors, and $r$ is the actual food ration, then $r$ varies with $P$ when $r \ll R$, but remains virtually unchanged if $r \to R$. To a first approximation, then,

$$dr/dP = \xi(R - r),$$

where $\xi$ is some constant. The solution is given by

$$r = R(1 - e^{-\xi P}), \quad \text{or} \quad N_A = R(1 - e^{-\xi N_o}).$$

The form of the curve generated is that of the type II curve and describes fish data gathered by Ivlev quite well (Fig. 7.2).

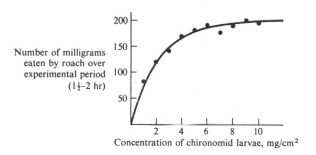

Number of milligrams eaten by roach over experimental period ($1\frac{1}{2}$–2 hr)

Concentration of chironomid larvae, mg/cm$^2$

**Fig. 7.2.** The feeding response of the roach (fish).

Type III curves, characteristic of mammals and sometimes other vertebrates, are merely type II curves with a dip (Fig. 7.1c). Holling (1965) generated a type III curve by offering mice varying numbers of buried sawfly cocoons. A standard biscuit food supply was always available in abundance. This experimental design then provided for abundant alternative food and thus a constant value of $a/\alpha$ (see above). A type II curve should then have been expected—and, in fact, was found—but with a dip in it. Why this dip occurs is not clear.

L. Tinbergen (1960) studied the feeding response of tits to changes in insect food abundance and discovered that as abundance of a given insect rose, the birds showed a delayed response—that is, the feeding rate on that insect rose with abundance in the manner of the type III curve. Tinbergen felt that animals (in this case, tits) did not learn to appreciate foods that were becoming progressively more abundant until their numbers reached some threshold value. This would explain the dip under conditions of alternative food and changing food abundance. Holling's data may be explained in the same way.

When individuals form mental images (*search images*) of a commonly encountered food type and subsequently feed disproportionately on that food (*apostatic selection*—see .5.2.5) the dip is explained. This argument applies to Tinbergen's results but not Holling's where one food was always superabundant. Holling reasons, since mammals (and other vertebrates) learn and forget, that they tend both to learn slowly and then forget the value of a food unless they encounter it fairly often.  If a food is scarce, therefore, and predators tend not to eat strange (unlearned or forgotten) food, the food will be eaten disproportionately seldom and a dip will appear in the type II curve.

Holling's learning and forgetting argument may be weak, however, for the following reason.  Invertebrates do learn.  Furthermore, they may forget more rapidly than vertebrates.  Landenberger (1966) managed to condition starfish (*Pisaster*) to move down aquarium walls in response to a conditioned stimulus of light when rewarded with mussels on the aquarium floor.  His subjects learned rapidly (less than eight trials) and showed rapid extinction.  He also showed that *Pisaster* in nature moved about actively but remained near food-rich areas around piers.  In a later paper (Landenberger, 1968) he reported that the starfish learned to concentrate their efforts in that part of an aquarium with the greatest relative abundance of their preferred food (mussels as opposed to gastropods).  Fisher-Piette (1935) showed that predatory intertidal snails, *Thais lapillus*, in France, had to learn to eat mussels after decimating their original diet of barnacles.  The switch in diet considerably lagged the crash in barnacle and rise in mussel numbers, and for several months the snails made all sorts of feeding errors—drilling through dead mussels, or drilling outward from the inside of empty shells.  By Holling's argument then, invertebrates should show an even more pronounced dip in the feeding curve than vertebrates—which they don't.

The cause of the dip in Holling's type III feeding curve thus remains inadequately explained.  For the moment, let us ignore the dip, and concentrate on the general shape of the curve (type II).  If we denote by $V$ the "value" (undefined for now) of a food item, its cost of acquisition by $C$, and let the metabolic rate during non-feeding activities be $M$, the proportion of the day feeding be $f$, and the length of a day $T_t$, then

$$\frac{\text{Net gain}}{\text{time}} = \frac{(V - C)N_A}{T_t} - M(1 - f)$$

and, substituting from Eq. (7.1),

$$\frac{\text{Net gain}}{\text{time}} = \frac{(V - C)aT_tN_of}{(1 + aT_HN_o)T_t} - M(1-f) = \frac{(V - C)afN_o}{1 + bN_o} - M(1-f). \quad (7.2)$$

For an animal to survive, the value of this expression must be positive.  For the animal to show a fitness of 1.0, it must exceed some minimal maintenance value.  Proper choice of food can maximize $(V - C)$ (see 7.2.1), and $a$ can be maximized and $b$ minimized by efficient foraging behavior.  It is the value $f$ that concerns us here.

It is, of course, theoretically possible to increase $f$ to 1.0.  This, however, leaves the animal no time for reproducing, sleeping, temperature regulating (in ectotherms), etc.  The value of $f$ must be a compromise between the various needs of the animal.  Homeotherms may raise $f$ close to 1.0 for brief numbers of days, to avoid starvation, but cannot hold it there long because of other demands on time. When $(V - C)$ falls to zero due to food shortage, inefficiency of foraging in cold temperatures, poor foraging conditions, or other reasons, they must either migrate, become temporarily poikilothermic (hibernate), or starve.

Poikilotherms, not having to maintain a constant, high body temperature, have more options open to them. Social concerns are generally of less importance to them than to, say, birds and mammals, except at breeding time (although there are many exceptions). Thus $f$ can be raised and held at high levels for considerable periods of time. Also, when $(V - C)$ drops below zero, it is possible for poikilotherms to gradually starve, and then to recover when $(V - C)$ again rises. Fish, for example, are capable of withstanding considerable periods of starvation, during which they may lose a large proportion of their body weight. Landenberger (1968) allowed starfish, *Pisaster*, to starve for up to five months with no apparent ill effects. This ability to withstand starvation allows such species to vary $f$ appropriately (to zero) when $(V - C)$ becomes negative. It is a better strategy to avoid exertion and starve slowly than to attempt to find food and, operating at a deficit, starve rapidly. Paine (1965) showed that the ectibranch mollusk *Navanax inermis* operated at an energy deficit in the winter months near San Diego, California, and cut way back in its feeding activities. The same situation seems to apply in the starfish *Pisaster ochraceus*, which nearly ceases feeding in the winter, off the coast of Washington (Mauzey, 1966; Paine, 1969). Also off the Washington coast, the intertidal snail *Thais emarginata* feeds on the barnacle *Balanus glandula*, whose larvae settle in great numbers in the spring and late summer, growing into good food items in early summer and autumn. By winter and early spring most have been eaten, or destroyed by other factors, and little remains for the snails to eat. The occasional, good food items are widely scattered and require considerable travel by the predatory snails. $C$ thus rises and eventually exceeds $V$. *Thais emarginata* feeds very little in the winter and early spring (Emlen, 1966b). The same may be true for the related and often sympatric species, *T. lamellosa*.

## 7.1.2 Hunger and Satiation

What controls the proportion of an animal's time spent feeding? An animal stops eating if interrupted, but if then left in peace may return to eating. An animal ceases to feed when the motivation to feed is overshadowed by the motivation to do something else. As more and more food is ingested the need for getting more food declines and other needs become relatively more important. Feedback mechanisms have evolved in which some measure of energy, or recently consumed nutrient, gradually decreases receptivity to new food stimuli so that a near-optimal balance is reached between feeding and other activities. When an animal ceases to respond to food, it is said to be satiated.

One problem surrounding the concept of satiation is that in the laboratory, with fewer non-nutritive stimuli and less cost of food acquisition, animals probably eat far more before stopping than in nature. To the extent that "hunger" means a feeding response to food stimuli under *some* set of conditions (say, laboratory conditions), a wild animal may always be hungry.  To avoid semantic confusion we define satiation to mean the state of nonacceptance of food under any situations whatever.  An animal may ignore food stimuli in nature when still hungry by laboratory standards.

We assume that a decision to cease feeding is related to the amount of nutrient taken in and available to the animal.  Of course it is difficult to know how closely an animal can recognize the amount of energy or various nutrients it has consumed, but there is some evidence to show that this recognition ability may be quite well developed in some species.  Richter (1953) claims that laboratory rats prefer foods which keep their caloric intake per day at a constant level.  Satiation appears to occur sooner on energy-rich than energy-poor diets.  Most humans would say the same about their own diets.  Barnett (1953) added supporting evidence when he noted that rats ate less when sugar was added to their water supply.  Holling (1965) found that his mice (mentioned above) ate more biscuits as the supply of sawfly cocoons was lowered, in such a way that

(number of cocoons eaten) + $B$(grams of biscuit eaten) = Constant.

Since one cocoon is roughly equivalent to $B$ grams of biscuit in energy content, "the overall consumption maintains a relative constant energy input per day" (Holling, 1965).  Smith, Pool, and Weinberg (1962) tested the hypothesis that food intake was limited by the calories ingested by feeding laboratory chow, either straight or thinned with 40% cellulose, to laboratory rats.  Those fed on straight laboratory chow ate 22.7 g per day (79.4 calories); those eating thinned laboratory chow ate 34.0 g per day (71.4 calories).  When food was available for a short time only, rather than being constantly available, however, the rats made no adjustments for energy differences, but ate all they could hold.  This suggests some mechanism acting to keep daily intake between 70 and 80 calories.

If recognition is poor, it might be necessary to fall back upon some cruder measurement of intake such as bulk, or energy expended, as a cue to stop feeding.  It is often assumed that animals feed to satiation and that satiation is based on food bulk in the stomach or food bulk consumed over some past period.  Pearse (1924), for example, cites the fish *Micropterus*, *Eupomotis*, *Amblophites*, and *Ameirus*, in Lake Mendota, Wisconsin, as eating a diet regularly 6% of their own weight per day at 18.7° to 19.6°C.  Alpers (1932, in Thorson, 1958) states that the marine snail *Conus mediterraneus* eats 50% of its own volume per meal; Thorson (1958) reports that the Japanese oyster drill, *Tritonalia japonica*, in Puget Sound, eats about 15% of its own weight in oysters per day, and the crab *Carcinus maenas*, more than 20% of its living weight per day.  Matthews (1955) estimates that migratory locusts eat their own weight in food every day.  Of course, since an animal's diet is fairly consistent from day to day, a standard weight or volume may not imply that bulk is the limiting factor on food intake— calories and nutrient content are fairly proportional to bulk.  Also, since daily activities do not vary greatly if many individuals are observed and their activities averaged, it is hardly surprising that their diets are consistent in amount.  Constant bulk intake does not necessarily imply satiation.

### 7.1.3 Animal Size

Many animals spend only a small proportion of their day feeding and give every impression that their food needs are easily met. Others seem to operate on very tight budgets indeed.

Gibb (1960) noted a dramatic increase in yearly survivorship of the coal tit (*Parus ater*) from 30% to 75% with an increase in the dry weight per square meter of invertebrate food stock. This could only happen under tight food budget conditions. Gibb also reported, in the same article, that tits (*Parus*) and goldcrests (*Regulus regulus*) must eat about 5 mg dry weight, or 24 "average-sized" insects every minute, to meet their basic energy needs. To do this, the birds must search 1100 trees every day. The anna hummingbird (*Calypte anna*) is reported to have to visit the equivalent of 1022 fuchsia blossoms every day to extract the nectar it needs for maintenance. This comes to one every 6.8 seconds (Pearson, 1954).

Regardless of the effort an animal appears to put into its daily eating activities, it is probably safe to say that satiation, in the laboratory sense, is rarely reached. Large predators who have glutted themselves on a large prey item are the primary exceptions.

Homeotherms might be expected to eat enough to satisfy their basic needs for maintenance—temperature regulation, normal growth (including reproduction), heat of activity, respiration—switching to other activities thereafter. This expectation is quite well borne out. Nice (1938) has compiled a list of bird species, living under similar temperature conditions, along with their average weights and daily food intake (by weight). Since heat is lost through the skin, we might assume (and physiological work bears this out) that basal metabolism is proportional to an animal's surface area (roughly proportional to the $\frac{2}{3}$ power of the weight). The birds listed are all of roughly the same shape and possess roughly equal insulation by feathers, metabolic needs are roughly proportional to basal metabolism, and equal food weights contain roughly equal numbers of calories, so there should be a linear relationship between $W^{2/3}$ and $f$ ($W$ = weight of bird and $f$ = weight of food eaten/day); that is, we expect $\frac{2}{3}\log W = \log f + k$, where $k$ is some constant. Nice's data and the resulting graph of $\log W$ against $\log f$ are shown in Table 7.1, and Fig. 7.3; the slope can be seen to approximate $\frac{2}{3}$.

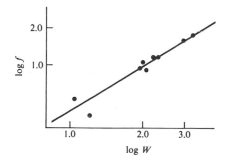

**Fig. 7.3.** The relation between body weight and metabolism in birds.

The "existence energy" (as opposed to basal metabolism) was measured by Kendeigh (1970) for a number of bird species. The animals were kept in cages so that the values are only rough approximations to those obtaining under natural conditions. Kendeigh generated the following equations

$$\log M = 0.1965 + 0.6210 \log W \pm 0.0633 \text{ (passerines)}$$

$$\log M = -0.2673 + 0.7545 \log W \pm 0.0630 \text{ (nonpasserines)},$$

where $M$ is kilocalories/bird/day. Since the amount of food ingested will be roughly proportional to the number of calories ingested, $\log f$ and $\log W$ again bear an approximately linear relationship to each other.

**Table 7.1.** The relation between body weight and metabolism in birds (from Nice, 1938)

| Bird species | Bird weight (grams) | Dry weight food eaten/day (grams) |
|---|---|---|
| *Parus caeruleus* | 11 | 3.3 |
| *Erithacus rubecula* | 16 | 2.4 |
| *Turdus ericetorum* | 89 | 8.7 |
| *Zenaidura macroura* | 100 | 11.2 |
| *Turdus merula* | 118 | 8.6 |
| *Colinus virginianus* | 170 | 15.0 |
| *Falco tinnunculus* | 200 | 15.4 |
| *Buteo buteo* | 855–900 (880) | 39.5 |
| *Gallus gallus* | 1800 | 61.2 |

There seems no such simple argument relating body size to amount ingested for poikilotherms.

## 7.1.4 Hoarding

When the amount of food falls, or its cost of acquisition rises sufficiently, animals must operate at energy deficits. As stated before, poikilotherms may shut down their feeding operations and homeotherms may migrate or hibernate. Both groups have an additional option: They may collect and store food during better times, thus buffering deleterious changes in the environment. The storing, or hoarding, of food is widespread. The popular, familiar example is that of squirrels burying their acorns. It also occurs in pikas, *Ochotona princeps*, which store "hay"—a mass of grass and weeds, often with marmot or coyote scats (see Broadbooks, 1965). The hay, which is stored in piles among rocks, is so important that its gathering and guarding figure prominently in pika social life all the year. Beavers hoard food and other objects, as do rats and packrats. Jays and acorn woodpeckers, *Melanerpes formicivorus*, commonly store food items and will attack other birds which attempt to steal from the food cache (Bent, 1939; Goodwin, 1956;

MacRoberts, 1970). Titmice, *Parus*, store both seeds and dead insects in bark crevices for winter consumption (Haftorn, 1933, cited in Welty, 1962). Many lizard species, such as the gecko, *Coleonyx variegatus*, of arid North and Central America, possess fat deposits in the tail. The animals are capable of eating fantastic amounts of food and storing excess energy in these fatty deposits. Four days of voracious eating may result in enough stored fat to sustain life for six to nine months (Bustard, 1967). In some of the formicine ants, food is stored in tremendously expanded crops in the abdomen of certain workers, known as *repletes*. These individuals are commonly so large they are unable to move, but in hot, dry weather, when food is scarce, can regurgitate enough food juices to feed the colony. The proportion of a colony in replete condition changes with the season. In the winter, after food juices are collected, it may be as high as 80%. It dwindles, thereafter, as the stored juices are used up (Talbot, 1943).

## 7.2 FOOD PREFERENCES

### 7.2.1 General Considerations

Many species appear to be true food generalists, merely eating different foods in proportion to their availability. Unfortunately, a vast bulk of the literature contains information only on stomach contents and not food availability, so that no conclusions as to food preference are possible.

Most complete studies of animal diet, though, indicate preferences, often marked, for certain food species within the class of foods normally eaten. Given a choice, the starfish *Pisaster gigantea* prefers (eats proportionately more in relation to number available), in order, the mussels *Mytilus edulis*, *M. californianus*, *Septifer bifurcata*; then the gastropods *Thais emarginata*, *Acanthina spirata*. *Tegula funebralis*; then the chiton, *Nutalina californica*. Another starfish, *Pisaster ochraceus*, prefers, in order, *Mytilus edulis*, *M. californianus*, *Septifer bifurcata*, *Thais emarginata*, *Acanthina spirata*; and then, equally, *Tegula funebralis* and *Nutalina californica* (Landenberger, 1968). Why should such preferences exist? And why should the order of preference be almost identical?

One factor determining food preference is differences in palatability. This answers only half the question, though: why do foods differ in palatability? It is possible, of course, that different palatabilities exist fortuitously, so that food preferences are totally unpredictable. But since populations are under strong selective pressure to choose and eat those foods in the proportion that will yield maximum "value" per unit time, this seems unlikely. Palatability, after all, is an animal's interpretation of a food's taste, and the interpretation of taste is at least in part the result of natural selection. Those individuals that interpret food of high "value" as being highly palatable and those foods of low "value" or toxic nature to be noxious, are the ones that pass on the most genes. We conclude, therefore, that natural selection has labeled different-quality foods with different palatabilities; that palatability is the mechanism by which animals choose their foods so as to maximize the net "value" per time of what they eat.

We now present a simple model of food preference reflecting the above conclusion (see also Emlen, 1968a). The "value" (still purposely undefined) of individual food items within one food species will show some kind of distribution curve. If the distribution curve plots "value" against number, rather than frequency, of items, the area under the curve can be drawn so as to give the total "value" of all items encountered by a feeding animal over some defined period of time. Fig. 7.4 shows such distribution curves for two hypothetical food species. Figs. 7.4(a), (b), (c) represent situations in which the total abundance of food encountered is the same but in which the ratio of food $j$ to food $i$ varies from 3:1 to 1:1 to 1:3. Fig. 7.4(d) corresponds to 7.4(b) except that both foods are now half as abundant. If individuals are capable of assessing the "value" distributions of the foods in their environments, and are attuned to choosing the best items, they should choose only those items to the right of the vertical bars (shaded portions of the curves). The total shaded area under both curves gives the total amount of "value" ingested over the specific period of time. The proportion of the shaded part to the whole area under both curves gives the "optimal" proportion of encountered foods actually taken.

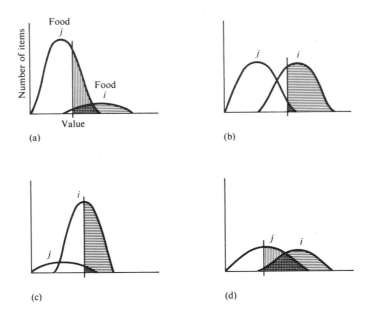

**Fig. 7.4.** The distribution of food value for two hypothetical food species.

This conceptual model leads to several interesting conclusions. First, comparison of 7.4(b) and 7.4(d), assuming the total amount consumed in the corresponding environments (shaded area) to be equal, shows that when food is less common (7.4d), less preference is found. That is $P_{is}$ and $P_{js}$ (the proportions of encountered foods *i* and *j* taken) converge. Animals should be more particular when food is abundant, less discriminating when food is scarce. Young (1945) has shown that rats become less particular in what they eat as food deprivation becomes increasingly severe, and Emlen (1966b) has done the same for snails of the genus *Thais*, feeding on barnacles in Washington state.

A good deal of information exists on dietary changes with seasonal or other changes in absolute and relative food abundance, but unfortunately the information necessary to an evolution of dietary specialization is usually lacking. An animal which eats equal amounts of two foods is *not* necessarily showing less preference than one which eats the two in the ratio 10:1; perhaps the relative availabilities are in the ratio 10:1.

Animals which are hungry and which may have to eat in spurts of uncertain duration and number (small prey species which are likely to be disturbed, for example) would be expected to feed as if food were scarce (little discrimination) while satiated individuals can afford to choose only the best items. Ivlev (1961) showed that as carp approached satiation, their food preferences became more pronounced (see Table 7.2. "Electivity", *E*, is a measure of food acceptance, and the degree to which preferences are shown can be described by the variance in *E*.) Preying mantids respond to drops in hunger level in the same way (Holling, 1966), and attack prey individuals only when they are very close by and easily caught.

**Table 7.2.** Electivity (E) of various food classes by carp (data from Ivlev, 1961)

| Food type | First food portion | Second | Third | Fourth | Fifth |
|---|---|---|---|---|---|
| Chironomids | 0.12 | 0.20 | 0.32 | 0.46 | 0.54 |
| Amphipods | 0.02 | 0.11 | 0.24 | 0.05 | −0.29 |
| Fresh water isopods | −0.03 | −0.03 | −0.33 | −0.67 | −0.74 |
| Mollusks | −0.15 | −0.52 | −0.95 | −1.00 | −1.00 |
| Variance (*E*) | 0.0126 | 0.1030 | 0.3473 | 0.4422 | 0.4588 |

Where $P_{is}$ is the proportion of encountered items of food *i* ingested, and $\mu_i$ the availability of food *i* in the environment, the proportion of food *i* in the diet $P_i$, is given by:

$$P_i = \frac{\mu_i P_{is}}{\sum_j \mu_j P_{js}}.$$

As the relative values of $\mu_i$ and $\mu_j$ change (Fig. 7.2a, b, c), then the changes in $P_{is}$, $P_{js}$ (hold total "value" constant) and $P_i$ can be calculated. Rough plots of $P_i$ against $\mu_i$ for the two frequency distribution arrangements shown in Fig. 7.5 and for data from Ivlev (1961) are given in Fig. 7.6. Note first that the rise in $P_i$

(and thus the amount of food $i$ eaten) with $\mu_i$ (the amount of food available) shows a dip characteristic of Holling's type III curve. This dip means that the ratio of food $i$ eaten to food available is not constant. This fact eliminates virtually all artificial scales of preference as measures valid for the whole range of $\mu_i$ (from 0.0 to 1.0), assuming animals approach the sort of optimal feeding described here. Second, note that when the food which, on average, is more "valuable" has a sufficiently small variance in value (Fig. 7.5b), preference may be shown for the generally less "valuable" food (Fig. 7.6b).

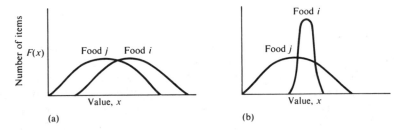

**Fig. 7.5.** The distribution of food value for two hypothetical food species.

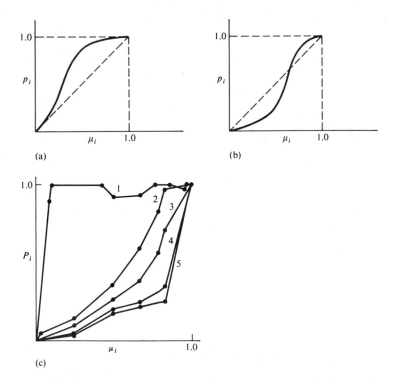

**Fig. 7.6.** The relation between the proportion of food in an animal's diet and its relative availability.

## 7.2.2 The Components of "Value"

*Chemical Nutrients.* It is a popular belief that, given a choice, animals and people eat according to their biochemical needs. Notwithstanding the candy-and-coke cult, this belief has some validity. Richter *et al.* (1938) tested the notion by offering rats olive oil, casein, sucrose, cod liver oil, wheat germ oil, yeast, salt, calcium lactate, sodium phosphate, and potassium chloride. The rats chose their diet efficiently enough to grow and reproduce as well as control rats on standard, enriched laboratory chow—and did it on fewer calories per day. Richter (see also Richter, 1942) concluded that rats possess "special appetites" for the different nutrients required for life. Darling (1937) reports that red deer, *Cervus elaphus*, keep up their calcium intake during the period of greatest need—when growing antlers—by chewing on shed antlers. When calcium is scarce, antlers may be nibbled while still on the heads of neighbors. Bone chewing is common in the eastern grey squirrel and fox squirrel (*Sciurus carolinensis, S. niger*), especially in pregnant and lactating females (Bakken, 1952). Blowfly (*Calliphora erythrocephala*) females can live on a pure sugar diet, but cannot reproduce unless also fed protein. They cannot survive, however, on protein alone. Given a choice of sucrose and yeast extract with protein, the food preference varies with the season. During the period of early egg growth the flies choose strongly in favor of protein. This preference wanes during the later stages of yolk formation (Strangeways-Dixon, 1961).

The ability to choose needed nutrients properly is not always shown, though. Rats fed sugar and casein prefer sugar, and if later fed a protein-deficient diet and then again offered the choice they may switch to casein. This switch in preference occurs, however, only if the new choice situation involves a change in the experimental milieu (Young, 1945). Rats form dietary habits which are changed only very slowly.

In nature, a certain characteristic appearance may be highly correlated with a certain nutrient content. In such cases an animal may "know" what nutrients a given food contains. In most cases, however, the nutrient value of a food is not so easily ascertained. Because of this fact, nutrient deficiencies may be most efficiently remedied by experimental feeding. That is, an animal might be expected, under conditions of nutrient deficiency, to show disproportionate interest in novel foods. The feeling of physiological well-being accompanying intake of the needed nutrient (those animals interpreting the new feeling as good and therefore as a positive reinforcement are selected for) tells the animal that the nutrient is present. Rodgers and Rozin (1966) found that rats on a thiamine-poor diet chose novel foods over their normal foods regardless of whether they were thiamine-rich or not. If not, the rats gradually reverted to their usual dietary habits. Rodgers (1967), pursuing this line of inquiry further, found a preference for thiamine-fortified food by thiamine-deficient rats, but only when the alternative food was otherwise identical to the fortified food and both were familiar to the rat. No preference was shown for the enriched food over other, new and different foods, even when the latter were thiamine-deficient. Given a choice of two novel foods, one with and one without thiamine, no preference was noted. Rats appear to respond immediately to sodium content and, perhaps, to inorganic salts in general, in their food (Barnett, 1963), but need to learn proper responses to foods containing biochemical nutrients.

*Energy Content.* The calories per ingested gram of food is obviously one component of a food's "value". All else being equal, we would expect animals to show preferences for energy-rich foods. Many animals show preferences for sweet or fat-rich foods, both of high caloric content. In western man, where there is no longer a struggle to obtain sufficient food energy, this preference for rich foods has become maladaptive. Prey species have occasionally capitalized on the almost universal preference for energy-rich food items by their predators, and increased their "bones to meat" ratio. Calcareous particles of every description are distributed through the bodies or skin of many mollusks, calcareous algae, and other organisms, making their caloric content per unit weight very low. Insect larvae may be generally preferred by birds to their adult forms largely because of the lower ratio of chiten to haemolymph. R. T. Paine (personal communication) notes that of 230 food items (molluskan prey with more than 4 kilocal/g) offered to tectibranch mollusks, *Navanax inermis*, all 230 were eaten. Of prey species with energy content between 3 and 4 kilocal/g, 141 of 142 offered items were eaten. But of those species with caloric contents below 3 kilocal/g, only 3 of 52 were taken.

Of course, what constitutes an energy-rich food to one species may be an energy-poor food to another. Grass has little energy content for carnivores who lack the intestinal flora to break down cellulose. A grazing animal will assimilate meat less efficiently than a cat. Few animals besides porcupines would find bark a nourishing food. Assimilation efficiencies—energy assimilated divided by energy ingested—have been calculated for a number of animals feeding on various foods, and some of the results are given below. The grazing beetle, *Chrysochus auratus*, eats apocynum (*Cannabinum*) leaves and digests about 43% of the ash-free matter by weight, retains 71% of the ash, and assimilates 56% of the available energy (Williams and Reichle, 1968).

Ricker (1946) figures the assimilation efficiency of fish on their natural diet to vary between 80% and 90% unless lots of hard parts of prey are ingested, and Emlen (in preparation) estimates laboratory mice to assimilate the energy in oats and millet seed with an efficiency approaching 90% to 95%. Welch (1968) gives assimilation efficiencies of a number of animals on their natural diets.

A final note of possibly great ecological importance: Kitchell and Windell (1969) have noticed that the growth rate—and thus, presumably, the assimilation efficiency—of blue gills, *Lepomis macrochirus*, is highest on a mixed diet. Offered unlimited amounts of algae, the fish lost 11.53% of their weight over a six-week period. Unlimited numbers of oligochaetes resulted in a 2.06% weight increase. When algae and oligochaetes were offered together, however, the weight gain was 3.86%. Tenebrio larvae allowed a weight increase of 0.95% while tenebrio larvae plus algae resulted in weight gains of 3.71%. Of course, nutrient differences in these two foods are great and one may possess trace nutrients, useful to the predator, not present in the other food. If this is so, increased growth might result from the nutrient balance offered by the mixed diet and may not necessarily imply increased assimilation efficiencies. Emlen (in preparation), however, has shown that assimilation efficiencies of laboratory mice feeding on a mixture of oats and millet are at least as high as those of the same mice feeding on oats or millet alone, and R. T. Paine (personal communication) believes the assimilation efficiencies of slugs to be higher on mixed than simple diets.

*Food Size.* Larger food items contain more energy than similar but smaller food items. Unless the cost of acquisition and the time to acquire and eat increases with item size at a proportionately even faster rate, larger items will provide a greater number of calories per unit time than smaller items. Thus, the value of food items of any food species increases with the size of the items until such a size is reached that the predator encounters some difficulty in catching or handling its prey. Food preferences, ideally, should reflect this change in food "value." Bimodal preferences with respect to prey size should seldom occur. Gibb (1956) examined the distribution, by size, of the intertidal snail *Littorina neritoides* both as it occurred on the beach and in the diets of three birds. Table 7.3 gives

**Table 7.3.** Prey size preference in birds (from Gibb, 1956)

Normalized ratio of percent taken to percent available—a measure of preference

| Size of snails (*Littorina neritoides*) | Percent available in environment | Percent taken by | | |
|---|---|---|---|---|
| | | Pipits | Purple sandpipers | Turnstones |
| < 1.5 mm | 3 | 0 ⎫ | 0 ⎫ | 0 ⎫ |
| 1.5 | 8 | 3 ⎬ 0.08 | 0 ⎬ 0.01 | 0 ⎬ 0.00 |
| 1.7 | 14 | 11 ⎭ | 3 ⎭ | 0 ⎭ |
| 2.0 | 25 | 25   0.14 | 6   0.01 | 3   0.00 |
| 2.3 | 28 | 31   0.16 | 20   0.04 | 8   0.01 |
| 2.6 | 17 | 23   0.28 | 35   0.12 | 30   0.07 |
| 3.0 | 3 | 7   0.34 | 24   0.47 | 27   0.33 |
| 3.4 | 1 | 0 ⎫ | 6 ⎫ | 11 ⎫ |
| 3.9 | 0 | 0 ⎬ 0.00 | 3 ⎬ 0.35 | 12 ⎬ 0.59 |
| 4.5 | 1 | 0 ⎪ | 2 ⎪ | 9 ⎪ |
| > 4.5 | 0 | 0 ⎭ | 1 ⎭ | 0 ⎭ |
| Total number observed | 2353 | 1087 | 189 | 89 |
| Mean size | 2.14 | 2.25 | 2.72 | 2.98 |

the results. All three bird species show a preference for one size range. Since larger predators might be expected to handle larger prey more efficiently, we might expect the size class of food preferred to increase with predator size. Purple sandpipers, *Erolia maritima*, are larger than pipits, *Anthus spinoletta*, and turnstones, *Arenaria interpres*, are larger than purple sandpipers. The preferred food size varies as expected (Table 7.3).

In seed-eating birds, the relation between the size of the bill and the range in food size has been studied. Newton (1967), for example, found the expected positive correlation (see Table 7.4). Myton and Ficken (1967) fed large and small sunflower seeds to chicadees (small bill) and titmice (larger bill). At temperatures above 32°F, chicadees ate 346 small to 98 large seeds when offered both sizes in equal numbers. The titmice ate 83 small and 301 large seeds. The results bear out our expectations. Below 32°F the figures were, respectively, 352 and 333, and 331 and 785. The same preferences exist but are considerably less pronounced. It seems reasonable to conclude that the birds associated low temperatures with low food levels (the usual relation in nature) and in the latter experiment ate as if food were scarce (i.e., showed less pronounced preferences).

**Table 7.4.** The relation between food size and bill size in finches (from Newton, 1967)

| Bird species | Mean bill depth (mm) | Percentage composition of diet, by weight, of individual seeds of size: | | | | |
|---|---|---|---|---|---|---|
| | | <0.5 | 0.5–1 | 1–10 | 10–100 | >100 (mg) |
| Lesser redpoll (*Carduelis flammea*) | 6.6 | 80 | 18 | 2 | — | — |
| Linnet (*C. cannabina*) | 7.6 | 15 | 31 | 52 | 2 | — |
| Greenfinch (*Chloris chloris*) | 11.5 | 8 | 9 | 27 | 54 | 2 |
| Hawfinch (*Coccothraustes coccothraustes*) | 17.7 | — | — | 1 | 30 | 68 |

*Novelty.* Among wild rats there is a tendency to avoid new situations. A change in familiar surroundings may lead to a temporary cessation of feeding (Barnett, 1963). In birds there seems to be a tendency to ignore new and scarce foods (L. Tinbergen, 1960; Allen and Clarke, 1968) perhaps as a result of the bird's forgetting the value of foods not often encountered (Holling, 1965). On the other hand, laboratory rats often show disproportionate interest in novel stimuli. This *novelty effect* has already been mentioned above with respect to nutrient needs. It is well known to psychologists and extends well beyond response to food stimuli. For example, Barnett (1963) reports that if rats are allowed to run through a T-maze, given no reward, they are likely on their second passage to turn opposite to the direction they turned on their first passage. Whelker and King (1962) offered novel foods and other items to laboratory rats and found strong preferences, which gradually waned, for the novel foods. The novelty effect was even evident in the rats' great interest in such non-food items as rubber erasers, which the rats busily gnawed at and occasionally ingested.

Scarce foods are often preferred simply by virtue of their being scarce. It is possible that this is because scarcity, since it leads to rare encounters for the feeding animal, renders a food effectively novel. It is also possible that animals tire of their normal foods and, when satiated on these, switch to other foods. Young (1945) noted that laboratory rats, when "satiated" with their preferred food, switched to a temporary preference for others. Emlen (unpublished), however, found that laboratory mice generally show most pronounced preferences for scarce foods when hungry and shy away from all but the preferred food as they approach satiation. Siegal and Pilgrim (1958), working with human subjects, asked 79 men to rate the palatability of a number of foods on a scale of one to nine. They found that the rating of those foods offered most frequently declined (indicating specific satiation), while also the rating of seldom-offered foods increased. On the basis of this evidence, preference for scarce foods seems to involve both the novelty effect and food-specific satiation and probably a host of other factors.

Reaction to novel and to scarce items has been studied extensively in laboratory animals and to a lesser extent in wild animals. It seems likely, in view of the differences in response by these two groups, that some animals have evolved tendencies for both neomania and neophobia. In laboratory situations, where dangers accruing from too much exploratory sampling of foods are slight, the first tendency is dominant. Wild animals, on the other hand, learn to be suspicious, but do not become entirely rigid in their rejection of new objects. Ecologically, of course, the question of interest involves the selective advantages of neomania: Why have such tendencies evolved and where and when are they most likely to show themselves in wild animals? The most reasonable argument for the evolution of neomania is that animals in patchy or temporally heterogeneous environments may, by showing interest in new or seldom-encountered foods, both keep the door open to discoveries of richer food resources and stave off starvation imposed by sudden drops in the abundance of their normal food supply. If this argument is correct, then, all else being equal, animals inhabiting unstable or patchy areas should more likely display the novelty effect and food-specific satiation than animals inhabiting stable, homogeneous areas. The relationship between animal size and environmental homogeneity is clear: Changes, both spatial and temporal, affect large animals less than small. Thus small animals such as mice and rats should more often show interest in new and scarce foods than large animals. Whether or not this is true is unknown.

*Availability.* One of the first things an ecologist interested in feeding must consider is that the availability of a food to a species may bear little relation to its abundance. A trivial illustration: The number of leaves above the upper browse line of deer may be very large, but their availability to the deer is zero. The codistribution of a potential food species with another species may affect its availability. What is not available to a coyote because it is hidden among thick brambles may be easy prey for a weasel. Landenberger (1968) notes that starfish, *Pisaster*, forage by sweeping their arms as they move. This means of searching for food makes it difficult to detect prey hiding in crevices. Thus when given a choice between turban snails, *Tegula*, and mussels, *Mytilus*, in aquarium tests, the proportion of mussels eaten declines when bricks are placed on the aquarium floor; the mussels tend to get wedged between the bricks while the snails perch on top of the bricks.

What is easy to catch becomes, to the predator, more available. The change in diet of the European white stork, *Ciconia ciconia*, from fish and frogs in wet periods to mice in dry periods may reflect only the changes in the food species' relative abundance; the fact that *Falco subbuteo* eats more swifts in cold-wet weather is not so simple. In the latter case, cold-wet weather reduces the number of insects, cutting the food supply and thus weakening the swifts which, therefore, become easier to catch (Welty, 1962). Ivlev (1961) has investigated the role of ease of capture on diet in fish. In one experiment, he clipped parts from the tail fin of roaches (prey species) so as to vary the speed at which they could flee predators, and then offered four speed classes, in equal numbers (all prey individuals of the same size), to pike and perch. The results (Table 7.5) show a clear tendency for the easier (slower) caught prey to be eaten more readily.

**Table 7.5.** The relation between speed of prey and numbers of prey eaten (the numbers give the relative numbers of prey taken)

| Predator | Speed of prey (cm/sec) | | | |
|----------|------|------|------|------|
|          | 40   | 70   | 90   | 105  |
| Pike     | 32   | 26   | 24   | 18   |
| Perch    | 38   | 28   | 18   | 16   |

When dealing with animals which spend large portions of their lives feeding on a single, individual plant or on a local clump of plants, another factor affecting preference is involved. The same factor may be important also in species which adjust their diets to changes in food species availability only with difficulty. Paine and Vadas (1969) have found that four sea urchin species, a type of abalone, and eight other invertebrate species all show food preferences for algae of neither very high nor very low caloric value. It is reasonable to expect those algal species with low caloric content, such as the corallines, to be eaten less than those with medium energy content, but why high-energy species should be avoided is not clear until we note that these latter types are mostly the annuals or ephemerals. They are not around for long enough periods of time to allow colonization on them or to allow feeding animals to readjust their dietary habits (and, perhaps physiology). The algae generally eaten are the more energetically rich perennials.

Finally, changes in relative food abundances, and hence diets, often result in physiological adaptations to new feeding regimes, thereby exaggerating the changes in relative availability. In the summer, herring gulls, *Larus argentatus*, move inland and eat a lot of grain. During the rest of the year, fish are much more important and grain much less important in their diets. By moving inland, the gulls are responding to increased grain supplies, and they are also putting themselves into a position where the relative abundance of grain is maximized. Furthermore, the stomach wall turns hard in response to the diet of grain, increasing digestive efficiency on grain and probably lowering digestive efficiency on fish (Thomson, 1923, cited in Welty, 1962).

## 7.3 SPECIALIZATION AND GENERALIZATION

When a species is broadly adapted so that it fits successfully into a wide variety of habitats or eats a wide variety of foods, etc., it is known as *eurytopic* and often referred to as a *generalist*. *Specialists* are *stenotopic*. A large number of animals will eat practically everything of conceivable food value and are thus eurytopic with respect to food, or *euryphagic*—food generalists. This is true of rats, bears, oppossums, some mammalian herbivores, some primates, and many others. Many smaller species are also generalists. The food of wood ants includes honeydew, other insects, millipedes, sap and resin, fungi, carrion, and seeds (Sudd, 1967). Not all animals behave in such a manner, however. Koalas, *Phascolarctos cinereus*, eat only the more tender leaves of eucalyptus trees, the three-toed sloth, *Tamandua tetradactyla*, eats only the leaves, buds, and shoots of the tropical mulberry, *Cecropia palmata*, or *Spondia lutea*. The giant panda, *Ailuropoda melanoleuca*, eats only bamboo shoots. The anteaters and aardvarks are also specialists, as are the southern elephant seal, *Mirounga leonina*, and the sperm whale, *Physeter catodon*, which exist almost exclusively on cephalopods. Among birds, hummingbirds eat nectar; sage grouse, *Centrocercus urophasianus*, eat only sagebrush foliage; and the everglades kite, *Rostrhamus sociabilis*, eats almost nothing but freshwater snails. The brant, *Branta bernicla*, underwent a population decline of about 80% when its one important food source, eel-grass, suffered a blight in 1931 to 1933 (Moffit and Cottam, 1941).

Lists of generalists, specialists, and their modes of specialization could fill a book. The interesting question is why some species should possess broad diets, and others very narrow diets. One important consideration is food availability. Consider two foods of the same "value", but suppose that one is much more available than the other. The available food will be eaten more frequently at first simply because of its availability. Because it is eaten more often, the predators feeding on it will undoubtedly become more practiced in its capture and ingestion and the predators' digestive efficiency may increase on it at the expense of efficiency on the alternative food. Thus, it becomes a more "valuable" food and preferences should appear. This argument applies both on a somatic and a genetic level. In the latter case, those individuals that possess foraging, ingesting, and digestive mechanisms which increase their efficiency on the most available food (all else being equal) will leave the most offspring. Thus efficiency and preference interact in a positive feedback manner, leading to specialization on the most available food. Of course, if the less-available food is more "valuable" initially, the direction of specialization may be reversed. The fact that specialization is not universal, as the above argument implies it should be, is due to three—and perhaps more— factors: overall food abundance, heterogeneity in the environment, and nutrient imbalances. These will be discussed one at a time.

Adaptive change can raise the relative "value" of one food only so far. It is seldom that the distribution curve (see Fig. 7.5) of one food will be so far to the right of those of all other foods so as not to overlap them. However, even if one food does not overlap, a sufficient shortage of it—low overall availability— will force predators to eat second-best food species. Where animal populations are prevented from further increase by available food, then, by definition, food is scarce and specialization is a luxury that cannot be afforded. The species, in such cases, is said to be food-limited (see Chapter 10). Where animal numbers are held down by predation (predator-limited) or physical environmental factors (see Chapter 10), food is, by definition, more plentiful, and food specialization is more likely.

There seems little doubt that most insect populations are held at low levels by inclement weather and predation (see Chapter 10) and only occasionally reach levels at which food is scarce. Terrestrial vertebrates, on the other hand, are a group more often affected by food shortages. Dietary specialization is far more common among insects than in most terrestrial vertebrates.

If we accept Hairston, Smith, and Slobodkin's (1960) argument that terrestrial carnivores are population limited by their food supply, and that terrestrial prey populations are limited by predation (Chapter 10), then we should expect ground predators to be food generalists and ground herbivores to be food specialists. Among large mammals, at least, this seems the general rule. Such a carnivore as the pine marten, *Martes martes*, eats almost anything it comes across that could be called food. In the summer, berries form the bulk of the diet (Lockie, 1961). Bears and foxes are also food generalists, as is the badger, *Meles meles*, which eats young, small mammals, insects, honey, mollusks, earthworms, grass, bark, tubers, rhizomes, fruits, and berries (Bourliere, 1956). Herbivores, on the other hand, usually stick to one class of foods, such as leaves, and only rarely branch out to take insects or other food in significant amounts. In the summer, the golden-mantled ground squirrel, *Citellus lateralis*, feeds almost exclusively on the dandelion, *Taraxacum officionale* (80% of its diet) and further specializes by showing a strong preference for the stems as opposed to the leaves or head (Carleton, 1965).

Spatial and temporal heterogeneity of the environment both affect dietary behavior. As noted in Chapter 4, large animals tend to view patchiness of the environment more often as fine-grained, while small species view the same patches as coarse-grained. In other terms, large animals wander over many different microhabitats, barely noticing the changes while small animals may live long periods of time, even their whole lives, in one microhabitat. A browsing deer and a leaf-mining insect look on individual leaves from an entirely different perspective. Refer now to the optimal choice model in Fig. 7.5. The distribution curves of food species occurring in different patches will undoubtedly differ from patch to patch. Thus, to an animal feeding in several patches, the effective shape of the distribution curves will be broadened and flattened relative to their shape in any one patch. That is, the variability in food value is effectively increased. A short glance at Fig. 7.5 will make it clear that any flattening of the curves will dictate less specialization. Large animals, then, should become more food generalized as their environment becomes patchier. Small animals, on the other hand, which are capable of spending great amounts of time in one patch (on one leaf, perhaps) cannot be anything but more specialized than larger species, and may be highly specialized (on one kind of leaf). Thus, food specialization should be more common among small rather than large animal species. This seems generally true.

A very important deterrent to the evolution of food specialization is temporal heterogeneity. The argument is simply an application of the principles spelled out in Chapter 4. Where foods become abundant or rare over time, specialists may survive by aestivating, hibernating, encysting, or producing resistant eggs. Invertebrates which can afford to fast over fairly long periods can also "escape" food shortages, as can species which are capable of migrating, but neither strategy is efficient in comparison with adaptation to a variety of foods. Of course, there seems no reason why a species should not become specialized to different diets at different seasons. Extreme, consistent dietary specialization, however, should occur only in such highly stable areas as the deep marine benthic or, to a lesser extent, tropical rain forests. It is often assumed (perhaps correctly) that diets are more narrow in the tropics than in temperate zones.

Perhaps the most important factor enhancing broad dietary habits is nutrient imbalance. Most animal species cannot produce all the amino acids and other macromolecules necessary to their survival and must acquire them directly in their food. Since not all foods possess the full spectrum of chemical needs of an animal, that animal may have to eat at least small amounts of several foods to survive. In addition, where increased assimilation efficiencies (for reasons of nutrient imbalance or other reasons) occur on mixed diets, it is advantageous to eat a broad diet. Avoidance of a specialized diet on these bases would appear as novelty effects (see section on food preference). Other explanations given previously for the novelty effect would be categorized under the effects of environmental heterogeneity.

In conclusion, there is a general selective advantage to food specialization which is opposed by a number of other selective pressures relating to food abundance, patchiness, change over time, and nutrient factors. Differences between environments and between animal sizes impinge directly on the latter selective pressures, and allow predictions of a very rough sort to be made *a propos* dietary habits.  Species whose populations are food limited—generally large predators—should seldom be food specialists; large species should be more generalized than small; large species should broaden their diet as their environment becomes more spatially heterogeneous; and animals inhabiting changing environments are less likely to be food specialists than animals inhabiting stable environments. Consistent narrow specialization such as that found in anteaters should be rare, because of nutrient imbalances, relative to the predator's needs, in most foods.

## 7.4 FORAGING

The techniques used by animals in foraging for their food are tremendously diverse. We shall touch upon only a few highlights here; the intake of dissolved organic substances has already been mentioned and endoparasitism is not discussed here.

### 7.4.1 Techniques of Food Gathering

Grazers, detritus, deposit, and filter feeders simply move about sampling the environment around them, or remain in one place waiting for (and occasionally producing water currents to encourage) prey to come to them.  Nevertheless, the nuances of these simple behaviors are diverse.  Molluskan grazers and detritus feeders, such as littorines, use a radula to scrape food from the rocks or rasp off parts of plants.  Predatory muricid snails use the radula to drill through barnacle or bivalve shells and often supplement this action with a chemical-producing accessory boring organ (Carriker, 1961).  A few (*Acanthina*, for example) possess spines used as a brace to aid in drilling (Paine, 1966).  Vertebrates and crustaceans simply use their mouthparts to cut, tear, or occasionally scrape, although interesting additional patterns are sometimes observed.

Gulls and crows, which pick up slow-moving or stationary food objects, and thus fit in this category, are often observed to open bivalves or snails by carrying them high in the air and dropping them on rocks.  Oyster catchers insert their laterally flattened bill between the valves of mussels and twist their heads to open their food.  Bivalves may either take in mud from which food is sorted (deposit feeders) or force water through their siphons, filtering particles out with their gills.

Some predators feed on relatively slow-moving prey and treat them essentially as herbivores would treat their plant food.  Anteaters fit in this category along with a number of insectivorous birds which prey upon marching driver ants, emerging naiads, or earthworms.  The pangolin, *Manis sp.*, allows ants to crawl among its scales and then licks them off, and the mongoose, *Mungos mungo*, has been reported in captivity to pick up hard-shelled millipedes and break them open by hurling them between the hind legs against trees or rocks (Eisner and Davis, 1967).

Scavenging is a common form of foraging among all predators and some, such as vultures, rely primarily on carrion.  White foxes, *Alopex lagopus*, may follow polar bears onto sea ice to feed on the remains of bear kills, or follow caribou herds to feed on whatever the wolves leave behind (Chesemore, 1968). Some predatory diptera wait along columns of army and driver ants, stealing their booty (Bequaert, 1922). Scavenging, or more accurately stealing, takes place also among aerial predators. Bald eagles commonly harrass osprey, forcing them to drop their kill and then swoop down and carry it off. Frigate birds, skuas, and jaegers do the same with a variety of victims (Friedmann, 1967).

Predators operating in the air or water must cope with an environment where the prey has an extra dimension in which to escape. They are thus more often equipped than terrestrial predators with structures designed to sweep up or hang on to their prey. The *Caprimulgidae* (night hawks and their relatives) are equipped with very broad mouths with a comblike fringe of feathers. Their mouths act like scoops, enabling them more easily to sweep up insects as they go. A bat uses its tail pouch and occasionally a wing tip to net insects which are relayed to the mouth in flight (Webster and Griffin, 1962), and some birds and fish-eating bats skim the surface of the water, the latter gaffing fish with sharp claws on specialized hind feet (Bloedel, 1955).

Dolphins, *Tursiops truncatus*, possess sharp, conical teeth pointing backward, presumably designed to hold slippery prey, and many sea birds possess similar structures. Pouncers may either trap their prey in the medium (air or water) or on the substrate. Examples of predators of the former type are flycatchers which fly from a perch, grab their prey, and immediately return to the perch, and morays, which dart from crannies in rocks, and likewise return to wait again.

The leaf fish, *Monocirrhus polyacanthus*, is greatly flattened laterally, and lies flat on the bottom or, head down, hangs from the water surface, mimicking a leaf. It remains motionless until a prey animal makes its appearance, and then pounces (Portmann, 1959). The asilid fly *Mallophora bomboides* mimics the bumblebee, *Bombus americanorum*, on which it preys. Even the low buzz produced by its wingbeat is a copy of the bee's drone. This robber fly waits on stems roughly three feet from the ground, takes off at the appearance of a stationary bee, and approaches its prey rapidly from above and behind. As the bee takes off, the fly grabs it and injects it with a paralyzing substance (Brower *et al.*, 1960).

Terrestrial pouncers generally feed on species smaller than themselves. They wait in ambush, occasionally stalk, and then attack rapidly. Cats use this strategy, and so do frogs and toads, which catch their prey with a flash of sticky tongue. Herons, which wait in shallow water and snatch prey with their long bills, are another example, as are preying mantids which stalk, fixate, and grab.

Terrestrial predators attacking prey species larger or faster than themselves have two basic strategies open to them. They may run in groups to exhaust their quarry, and then move in for the kill or, if they are strong enough and fast enough, they stalk and then attack very rapidly and alone. Among the large mammalian predators these two strategies are best exemplified by the pack-hunting dogs and large cats. Wolves work in packs, chasing an animal down and generally first attacking the rump, the safest end. The adaptations associated with this hunting technique are far-reaching. Wolves form close-knit social groups with well-defined dominance and leadership hierarchies and long-term family bonds. They possess long, lightweight legs and fairly slender bodies, an ideal shape, not for great speed and power, but for tremendous endurance. Coyotes and foxes are similar in shape, but lack the elaborate social structure. They hunt smaller prey, usually alone or in pairs. In contrast to the large dogs, the large cats—with the exception of the cheetah—have shorter, more massive legs and a highly flexible spine. The tremendous bulk of muscle makes for poor endurance, but with the extra joint supplied by the spine, these cats are extremely powerful and capable of almost unbelievable rates of acceleration. They depend on surprise, and hit their prey suddenly and hard, relying on their great strength. Unlike the weaker dogs, the large cats usually attack the neck or head of their prey, wrenching the animal down and holding it there for several minutes. This suggests suffocation of the prey, although in falling the prey's own weight may snap its neck (see, for example, Schaller, 1967).

A fascinating array of foraging techniques exists among predatory arthropods. Many ants attack creatures their own size or larger, and often find it necessary to cover considerable distances in groups. *Formica fusca*, in England, may have to travel up to 26 feet from its nest to satisfy its food needs and *Myrmica gulosa* may forage over an area of two million square meters (Haskins and Haskins, 1950). Wood ants are attracted by the struggles of one of their own and go to aid the comrade in pulling down its prey (Sudd, 1965). Many ants form raiding parties which range in size from only a few individuals to several hundred thousand. In the African species *Megaponera foetans*, up to a hundred ants move through the forest in single file to raid termite nests. Such small-scale raids are very common and earn their participants the name *company ants*. The *driver ants* of the old world and *army ants* of the new world are less common and forage in parties of up to several hundred thousand in columns anywhere from less than a millimeter (*Annoma sp.*) to over 25 meters (*Eciton sp.*) in width (Sudd, 1967). These columns move forward as a wave, those individuals at the head moving then falling back and being overtaken by the next rank. The columns advance at rates of about two meters an hour and may take hours to pass a given point. Generally the raids of driver and army ants seem to be aimed at no particular goal. Food matter is seized as encountered and the ants sweep the area clean as they move. The paths followed are usually quite regular, with definable widths and speed of the advancing front. Occasionally, soldier ants line the edges of the columns where they appear to be keeping the ranks in line. Such raiding parties, as one might expect, are taken advantage of by ant birds, anteaters, and parasites which lay eggs in captured larvae carried by returning raiders. As prey, these raiders are highly vulnerable, but as predators on other small arthropods they are almost invincible.

Spiders have evolved a number of foraging techniques. Unlike the ants discussed above, spiders, like the large cats, rely on speed and strength—as well as poison—to subdue their prey by themselves. The classic example is the familiar web which either entangles or adheres. Kaston (1965) gives these additional examples. Spiders of the family *Dinopidae* spin sheets which are flung over the prey, thereby trapping it. *Theridiosoma* pulls its web into an elastic funnel shape and holds it taut. When a food item appears the web is released, snaps, and hits and entangles the prey. The web need not always be a sheet. One species spins a single thread and captures its victims with a mad circular dash in which it lashes the prey to a twig. Other spiders produce a bola, a sticky knob on the end of a thread, which is thrown at the prey. Lycosid and salticid spiders are nomadic predators which run or hop about over the ground, and the huge ground predatory *Avicularia* (tarantulas) gain their name from the fact that they have been known to attack and eat birds (as well as bats, mice, lizards, and smaller prey).

## 7.4.2 Tools and Cooperation

Tools are often used in food gathering. The example of the Galapagos woodpecker finch, *Cactospiza pallida*, which uses a twig or thorn to pry food from beneath bark, is well known (Gifford, 1919; Lack, 1945; Bowman, 1961). Sea otters, *Enhydra lutris*, open mussels and occasionally sea urchins and crabs by setting the food item on their chests and pounding them, often many blows, with a stone. The same stone used once by an individual is often saved for future use (Hall and Schaller, 1964).

Tool-using of a sort also occurs in one group of Japanese macaques, *Macaca fuscata*. Given wheat, these monkeys have learned to separate the grains from sand with which they invariably get mixed, by placer mining. A mixture of the seed and sand is dipped in water and the seed, which floats, is skimmed from the surface. This trait has been passed on culturally in this particular troop of monkeys (Kawai, 1955).

Somewhat further removed from real tool use, but still related, is the use of tree crotches by leopards and thorns by shrikes, *Lanius sp.*, to brace or store carcasses of their prey. The use of perches by bee-eaters, *Merops sp.*, first to smash the heads of venomous hymenopteran food, then to rub off the tip of the abdomen, voiding the poison, falls into this same category (Fry, 1969).

The most interesting aspects of foraging behavior involve cooperative effort. The case of pack hunting in wolves has already been mentioned, and cooperation between coyotes, *Canis latrans*, and badgers, has a long folk history, as well as being reported by serious scientists (Galati, personal communication). In the latter case coyotes are said to drive the prey into dense shrubbery. The coyote cannot follow, but can prevent escape while the badger hunts the quarry down in the thicket.

It has been reported that white pelicans, *Pelicanus erythrorhynchos*, may form arcs of several swimming individuals, moving slowly toward shore and driving fish before them. Cormorants (Bartholomew, 1942) and red-breasted mergansers (Emlen and Ambrose, 1970) have also been observed driving fish. Harbor porpoises, *Phocaena vomerina*, have been seen surrounding a school of sardines, forcing it into a long, narrow shape and then diving in rotating groups of five or seven lengthwise through the school, scooping up fish, then diving down and back, forcing the school to remain at the water's surface (Fink, 1959).

Perhaps the most fascinating example of all is reported by Friedmann (1967) who describes wattled starlings, *Creatophora carunculata*, feeding on large flocks of locusts. The birds form a hollow, flying cylinder about part of the swarm, thus entrapping the locusts, and gradually close in from the sides, eating as they do so.

### 7.4.3 The Effect of Prey Dispersion

Successful foraging depends not only on efficient techniques of prey capture but also on the efficient localization of time and effort. Where food occurs in clumps, it is trivially apparent that an efficient predator wastes as little time in poor areas and spends as much time in rich areas as possible. The pattern of foraging, whether random, criss-cross, spiral, or other may have a profound effect on efficiency of food gathering or the type of food gathered, depending on the nature of food distribution. Unfortunately, the relation between food dispersion (distribution pattern) and foraging pattern is poorly known. Tinbergen *et al.* (1967) reasoned that an animal which finds one prey item in an area will search further in that area. This should be adaptive behavior where food items are clumped, and would lead to higher rates of exploitation in more food-dense areas.

To test this hypothesis, eggs were placed in grids of 3 × 3, with differing inter-egg distances. Each test consisted of two grids, and the number of eggs taken by carrion crows from each grid was noted. Predators nearly always took the eggs in the more dense grid at more rapid rates.

Random movements between and concentrated efforts within food clumps might be expected to make patchily distributed food more readily obtainable than randomly or evenly distributed food. Ivlev (1961), working with fish, found that, in fact, evenly distributed food was eaten at lower rates than food scattered about in patches. Ivlev derived the empirical equation

$$r \propto 1 - e^{-x\zeta},$$

in which $r$ is food ration per time, $\zeta$ is the standard deviation of food densities about the mean (measured over some standard grid size), and $x$ is a constant related to the grid size and the species of fish involved.

One can take a more theoretical approach to foraging pattern. Suppose that food occurs in clumps and that upon discovering a clump the predator concentrates its efforts in searching. Clearly, the search for food items in any one clump eventually brings diminishing returns; when intake falls to some critical level, determined by the time required to find individual food items, it is best (average rate of food intake is maximized) to move on to another clump. Let $\theta$ be the average probability that any given food item will be found in a given clump in a given unit of time, $x'$ be the number of food items available per clump (a function of time), and $\kappa$ be the threshold rate of food discovery at which the predator moves to the next clump. Then $x'\theta$ is the rate of food capture, and the predator leaves the clump when $x'\theta \to \kappa$. Now it is likely that some of the food items are either unfit to eat (for any of a variety of reasons) or so well hidden that they are effectively not available. Thus $x'$, the number of items available, may be somewhat less than the number of items present, $x$, and we write (as a first approximation) $x' = x - cx_0$. Then

$$\kappa \to x'\theta = (x - cx_0)\theta,$$

so that

$$x \to cx_0 + \kappa/\theta,$$

and the proportion of all items present that are eaten is

$$p = 1 - x/x_0 \to 1 - \frac{cx_0 + \kappa/\theta}{x_0} = 1 - c - \kappa/x_0 = a - b/x_0, \qquad a < 1. \qquad (7.3)$$

Where several clumps are taken together, we must write

$$E(p) = \overline{a - b/x_0} = \bar{a} - \bar{b}\overline{(1/x_0)} + \text{Cov}\,(b, 1/x_0) = u - v\overline{(1/x_0)}. \qquad (7.4)$$

A preliminary test of the simple hypothesis above is afforded us by the data of Gibb (1966) on the feeding of tits, *Parus*. These birds feed on the larvae of *Ernarmonia conicolana*, an insect which lays its eggs between the scales of Scots pine cones. In the above terminology, the cones represent areas of food clumps, and the larvae represent individual food items. The eggs of the insect are all laid during a short portion of the year, and the hatching larvae leave scars as signs of their presence; because the birds do not revisit used cones, it is an easy matter to examine cones and to determine both the total number of larvae before feeding ($x_0$) and the number eaten ($x_0 - x$). Gibb made several collections of data. The one reported here is presented along with predicted results, in Table 7.6 (see also Fig. 7.7). Lots of 50 cones, each from one tree, were examined and $x_0$ was divided into several categories: 0 to 5 larvae per 50 cones, 6 to 10 larvae per 50 cones, etc. The fit in Table 7.6 was calculated by the least-squares method, yielding

$$p = 0.39 - 0.89\overline{(1/x_0)}.$$

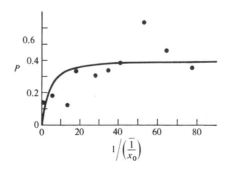

**Fig. 7.7.** Predicted and observed thoroughness (proportion of prey taken) of tits feeding on larvae in pine cones (data from Gibb, 1966).

**Table 7.7.** Predicted and observed thoroughness (proportion of prey taken) of tits feeding on larvae in pine cones (data from Gibb, 1966)

| Number of larvae per 50 cones | Number of 50-cone lots | $\overline{(1/x_0)}$ | Number of larvae eaten $(x_0 - x)$ | Fraction of larvae eaten $\overline{(1 - x/x_0)}$ | Expected fraction eaten $(P) = u - v(\overline{1/x_0})$ |
|---|---|---|---|---|---|
| 0 | 0 | | — | | |
| 1 | 0 | | — | | |
| 2 | 3 | | 2, 0, 0 | | |
| 3 | 5 | 0.357 | 0, 0, 0, 0, 0 | 0.14 | 0.07 |
| 4 | 0 | | — | | |
| 5 | 2 | | 1, 1 | | |
| 6 | 3 | | 1, 2, 0 | | |
| 9 | 2 | 0.133 | 0, 1 | 0.18 | 0.27 |
| 10 | 2 | | 2, 1 | | |
| 12 | 1 | | 1 | | |
| 13 | 2 | | 0, 0 | | |
| 14 | 1 | 0.073 | 4 | 0.13 | 0.325 |
| 15 | 4 | | 2, 1, 1, 6 | | |
| 17 | 1 | | 4 | | |
| 18 | 2 | 0.055 | 4, 6 | 0.33 | 0.34 |
| 19 | 2 | | 13, 3 | | |
| 21 | 1 | | 11 | | |
| 22 | 3 | 0.041 | 10, 0, 5 | 0.30 | 0.355 |
| 23 | 2 | | 5, 1 | | |
| 24 | 1 | | 7 | | |
| 26 | 3 | | 5, 7, 5 | | |
| 29 | 2 | | 10, 10 | | |
| 30 | 1 | | 23 | | |
| 32 | 1 | | 9 | | |
| 37 | 2 | 0.028 | 8, 12 | 0.34 | 0.365 |
| 38 | 1 | | 21 | | |
| 42 | 1 | | 20 | | |
| 43 | 2 | 0.023 | 7, 14 | 0.38 | 0.37 |
| 44 | 1 | | 17 | | |
| 48 | 1 | | 26 | | |
| 51 | 1 | | 33 | | |
| 56 | 1 | 0.018 | 33 | 0.62 | 0.375 |
| 58 | 1 | | 30 | | |
| 61 | 1 | | 41 | | |
| 67 | 1 | 0.015 | 25 | 0.47 | 0.375 |
| 69 | 1 | | 25 | | |
| 71 | 1 | 0.013 | 20 | 0.34 | 0.38 |
| 79 | 1 | | 31 | | |

The fit is far from perfect, but is an approximate description of the behavior of the birds. Of course, $\text{Cov}(b, 1/x_0)$ may not be invariant with $x_0$ and $\kappa$ may change with $x_0$. If there is a positive correlation between values of $\overline{1/x_0}$ in successive lots of 50 cones, for example, the birds might lower $\kappa$ in poor food areas or raise it in good food areas, producing a curve which rises less steeply at lower values of $1/(\overline{1/x_0})$ (as observed). This kind of approach is both interesting and potentially useful in describing food exploitation patterns and the distribution of feeding animals in a population. Assessments of its potential accuracy await badly needed field experimentation.

# 8

# Spacing Systems

It is clear from the discussions in Chapters 5 through 7 that proximity to other individuals of one's own species may at times be selectively advantageous. There must also be times when proximity is very disadvantageous. The purpose of this chapter is to examine the various advantages and disadvantages of proximity and to apply the findings to the nature of animal dispersion patterns.

Mathematicians and ecologists, particularly plant ecologists, have produced a great deal of literature on measures of dispersion. Organisms are said to be randomly dispersed when one unit area (or volume) is as likely to contain an individual as any other. Clumped distributions are generally referred to as *contagious* or *underdispersed*, while more even spacing is called *overdispersal*. While these terms, and the measurements behind them, are interesting and useful in describing populations, their ecological meaning is related to the degree to which the distributions they describe depend upon individual behavior—the tendency of organisms to avoid or associate with one another. The thrust of the following discussion then, deals with individual behavior and its effect on spacing patterns, rather than with grouping or overdispersion *per se*. Nonsocial grouping, arising from physical environmental factors (the accumulation of seeds along wind rows, the concentration of various organisms in tide pools) or preferenda (the aggregation of forest arthropods under rotting logs, the collecting of ribbon worms under algal mats) will not be considered.

## 8.1 THE ADVANTAGES AND DISADVANTAGES OF PROXIMITY

Because surface area varies with the square, and volume with the cube of the linear dimension, large organisms lose a smaller proportion of their total body heat per unit time than small organisms. For this reason small animals more often than large huddle together to retain heat and thus lower their metabolic needs (Prosser and Brown, 1961; Sealander, 1952; Wiegert, 1961). Temporary groupings, or huddles, of quail, mice, and other small animals are often found, particularly in cold situations. These huddles may also serve social functions.

A number of species cooperate to build refuges from the fluctuations in temperature, humidity, or to escape from harsh situations and, as such, form cooperative groupings. Termite mounds act as environmental buffers as well as centers for social activity, as do subterranean ant colonies.

In addition to physical environmental factors, proximity also affects fitness through its role in defense, reproduction, and feeding efficiency. These factors are considered below. A possible reproductive consideration may be neglected in an example illustrating defensive advantage, but this should not be taken as an indication of its insignificance. A predictive scheme must consider all factors; but for simplicity of organization the following discussions cover just one factor at a time.

### 8.1.1 Defense

It has already been noted (Chapter 5) that aposomatic prey species might find grouping advantageous as an advertisement of their noxious qualities. The individual which associates closely with others of its species is more likely to be recognized by the predator for what it is, and bypassed.

It was also noted that group mobbing by birds acts as a predator deterrent. Thus it may be selectively advantageous for a nesting bird to place its nest close to that of another bird of the same species. Of course, there are countering disadvantages: Where nests are placed in a line the predator is led to them (see 5.3.3), and close proximity allows for mutual infestation with mites, botflies, or disease.

The size of a breeding bird colony is limited by the failure of new individuals to join. This occurs when the disadvantages of proximity to a colony override the advantages—perhaps because increased colony size requires ever more far-reaching foraging flights (Hamilton et al., 1967), or gives greater chances for the spread of contagious diseases.

Bison, *Bison bison*, and musk oxen, *Ovibos moschatus*, are capable by virtue of their size of jointly defending themselves and their young; they are good examples of species in which proximity is of defensive value. Where defense is aided by proximity, selection should act to increase both the density and extent of the herd. It is not surprising that the large herbivores (bison, musk oxen, African buffalo) which use shoulder-to-shoulder defense tactics have larger herds than the smaller related herding herbivores which don't (pronghorn, most small African antelopes). The factor limiting herd size in those species is not known. It may be related to foraging effort, disease, or mutual parasitic infection, or it may be that social structure breaks down in large herds with detrimental effects on their members.

Grouping, if sufficiently dense, may lower the frequency of contacts between prey and predator. It has already been suggested that this is one cause of schooling in some fish (5.3.3). In some fish it is likely that group size is limited by energetic problems: Individuals will become increasingly hesitant to pack into a school as its size and density lower the oxygen tension in the surrounding water (McFarland and Moss, 1967).

### 8.1.2 Reproduction

It is fairly obvious that, except among self-fertilizing hermaphrodites, reproduction requires a close association of two individuals, and that groupings of individuals will make it more likely that their gametes form zygotes. A lone oyster is, for all practical purposes, a dead oyster, since it passes on no genes. Acorn barnacles face the same problem as oysters, and selection has favored a behavior pattern in the larvae which maximizes the chance of reproducing. The barnacle *Balanus balanoides* settles only among others of its own kind (but preferentially in neither very sparsely nor very densely populated areas) unless, after two weeks, it has still to find an appropriate place. Then it becomes less particular. Further procrastination is rewarded by bacterial infection of its tissues (Knight-Jones, 1953). *Elminius modestus*, another barnacle, has much the same behavior (Crisp, 1961).

In sessile or very slow-moving species such as oysters, barnacles, mussels, tunicates, some annelids, sea urchins, and others, the social stimulation that comes with grouping is essential to reproduction. Thus the shedding of sperm by male oysters stimulates the females to release eggs. In urchins it is the release of the eggs which stimulates the shedding of sperm. Social stimulation may be psychological as well as chemical. In a number of fish species such as salmon and white fish, large, thrashing schools may stimulate spawning behavior. In the European minnows *Leuciscus erythrophthalamus*, *Abramis brama*, and *Esox lucius*, the spawning of one pair excites the same behavior by other pairs (Svärdson, 1949.) Fraser Darling (1938) has suggested that highly synchronous breeding results in better reproductive timing and hence greater reproductive success in birds. The accurate timing, he claims, results from the *social stimulation* which arises when many pairs in close proximity undergo together the physiological changes in preparation for nesting. Horn (personal communication) reports that the precopulatory display of one female Brewers blackbird triggered similar displays by other, nearby females.

Laboratory evidence for social stimulation is somewhat more clear-cut. Lott *et al.* (1967), for example, found ovarian development in ring doves to be stimulated after seven days of seeing males through a glass partition. Behavior, sound, and perhaps appearance, are also important stimulants, as evidenced by the fact that the stimulation is weaker if the male is castrated and stronger if the female is allowed to hear his voice.

There does seem to be greater synchrony of reproduction within rather than on the periphery of bird colonies. Horn (1966), for example, working with Brewers blackbirds, *Euphagus cyanocephalus*, found a significant correlation ($r = 0.91$) between the mean distance to colony center and the mean square difference in dates of first egg laying for all nests in several colonies. In black-headed gulls, *Larus ridibundus*, 157 of 1703 eggs in one colony hatched, while of 157 eggs in outlying nests none hatched. In another colony egg losses were earliest and most frequent at colony edges (except on the seaward side) and decreased toward the colony center (Patterson, 1965). The increased success of those nesting on the seaward margin may be due to the increased proximity to food there.

The spacing pattern followed by individuals in a breeding colony will reflect a balance of selective forces acting both to draw them together and spread them apart. The size and density of the group may depend ultimately on the likelihood of disease, parasitic infection, or perhaps by the increasing risk to males of stolen copulations between their females and other males in large, crowded colonies. Polygynous species in which the nesting periods of one male's several females overlap may suffer most from stolen copulations; thus we might expect them to nest at widely spaced intervals more often than monogamous species. Perhaps this is one reason why the polygynous redwing (*Agelaius phoeniceus*) and yellow-headed (*Xanthocephalus xanthocephalus*) blackbirds are territorial, while tri-colored (*A. tricolor*) and Brewers blackbirds, mainly monogamous, nest colonially.

### 8.1.3 Feeding

The larvae of the sawfly, *Neodiprion pratti*, hatch on pine needles which serve as their source of food. It is clear, after some observation, that the clusters of larvae found are too large and frequent to be accounted for by chance. It turns out that these are, in fact, local clumps which arise when larvae migrate toward each other, attracted by the smell of pine and the saliva of other, feeding larvae. The ultimate cause of this grouping is that once one larva eats through the cuticle of a pine needle, others can eat from the same opening without spending the time and energy to cut through the cuticle somewhere else. Thus it is always advantageous to be drawn to an already "opened" source of food (Ghent, 1960). Similar obser-vations have been made (Emlen, 1966b) on the intertidal snail *Thais emarginata*. Generally these creatures do not attack the large individuals of the barnacle *Balanus cariosus*, because they cannot drill through to the food inside during one period of tidal immersion. During subsequent exposure they may move to shelter or "forget" what they began and, upon reimmersion, move on without feeding on the partly drilled prey. When one snail is found feeding on such a large barnacle, however, it is very often accompanied by at least one and often two or three other snails. If the barnacle is already drilled and large enough for more than one snail to perch on, it represents a free meal for passers-by.

It has been suggested that one possible, ultimate factor behind bird flocking is *local enhancement*. When one bird finds a food source, others observe its dis-covery and join in feeding. This advantage to flocking is minimal—or negative—if food occurs spaced out or in tiny clumps, but may be substantial when food occurs in large groups. Where small fish travel in schools, their gull and tern predators form feeding flocks. In open fields, seed distribution may be clumped around seed sources, or form wind rows. Thus seed eaters might be expected to flock more often than species which feed on some more motile and generally overdispersed prey such as insects (see 5.3.3). Newton (1967) notes that cardueline finches (seed eaters) always feed in flocks and believes local enhancement to be the reason. Lack, in the appendix of his 1968 book, gives data on the number of solitary and flock feeders (by family) in relation to their food. The figures for passerines and other nidicolous or partly nidicolous bird species, are shown in Table 8.1. Fractions indicate that only some species of one family do as indicated. Fractions are added so that if one-half of each of two families fits one category, they are scored as $\frac{1}{2} + \frac{1}{2} = 1$ family. The difference in spacing (solitary versus flock) between seed and insect eaters is significant, with $p < 0.0001$ ($\chi_1^2 = 17.09$).

Grouping during reproduction—coloniality—may be selected for partly on the basis of local enhancement. Horn (1968) divided his Brewers blackbird colonies into outer half and inner half (central) nests and reasoned that the latter group, benefiting from more visual contact with incoming and outgoing foragers, might be more successful in food gathering. He found weight gain in central colony nestlings to be 27·5 g per nest per day, while that of peripherally located (outer half) nestlings was 23.8 g per nest per day. The objection which might be raised to the application of these data to the notion of local enhancement is that the central colony birds may simply be more adept at foraging, and occur centrally because their general superiority (experience?) allows them to usurp this position in adaptive response to other advantages (lower predation, for example).

**Table 8.1**. The number of bird families flocking or not flocking as a function of diet

| Solitary feeders | Flock feeders | Food |
|---|---|---|
| $48\frac{1}{6}$ | $17\frac{1}{4}$ | Insects *only* |
| 5 | $15\frac{11}{12}$ | Seeds *only* |
| $25\frac{1}{2}$ | $23\frac{2}{3}$ | Combination of above, or others |

## 8.2 ASSOCIATION WITH AN AREA

The above discussion provides us with at least the skeleton of a theory for predicting whether individuals will move alone or in groups. Where animals are associated with a piece of ground (or space) we need additional information to understand their spacing.

### 8.2.1 Dominance Gradients

There are a number of advantages to an animal in restraining the amount of area over which it wanders (see 8.2.8). One is that familiarity with a locality and its inhabitants leaves an individual with more certainty of its escape routes. Possibly for this reason, and undoubtedly for many others, too, familiarity seems to impart to individuals a sense of "confidence"—a psychological state conducive to winning skirmishes with conspecifics. Because of the disadvantages of displacement from the area there must also be a strong motivation to fight aggressively, if necessary, to maintain ground. The result of these facts is that its relative ability to win fights, *relative dominance*, gradually shifts with a transect across an individual's range.

In the fish *Platypoecilius sp.*, residence in an area confers greater potential for dominance in both sexes (Braddock, 1949). Permanent resident juncos and tree sparrows tend to be dominant in their home areas over part-year residents and stray birds (Sabine, 1949, 1955, 1956). In stellars jays, *Cyanocitta stelleri*, birds are dominant, during the winter, in the area of their previous summer territories (Brown, 1963).

A plot of dominance along a transect might be as shown in Fig. 8.1. Each peak represents one individual. At point *x*, individual rankings would be $5 > 6 > 7 > 4$.

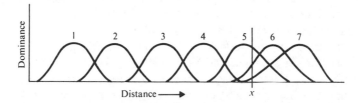

Figure 8.1

Were the advantages of remaining within a home range related merely to the advantage of familiarity with an area of ground or space, Fig. 8.1 might be expected to show broadly overlapping curves. Suppose, however, that the individuals in question require some resource which is in short supply. If the resource is strongly clumped in its distribution, the population utilizing that resource must also be clumped in its distribution, for those individuals whose home ranges do not include part of a resource clump will die. If the resource is very homogeneously distributed the population might also be expected to be homogeneous in distribution. This much is obvious. But suppose now that some individual acquires a genetic recombination or mutation that predisposes it to spend extra time and energy in aggressive encounters near the boundary of its home range. This has the effect of greatly narrowing the range of uncertainty in dominance and broadening the area held exclusively or almost exclusively by that individual. If such behavior were advantageous with respect to the critical resource (or for some other reason—see 8.2.3, 4, 5) we should expect the new genetic innovation to spread until Fig. 8.1 looked like Fig. 8.2. The individuals are now said to be *territorial*.

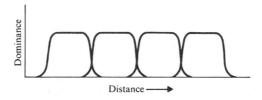

Figure 8.2

## 8.2.2 Territoriality: A Survey

A number of definitions of "territoriality" exist. The two most widely used are "an exclusive area", and "a defended area". Since a territory usually houses not only the resident male who defines it but also his family, the term "exclusive" is poor; most workers prefer the second definition. But the description "defended area" is teleological (Emlen, 1957). The term has become far too ingrained in the literature to deny its continued use, but the reader should be aware of the difficulties and errors in thinking into which teleological definitions such as "defended area" may lead him.

Territoriality is amazingly widespread. It has been little studied in invertebrates, but a fair number of known cases exist. Connell (1963) observed that the marine amphipod *Erichthonius braziliensis*, which lives in tubes, scours the area immediately around the opening of its tubes for food. Amphipods settling in this area are driven out or destroyed. The fiddler crabs, *Uca sp.*, are territorial for a brief period in the summer, and have evolved elaborate territorial–aggressive displays in which the larger chela is waved in species-characteristic patterns and rhythms. Some grasshoppers also "defend" territories, and cicada killers, *Sphecius speciosis*, fiercely territorial, attack intruding male cicada killers, other insects, and even pebbles tossed into their domain (Lin, 1963). The California limpet, *Lottia gigantea*, also ranges over a limited area from which it repels intruders both of its own and other species (Stimson, 1970).

Dragonflies are the invertebrates whose territorial behavior is best documented. These animals may "defend" not areas, but linear distances along pond shores. When the tenerals emerge, the males simply stay on the pond's edge, while the females fly inland. Later, when the females return to breed, they are fertilized by the male whose territory they cross. Generally, the number of individuals per length of shore line is constant for a species (Moore, 1964).

In fish, territories vary from the area immediately about the nest in sticklebacks, to the sleeping places of some subtropical fish. Bluegills (*Lepornis*) possess territories during the breeding season, as do many other fish, and the mullet, *Mugil labrosus*, is territorial outside of the breeding season along the English intertidal (Carlisle, 1961). The garibaldi, *Hypsypops rubicunda*, is territorial all year off Southern and Baja California (Clarke, 1971). Among amniotes, some amphibians, including the South American frog *Phyllobates trinitatis*, are territorial (Test, 1954), and S. T. Emlen (1968) gives a good description of the "defended" areas and displays of bullfrogs, *Rana catesbiana*, which space themselves out at distances of about 18 feet. Among reptiles, alligators, the gecko *Coleonyx variegatus*, and several skinks and iguanids, including the American anole, *Anolis carolinensis*, are territorial (Oliver, 1955). In mammals territorial behavior is found primarily among carnivores. But it is among birds that territoriality has been most studied and debated, and it is primarily to this group that we look for information in trying to construct general hypotheses concerning territoriality.

## 8.2.3 Territoriality: Reproduction

Before proceeding it is important to point out that territories may be "defended" all year round or only seasonally. The arguments presented below must be applied on a season-to-season basis. The coot, *Fulica atra*, may hold its territory all year (Gullion, 1953), as does the European robin, *Erithacus rubecula*, in England (Lack, 1943). Commonly when birds give up their territories during the winter they return to them—or to the same general areas—next spring. In view of earlier comments on the value of familiarity with an area, the above strategies make good sense.

Territoriality has evolved—more accurately, territoriality is retained—by a variety of ultimate factors. In some species the advantage of "defending" a food cache may be the cause. In others both food and cover may be involved. We discuss first those cases in which the selective forces behind territoriality seem to be sexual in nature.

Female long-billed marsh wrens, *Telmatodytes palustris*, appear to choose a male on the basis of his territory (Verner, 1964) as, also, do redwing blackbirds (Orians, personal communication; D. G. Smith, personal communication). In the long-billed marsh wren, bigamous males possess the largest territories, monogamous males the next largest, and bachelors the smallest. These cases suggest that mating success and success in pair formation are the ultimate factors in territoriality. Of course an advantage to territoriality relating to mating or pair formation (sexual considerations) need not exclude food supply as another ultimate factor. In fact food and sexual factors may be inseparable. If redwing females choose a territory on the basis of its food value, then the maintenance of a food-rich territory has a sexual benefit to males.

Occasionally, however, a territory appears to function strictly as a vehicle for attracting females. This occurs commonly in birds. In the black-headed gull, males "defend" special areas, 10 to 12 square yards in size, on or near the periphery of the nesting colony. After a pair bond is formed—on this territory—the pair moves to a small nesting territory in the colony (Moynihan, 1955; Tinbergen, 1956). The male anna and Allen's hummingbirds, *Calypte anna* and *Selasphorus sasin*, use certain well-defined areas to perform spectacular diving courtship displays leading to pair formation (Pitelka, 1951), as do many other birds. In the case of dragonflies, one advantage to territoriality seems quite clearly to be sexual since females, at least in some species, when returning to the water's edge appear to hit the shoreline at random points. The larger a male's territory the more females he will be able to inseminate.

Some bird species gather together in *arenas* where they "defend" small territories known as *courts* and perform *lek* displays. These displays attract females and copulation occurs on the court. Nesting occurs elsewhere. The courts may have abutting boundaries or be quite widely separated. Arenas of the capercaillie, *Tetrao urogallus*, are comprised of courts barely within earshot of each other, while in blue-backed manakin (*Chiroxiphia pareola*) arenas, individuals display together before females (Armstrong, in Thomson, 1964, pp. 431–433).

There is evidence that mating success is related to both the size and position of the court in the arena. Robel (1966) found that of six male prairie chickens, *Tympanuchus cupido*, on one "booming ground" (collection of courts named for the display performed) in 1964, 21 of 23 copulations were performed by the male with the largest court. In 1965, 29 of 31 copulations were achieved by that male of four with the largest court. In mammals, leks occur in the Uganda kob (*Adenota kob*) and other antelopes. In the case of the kob, 30 or 40 individual males space themselves out over a total area of roughly 400 square kilometers, and form clusters of territories each 11 to 30 meters in diameter. The estrous females move about freely over these arenas and mate with their choice of territorial males (Buechner, 1961; Buechner *et al.*, 1966). Lek species clearly form only temporary pair bonds and must be considered promiscuous (6.5).

### 8.2.4 Territoriality: Defense and Population Control

Territories may be useful in "defending" a nest from attack or pilfering by adjacent nesting birds.

They might serve as buffer zones around the nest and the female, thus cutting down on stolen copulations and the stealing of nesting material. The mean size of territories in the black-capped chickadee, *Parus atricapilla*, varies through the breeding season, reaching its maximum during nest building and its minimum when food is most needed during incubation and the nestling period (Stefanski, 1967). Furthermore, this species tends to establish territories after it has already formed pair bonds (Odum, 1941). Thus neither food supply nor mating success seems the ultimate cause of territoriality.

Where territories are very small and adjacent, the spacing system is known as *colonial*. The advantages of coloniality in defense have already been discussed (5.3.3). Where territories are large, territoriality may be valuable in defense in another sense. Recall (from 7.4.3) that, where food is spread out, predators spend less time foraging and eat less. Where territories are large, or separated, with neutral zones between them, the value of territoriality as a protection against the spread of infectious diseases is also a possibility, but no evidence is available; whether the incidence of infectious disease is high enough during the nesting season to result in significant selection intensities for territoriality is not known.

It has been suggested, particularly by Wynne-Edwards (1962), that territoriality has evolved as a mechanism for holding populations below starvation levels. Even assuming that starvation of some of its members is "unfit" for a population—which is by no means certain—the evolution of territories for this purpose requires a rather heady dose of large-group selection (3.1). Also, while territoriality limits the number of breeding pairs, it maximizes fitness for the territorial individuals—otherwise it would not have evolved. It thus may be that territoriality *increases* the size of the population which the environment can support.

### 8.2.5 Territoriality: Resources

During the breeding season, food is usually a critical factor in avian reproductive success. Thus it is possible that an area is "defended" for its food supply. This would seem to be the ultimate cause of territoriality in hummingbirds (Pitelka, 1942), as well as many other birds.

The pika, *Ochotona princeps*, leaves no doubt as to the food value of its territory. Piles of hay and other food matter form the center of interest for each pika during much of the year and incursions near the area of the food piles are vigorously rebuffed (Broadbooks, 1965). In otters, females with their young appear to inhabit exclusive areas, probably within male territories, with no other obvious purpose than to assure an exclusive source of food (Erlinge, 1968). Eastern kingbirds, *Tyrannus tyrannus*, generally establish territories after mating. Thus territories are not "defended" for the purpose of sex and would appear to be food-related (Davis, 1952).

While there is much indirect evidence, there exist only a few cases (such as the pika) in which food can be shown rigorously to be the ultimate factor in territoriality. Schoener (1968) ran regression analyses on territory size, bird size (and thus energy needs), and clutch size for birds feeding primarily on their territories and found

$$\log_{10} \text{(territory area)} = (1.09 \pm 0.11) \log_{10} \text{(body weight)}.$$

For predators alone, the coefficient rose to 1.31. This suggests that birds with greater metabolic needs—large birds—and birds which must forage far to get adequate food—predators—choose their territory size to match their food requirements. However, the crucial food factor probably relates to feeding of the young. Larger clutches, all else being equal, should require larger territories. Yet Schoener found, in a partial regression analysis, that clutch size accounted for very little (insignificant regression coefficient) of the variance in territory size.

Whether territoriality exists or not depends first upon the magnitude of the selective advantage to possessing a territory and second upon the cost (in terms of diminished fitness) of using and maintaining it (Brown, 1964). A simple, beginning analysis of cost in species which feed, or make use of some resource, on the nesting area, is provided by Horn (1968). The following discussion is paraphrased from his model.

We consider the three possible resource distributions: uniform in space (or at least not highly clumped) and predictable in time; clumped and predictable in time; clumped but unpredictable in time.

**a) Resources are uniform and predictable.** We picture the resource as spread over a large area and the nests as either evenly spaced (large territories) or clumped (small territories—coloniality) at even intervals (Figs. 8.3a, b). We draw the figures in such a way that each nest requires the use of four resource sources (points in the figure). In the first case (8.3a) the average expenditure in time and energy (cost) in harvesting the resource, where $k$ is the separation of the resource sources, and assuming cost to be proportional to distance traveled in harvesting, is

$$\sqrt{(\tfrac{1}{2}k)^2 + (\tfrac{1}{2}k)^2} = 0.71k.$$

In the second case (8.3b)

$$\frac{4\sqrt{(\tfrac{1}{2}k)^2 + (\tfrac{1}{2}k)^2} + 8\sqrt{(\tfrac{1}{2}k)^2 + (\tfrac{3}{2}k)^2} + 4\sqrt{(\tfrac{3}{2}k)^2 + (\tfrac{3}{2}k)^2}}{16} = 1.50k.$$

Time and energy are clearly saved in the first case. We conclude that so long as the added time and energy required to "defend" a territory does not overbalance the time and energy saved in foraging, territoriality should be selected for.

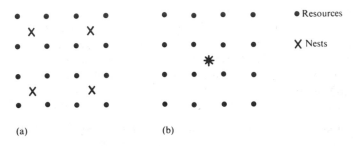

**Fig. 8.3.** Diagrammatic representation of nest placement among resources: two possibilities (a) animals nest on territories, (b) animals nest colonially

Note that if the resource is particularly plentiful, $k$ is small so that the net advantage (in terms of time and energy spent foraging) of territoriality is reduced. Since "defense" of the territory still requires time and energy, there may now be a net advantage to coloniality.

**b) Resources are clumped and predictable.** In this case all breeding individuals crowd into the resource clumps and territories must be small. It is a matter of semantics at what point we speak of coloniality as opposed to territoriality. Where nesting does not occur where the resource is found, other factors will determine whether nests are clumped or widely spaced.

**c) Resources are clumped and unpredictable.** We again make use of Fig. 8.3, but now view the points as locations where resource clumps appear now and then with roughly equal frequency. The nesting animals must now visit all of these points to acquire the resource. Territorial spacing requires traveling a distance of

$$\tfrac{1}{16}[4\sqrt{(\tfrac{1}{2}k)^2 + (\tfrac{1}{2}k)^2} + 4\sqrt{(\tfrac{3}{2}k)^2 + (\tfrac{1}{2}k)^2} + 4\sqrt{(\tfrac{5}{2}k)^2 + (\tfrac{1}{2}k)^2}$$
$$+ \sqrt{(\tfrac{3}{2}k)^2 + (\tfrac{3}{2}k)^2} + 2\sqrt{(\tfrac{5}{2}k)^2 + (\tfrac{3}{2}k)^2} + \sqrt{(\tfrac{5}{2}k)^2 + (\tfrac{5}{2}k)^2}] = 1.93k.$$

Colonial spacing requires, as before, $1.50k$ units of travel. Coloniality is advantageous where resources are clumped and unpredictable, even if the cost of territorial "defense" is negligible.

We recall that most insects are cryptic and that cryptic insects tend to overdisperse, at least in woodland (5.3.3). On the other hand, seeds and insects in open country both tend to blow and to become relatively concentrated in certain areas. If food is the critical factor in determining selective pressures for or against territoriality, then forest birds should more often be territorial than field birds. Coloniality, at least among temperate birds species, is primarily a phenomenon of open country. Carnivorous sea birds, by virtue of the unpredictability of one prime food source (fish, for example), and the impossibility of nesting in areas predictable with respect to another prime food source (the intertidal zone), are generally colonial.

### 8.2.6 Interspecific Territoriality

Where ecologically similar species live sympatrically (where their geographic ranges overlap) they often exclude each other from their territories. Thus redwing and yellow-headed blackbird males "defend" territories, excluding males of both their own and the other species (Orians and Willson, 1964). Red-tailed hawks (*Buteo jamaicensis*) and eagles are also interspecifically territorial (Fitch *et al.*, 1946) as are eastern (*Sturnella magna*) and western (*S. neglecta*) meadowlarks (Lanyon, 1956), and the two woodpeckers *Centurus carolinensis* and *C. aurifrons* where their ranges overlap in Texas (Selander and Giller, 1959). In the case of the meadowlark, visual appearance of the two species is almost identical, and interspecific territories may result from the need for species recognition by potential mates. In addition to this, the food needs of the two species, and the food needs of the members of the other bird pairs mentioned, are very similar. In general, interspecific territoriality is probably evolved to alleviate competition for food. An exception probably occurs in the exclusion by redwings of marsh wrens. Marsh wrens do not compete with the blackbirds for food, nor is there any danger of species misidentification. Marsh wrens have the rather provocative habit of poking holes in redwing eggs.

### 8.2.7 Home Ranges

A *home range* has been defined as the area to which an animal normally confines its movements (Burt, 1943). Most animals appear not to wander freely about but rather to stay in some area. The cow water buffalo, *Bubalus bubalis*, with her young is so stubbornly tenacious to one general area that she will refuse to leave it for food and water even if starving to death (Tullock, 1969). Young *et al.* (1950), keeping track of the movements of house mice, *Mus musculus*, in two buildings, found the distance between successive live trappings of individuals to be only a few feet. Over an eight-month period most mice had never wandered more than about 20 feet from their original point of capture. It is by no means certain that all animals possess restricted ranges, and certainly some do not. However, most workers interested in animal movements, particularly those working with mammals, believe that limited movement in a home range is the rule. Throughout the rest of this chapter we shall refer to *home range* in perhaps a more limited way than Burt originally intended, as a restricted area.

Evidence for the existence of home ranges is of two basic sorts. Extensive trapping, with live traps set in grids, is the usual way. If a mouse, say, is trapped in positions 12, 13, 14, 16, and 17 along a transect, but not 11 and 18, its home range may be estimated as extending, linearly, from halfway between 11 and 12 to halfway between 17 and 18 (see Fig. 8.4). In this manner Blair (1940a) has estimated the prairie deer-mouse, *Peromyscus maniculatus*, in southern Michigan to have home ranges of 2.31 acres (male) and 1.39 acres (female) when in woodland, and 0.63 acres (male) and 0.61 acres (female) when in meadows. Hamilton (1937) gives the home ranges of the vole *Microtus pennsylvanicus* to be 0.7 acres. The male wood mouse, *Apodemus sylvaticus*, ranges over 0.02 to 5.67 (mean = 0.72) acres and the female covers 0.004 to 1.26 (mean = 0.24) acres (Miller, 1958). The red-backed vole, *Clethrionomys glareolus*, wanders over an area roughly 30 to 40 yards in diameter (Jewell, 1966). Using a similar technique, Tinkle (1965) estimated the range of the lizard *Uta stansburiana* at 0.10 acres for males, 0.02 acres for females, and 0.01 acres for juveniles. Other techniques give somewhat different results (Jorgenson and Tanner, 1963).

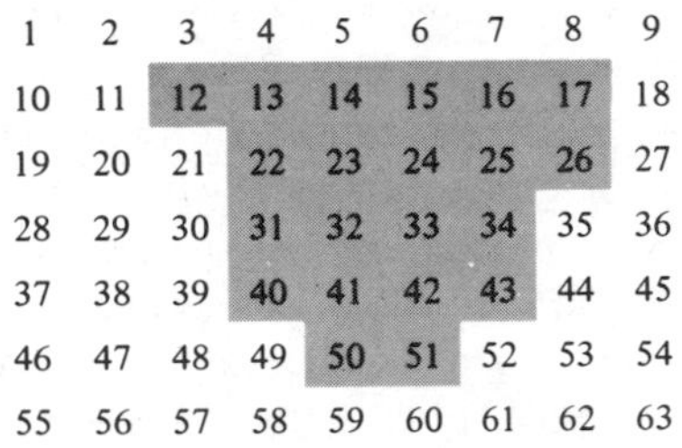

**Fig. 8.4.** Home range as determined by trapping data.

The other source of evidence for home ranges comes from homing experiments. Here animals are released some distance from their point of capture and their movements observed. If the distance is not too great, the animals usually return to the general area in which they were picked up (their home range, presumably). As an example of the results from these tests, Fisler (1962) discovered that the California vole, *Microtus californicus*, homed from distances of up to 600 feet from its point of capture. Thus the home range seems to exert a strong attraction for animals.

Unfortunately in spite of the implication of these figures, it is not always clear just what the home range is. Blair (1940a) found that *Peromyscus maniculatus* might wander over one fairly well-defined area for a while and then move to another. Miller (1958) found the same phenomenon in *Apodemus sylvaticus*, and it appears to occur also in the harvest mouse, *Reithrodontomys humulis* (Dunaway, 1968). If individuals possess several ranges over the course of the year, these movements present no problem, but the concept of a limited area of wandering becomes a little strained. In fact the situation may often be even more bleak. Stickel (1960) claims that the area over which a *Peromyscus leucopus* is trapped increases insignificantly with further trapping beyond that indicated after four captures. On the other hand Stoecker (personal communication), working with *Microtus pennsylvanicus* and *M. montanus*, often finds no leveling off of the total area over which a mouse is caught after 8, 10, or even 30 recaptures. Kaye (1961) wire tagged individuals of *Reithrodontomys humulis* with radioactive gold and followed movements by walking his study area with a geiger counter. He discovered that individuals spent most of their time occupying two or more nests at the periphery of the areas where they were found. This suggests that most of what is generally considered home range is merely along the path between where an individual is at any one time and where it "wants" to go in the future.

One way of viewing home ranges is to consider them as probability-of-capture distributions. Harrison (1958), following the original suggestion of this approach by Hayne (1949), has compiled live-trap data on rats, calculated an *activity center*, and looked at the fall off in probability of capture with distance from that center. Davis (1945) has done the same for several small Brazilian rodents. The data of Dasmann and Taber (1956) using a similar approach with black-tailed deer, *Odocoileus hemionus*, are shown in Table 8.2.

**Table 8.2.** Home range as measured by capture frequency at increasing distances from an activity center

|  | \multicolumn{8}{c}{Distance from activity center in yards} |
|---|---|---|---|---|---|---|---|
|  | 0–250 | 250–500 | 500–750 | 750–1000 | 1000–1250 | −1500 | +750 |
| No. of sight records | 410 | 157 | 46 | 14 | 4 | 0 | 2 | 4 |
| Percentage | 64 | 25 | 7 | 2 | 1 | 0 | 1 | 1 |

Returning to Fig. 8.1, we note that home range, by the preceding definition, can reflect any kind of overlap except that extreme case indicating territoriality (Fig. 8.2). "Home range" represents the entire continuum between a "defended" area with clear-cut boundaries, and a system with freely overlapping movement limited only by considerations of possible contact avoidance. But this does not imply lack of social structure. When home ranges exist they often include a central *core area* which is "defended" and thus meets the definition of a territory. The American eastern chipmunk, *Tamias striatus*, "defends" a sphere around its burrow (Burt, 1940), as does the golden-mantled ground squirrel, *Citellus armatus*, (Gordon, 1936). The territories of the pika, *Ochotona princeps*, surround the vital hay piles, but foraging extends beyond these territories into overlapping home ranges (Broadbooks, 1965).

### 8.2.8 Home Range Size

All else being equal, that species with the greatest speed of movement might be expected to exhibit the largest home range. But there is clearly more to the matter of home range size. Let us examine some of the facts.

In 1963, McNab proposed that home range size could be accounted for largely on the basis of metabolic needs. The reasoning was that if one animal's metabolism requires twice as much food as that of another, the first would have to scour twice as much area as the second to get that food. Since metabolism rate varies roughly with the surface area of a homeotherm, it should vary roughly with the $\frac{2}{3}$ power of that animal's weight (Lesiewski and Dawson, 1967). Accordingly, McNab collected data on weights and home range sizes of a number of mammals and plotted the relation. He obtained a graph in which the log of home range size increased linearly with the log of weight and showed a slope of 0.63, not significantly different from $\frac{2}{3}$. Furthermore, *hunters*, which would be expected to forage farther for equal amounts of food than *croppers*, fitted a somewhat higher curve. Schoener (1968), using data from Burt and Grossenheider (1964), and applying the same ideas to predatory mammals arrived at the expression

$$\log (\text{home range size}) = (1.41 \pm 0.16) \log (\text{body weight}),$$

and found body weight to account for 92% of the variation in home range size. The slope of this equation seems at first glance too high to fit the McNab hypothesis. but it nevertheless supports the view that food needs play an important role in determining the size of the area over which an animal wanders.

Suppose that some ranges never overlapped—that they were, in fact, territories. Then if home range size doubles, each individual has at its command twice the resources and one-half of the population must have been driven from the area. If there is unlimited overlap, a doubling in home range size doubles both the resources available per individual and the competition for those resources. No animals have been driven from the area and the amount of resource per individual remains unchanged. The usual situation in mammals, at least, may be intermediate: A degree of aggression may limit range overlap but does not result in strict territoriality. Thus a doubling of resources per individual should require somewhat more than a doubling of range size and the slope of regression equations of the sort shown above should exceed $\frac{2}{3}$. It is therefore satisfying that the figure from Burt and Grossenheider's data, 1.41, is greater than $\frac{2}{3}$. For terrestrial lizards the figure is 0.95 (Turner *et al.*, 1969). It is strange, however, that McNab found a value of only 0.63, and that birds, which are truly territorial, should show figures way in excess of $\frac{2}{3}$. Schoener (1968) and Armstrong (1965) give the values 1.09 and 1.23, respectively, for territorial birds.

The prime objection to the McNab hypothesis is that it is simplistic. If food were the only criterion in determining home range size, why are some small mammal ranges smaller in the winter months when food is most scarce (Jewell, 1966; Dunaway, 1968)? And is it possible that such creatures as meadow voles ever really run short of grass to eat? Let us examine the question of home range size further by asking the following two questions:

1.   What are the factors causing individuals to wander at all,

2.  What are the factors limiting the extent of their wanderings?

There are at least two factors which might induce wandering.  Since most habitats are spatially heterogeneous in some respect, it may be that wandering increases the chances of an individual's finding clumps of superior food or micro-habitat.  There would, by this criterion, be less advantage to moving widely in a relatively homogeneous environment and an equilibrium between forces causing wandering and forces preventing wandering should occur at a relatively small home range size in such an environment.  No data pertaining to this hypothesis exist.

A second possible cause of wandering is social.  Home ranges in many small mammals increase in size during the breeding season (Brown, 1966; Jewell, 1966; Dunaway, 1968), indicating a wider search for mates.  Since mate seeking is generally the male's role, male ranges should—and do—exceed female ranges in size (Chitty, 1937; Blair, 1942; Jenkins, 1948; Erickson, 1949; Brown, 1966).  The same appears true in the lizard *Uta stansburiana* in Texas (Tinkle, 1965).  This indirect support of the hypothesis is clouded by the fact that males also hold larger ranges than females in the winter.  The fact that in *Apodemus sylvaticus*, at least, a social-dominant male possesses a larger range than subordinate males (Brown, 1966) may indicate some social advantage in large range size.

There are several factors limiting movement.  One of the most obvious is that mice are more prone to predation when they are in strange surroundings and thus limited in range by a tendency to stay within familiar haunts (Davis and Emlen, 1948).  This conjecture is supported by laboratory experiments in which it was found that *Peromyscus* unfamiliar with a large test room were captured in that room by screech owls significantly more often than individuals which had been allowed first to explore the room (Metzgar, 1968).  Blair (1940a) remarked that *Peromyscus maniculatus* seemed to know the terrain at its point of capture very well and that upon release mice made direct dashes for the nearest holes.  The fact that beach mice, *Peromyscus polionotus*, possess smaller ranges in open country than in regions of dense cover (Blair, 1951) also supports the contention that danger restricts movements.

Social factors undoubtedly are also important in limiting range size.  The possibility that aggressive encounters influence the extent of wandering is supported by the fact that high population densities—and thus increased numbers of contacts with the distance covered—lead to smaller home ranges in *Microtus pennsylvanicus* (Blair, 1940b), *Peromyscus leucopus* (White, 1964), and the raccoon *Procyon lotor* (Ellis, 1964).

It is also known that dominant males tend to restrict the movements of subordinates (Brown, 1966).  It is likely that subordinate males will not risk wandering into unfamiliar areas if for no other reason than that doing so brings them into contact with strange males and consequent strife.  Within a familiar area dominance relations (8.3.1) are recognized and fighting is at a minimum—not so in new surroundings.  This may be one reason why socially dominant males wander farther and it may profoundly influence range size.  If a female is to have a choice of males, her home range must overlap that of several males. A male's problem is more complex. If he is to mate, he must not only cover enough area to contact unmated females, but he must exhibit a sufficient degree of dominance in that area to insure a chance of being accepted by a female. As his range increases in size the first criterion is more likely to be met, but in making contacts with more males, and covering a larger area—and therefore spending less time developing familiarity with any one part of it—the second criterion is likely to be increasingly lost. A male extending his range into unfamiliar area with unfamiliar residents, some of which may be stronger or more aggressive, runs real risks. Brown (1966) reports that adult male *Apodemus sylvaticus* drive away or even kill new males in their home range. Young animals of almost all mammalian species, once they have left the nest and established themselves in one of a few places, remain there for life. The size of a male's home range should reflect a balance of needs and risks.

Two possible pictures emerge. The first is one of an area divided into partially overlapping spheres of influence, each occupied by a dominant male. Within these spheres are the overlapping home ranges of subordinate males. It is likely that a subordinate animal gains a measure of protection from staying within the confines of a local dominant male's range and not wandering into the range of another top male. A hypothetical diagram of male ranges is given in Fig. 8.5. The solid lines indicate the home range boundaries of the largest, most aggressive males, and the shaded areas show the overlapping ranges of subordinate males. Females, because they seldom display either the level of aggressiveness of the well-developed dominance hierarchies of males, would be expected to possess less rigidly defined, smaller ranges with less influence by dominant individuals and thus less spatial structure. Brown (1966) found exactly this system in the wood mouse, *Apodemus sylvaticus*.

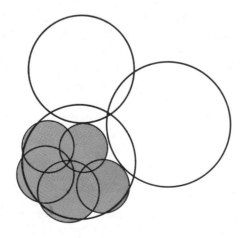

Fig. 8.5. Representation of home range in a system of dominant and subordinate individuals.

The alternative picture is one in which the dominant males choose and at least partly exclude subordinates from a highly desirable area—food, location of females—and the subordinates occupy the area left. The darkened areas in Fig. 8.6 represent the ranges of five male swamp rabbits, *Sylvilagus aquaticus*, in a single study pen (represented by the rectangle). The blackened areas are portions of the ranges also occupied by females (Marsden and Holler, 1964). $M$ denotes male, and dominance is given by $M1 > M2 > M3 > M4 > M5$.

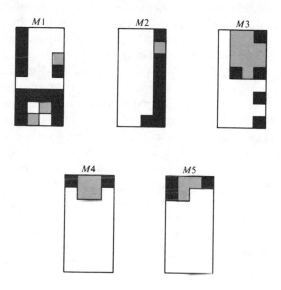

**Fig. 8.6.** Diagram of home ranges in the swamp rabbit (after Marsden and Holler, 1964).

## 8.3 GROUPS

Except in territorial systems, there is always some interaction between individuals of overlapping ranges. We may thus define any collection of individuals with overlapping home ranges as a group. Where individuals move in flocks or herds their ranges clearly overlap. The following discussion applies to intragroup and intergroup relations.

### 8.3.1 Internal Structure

The proximity of individuals is governed by psychological tendencies to join or leave groups, to passively mingle or aggressively drive others away, tendencies which may have evolved for any number of reasons. J. T. Emlen (1952) speaks of positive and negative social "forces", noting that animals are generally drawn to their own kind, but only to a certain distance. He observed, for example, cliff swallows, *Petrochelidon pyrrhonota*, pass up thousands of feet of otherwise perfectly acceptable telephone wire to sit within 4 to 10 inches of one another. However, this distance rarely fell below 4 inches. If a space of 8 to 10 inches existed between perched birds, another might alight, but this was unlikely in spaces less than 8 inches. If a newcomer lit in a space of 6 inches, he was either immediately driven away or stayed his ground, causing a major reshuffling of position all up and down the line of swallows. Four inches is roughly the distance at which a swallow can jab its neighbor without moving its feet. (It is also about the distance its wings reach during a quick take-off).

It seems reasonable to assume the existence of both positive and negative social forces at work here, the former acting over greater distances, the latter strongest at small distances. Where the two are equal is the average distance between individuals. This equilibrium distance generates an area around an individual which is known as its *personal space*.

The personal space clearly changes with an animal's mood and with the identity of the next individual. In many species individuals spend some of their time fighting and some of their time huddling. In most birds and mammals closer distances are tolerated between mates or parents and offspring than between adult males. Size of the personal space also should change with the aggressive level of individuals over the year. Thus in the breeding season, redwing blackbird males are highly aggressive toward other males, and the equilibrium distance is large. This plus site fixation (on a nest site) results in territoriality. In the winter, aggression is low and equilibrium distance falls, allowing for the formation of flocks. Emlen and Lorenz (1942) found that injections of male hormone into wild California quail resulted in increased levels of aggression and subsequent withdrawal from the flock. The proximate cues for aggression vary, but two seem to be humidity and temperature. In dry periods, coveys of California quail fail to break up during the nesting season. In bouts of cold weather during the early part of the nesting season, redwings and cliff swallows abandon their aggressions and re-form flocks.

Since aggression level plays such an important role in the cohesiveness of groups it is clear that its control will increase fitness within a group. This control is often brought about through the operation of *dominance hierarchies*.

A dominance hierarchy is an ordering of individuals such that each animal can usually successfully intimidate those individuals below it in the order, but is usually successfully intimidated by individuals above it in the order. The top individual is denoted the alpha individual. In rhesus monkey troops, the top males are the ones who keep discipline. Generally, dominance hierarchies are more marked in males than in females. Females may have hierarchies independent of the males or not at all; occasionally a female takes the rank (position in the hierarchy) of her mate. If conflict arises, the top-ranking individuals, in order of rank if necessary, get first choice at the object of conflict and, when satisfied, give way to lower-ranking individuals.

Hierarchies are highly developed among the higher-ranking males in most primate groups. In wolves, one alpha individual is dominant over all other, lower-ranking individuals. In most species organization is less well defined in the lower echelons of the hierarchy.

Where food is plentiful and spaced so that conflicts—at least involving food—seldom arise, hierarchies are poorly developed.

Near Ishasha forest in Uganda, baboons have an unusually rich food supply and plentiful cover relative to their abundance. This diminishes the need for a group to "defend" its food, water, and shelter and leads to considerable mixing of troops as compared with baboons in most other areas. The baboons of Ishasha forest show no apparent dominance hierarchy, while in other areas baboons show highly developed hierarchies (Rowell, 1967).

The evolution of dominance hierarchies has, in the past, usually been explained on the basis of group selection (3.1). Those groups which did not possess hierarchies were internally disrupted and ceased to exist. Where groups are small and disruption leads to death, or failure to reproduce, group selection is undoubtedly a factor to consider. As we have pointed out previously, however, group selection is probably a very weak force in nature. A more satisfactory argument for the evolution of hierarchies is that certain individuals naturally have the ability to win in conflict situations and that the losers learn that trying to win costs them more in terms of time, energy, and injury than it gains them. (This argument is supported by the fact that individuals of species which are territorial—and display no hierarchy in nature—quickly develop peck orders when crowded in captivity.)

The loser might never put up a struggle even if his chances of feeding or reproducing thus become vanishingly small—so long as his chances are not actually zero.

Clearly, the top-ranking individuals would seem to get first choice of food and females and thus pass disproportionate numbers of genes. Research on a number of species indicates this to be so and in some cases we emerge with a picture of despot males ruling over their psychologically castrated underlings. Davis (1952) has suggested that the ultimate cause of hierarchies is to insure that the "best" genes are passed on. This argument, however, because it draws too heavily on group selection and smacks of teleologic thought, is totally unacceptable.

To some extent dominance is a function of age. The fact that only a few males mate need not mean that other males will not also have the opportunity to mate as they grow. But dominance is a function of more than just age. Thus the notion that dominant individuals contribute most of the genes to future generations carries some interesting implications. It suggests that whatever traits characterize dominant individuals—size, aggressiveness—will be selected for. Males generally have the best-developed hierarchies so that such selection should generally act most strongly in males. Selection for social rank, then, is a form of intrasexual selection and like other forms of intrasexual selection leads to increasing size and aggressiveness.

If size and aggressiveness are selected for, then there must be opposing selective forces, unless we suppose that hierarchical animals are, always have been, and always will grow in size and aggressiveness. Let's look at some possible disadvantages of high social rank.

Highly aggressive males may benefit by getting first rights at food, but waste time and energy by "defending" too large a territory in the breeding season or being too aggressive to mate properly. Where the dominant males have first rights with the females it is possible that their aggressive habits decrease their chances of living to reproduce. It is suspected, though not proven, that high-ranking lizards by displaying in open spaces are highly prone to predation. Usually, dominant male macaques lead their troops and form the final line of defense against males of opposing troops or predators. We are uncertain how dangerous this is and, furthermore, in some groups it is not the dominant male that leads the troop. In baboons, *Papio ursinus*, the alpha male is usually in the middle of the troop, surrounded by females, and the vanguard is formed by peripheral males. Salomon and Schein (1965) show that top male cockerels do a disproportionate amount of crowing and usually experience more aggressive encounters than lower ranking individuals. But is their mortality higher as a result?

Perhaps it would be wise also to reassess the popular picture of sexual hoarding by high-ranking males. Hall (1962) reports that in wild chacma baboons, the alpha male mated with fully estrous females 93% of the time. Other males usually never had the opportunity to contact females more than $\frac{2}{3}$ of the way towards full estrous. The beta male mated with fully estrous females 90% of the time and mated only half as often as the alpha male. In the northern and southern elephant seals, *Mirounga augustirostris* and *M. leonina*, high-ranking males spend their time with sexually receptive females and drive other males away. In *M. augustirostris* the top 4 males of one colony containing 71 males inseminated 88% of 120 females. In another colony, the alpha male alone performed 73% of the observed copulations (LeBeouf and Peterson, 1969). On the other hand, Rowell (1967) reported no dominance hierarchy in her troop of Ishasha forest baboons, and recent work by others indicates that Hall and Washburn may have overstated their case (for recent data see DeVore, 1965b). Jolly (1967) reports that although *Lemur catta* shows dominance hierarchies, these hierarchies break down as regards sexual rights. Subordinate males can and do successfully chase dominant males from females. It has always been assumed that high-ranking males of the rhesus monkey accounted for the vast bulk of gene passing. Conaway and Koford (1965), however, have disputed this notion. They acknowledge the greater activity of dominant males with fully estrous and dominant females but suggest that these males may be wasting their time with already fertilized females.

There are even more complications. Among some primates two or more males may band together to depose higher ranking individuals. DeVore (1965a) observed one male stand guard to head off attack or harassment while his colleague copulated.

We know that, at least in some species, a tendency toward aggressiveness can be inherited. This is true of chickens, for example (Craig *et al.*, 1965). On the other hand the heritability (see Chapter 2) of social rank or even aggression may be low. If so, the above-mentioned disadvantages might easily provide sufficient balancing selective forces to stop an increase in size and aggressiveness.

It is instructive to look at the ontogeny of dominance. Rhesus monkey males begin to fight first as old infants or yearlings. At this stage they are still closely associated with their mothers and their rank at this time is often decided by the interference, and thus the rank, of their mothers (Sade, 1967; Marsden, 1968). Their ranks attained at this stage are good indicators of their later ranks as adults. To the extent that their mothers' ranks are inherited, the ranks of young males are also partly heritable. However, it is almost certain that chance, the rate of early maturation, and possibly other factors play a role in determining the eventual social status of a male. It is true of mice, and probably also in rhesus monkeys, that aggressiveness may be raised by a systematic series of victories (Ginsburg and Allee, 1942), and increased aggressiveness will generally lead to continued victories. There is a need to measure heritability values of social rank.

### 8.3.2 Group "Defense"

Groups with little internal organization have little to be disrupted by the coming of new and the going of old members. They tend to vary in size and composition. Most bird flocks split, mix, and move freely about. Highly structured groups on the other hand, presumably have become highly structured for some purpose and have much to lose from disruption. Such groups tend to "defend" their integrity. Social insects, for example, attack all intruders to their group. In ants, group territorial combat occurs particularly in the spring when new territorial boundaries are established after partial contraction of each colony's sphere of influence during the winter (Brian et al., 1966). In the Galapagos mockingbird, *Nesomimus macdonaldi*, 4 to 10 individuals jointly "defend" a good nesting area, excluding others (Hatch, 1966). Such group territories also occur in *Crotophaga ani*, a bird of the American tropics, which forms communal nesting groups of 2 to 24 individuals. Mutual preening seems to serve as a cohesive force in this species (Davis, 1940).

Among mammals, the European rabbit, *Oryctolagus cuniculus*, lives in groups of 8 to 10 in distinct areas with a central burrow and a number of entrances. This rabbit *warren* is demarcated with urine, feces, and a special scent gland. Little fighting of intruders seems to occur but outsiders "respect" the boundaries. Individuals are recognized as members of the in-group by their smell, partly acquired through these rabbits' habit of urinating on each other (Mykytowycz, 1968). Prairie dogs, *Cynomys ludovicianus*, also possess group territories (King, 1955), as do beavers and a number of mammalian carnivores. Vicuna, relatives of the llama, in groups of one male, usually four females, and two juveniles, "defend" joint territories of 20 to 100 acres all year round (Koford, 1957).

Groups which "defend" their immediate area from intruders, but whose group ranges overlap, are quite common. Such group areas are counterparts of home ranges and we may more properly speak of group ranges. The wolves of Isle Royale, in Lake Superior, have no clear-cut pack boundaries, but regularly chase "intruders" (read "nonmembers" of the pack) wherever they are encountered (Jordan et al., 1967). Similar behavior has been repeatedly observed in northern Minnesota by Stenlund (1955) and Mech et al. (1971).

Primates commonly "defend" their group integrity and occasionally group territories.  In prosimians, *Propithecus* (diurnal lemur) forms groups of 2 to 4 or 5 consisting of a monogamous pair with their young.  The family "defends" a group territory.  *Lemur macaco* forms discrete groups of 4 to 15 individuals during the day, but group integrity breaks down at night when members of different groups come together to sleep (Bourliere, 1961b).  Gibbons (*Hylobates*) behave much as *Propithecus* (Carpenter, 1940).  In rhesus monkeys, *Macaca mulatta*, groups of up to 100 are often found.  These *troops* move through a forest, displaying a propensity to stay within certain boundaries, but not defending any particular area.  However, except for an occasional *peripheral male*—usually young males who participate little in affairs of the troop except to move on its periphery, often acting as lookouts for other troops—the integrity of these troops is maintained.  Several older males working alone or in combination generally oversee each troop's activities, mediate disputes, deal out discipline, and form the muscle behind intimidation of other troops.  When two troops meet, there are elaborate displays of aggression and the smaller troop usually withdraws to let the larger pass.  In mountain gorillas (Schaller, 1963), meetings between independent troops—consisting of one adult male, his females, juveniles, and young —are infrequent.  Troops present little competitive threat to one another and intertroop aggression is slight.  Baboons, *Papio ursinus*, possess roughly the same social organization as rhesus monkeys.  Here, however, there is a "defended" *core area*, central in the range over which wandering takes place.  Vervet monkeys, *Cercopithecus aethiops*, also "defend" core areas within group home ranges, the latter overlapping those of other groups (Struhsaker, 1968), and a similar situation probably existed in preneolithic man (Hiernaux, 1963).

When is it advantageous for individuals to jointly rather than individually "defend" an area?  Some of the criteria involved in the selective advantages and disadvantages of proximity and territoriality have already been discussed—it seems reasonable that where both are advantageous group territoriality will evolve, that where only the latter is advantageous, individual territories will appear.  There is a further energetic argument that must be considered.

G. C. Smith (1968), in his study of red squirrels, *Tamiasciurus spp.*, found that both males and females "defended" separate territories and shared joint territories only in the breeding season.  Why should joint territories not exist all year?  In an attempt to explain this, Smith constructed the following model. Red squirrels tend to cache their food at some central location on their territory, and their *raison d'être* for "defense" in the first place is to assure an adequate food supply.  Suppose that food is homogeneously (randomly) distributed over the territory.  Then, if we define our units of measurement so that the radius of a territory with sufficient food for one squirrel is 1, the average distance the squirrel must move in foraging (average distance from the territory center to all points in the territory) is given by

$$\frac{\int_0^1 r'\, 2\pi r'\, dr'}{\int_0^1 2\pi r'\, dr'} = \frac{2}{3}$$

and the area covered is $\pi(1)^2 = \pi$. If two squirrels were to occupy a joint territory, its area would have to be $2\pi$. The radius of this territory must, then, be $r = 1.41$, and the average distance from territory center to all points in the territory is now

$$\frac{\int_0^{1.41} r' 2\pi r' \, dr'}{\int_0^{1.41} 2\pi r' \, dr'} = \frac{2.82}{3}.$$

This distance must be traveled even if one squirrel took one side and the second took the other side of the territory. The perimeter of the joint territory is 1.41 times that of a single territory, and so if the squirrel defended one-half of the perimeter, the other squirrel the other half, some energy and time would be saved through this cooperative effort. But it is doubtful that such perfect cooperation would occur and, even if it were to occur, the energy saved would not compensate for the extra foraging distance needed to acquire food (unless all food energy went into territorial "defense" efforts—which, of course, it doesn't). Joint territories, except in the breeding season, would be disadvantageous to red squirrels unless cooperative "hunting" significantly increased their food intake—which it doesn't. Group territories, except perhaps in such social carnivores as dogs, do not pay energetically and are not likely to occur unless energy deficits are more than compensated by other considerations. Prairie dogs are undoubtedly group territorial because they stay in one place and cannot "defend" the integrity of the group without being territorial. Gibbons and vicuna may stick together over the year to insure successful growth of their offspring. If "defense" of a family territory insures adequate food and if family cohesiveness is advantageous, group territories should exist.

As a closing comment it is interesting that, following the food arguments behind territoriality (8.2.5), "defended" core areas seem to occur in situations where food is evenly distributed and dependable but not otherwise. In primitive man, for example, the Yaghans of Tierra del Fuego live on fish—a clumped, moving source of food—and do not possess a home, core area. The only aggressive, exclusive behavior is exhibited toward non-Yaghans (Service, 1958). Food is also changing in its distribution in the Arctic and in desert regions where hunters must follow the game to survive. Neither the Eskimos nor the Great Basin Sho-shone were (or are) territorial (Sahlins, 1959). On the other hand, in parts of Australia where food is evenly distributed and dependable, the Ona and most Australian aboriginal groups possess group territories. Furthermore, where food is dependable only seasonally, group territories appear only seasonally (Sahlins, 1959).

# Discussion II

## II.1 THE NICHE

There are almost as many definitions of "niche" as there are ecologists. Niches are variously thought of as the habitats occupied by a species, the resources utilized by a species, or the set of all characteristics of a species (behavioral, physiological, etc.). Hutchinson, in his now famous 1957 "Concluding Remarks," introduced the concept of a niche as a geometric representation, in $n$-space, of all conditions in combination under which a species can perpetuate itself. Since species, at least in local populations, do not necessarily utilize all resources they are capable of using, nor live under all conditions they are capable of tolerating, we distinguish between a "fundamental niche" and a "realized niche." This concept of niche has attracted many adherents (perhaps because ecologists, who are historically not mathematically inclined, are impressed with $n$-dimensional spaces), but has yet to prove really useful. The fault, however, may lie not in the concept, but in most ecologists' approach to it. Below is presented a slightly altered concept in somewhat different terms. The different terms may prove to open some heuristic doors (for example, see Chapter 15).

Suppose we define two axes of a Cartesian coordinate system, in the plane of this paper, with two environmental parameters (or stimuli)—say foods $A$ and $B$. The distance along the axes from the origin represents the response to (degree of use of) these stimuli by an individual—or time spent in the situation defined by the parameter, if the parameter does not represent a resource.

For every combination of use of the two parameters, given the same response to other parameters, there will accrue some net benefit to the individuals responding. The net benefit we may define as a net increment in fitness. The net increment in fitness is given by the fitness increment accruing from the use minus the fitness lost due to time and energy spent, or risk involved in the use. If an individual feeds, on a certain day, on one food combination, its fitness will be different than if it had fed on some other combination. The difference defines the difference in fitness increments of the two responses. Possible nonlinear interactions between behavior on this and other days is clearly ignored here for the sake of simplicity. Of course, the benefit accruing from some response combination to the two stimuli will be genotype-specific, so we shall think, for the moment in terms of individuals in a population of one genotype.

If net increment is plotted on a third axis rising up from the plane of the paper, it is seen that the net increments of all combinations of response efforts are described by a surface. This surface is close to the plane of the paper when the net increment is small, but far above it when the net increment is large. Suppose we now cut this *adaptive surface* with a series of planes regularly spaced above and parallel to the surface of this paper. The lines of intersection of these planes and the adaptive surface give isoclines of fitness and, when projected onto the plane of the paper, form a topographic map of fitness increment (Fig. II.1).

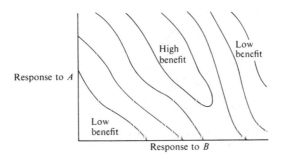

**Fig. II.1.** Topographic map of fitness (the adaptive function).

This is the general case of an adaptive function (Chapter 4).

Let us now construct a fitness set, superimpose it on Fig. II.1 (Fig. II.2), and suppose that the isocline marked $x$ corresponds to the $W = 1$ isocline. We see then that there is a certain range in response under which individuals of the genotype in question can maintain themselves. This range is more clearly shown in Fig. II.3. The upper line is the boundary of the fitness set, the lower line the $W = 1$ isocline. In the shaded area between the lines, $W > 1$, so that a population consisting of individuals of this genotype can afford some sloppiness of response (shaded area below the $W = 1$ isocline) and still maintain itself. The shaded area is the *niche*.

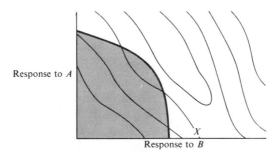

**Figure II.2**

Now, of course, the shaded area in any two dimensions depends on the responses to all other stimuli not depicted in the above figures. Thus we must examine not Fig. II.3 but its equivalent in $n$-space. Each discrete stimulus can be treated as $A$ and $B$, above. Whenever a stimulus can be graded—for example a food item may vary in size, temperature changes will vary in degree—one may depict the stimulus by two axes, one for response effort and one for value along the appropriate gradient scale. Each kind of response to a stimulus calls for another axis. The $n$-dimensional space in which niches lie is called the *niche space*.

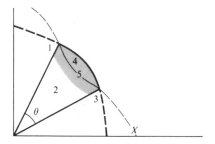

**Fig. II.3.** Fitness set—Adaptive function depiction of the niche.

We now consider the union of niches for all genotypes in a population—i.e. the niche for the entire population in one locality. Note that since selection tends to minimize the frequency of genotypes most divergent from the mean, the upper boundary in Fig. II.3 is now no longer a well-defined line, and the niche becomes a probability cloud, most individuals responding most of the time as depicted by the niche center, and progressively fewer responding progressively less of the time at increasing distances from the center: The probability of observing response 5 (Fig. II.3) is greater than that of observing responses 1, 2, 3, or 4. This probability cloud is the *realized niche*. It's density reflects population distribution. It may be drawn as in Fig. II.3 but with the "boundaries" now denoting the standard deviation in response.

Suppose the population expands. Then there is less food per individual and the net increment of a given response drops due to the increased cause of food acquisition. In addition, the fitness set contracts toward the origin. The realized niche therefore shrinks. (Note that the degree of shrinkage depends on the degree to which food is initially available.) The situation is exactly similar if food becomes generally scarce. If one food (say $B$) becomes more available, the fitness set expands to the right (refer to Fig. II.2) and the net increments of any given response rise. The realized niche thus expands and moves to the right. If food is critical to population maintenance, however, the increased food brings on a population increase, resulting in a general contraction of the fitness set and a drop in fitness increments; the realized niche shrinks again. Note, then, that the size and position of the realized niche in $n$-space varies from locality to locality. The union of all possible realized niches for a species we define as the *fundamental niche*.

In a local population, some genotypes exhibited by the species will undoubtedly be missing and the options for response open or necessary may be less than for the whole species. Thus the realized niche lies within the fundamental niche in all dimensions and is of equal or smaller dimensionality.

Notice that this concept of niche differs from that of Hutchinson in being dynamic. That is, it is a measure of an organism's response to all environmental stimuli. Hutchinson's concept refers to an organism's tolerance to all environmental stimuli.

Three meanings of niche breadth are appropriate. One is simply a measure of range in response and may be given by the variance in $x$, where $x$ is the distance of a given response from the niche center (see 15.3). The range in response relative to the total response, a measure of phenotypic plasticity, may be given by $\theta$, or more practically, by $\mathrm{Var}\,(x)/r$, where $r$ is some measure of the distance of the niche center from the origin. Still another measure of niche breadth, in this case a measure of eurytopy (7.3), refers to the dimensionality of the niche. An appropriate measure is $-\sum_i p_i \log p_i$ (the information theoretic index) where $p_i$ defines the relative mean response to the $i$th niche parameter. For example, if niche center lies at the point $A = 3$, $B = 6$, then $p_A = \frac{1}{3}$, $p_B = \frac{2}{3}$, and niche breadth is

$$-(\tfrac{1}{3} \log \tfrac{1}{3} + \tfrac{2}{3} \log \tfrac{2}{3}).$$

(For a discussion of this measure of niche breadth, see Colwell and Futuyma, 1971). Of course other measures of $\theta$ or dimensionality may be more appropriate under some circumstances. Levins (1968), who seems to be speaking of "niche" in the above sense rather than in Hutchinson's terms, uses either

$$\sum_i 1/p_i^2, \quad \text{or} \quad \exp\left(-\sum_i p_i \log p_i\right)$$

to describe niche breadth in fruit flies. Since the niche parameters are merely different kinds of bait to which the flies are attracted, however, it is not certain whether $\theta$ or dimensionality is being measured.

Note that there must (at least over a wide range of $\theta$) be a positive correlation between the solid angle, $\theta$, and the degree to which the $x_i$'s are equal. Thus there is a positive correlation between the two meanings of niche breadth: genetically and somatically plastic populations will tend to be more eurytopic.

Let us see what else this approach tells us.

**1.** The shape and size of the realized niches vary with the genetic structure of the populations they describe. Since genetic structure of a population may change over the lifetime of any one individual, it is impossible to view the niche as an unchanging entity, even in the very short run.

**2.** Since selection acts to increase fitness, it will, in the absence of environmental change, act to decrease $\theta$: those individuals closest to the number 5 in Figure II.3 pass the most genes. In environments that are variable in time or space with respect to those stimuli, however, genetic variability will be maintained (Chapters 3, 4 and 7), preventing $\theta$ from shrinking beyond some point. Where food is in short supply, the fitness set contracts toward the origin and fitness increments, for any response, fall. Therefore $\theta$ diminishes. Thus, while specialization will develop to a smaller degree when resources are in short supply than when resources are plentiful (7.3), individuals with limited food can less afford to be careless in their response to poor foods, and should follow the optimal feeding model (7.2.1) more closely.

3.    The narrowing of the ($\theta$) niche implies less plasticity of response to changes in the relative abundance of resources. Populations for which food is in short supply should be less plastic in their responses to food change than ecologically and neurologically equivalent populations with an easy food supply.

4.    The adaptive function and fitness set associated with stimuli $A$ and $B$ may vary over time with changes in other stimuli such as temperature, humidity, and amount of shelter. Selection may thus act to narrow $\theta$ with respect to $A$ and $B$ to one range under one set of circumstances, to some other range under other circumstances.

Change in the range of $\theta$ with change in circumstance in one individual is usually accomplished by genetic guidance of physiological and/or learned response changes. That is, selection may favor directed somatic plasticity and directed learning (see below).


## II.2 DIRECTED PLASTICITY AND LEARNING

Point 4 above serves to introduce a subject which still disturbs and confuses a number of diehard, old-school ethologists and psychologists: the nature–nurture controversy. No animal learns without genetic guidance. To do so would be to change phenotype without regard to its consequences for fitness. Most such un-directed changes would be lethal or, at best, deleterious. An animal is constructed in such a way that the chances that what it learns is beneficial are maximized. Dogs learn with great difficulty, if at all, not to chase small animals moving away from them. In nature such animals often are easily caught prey items and it is beneficial to respond by chasing them, detrimental to learn not to chase them. A single bad experience by a dog involved in such a chase should not act as a strong negative reinforcement and natural selection has eliminated those individuals which in the past treated it as such. As stated by Emlen (1968e): "While the social behavior of higher vertebrates may be largely, or almost entirely, based on learning, that learning is to a large extent genetically directed. Evolution has modified the nervous and endocrine systems in such a way that:

1.  Stimulus situations "motivate" animals into certain physiological states;

2.  In response to any one such physiological state, and depending in part on previous experience, certain motor patterns and associations are more easily learned than others; and

3. The positive or negative reinforcing aspect of a given behavior pattern under a given stimulus situation is genetically as well as environmentally determined.

"For example, a rat on a protein-deficient diet learns to show preference for protein foods when these foods are offered. The reinforcement for such learning can only be the sensation associated with proper physiological balance in the body, a sensation to which natural selection could have assigned either a positive or a negative reinforcement valence."

The degree to which behavioral plasticity allows genetically guided learning varies between species and is correlated both with a species' niche breadth and the nature of its habitat. Broad niches allow for a greater range of behavioral responses. A species occupying an unchanging habitat will be maximally fit if little plasticity exists in its responses to stimuli, since such plasticity may occasionally lead to errors in a situation where stereotyped, rigid responses are adequate. In habitats where changes occur, especially unpredictable and thus "unexpected" changes, a species must either live in such a way as to buffer itself from the changes (pick a stable microhabitat and stay there—display a narrow niche with respect to habitat), suffer high mortality, or possess the ability to appropriately vary its responses to such changes. There should be—and is—a high correlation between opportunism (6.3) and rigidity of behavior, long life and behavioral plasticity.

We must be cautious, though, in carrying these arguments too far in application. An animal's behavioral predilections may result from selection pressures of the past and are thus no longer predictable. They may result from selection pressures which recur at wide intervals in time. We cannot assume that an animal doesn't do something simply because, fortuitously, it "enjoys" doing it—the "enjoyment" being a mechanism by which selection may have met quite unrelated needs.

## BIBLIOGRAPHY, SECTION II

Allee, W. C., A. E. Emerson, O. Park, T. Park, and K. P. Schmidt, 1949, *Principles of Animal Ecology*, Saunders, Philadelphia.
Allen, J. A., and B. Clarke, 1968, Evidence for apostatic selection by wild passerines, *Nature* 220: 501–502.
Alpers, F., 1932, Über die Nährungsaufnahme von *Conus meditteraneus* Brug, eines toxoglossen Prosobranchier, *Pubb. Staz. Napoli* 11: 426–445.
Aneshansley, D. J., and T. Eisner, 1969, Biochemistry at 100°C: explosive secretory discharge of bombardier beetles (*Brachinus*), *Science* 165: 61–63.
Armstrong, E. A., 1955, *The Wren*, Collins, London.
Armstrong, J. T., 1965, Breeding home range in the nighthawk and other birds; its evolutionary and ecological significance, *Ecol.* 46: 619–629.
Aronson, L. R., 1965, Environmental stimuli altering the physiological condition of the individual among lower vertebrates, in F. A. Beach (Ed.), *Sex and Behavior*, Wiley, New York.
Asdell, S. A., 1946, *Patterns of Mammalian Reproduction*, Comstock Publication, Vail-Ballon Press, Binghamton, New York.

Ashmole, N. P., 1961, The biology of certain terns, Ph.D. dissertation, Oxford University, Oxford.

Baerands, G. P., and J. M. Baerands van Roon, 1950, An introduction to the study of the ethology of cichlid fishes, *Behaviour suppl.* No. 1, pp. 1–242.

Bakken, A., 1952, Interrelationships of *Sciurus carolinensis* (Gmelin) and *Sciurus niger* (Linnaeus) in mixed populations, Ph.D. thesis, University of Wisconsin, Madison.

Barnes, H., and M. Barnes, 1964, Egg size, nauplius size, and their variation with local geographic and specific factors in some common cirripedes, *J. Anim. Ecol.* **33**: 391–402.

Barnett, S. A., 1953, Problems of food selection by rats, *Anim. Beh.* **1**: 159.

Barnett, S. A., 1963, *The Rat, A Study in Behavior,* Methuen, London.

Bartholomew, G. A., Jr., 1942, The fishing activities of double-crested cormorants on San Francisco bay, *Condor* **44**: 13–21.

Benson, C. W., 1963, The breeding seasons of birds in the Rhodesias and Nyasaland, *Proc. XIII Inter. Ornith. Congress* pp. 623–639.

Bent, A. F., 1939, Life Histories of North American woodpeckers, *U.S. Nat. Mus. Bull.* No. 174.

Bequaert, J., 1922, The predaceous enemies of ants, *Bull. Amer. Mus. Nat. Hist.* **45**: 271–331.

Blair, W. F., 1940a, A study of prairy deer-mouse populations in southern Michigan, *Amer. Midl. Natur.* **24**: 273–305.

Blair, W. F., 1940b, Home ranges and populations of the meadow vole in southern Michigan, *J. Wildl. Mgt.* **4**: 149–161.

Blair, W. F., 1942, Size of home range and notes on the life history of the woodland deer-mouse and eastern chipmunk in northern Michigan, *J. Mammal.* **23**: 27–36.

Blair, W. F., 1947, Estimated frequencies of the buff and gray genes ($G$, $g$) in adjacent populations of deer mice (*Peromyscus maniculatus bairdi*) living on soils of different colors, *Contr. Lab. Vert. Biol., Univ. Mich.* **36**: 1–16.

Blair, W. F., 1951, Population structure, social behavior, and environmental relations in a natural population of the beach mouse, (*Peromyscus polionotus leucocephalus*), *Contr. Lab. Vert. Biol., Univ. Mich.* **48**: 1–47.

Bloedel, P., 1955, Hunting methods of fish-eating bats, particularly *Noctilio leporinus,* *J. Mammal.* **36**: 390–399.

Bourliere, F., 1956, *The Natural History of Mammals,* Knopf, New York.

Bourliere, F., 1961a, Calling and spawning seasons in a mixed population of anurans, *Ecol.* **42**: 99–110.

Bourliere, F., 1961b, Patterns of social grouping among wild primates, in S. L. Washburn (Ed.), *Social Life of Early Man,* Aldine, Chicago.

Bowman, R. I., 1961, Morphological differentiation and adaptation in the Galapagos finches, *Univ. Calif. Publ. Zool.* **58**: 1–326.

Braddock, J. C., 1949, The effect of prior residence upon dominance in the fish, *Platypoecilius maculatus, Physiol. Zool.* **22**: 161–169.

Brenner, F. J., 1967, Seasonal correlations of reserve energy of the redwing blackbird, *Bird Banding* **38**: 195–211.

Brian, M. V., J. Hibble, and A. F. Kelley, 1966, The dispersion of ant species in a southern English heath, *J. Anim. Ecol.* **35**: 281–290.

Brisbin, I. L., Jr., 1969, Bioenergetics of the breeding cycle of the ring dove, *Auk,* **86**: 54–74.

Broadbooks, H. E., 1965, Ecology and distribution of the pikas of Washington and Alaska, *Amer. Midl. Natur.* **73**: 299–335.

Brock, V. E., and R. H. Riffenburg, 1963, Fish schooling: a possible factor in reducing predation, *J. du Conseil, Intern. Council for the Explor. of the Sea* **25**: 307–317.

Brower, J. V. Z., 1960, Effects of mimicry at different proportions of models and mimics, *Amer. Natur.* **94**: 271–282.

Brower, J. V. Z., and L. P. Brower, 1966, Experimental evidence of the effects of mimicry, *Amer. Natur.* **100:** 173–187.

Brower, L. P., 1969, Ecological chemistry, *Sci. Amer.* **220:** 22–29.

Brower, L. P., and J. V. Z. Brower, 1962, The relative abundance of model and mimic butterflies in natural populations of the *Battus philenor* mimicry complex, *Ecol.* **43:** 154–158.

Brower, L. P., and J. V. Z. Brower, 1964, Birds, butterflies and plant poisons; a study in ecological chemistry, *Zoologica,* **49:** 137–159.

Brower, L. P., J. V. Z. Brower, and J. M. Corvino, 1967, Plant poisons in a terrestrial food chain, *Proc. Nat. Acad. Sci.* **57:** 893–898.

Brower, L. P., J. V. Z. Brower, F. G. Stiles, H. J. Croze, and A. S. Hower, 1964, Mimicry: differential advantages of color patterns in the natural environment, *Science* **144:** 183–185.

Brower, L. P., J. V. Z. Brower, and P. W. Westcott, 1960, Experimental studies of mimicry, 5, The reaction of toads (*Bufo terrestris*) to bumblebees (*Bombus americanorum*) and their robberfly mimics (*Mallophora bomhoides*), with a discussion of aggressive mimicry, *Amer. Natur.* **94:** 343–356.

Brower, L. P., L. M. Cook, and H. J. Croze, 1967, Predator response to artificial mimics released in a neotropical environment, *Evol.* **21:** 11–23.

Brower, L. P., W. N. Ryerson, L. L. Coppinger, and S. C. Glazier, 1968, Ecological chemistry of the palatability spectrum, *Science* **161:** 1349–1351.

Brown, J. L., 1963, Aggressiveness, dominance, and social organization in the stellar jay, *Condor,* **65:** 460–484.

Brown, J. L., 1964, The evolution of diversity in avian territorial systems, *Wilson Bull.* **76:** 160–169.

Brown, L. E., 1966, Home range and movements of small mammals, *Symp. Zool. Soc. Lond.* No. 18, pp. 111–142.

Buechner, H. K., 1961, Territorial behavior in Uganda kob, *Science* **133:** 698–699.

Buechner, H. K., J. A. Morrison, and W. Leuthold, 1966, Reproduction in Uganda kob with special reference to behavior, *Symp. Zool. Soc. Lond.* No. 15, pp. 69–88.

Burt, W. H., 1940, Territorial behavior and populations of some small mammals in southern Michigan, *Misc. Publ. Mus. Zool. Mich.* **45:** 1–58.

Burt, W. H., 1943, Territoriality and home range concepts as applied to mammals, *J. Mammal.* **24:** 346–352.

Burt, W. H., and R. P. Grossenheider, 1964, *A Field Guide to the Mammals,* second edition, Houghton Mifflin, Boston.

Bustard, H. R., 1967, Gekkonid lizards adapt fat storage to desert environments, *Science* **158:** 1197–1198.

Butler, C. G., 1967, Insect pheromones, *Biol. Rev.* **42:** 42–87.

Buxton, P. A., 1923, *Animal Life in Deserts,* Edward Arnold, London.

Carleton, W. M., 1965, Food habits of two sympatric Colorado sciurids, *J. Mammal.* **47:** 91–103.

Carlisle, D. B., 1961, Intertidal territory in fish, *Anim. Beh.* **10:** 106–107.

Carpenter, C. R., 1940, A field study in Siam of the behavior and social relations of the gibbon (*Hylobates lar*), *Comp. Psych. Monogr.* **16:** 1–212.

Carr, A., 1952, *Handbook of Turtles,* Comstock, Ithaca, New York.

Carr, A., and L. Giovanelli, 1957, The ecology and migrations of sea turtles, 2, Results of field work in Costa Rica, *Amer. Mus. Novit.* No. 1835, pp. 1–32.

Carrick, R., 1963, Ecological significance of territoriality in the Australian magpie, *Proc. XIII Intern. Ornithol. Cong.* pp. 740–753.

Carriker, M. R., 1961, Comparative and functional morphology of boring mechanisms in gastropods, *Amer. Zool.* **1:** 263–272.

Chan, S. T. H., 1970, Natural sex reversal in vertebrates, *Phil. Trans. Roy. Soc. Lond.* (*B*) **259**: 59–71.

Chesemore, D. L., 1968, Notes on the food habits of arctic foxes in northern Alaska, *Can. J. Zool.* **46**: 1127–1130.

Chitty, D., 1937, A ringing technique for small mammals, *J. Anim. Ecol.* **6**: 36–53.

Clarke, T. A., 1971, Territory boundaries, courtship, and social behavior in the garibaldi, *Hypsypops rubicunda* (Pomacentridae), *Copeia* pp. 295–299.

Cody, M. L., 1966, A general theory of clutch size, *Evol.* **20**: 174–184.

Cole, L. C., 1954, The population consequences of life history phenomena, *Quart. Rev. Biol.* **29**: 103–137.

Colwell, R. K., and D. K. Futuyma, 1971, On the measurement of niche breadth and overlap, *Ecol.* **52**: 567–576.

Conaway, C. H., and G. B. Koford, 1965, Estrous cycles and mating behavior in a free-ranging land of rhesus monkeys, *J. Mammal.* **45**: 577–588.

Conder, P. J., 1948, The breeding biology and behavior of the continental goldfinch, *Carduelis carduelis carduelis, Ibis* **90**: 493–525.

Connell, J. H., 1963, Territorial behavior and dispersion in some marine invertebrates, *Res. Pop. Ecol.* **2**: 87–101.

Cook, L. M., L. P. Brower, and J. Alcock, 1969, An attempt to verify mimetic advantage in a neotropical environment, *Evol.* **23**: 339–345.

Coppinger, R. P., 1969, The effect of experience and novelty on avian feeding behavior with reference to the evolution of warning coloration in butterflies, I: reactions of wild-caught adult blue jays to novel insects, *Behaviour* **35**: 45–60.

Coppinger, R. P., 1970, The effect of experience and novelty on avian feeding behavior with reference to the evolution of warning coloration in butterflies, II: reactions of naive birds to novel insects, *Amer. Natur.* **104**: 323–335.

Cott, H. B., 1940, *Adaptive Coloration in Animals,* Methuen, London.

Coulson, J. C., 1966, The influence of the pair-bond and age on the breeding biology of the kittiwake gull, *Rissa tridactyla, J. Anim. Ecol.* **35**: 269–279.

Coulson, J. C., and E. White, 1956, A study of the kittiwake, *Rissa tridactyla, Ibis* **98**: 63–79.

Coulson, J. C., and E. White, 1958, The effect of age on the breeding biology of the kittiwake, *Rissa tridactyla, Ibis* **100**: 40–51.

Craig, J. V., L. L. Ortman, and A. M. Guhl, 1965, Genetic selection for social dominance in chickens, *Anim. Beh.* **13**: 114–131.

Crisp, D. J., 1961, Territorial behavior in barnacle settlement, *J. Exper. Biol.* **38**: 429–446.

Crook, J. H., 1964, The evolution of social organization and visual communication in the weaver birds (*Ploceinae*), *Beh. suppl.* No. 10.

Crook, J. H., 1965, The adaptive significance of avian social organization, *Symp. Zool. Soc. Lond.* **14**: 181–218.

Darling, F., 1937, *A Herd of Red Deer,* Amer. Mus. Nat. Hist. (1964), Garden City, New York.

Darling, F., 1938, *Bird Flocks and the Breeding Cycle,* Cambridge University Press, Cambridge, England.

Dasmann, R. F., and R. D. Taber, 1956, Behavior of Columbian black tailed deer with reference to population ecology, *J. Mammal.* **37**: 143–164.

Davis, D. E., 1940, Social nesting habits of the smooth-billed Ani, *Auk* **57**: 179–218.

Davis, D. E., 1945, The home range of some Brazilian mammals, *J. Mammal.* **26**: 119–127.

Davis, D. E., 1952, Social behavior and reproduction, *Auk* **69**: 171–182.

Davis, D. E., and J. T. Emlen, 1948, Studies on home range in the brown rat, *J. Mammal.* **29**: 207–225.

De Ruiter, L., 1952, Some observations on the camouflage of stick caterpillars, *Behaviour* **4**: 222–232.

DeVore, I., 1965a, Male dominance and mating behavior in baboons, in F. A. Beach (Ed.), *Sex and Behavior,* Wiley, New York.

DeVore, I., 1965b, *Primate Behavior: Field Studies of Monkeys and Apes,* Holt, Rinehart, and Winston, New York.

Dice, L. R., 1947, Effectiveness of selection by owls of deer mice (*Peromyscus maniculatus*) which contrast in color with their background, *Cont. Lab. Vert. Biol. Mich.* **34**: 1–20.

Dunaway, P. B., 1968, Life history and populational aspects of the eastern harvest mouse, *Amer. Midl. Natur.* **79**: 48–67.

Duncan, C. J., and P. M. Sheppard, 1965, Sensory discrimination and its role in the evolution of Batesian mimicry, *Behaviour* **24**: 269–282.

Dunning, D. C., 1968, Warning sounds of moths, *Z. Tierpsychol.* **25**: 129–138.

Ehrlich, P. R., and P. H. Raven, 1965, Butterflies and plants: a study in coevolution, *Evol.* **18**: 586–608.

Eisenberg, J. F., 1966, The social organizations of mammals, *Handbuch der Zoologie* **8**: 1–92.

Eisner, T., and J. A. Davis, 1967, Mongoose throwing and smashing millipedes, *Science* **155**: 577–579.

Ellis, R. J., 1964, Tracking raccoons by radar, *J. Wildl. Mgt.* **28**: 363–368.

Ely, C. A., 1957, Comparative success of certain south-central Oklahoma birds, Unpublished MS., University Oklahoma, Norman.

Emerson, A. E., 1958, The evolution of behavior among social insects, in A. Roe and G. G. Simpson (Eds.), *Behavior and Evolution,* Yale University Press, New Haven.

Emlen, J. M., 1966a, The role of time and energy in food preference, *Amer. Natur.* **100**: 611–617.

Emlen, J. M., 1966b, Time, energy, and risk in two species of carnivorous gastropods, Unpublished, Ph.D. thesis, University of Washington, Seattle.

Emlen, J. M., 1968a, Optimal choice in animals, *Amer. Natur.* **102**: 385–389.

Emlen, J. M., 1968b, Batesian mimicry: a preliminary investigation of theoretical aspects, *Amer. Natur.* **102**: 235–241.

Emlen, J. M., 1968c, A note on natural selection and the sex ratio, *Amer. Natur.* **102**: 95–95.

Emlen, J. M., 1968d, Selection for the sex ratio, *Amer. Natur.* **102**: 589–591.

Emlen, J. M., 1968e, Biological and cultural determinants in human behavior, *Amer. Anthrop.* **69**: 513–514.

Emlen, J. T., 1952, Flocking behavior in birds, *Auk* **69**: 169–170.

Emlen, J. T., 1957, Defended area? A critique of the territory concept and of conventional thinking, *Ibis* **99**: 352.

Emlen, J. T., and R. W. Lorenz, 1942, Pairing responses of free-living valley quail to sex hormone pellet implants, *Auk* **59**: 369–378.

Emlen, S. T., 1968, Territoriality in the bullfrog, *Rana catesbiana, Copeia* pp. 240–243.

Emlen, S. T., and H. W. Ambrose, III: 1970, Feeding interactions of snowy egrets and red-breasted mergansers, *Auk* **87**: 164–165.

Erickson, A. B., 1949, Summer populations and movements of the cotton rat and other rodents on the savannah River refuge, *J. Mammal.* **30**: 133–140.

Erlinge, S., 1968, Territoriality of the otter, *Lutra lutra L., Oikos* **19**: 81–98.

Feder, H. M., 1963, Gastropod defensive responses and their effectiveness in reducing predation by starfishes, *Ecol.* **44**: 505–512.

Fink, B., 1959, Observations of porpoise predation on a school of Pacific sardines, *Calif. Fish and Game Bull.* **45**: 216–217.

Fishelson, L., 1970, Protogynous sex reversal in the fish *Anthias squamipinnis* (Teleostei, Anthiidae) regulated by the presence or absence of a male fish, *Nature* **227**: 90–91.

Fisher, R. A., 1958a, *The Genetical Theory of Natural Selection,* second edition, Dover, New York.

Fisher, R. A., 1958b, Polymorphism and natural selection, *J. Ecol.* **46**: 289–293.

Fisher–Piette, E., 1935, Histoire d'une mouliere, *Bull. Biol.* **69**: 154–180.

Fisler, G. F., 1962, Homing in the California vole, *Microtus californicus, Amer. Midl. Natur.* **68**: 357–368.

Fitch, H. S., F. Swenson, and D. F. Tillotson, 1946, Behavior and food habits of the red-tailed hawk, *Condor* **48**: 205–237.

Ford, E. B., 1960, *Ecological Genetics,* Broadwater Press, Welwyn, Garden City.

Friedmann, H., 1967, Avian symbiosis, in S. M. Henry (Ed.), *Symbiosis,* Vol. II, Academic Press, New York.

Frings, H., and M. Frings, 1964, *Animal Communication,* Blaisdell, New York.

Fry, C. H., 1969, The recognition and treatment of venomous and non-venomous insects by small bee-eaters, *Ibis* **111**: 23–29.

Gershenson, S., 1928, A new sex ratio abnormality in *Drosophila pseudoobscura, Genetics* **13**: 488–507.

Gerstell, R., 1939, Certain mechanics of winter quail losses revealed by laboratory experimentation, *Trans. 4th North American Wildl. Conf.* pp. 462–467.

Ghent, A. W., 1960, A study of the group-feeding behavior of larvae of the jack-pine sawfly (*Neodiprion pratti banksiannae* Roh), *Behaviour* **16**: 110–148.

Gibb, J., 1956, Food, feeding habits, and territoriality of the rock pipit (*Anthus spinoletta*), *Ibis* **98**: 506–530.

Gibb, J., 1960, Populations of tits and goldcrests and their food supply in pine plantations, *Ibis* **102**: 163–208.

Gibb, J., 1966, Tit predation and the abundance of *Ernarmonia conicolana* (Heyl) on Weeting Heath, Norfolk, 1962–63, *J. Anim. Ecol.* **35**: 43–53.

Gifford, E. W., 1919, Field notes on the land birds of the Galapagos Islands and of Cocos Island, Costa Rica, *Proc. Calif. Acad. Sci.* series 4, **2**: 189–258.

Ginsburg, B., and W. Allee, 1942, Some effects of conditioning on social dominance and subordination in inbred strains of mice, *Physiol. Zool.* **15**: 485–506.

Goodwin, D., 1956, Further observations on the behavior of the jay *Garrulus glandarius, Ibis* **98**: 186–219.

Gordon, K., 1936, Territorial behavior and social dominance among *Sciuridae, J. Mammal.* **17**: 171–172.

Gullion, G. W., 1953, Territorial behavior of the American coot, *Condor* **55**: 169–186.

Haftorn, S., 1933, Contributions to the food biology of tits, part I. The crested tit (*parus c. cristatus* L.), *Det. Kgl. Norske Videnskabers selskabs Skrifter,* **4**: 1–124.

Hahn, W. E., and D. W. Tinkle, 1965, Fat body cycling and experimental evidence for its adaptive significance to ovarian follicle development in the lizard *Uta stansburiana, J. Exper. Zool.* **158**: 79–86.

Hairston, N., F. E. Smith, and L. B. Slobodkin, 1960, Community structure, population control, and competition, *Amer. Natur.* **94**: 421–425.

Hall, K. R. L., 1962, The sexual, agonistic, and derived social behavior patterns of the wild chacma baboon, *Papio ursinus, Proc. Zool. Soc. Lond.* **139**: 283–328.

Hall, K. R. L., and G. B. Schaller, 1964, Tool-using behavior of the California sea otter, *J. Mammal.* **45**: 287–298.

Hamilton, T. H., and G. H. Orians, 1965, The evolution of brood parasitism in altricial birds, *Condor* **67**: 361–382.

Hamilton, W. J., 1937, Activity and home range of the field mouse (*Microtus pennsylvanicus*), *Ecol.* **18**: 255–263.

Hamilton, W. J., III, W. M. Gilbert, F. H. Heppner, and R. Planck, 1967, Starling roost dispersal and a hypothetical mechanism regulating rhythmical and animal movement to and from dispersal centers, *Ecol.* **48**: 825–833.

Harrison, J. L., 1958, Range of movements of some Malayan rats, *J. Mammal.* **38**: 190–206.

Haskins, C. P., and E. F. Haskins, 1950, Notes on the biology and social behavior of the archaic ponerine ants of the genera *Myrmecia* and *Promyrmecia, Ann. Ent. Soc. Amer.* **43**: 461–491.

Hatch, J. J., 1966, Collective territories in Galapogos mocking birds, with notes on other behavior, *Wilson Bull.* **78**: 198–207.

Hayne, D. W., 1949, Calculation of the size of home range, *J. Mammal.* **30**: 1–18.

Heatwole, H., 1968, Relationship of escape behavior and camouflage in anoline lizards, *Copeia* pp. 109–113.

Herter, K., 1929, Vergleichende bewegungsphysiologische studien an deutschen Egeln, *Z. Vergl. Physiol.* **9**: 145–177.

Heydweiler, A. M., 1935, A comparison of winter and summer territories and seasonal variations of the tree sparrow, (*Spizella a. arborea), Bird Banding* **6**: 1–11.

Hiernaux, J., 1963, Some ecological factors affecting human populations of sub-Saharan Africa, in F. C. Howell and F. Bourliere (Eds.), *African Ecology and Human Evolution,* Wenner–Gren Foundation, Aldine, Chicago.

Hoddenbach, G. A., and F. B. Turner, 1968, Clutch size of the lizard *Uta stansburiana* in southern Nevada, *Amer. Midl. Natur.* **80**: 262–265.

Holcomb, L. C., and G. Twiest, 1968, Ecological factors affecting nest building in red-wing blackbirds, *Bird Banding* **39**: 14–22.

Holling, C. S., 1965, The functional response of predators to prey density and its role in mimicry and population regulation, *Mem. Ent. Soc. Can.* No. 45, pp. 5–60.

Holling, C. S., 1966, The functional response of invertebrate predators to prey density, *Mem. Ent. Soc. Can.* No. 48.

Horn, H. S., 1966, Colonial nesting in the brewers blackbird (*Euphagus cyanocephalus*) and its adaptive significance, Ph.D. thesis, University of Washington, Seattle.

Horn, H. S., 1968, The adaptive significance of colonial nesting in the brewers blackbird (*Euphagus cyanocephalus), Ecol.* **49**: 682–694.

Hornocker, M. G., 1970, An analysis of mountain lion predation upon mule deer and Elk in the Idaho primitive area, *Wildl. Monog.* No. 21.

Horvath, O., 1964, Seasonal differences in rufous hummingbird nest height and their relation to nest climate, *Ecol.* **45**: 235–241.

Hubbs, C., 1958, Geographic variation in egg complement of *Percina caprodes* and *Etheostoma spectabile, Copeia* pp. 102–105.

Hunt, E. E., Jr., W. A. Lessa, and A. Hicking, 1965, The sex ratio of live births in three Pacific Island populations, *Hum. Biol.* **37**: 148–155.

Hutchinson, G. E., 1957, Concluding remarks, *Cold Spring Harbor Symposium Quant. Biol.* **22**: 415–427.

Iersel, J. van, 1953, An analysis of the parental behavior of the male three-spined stickleback, *Behaviour suppl.* No. 3, pp. 1–159.

Inger, R. F., and T. P. Bacon, Jr., 1968, Annual reproduction and clutch size in rain forest frogs from Sarawak, *Copeia* pp. 602–606.

Inger, R. F., and B. Greenberg, 1966, Annual reproductive patterns of lizards from a Bornean rain forest, *Ecol.* **47**: 1007–1021.

Ivlev, V. S., 1961, *Experimental Ecology of the Feeding of Fishes,* Yale University Press, New Haven.

Janzen, D. H., 1969, Seed-eaters versus seed size, number toxicity and dispersal, *Evol.* **23**: 1–27.

Jenkins, O., 1948, A population study of meadow mice (*Microtus*) in three Sierra Nevada meadows, *Proc. Calif. Acad. Sci.* (4th series) **26**: 43–67.

Jewell, P. A., 1966, The concept of home range in mammals, *Symp. Zool. Soc. Lond.* No. 18, pp. 85–109.

Johnston, R. F., 1960, Variation in breeding season and clutch size in song sparrows of the Pacific coast, *Condor* **56**: 268–273.

Jolly, A., 1967, Breeding synchrony in wild *Lemur Catta.* in S. A. Altmann (Ed.), *Social Communication Among Primates,* University of Chicago Press, Chicago.

Jordan, P. A., P. C. Shelton, and D. L. Allen, 1967, Numbers, turnover, and social structure of the Isle Royale wolf population, *Amer. Zool.* **7**: 233–252.

Jorgenson, C. D., and W. W. Tanner, 1963, The application of the density-probability function to determine the home ranges of *Uta stansburiana stansburiana* and *Cnemidophorus tigris tigris, Herpetol.* **19**: 105–115.

Joslin, J., H. Fletcher, and J. T. Emlen, 1964, A comparison of the responses to snakes of laboratory and wild-reared rhesus monkeys, *Anim. Beh.* **12**: 348–352.

Judy, K. J., 1969, Toxin from skin of frogs of the genus *Atelopus:* Differentiation from dendrobatid toxins, *Science* **165**: 1376–1377.

Kaston, B. J., 1965, Some little-known aspects of spider behavior, *Amer. Midl. Natur.* **73**: 336–356.

Kawai, M., 1965, Newly acquired pre-cultural behavior of the natural group of Japanese macaques on the Koshima inlet, *Primates* **6**: 1–30.

Kaye, S. V., 1961, Movements of harvest mice tagged with gold-198, *J. Mammal.* **42**: 323–337.

Kendeigh, S. C., 1970, Energy requirements for existence in relation to size of bird, *Condor* **72**: 60–65.

King, J., 1955, Social behavior, social organization and population dynamics in a black-tailed prairie dog town in the Black Hills of South Dakota, *Contr. Lab. Vert. Biol. Univ. Mich.* No. 67.

Kitchell, J. F., and J. T. Windell, 1969, Nutritional value of algae to bluegill sunfish, *Lepomis macrochirus, Copeia* pp. 186–189.

Knight–Jones, E. W., 1953, Laboratory experiments on gregariousness during settling in Balanus balanoides and other barnacles, *J. Exper. Biol.* **30**: 584–598.

Koford, C. B., 1957, The vicuna and the puna, *Ecol. Monog.* **27**: 153–219.

Kohn, A. J., 1961, Chemoreception in gastropod mollusks, *Amer. Zool.* **1**: 291–308.

Kohn, A. J., and V. Waters, 1966, Escape responses of three herbivorous gastropods to the predatory gastropod, *Conus textile, Anim. Beh.* **14**: 340–345.

Koskimies, J., 1950, The life of the swift, *Micropus apus* (*L*) in relation to the weather, *Annales Academide Scientiarum Fennicae,* series A, IV, *Biologica* **12**: 1–151.

Kruuk, H., 1964, Predators and anti-predator behavior of the black-headed gull (*Larus ridibundus* L.), *Behaviour suppl.* **11**: 1–130.

Lack, D., 1943, *The Life of the Robin,* Cambridge University Press, London.

Lack, D., 1945, The Galapagos finches (*Geospizinae*): a study in variation, *Occas. Papers,* No. 21, *Calif. Acad. Sci.*

Lack, D., 1948a, The significance of clutch size, part III. Some interspecific comparisons, *Ibis* **90**: 24–45.

Lack, D., 1948b, Selection and family size in starlings, *Evol.* **2**: 95–110.

Lack, D., 1954, *The Natural Regulation of Animal Numbers,* Oxford University Press, London.

Lack, D., 1966, *Population Studies of Birds,* Clarendon Press, Oxford.

Lack, D., 1968, *Ecological Adaptations for Breeding in Birds,* Methuen, London.

Lack, D., and R. E. Moreau, 1965, Clutch size in tropical passerine birds of forest and savanna, *L'oiseau et la Revue Francaise d'Ornithologie* **35**: 76–89.

Landenberger, D. E., 1966, Learning in the Pacific starfish, *Pisaster giganteus, Anim. Beh.* **14:** 414–418.

Landenberger, D. E., 1968, Studies on selective feeding in the Pacific starfish, *Pisaster*, in southern California, *Ecol.* **49:** 1062–1075.

Lanyon, W. E., 1956, Territory in the meadowlarks, genus *Sturnella, Ibis* **98:** 485–489.

Lavenda, N., 1949, Sexual differences in normal protogynous hermaphroditism in the Atlantic sea bass, *Centropristes striatus, Copeia* pp. 185–194.

LeBeouf, B. J., and R. S. Peterson, 1969, Social status and mating activity in elephant seals, *Science* **163:** 91–93.

Leigh, E. G., Jr., 1970, Sex ratio and differential mortality between the sexes, *Amer. Natur.* **104:** 205–210.

Lesiewski, R. C., and W. R. Dawson, 1967, A reexamination of the relation between standard metabolic rate and body weight in birds, *Condor* **69:** 13–23.

Levin, D. A., 1971, Plant phenolics: an ecological perspective, *Amer. Natur.* **105:** 157–182.

Levins, R., 1968, *Evolution in changing environments,* Princeton University Press, Princeton.

Lin, N., 1963, Territorial behavior in the cicada killer wasp *Sphecius speciosus* (Drury) (Hymenoptera: Sphecidae), *Behaviour* **20:** 115–133.

Lloyd, M., 1964, Mean crowding, *J. Anim. Ecol.* **36:** 1–30.

Lloyd, M., and H. S. Dybas, 1966, The periodical cicada problem, I, Population ecology, *Evol.* **20:** 133–149.

Lockie, J. D., The food of the pine marten, *Martes martes,* in west Ross-shire, Scotland, *Proc. Zool. Soc. Lond.* **136:** 187–195.

Loschiavo, S. R., 1968, Effect of oviposition on egg production and longevity in *Trogoderma parabile* (*Coleoptera: Dermestidae*), *Canad. Ent.* **100:** 86–89.

Lott, D., S. D. Scholz, and D. S. Lehrman, 1967, Exteroceptive stimulation of the reproductive system of the female ring dove (*Streptopelia risoria*) by the mate and by the colony milieu, *Anim. Beh.* **15:** 433–437.

MacArthur, R. H., 1958, Population ecology of some warblers in northeastern coniferous forests, *Ecol.* **39:** 599–619.

MacArthur, R. H., 1965, Ecological consequences of natural selection, in T. H. Waterman and H. J. Morowitz (Eds.), *Theoretical and Mathematical Biology,* Blaisdell, New York.

MacRoberts, M. H., 1970, Notes on the food habits and food defense of the acorn woodpecker, *Condor* **72:** 196–204.

Marsden, H. M., 1968, Agonistic behavior of young rhesus monkeys after changes induced in social rank of their mothers, *Anim. Beh.* **16:** 38–44.

Marsden, H. M., and N. R. Holler, 1964, Social behavior in confined populations of the cottontail and the swamp rabbit, *Wildl. Monog.* No. 13.

Marshall, A. J., 1959, Internal and environmental control of breeding, *Ibis* **101:** 456–478.

Matthews, G. V. T., 1955, Animal migration, *Nature* **176:** 772–773.

Mauzey, K. P., 1966, Feeding behavior and reproductive cycles in *Pisaster ochraceus, Biol. Bull.* **131:** 127–144.

Mayfield, H., 1965, Chance distribution of cowbird eggs, *Condor* **67:** 257–263.

McFarland, W. N., and S. A. Moss, 1967, Internal behavior in fish schools, *Science* **156:** 260–262.

McLaren, I. A., 1966, Adaptive significance of large size and long life of the chaetognath, *Sagitta elegans,* in the Arctic, *Ecol.* **47:** 852–855.

McNab, B., 1963, Bioenergetics and the determination of home range size, *Amer. Natur.* **97:** 133–140.

McWhinnie, M. A., and R. Johanneck, 1966, Utilization of inorganic and organic compounds by Antarctic zooplankton, *Antarct. J. U.S.* **1:** 210.

Meanley, B., and J. S. Webb, 1963, Nesting ecology and reproductive rate of the red-winged blackbird in tidal marshes of the upper Chesapeake Bay region, *Chesapeake Sci.* **4:** 90–100.

Mech, L. D., L. D. Frenzel, Jr., R. R. Ream, and J. W. Winship, 1971, Movements, behavior, and ecology of timber wolves in northeastern Minnesota, in L. D. Mech, and L. D. Frenzel, Jr., (Eds.), *Ecological Studies of the Timber Wolf in Northeastern Minnesota,* USDA Forest Service Res. Paper NC–52.

Metzgar, L. H., 1968, An experimental comparison of screech owl predation on resident and transient white-footed mice (*Peromyscus leucopus*), *J. Mammal.* **48:** 387–391.

Miller, R. S., 1958, A study of a woodmouse population in Wytham Woods, Berkshire, *J. Mammal.* **39:** 477–493.

Moffit, J., and C. Cottam, 1941, The eel grass blight and its effect on brant, *U.S, Fish and Wildlife Service Leaflet* **204:** 1–26.

Moment, G. B., 1962, Reflexive selection: a possible answer to an old puzzle, *Science* **136:** 262–263.

Moore, I. A., 1942, The role of temperature in speciation of frogs, *Biol. Symp.* **6:** 189–213.

Moore, I. A., 1949, Geographic variation of adaptive characters in *Rana pipiens, Evol.* **3:** 1–24.

Moore, N. W., 1964, Intra and interspecific competition among dragonflies, *J. Anim. Ecol.* **33:** 49–71.

Moreau, R. E., 1944, Clutch size: a comparative study with special reference to African birds, *Ibis* **86:** 286–347.

Moreau, R. E., 1950, The breeding seasons of African birds, I. Land birds, *Ibis* **92:** 223–267.

Morgan, T. H., C. B. Bridges, and A. H. Sturtevant, 1925. The genetics of *Drosophila, Bibl. Genet.* **2:** 1–262.

Moynihan, M., 1955, Some aspects of reproductive behavior in the black-headed gull (*Larus ridibundus r.* L.) and related species, *Behaviour suppl.* No. 4.

Mutere, F. A., 1967, The breeding biology of equatorial vertebrates: reproduction in the fruit bat (*Eidolon helvum*) at latitude 0° 20′ N., *J. Zool.* (Lond.) **153:** 153–161.

Mykytowycz, R., 1968, Territorial marking in rabbits, *Sci. Amer.* **218:** 116–126.

Myton, B. A., and R. W. Ficken, 1967, Seed size preference in chicadees and tits in relation to ambient temperature, *Wilson Bull.* **79:** 319–321.

Neill, W. T., 1964, Viviparity in snakes, *Amer. Natur.* **98:** 35–55.

Newton, I., 1966, Fluctuations in the weights of bullfinches, *Brit. Birds,* **59:** 89–100.

Newton, I., 1967, The adaptive radiation and feeding ecology of some British finches, *Ibis* **109:** 33–98.

Nice, M. M., 1937, Studies in the life history of the song sparrow, I, *Trans Linn. Soc. N.Y.* **4:** 1–247.

Nice, M. M., 1938, The biological significance of bird weights, *Bird Banding* **9:** 1–11.

Nice, M. M., 1957, Nesting success in altricial birds, *Auk* **74:** 305–321.

Nikolskii, G. V., 1963, *The Ecology of Fishes,* Academic Press, London, New York.

Noble, G. K., and B. Curtis, 1939, The social behavior of the jewel fish, *Hemichromis bimaculatus* (Gill), *Bull. Amer. Mus. Nat. Hist.* **76:** 1–46.

Novitsky, E., 1947, Genetic analysis of an anomalous sex ratio condition in *Drosophila affinis, Genetics* **32:** 526–534.

Odum, E. P., 1941, Annual cycle of the black-capped chicadee, *Auk* **58:** 314–333.

Oliver, J. A., 1955, *North American Amphibians and Reptiles,* Van Nostrand, Princeton, New Jersey.

Orians, G. H., 1966, Food of nestling yellow-headed blackbirds, Caribou Parklands, British Columbia, *Condor* **68:** 321–337.

Orians, G. H., 1970, On the evolution of mating systems in birds and mammals, *Amer. Natur.* **103**: 589–603.

Orians, G. H., and M. F. Willson, 1964, Interspecific territories of birds, *Ecol.* **45**: 736–745.

Paine, R. T., 1963, Food recognition and predation on opisthobranchs by *Navanax inermis, Veliger* **6**: 1–8.

Paine, R. T., 1965, Natural history, limiting factors, and energetics of the opisthobranch, *Navanax inermis, Ecol.* **46**: 603–619.

Paine, R. T., 1966, Function of labial spines, composition of diet, and size of certain marine gastropods, *Veliger* **9**: 17–24.

Paine, R. T., 1969, The pisaster-tegula interaction, prey patches, predator food preferences, and intertidal community structure, *Ecol.* **50**: 950–961.

Paine, R. T., and R. L. Vadas, 1969, Caloric values of benthic marine algae and their relation to invertebrate food preferences, *Marine Biol.* **4**: 79–86.

Patterson, I. J., 1965, Timing and spacing of broods in the black-headed gull, *Larus ridibundus, Ibis* **107**: 433–459.

Patterson, J. T., 1913, Polyembryonic development in *Tatusia novemcincta, J. Morphol.* **24**: 559–684.

Payne, R. B., 1965a, Clutch size and numbers of eggs laid by brown-headed cowbirds, *Condor* **67**: 44–60.

Payne, R. B., 1965b, The molt of breeding Cassin's auklets, *Condor* **67**: 220–228.

Payne, R. B., 1967, Interspecific communication signals in parasitic birds, *Amer. Natur.* **101**: 363–375.

Pearse, A. S., 1924, Amount of food eaten by four species of freshwater fishes, *Ecol.* **5**: 254–258.

Pearson, O., 1954, The daily energy requirement of a wild anna hummingbird, *Condor* **56**: 317–322.

Perrins, C. M., 1965, Population fluctuations and clutch size in the great tit, *Parus major, J. Anim. Ecol.* **34**: 601–647.

Perrins, C. M., 1970, The timing of birds' breeding seasons, *Ibis* **112**: 242–255.

Petersen, B., 1964, Monarch butterflies are eaten by birds, *J. Lepid. Soc.* **18**: 165–169

Phillips, C. L., 1887, Egg-laying extraordinary in *Colaptes auratus, Auk* **4**: 346.

Pitelka, F. A., 1942, Territoriality and related problems in North American hummingbirds, *Condor* **44**: 189–204.

Pitelka, F. A., 1951, Ecological overlap and interspecific strife in breeding populations of Anna and Allens hummingbirds, *Ecol.* **32**: 641–661.

Popham, F. J., 1941, The variation in the color of certain species of *Arctocorisa* (Hemiptera, Coroxidae) and its significance, *Proc. Zool. Soc. Lond .A,* **111.** 135–172.

Portmann, A., 1959, *Animal Camouflage,* University of Michigan Press, Ann Arbor.

Preston, F. W., 1948, the cowbird (*M. ater*) and the cuckoo (*C. canorus*), *Ecol.* **29**: 115–116.

Prop, N., 1960, Protection against birds and parasites in some species of tenthredinid larvae, *Archiv. Neerl. de Zool.* **13**: 380–447.

Prosser, C. L., and F. A. Brown, Jr., 1961, *Comparative Animal Physiology,* Saunders, Philadelphia and London.

Rand, A. S., 1968, A nesting aggregation of iguanas, *Copeia* pp. 552–561.

Richter, C. P., 1942, Total self-regulatory function in animals and human beings, *Harvey Lect.* series **38**: 63–103.

Richter, C. P., 1953, Alcohol, beer, and wine as foods, *Quart. J. Stud. Alcohol* **14**: 525–539.

Richter, C. P., L. E. Holt, Jr., and B. Barelane Jr., 1938, Normal growth and reproduction in rats studied by the self-selection method, *Amer. J. Physiol.* **122**: 734–744.

Ricker, W. E., 1946, Production and utilization of fish populations, *Ecol. Monog.* **16**: 373–391.

Ricklefs, R., 1968, Patterns of growth in birds, *Ibis* **110**: 419–445.

Robel, R. J., 1966, Booming territory size and mating success of the greater prairie chicken (*Tympanuchus cupido pinnatus*), *Anim. Beh.* **14**: 328–331.

Rodgers, W. L., 1967, Specificity of specific hungers, *J. Comp. Physiol. Psychol.* **64**: 49–88.

Rodgers, W. L., and P. Rozin, 1966, Novelty food preferences in thiamine-deficient rats, *J. Comp. Physiol. Psychol.* **61**: 1–4.

Roeder, K. D., 1962, The behavior of free-flying moths in the presence of artificial, ultrasonic pulses, *Anim. Beh.* **10**: 300–304.

Roeder, K. D., 1966, Acoustic sensitivity of the noctuid tympanic organ and its range for the cries of bats, *J. Insect Physiol.* **12**: 843–859.

Rowell, T. E., 1967, A quantitative comparison of the behavior of a wild and a caged baboon group, *Anim. Beh.* **15**: 499–509.

Sabine, W. S., 1949, Dominance in winter flocks of juncos and tree sparrows, *Physiol. Zool.* **22**: 68–85.

Sabine, W. S., 1955, The winter society of the Oregon junco: the flock, *Condor* **57**: 88–111.

Sabine, W. S., 1956, Integrating mechanisms of winter flocks of juncos, *Condor,* **58**: 338–341.

Sade, D. S., 1967, Determinants of dominance in a group of free-ranging rhesus monkeys, in S. A. Altmann (Ed.), *Social Communication Among Primates,* Chicago University Press, Chicago.

Sahlins, M., 1959, The social life of monkeys, apes, and primitive men, in Spuhler, J. N. (Ed.), *The Evolution of Man's Capacity for Culture,* Wayne State University Press, Detroit.

Salisbury, E. J., 1942, *The Reproductive Capacities of Plants,* Bell, London.

Salomon, A. L., and M. W. Schein, 1965, The effect of social rank on the incidence of crowing in cockerels (abstract), *Amer. Zool.* **5**: 209.

Salthe, S. N., 1969, Reproductive modes and the number and size of ova in the urodeles, *Amer. Midl. Natur.* **81**: 467–490.

Sandler, L., and E. Novitsky, 1957, Meiotic drive as an evolutionary force, *Amer. Natur.* **91**: 105–110.

Schaller, G. B., 1963, *The Mountain Gorilla,* University of Chicago Press.

Schaller, G. B., 1967, *The Deer and the Tiger,* University of Chicago Press.

Schmidt-Nielson, K., and B. Schmidt-Nielson, 1952, Water metabolism of desert animals, *Physiol. Rev.* **32**: 135–166.

Schoener, A., 1968, Evidencef or reproductive periodicity in the deep sea, *Ecol* **49**: 81–87.

Schoener, T. W., 1968, Sizes of feeding territories among birds, *Ecol.* **49**: 123–141.

Sealander, J. A., 1952, The relationship of nest protection and huddling to survival of *Peromyscus* at low temperatures, *Ecol.* **33**: 63–71.

Selander, R. K., 1960, Sex ratio of nestlings and clutch size in the boat-tailed grackle, *Condor* **62**: 34–44.

Selander, R. K., 1961, Supplemental data on the sex ratio of nestling boat-tailed grackles, *Condor* **63**: 504.

Selander, R. K., 1965, On mating systems and sexual selection, *Amer. Natur.* **99**: 129–141.

Selander, R. K., and D. R. Giller, 1959, Interspecific relationships of woodpeckers in Texas, *Wilson Bull.* **71**: 107–124.

Service, E. R., 1958, *A Profile of Primitive Culture,* Harper, New York.

Sexton, O. J., 1960, Experimental studies of artificial batesian mimics, *Beh.* **15**: 244–252.

Sheppard, P. M., 1959, The evolution of mimicry: a problem in ecology and genetics, *Cold Spring Harbor Symposium Quant. Biol.* **24**: 131–140.

Sheppard, P. M., 1960, *Natural Selection and Heredity,* Harper Torch book, New York.

Short, R. V., and M. F. Hay, 1966, Comparative biology of reproduction in mammals, *Symp. Zool. Soc. Lond.* **15:** 173–194.

Siegal, P. S., and F. J. Pilgrim, 1958, The effect of monotony on the acceptance of food, *Amer. J. Psychol.* **71:** 756–759.

Skutch, A. F., 1949, Do tropical birds rear as many young as they can nourish?, *Ibis* **91:** 430–455.

Skutch, A. F., 1950, The nesting seasons of Central American birds in relation to climate and food supply, *Ibis* **92:** 185–222.

Skutch, A. F., 1954, Life histories of central American birds, *Pacific Coast Avifauna* **31:** 1–448.

Skutch, A. F., 1960, Life histories of Central American birds, *Pacific Coast Avifauna* **34:** 1–593.

Smith, A. D., 1940, A discussion of the application of a climatological diagram, the hythergraph, to the distribution of natural vegetation types, *Ecol.* **21:** 184–191.

Smith, G. C., 1968, The adaptive nature of social organization in the genus of tree squirrels, *Tamiasciurus, Ecol. Monog.* **38:** 31–64.

Smith, M., R. Pool, and H. Weinberg, 1962, The role of bulk in the control of eating, *J. Comp. and Physiol. Psychol.* **55:** 115–120.

Smith, N. G., 1968, The advantage of being parasitized, *Nature* **219:** 690–694.

Sowles, L. K., 1966, Reproduction in the collared peccary, *Symp. Zool. Soc. Lond.* No. 15, pp. 155–172.

Springer, R., 1957, Some observations on the behavior of schools of fishes in the Gulf of Mexico and adjacent waters, *Ecol.* **38.** 166–171.

Stefanski, R. A., 1967, Utilization of the breeding territory in the black-capped chicadee, *Condor* **69:** 259–267.

Stenlund, M. H., 1955, A field study of the timber wolf (*Canis lupus*) of the superior National Forest, Minnesota, *Minn. Dept. Conserv. Techn. Bull.* No. 4.

Stephens, G. C., 1968, Dissolved organic matter as a potential source of nutrition for marine organisms, *Amer. Zool.* **8:** 95–106.

Stevenson-Hamilton, J., 1937, *South African Eden,* Cassell and Co., Ltd., London.

Stickel, L. F., 1960, *Peromyscus* ranges at high and low population densities, *J. Mammal.* **41:** 433–441.

Stimson, J., 1970, Territorial behavior of the owl limpet, *Lottia gigantea, Ecol.* **51:** 113–118.

Strangeways-Dixon, J., 1961, The relations between nutrition, hormones, and reproduction in the blowfly (*Calliphora erythrocephala*), I, Selective feeding in relation to the reproductive cycle, the corpus allatum volume, and fertilization, *J. Exper. Biol.* **38:** 225–235.

Struhsaker, T. T., 1968, Social structure among vervet monkeys, (*Cercopithecus aethiops*), *Behaviour* **29:** 83–121.

Sudd, J. H., 1965, The transport of prey by ants, *Behaviour* **25:** 234–271.

Sudd, J. H., 1967, *An Introduction to the Behavior of Ants,* St Martins, New York.

Sumner, F. B., 1934, Does "protective coloration" protect? Results of some experiments with fishes and birds, *Proc. Nat. Acad. Sci.* **20:** 559–564.

Sumner, F. B., 1935, Studies of protective coloration change, III, Experiments with fishes both as predators and prey, *Proc. Nat. Acad. Sci.* **21:** 345–353.

Svärdson, G., 1948, Natural selection and egg number in fish, *Rpt. Instit. Freshwater Res. Drottningholm* No. 29, pp. 115–122.

Svärdson, G., 1949, Notes on spawning habits of *Leuciscus erythrophthalmus* (L), *Abramis brama* (L), and *Esox lucius* (L); *Rpt. Instit. Freshwater Res. Drottningholm* No. 32, pp. 102–107.

Talbot, M., 1943, Population studies of the ant, *prenolepis imparis* (say), *Ecol.* **24:** 31–44.

Talbot, M., and C. H. Kennedy, 1940, The slave-making ant, *Formica sanguinea subintegra* (Emery), its raids, nuptual flights, and nest structure, *Ann. Ent. Soc. Amer.* **33**: 560–577.

Test, F. H., 1954, Social aggressiveness in an amphibian, *Science* **120**: 140–141.

Thompson, J. A., 1923, *The Biology of Birds,* MacMillan, New York.

Thomson, A. L., 1964, *A New Dictionary of Birds,* McGraw Hill, New York.

Thorson, G., 1950, Reproductive and larval ecology of marine bottom organisms, *Biol. Rev.* **25**: 1–45.

Thorson, G., 1958, Parallel level-bottom communities, their temperature adaptation, and their "balance" between predators and food animals, in A. Buzzati-Traverso (Ed.), *Perspectives in Marine Biology,* University of California Press, Berkeley.

Tinbergen, L., 1960, The natural control of insects in pine wood, I, Factors influencing the intensity of predation by songbirds, *Archiv. Neerl. de Zool.* **XIII**: 265–343.

Tinbergen, N., 1956, On the functions of territory in gulls, *Ibis* **98**: 401–411.

Tinbergen, N., M. Impekoven, and D. Franck, 1967, An experiment on spacing-out as a defense against predation, *Behaviour* **28**: 307–321.

Tinkle, D. W., 1965, Home range, density, dynamics, and structure of a Texas population of the lizard *Uta stansburiana,* in W. W. Milstead (Ed.), *Lizard Ecology, A Symposium,* University of Missouri Press. Kansas City.

Tinkle, D. W., 1969, The concept of reproductive effort and its relation to the evolution of life histories of lizards, *Amer. Natur.* **103**: 501–516.

Tinkle, D. W., H. M. Wilbur, and S. G. Tilley, 1970, Evolutionary strategies in lizard reproduction, *Evol.* **24**: 55–74.

Tullock, D. G., 1969, Home range in feral water buffalo, *Bubalus bubalis* (Lydekker), *Austr, J. Zool.* **17**: 143–152.

Turner, F. B., R. I. Jennrich, and J. D. Weintraub, 1969, Home ranges and body size of lizards, *Ecol.* **50**; 1076–1081.

Urquhart, F. A., 1957, A discussion of batesian mimicry as applied to the monarch and viceroy butterflies, *Contr. of Division of Zool. and Palaentol., Royal Ontario Museum,* University of Toronto Press, Toronto.

Verner, J., 1964, Evolution of polygamy in the long-billed marsh wren, *Evol.* **18**: 252–261.

Verner, J., and M. F. Willson, 1966, The influence of habitats on mating systems of North American passerine birds, *Ecol.* **47**: 143–147.

Verner, J., and M. F. Willson, 1969, Mating systems, sexual dimorphism, and the role of male north American passerine birds in the nesting cycle, *Ornithol. Monog.* No. 9.

Weber, H., 1924, Ein Umdreh und ein Fluchtreflex bei *Nassa mutabilis, Zool. Anz.* **60**: 261–269.

Webster, F. A., and D. R. Griffin, 1962, The role of the flight membranes in insect capture by bats, *Anim. Beh.* **10**: 332–340.

Welch, H. E., 1968, Relationships between assimilation efficiencies and growth efficiencies for aquatic consumers, *Ecol.* **49**: 755–759.

Weller, M. W., 1959, Parasite egg laying in the redhead (*Aythya americana*) and other North American Anatidae, *Ecol. Monog.* **29**: 333–365.

Welty, J. C., 1962, *The Life of Birds,* Saunders, Philadelphia and London.

Whelker, W. I., and W. A. King, 1962, Effects of stimulus novelty on gnawing and eating by rats, *J. Comp. and Physiol. Psychol.* **55**: 838–842.

White, J. E., 1964, An index of the range of activity, *Amer. Midl. Natur.* **71**: 369–373.

Wiegert, R. G., 1961, Respiratory loss and activity patterns in the meadow vole (*Microtus pennsylvanicus p.*), *Ecol.* **42**: 245–253.

Wiens, J. A., 1963, Aspects of cowbird parasitism in southern Oklahoma, *Wilson Bull.* **75**: 130–139.

Williams, E. C., Jr., and D. E. Reichle, 1968, Radioactive tracers in the study of energy turnover by a grazing insect (*Chrysochus auratus* Fab) (*Coleoptera, Chrysomelidae*), *Oikos,* **19:** 10–18.

Willson, M. F., 1966, Breeding ecology of the yellow-headed blackbird, *Ecol. Monog.* **36:** 51–77.

Willson, M. F., and E. R. Pianka, 1963, Sexual selection, sex ratio, and mating system, *Amer. Natur.* **97:** 405–407.

Wilson, E. O., 1963, The social biology of ants, *Ann. Rev. Ent.* **8:** 345–368.

Wynne-Edwards, V. C., 1962, *Animal dispersion in relation to social behaviour,* Oliver and Boyd, Edinburgh.

Young, H., R. L. Strecker, and J. T. Emlen, Jr., 1950, Localization of activity in two indoor populations of house mice, *Mus musculus, J. Mammal.* **31:** 403–410.

Young, P. T., 1945, Studies of food preference, appetite, and dietary habit, *Comp. Psych. Monog.* **19:** 1–58.

# Section III
# The Ecology of Populations

# 9
# Population Growth

Population is not an easy word to define. Ordinarily, biologists think of populations as units of interbreeding individuals, but this is vague. Are these units discrete? If not, do individuals interbreeding with two units belong to both? Do we include such individuals in counts of both "populations?" If breeding is nonrandom so that only a fraction of the group's members reproduce, does the population consist of the whole group, or those that breed, and how often must one breed to be considered part of the population? From a geneticist's viewpoint the effective size (with respect to genetic drift, inbreeding effects, see Discussion I) of a population is usually smaller than the number of individuals it contains. From an ecologist's viewpoint the ecological impact of a group may be only slightly correlated with its number of members. Use and production of energy, effect on other ecological factors in the environment, and even intrapopulation interactions depend on age distribution, physiological and psychological state, mobility, and spacing pattern of the individuals as well as their number or number per unit area. There are no easy answers to these questions.

An appreciation of "population" as it is used by ecologists should be gained from reading Chapters 9 through 12, even if no adequate definition can be invented. For the moment, we ignore the above complications and consider populations to be isolated units of interbreeding individuals. In part 9.1 all individuals are considered as entities drawn from the same random sample, independent of age. In the remaining sections of the chapter, age differences will be discussed. Throughout this chapter we assume that genetic changes are sufficiently slower than population changes that they can be safely ignored. This restriction will be eased in subsequent chapters.

## 9.1 EXPONENTIAL AND LOGISTIC GROWTH

For the sake of simplicity, we assume in the first half of this chapter that breeding occurs continuously throughout the year, with no peaks. This is seldom the case in nature (Chapter 6), and the conclusions reached below are limited in application except as very rough models of the real world. Nevertheless, the insight gained through the simple approach below is very helpful in understanding the more accurate and more complex treatment that will follow.

The number of individuals entering a continuously breeding population at any time is given by the average birth rate per individual, $b$, times the number of individuals, $n$, and the number leaving the population by the product of $n$ and the death rate, $d$. Thus the net rate of change in population size (number) is

$$\frac{dn}{dt} = bn - dn = (b - d)n.$$

The quantity $b - d$ is usually written $r$, so that

$$\frac{dn}{dt} = (b - d)n = rn. \tag{9.1}$$

Note that this may also be written

$$\frac{1}{n}\frac{dn}{dt} = \frac{d \ln n}{dt} = r.$$

This same simple equation might equally well describe some other measure of population size such as total weight (*biomass*) or potential energy content. In many cases, involving budding hydra, hydrozoan colonies, and clonally breeding plants, for example, it is not at all apparent what is meant by "individual." In such cases it is clearly preferable to use $n$ to mean something other than number. In this text, however, we shall continue to use $n$ to denote number, noting that other meanings generally require only minor changes in the following equations.

If birth rate and death rate are independent of population size, then $r$ is also, and Eq. (9.1) becomes $\int d(\ln n) = \int r \, dt$, or

$$n(t) = n(0) e^{rt}. \tag{9.2}$$

Where perturbations occur in $r$, so that $r$ takes the value $r(t)$ in the interval $t$ to $t + dt$, then

$$n(t) = n(0) \left(e^{r(0) \, dt} \cdot e^{r(dt) \, dt} \cdot e^{r(2 \, dt) \, dt} \ldots\right)$$

$$= n(0) \exp\left(\int_0^t r(t) \, dt\right) = n(0) e^{\bar{r}t},$$

where

$$\bar{r} = \frac{1}{t} \int_0^t r(t) \, dt$$

is the mean value of $r$. Clearly, if $\bar{r}$ is greater than zero the population increases exponentially to infinity. If $\bar{r}$ is less than zero, the population declines to extinction. If $\bar{r} = 0$, then the population stays at, or fluctuates about, some finite value. But it is obvious that while some populations go extinct, none reach infinite levels. Furthermore, while populations often fluctuate about some mean size, they also tend to grow when small and decline when very large. The value $r$, in fact, is nearly always correlated, at least slightly, with $n$ (see Chapter 10). For example, if a population increases in number without a compensatory increase in food supply, its members will eventually starve. Less energy is available for reproduction and the birth rate drops (Chapter 6); less energy is available for maintenance and the death rate increases. When $(b - d) = r$ drops to zero, population growth stops.

This rather simple approach was examined—in somewhat different terms—by Lotka, Volterra, and Gause, who independently published equations describing population growth under conditions of changing $r$. The most widely used equation of this sort will now be discussed.

We let $r_0$ be the *intrinsic rate of natural increase* of a population, the maximal rate attainable under some defined circumstances. Next we suppose that there is enough of some resource to support $K$ individuals in a given area. If this is so, and if $n < K$, then there is room for more individuals. Suppose, for example, that there are $K$ burrows and $n$ individuals. Then $K - n$ burrows go unused and we might expect the population to increase until $n$ reaches $K$. In the most simple instance, the rate of increase, $r$, will vary linearly with the amount of resource remaining: $r \propto K - n$. Since, in this scheme, $r$ approaches $r_0$ when $K - n$ is maximum $(n \to 0)$, we write

$$r = r_0\left(\frac{K - n}{K}\right).$$ 
(9.3)

An implicit assumption is that a change in $r$ *simultaneously* accompanies a change in $n$. Equation (9.1) now becomes

$$\frac{d \ln n}{dt} = r = r_0(1 - n/K),$$ 
(9.4)

whose solution is given by

$$n(t) = n(0) \frac{Ke^{r_0 t}}{K - n(0)(1 - e^{r_0 t})}.$$

This is the so-called *logistic growth equation*. A graph of the form of the curve generated is shown in Fig. 9.1. The value $K$ is the asymptote approached by $n$, and represents, therefore, the maximum number of individuals that the environment can support on a given area. It is known as the *carrying capacity* of

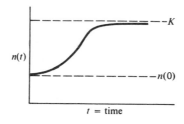

Fig. 9.1. Population (logistic) growth over time.

the environment and, in spite of its name, clearly varies with features of the population as well as with the environment. When $n < K$, $r$ is positive and the population rises, when $n > K$, $r$ is negative and the population falls. Thus $n = K$ represents a stable equilibrium.

Note here a source of confusion in the population literature. The term $r$ is used as a measure of population growth rate when such growth is continuous. It is occasionally written $m$ and called the *Malthusian parameter*. Since most populations cannot be said to change size continuously, the meaning of $r$ is not always clear. Instead, it may be preferable to use

$$R = \frac{n(t+1)}{n(t)}.$$

Note where growth is continuous and $r$ is independent of $n$, that

$$\frac{n(t+1)}{n(t)} = e^r,$$

so that

$$e^r = R, \quad \text{or} \quad r = \ln R.$$

When $r$ is not independent of $n$, we write

$$n(t+1) - n(t) = n(0)R^{t+1} - n(0)R^t = n(0)R^t(R-1),$$

so

$$\frac{\Delta n}{\Delta t} = n(0)R^t(R-1) = n(t)(R-1),$$

and

$$\frac{1}{n}\frac{\Delta n}{\Delta t} = R - 1.$$

But when $R$ is very close to 1, $R - 1$ is very nearly $\ln R$, and if $R$ is not changing greatly over time

$$\frac{1}{n}\frac{\Delta n}{\Delta t} \approx \frac{1}{n}\frac{dn}{dt} = \frac{d\ln n}{dt} = r.$$

Thus $\ln R \approx r$. It is seldom clear in the literature whether $r$ is taken to mean

$$\frac{d\ln n}{dt}, \quad \frac{1}{n}\frac{\Delta n}{\Delta t}, \quad \text{or} \quad \ln R,$$

and ecologists have compounded the confusion by occasionally also denoting $r_0$ by $r$. These facts should be borne in mind when reading the literature. In this book we shall use $r$ to mean $\ln R$, which, when $r$ is independent of $n$, is equal to $d(\ln n)/dt$. We shall let $r_0$ denote the maximum or *intrinsic rate of natural increase* in the same manner.

### 9.1.1 Discussion of the Logistic Growth Equation

The logistic growth equation has proved remarkably useful in population ecology despite the simplifying assumptions inherent in its derivation. First, the nature of population growth may differ from that predicted by the equation without affecting the equation's validity at equilibrium. Because of this it is possible to use the logistic equation to construct equilibrium models of competition. Second, the equation seems to describe growth reasonably well for a number of populations. Yeast and protozoan growth curves, for example, fit the model quite well, as to a less precise degree do those of some metazoans (Gause, 1934). Gause (1931, 1934) found reasonable fits for *Paramecium* and for flour beetles, *Tribolium*, and Pearl (1930) found the logistic equation to describe quite accurately growth in *Drosophila melanogaster* (fruit fly) populations. Even Tasmanian sheep seem to approximate logistic growth (Davidson, 1938a, b), and pheasants, *Phasianus colchicus*, may also. In the spring of 1937 two cock and six hen pheasants were released on Protection Island off the coast of Washington state. By spring of 1938, their number had swelled to 30. In spring 1939 it reached 81, in spring 1940, 282, in spring 1941, 641. By spring of 1942, the number had reached 1194 and continued to rise to 1898 in spring of 1943 (Fig. 9.2). Unfortunately, the army arrived in 1943 and started shooting pheasants, so the final phase of growth could not be observed. It appears, though, that the growth curve may have begun to level off by about 1942 (Einarsen, 1945a, b).

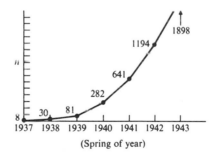

**Fig. 9.2.** Population growth of pheasants on Protection Island.

While growth curves similar to those predicted by the logistic growth model are occasionally found, most metazoans do not show very good fits to the equation. In fact, growing metazoan populations usually show quite regular, damped oscillations. Davidson's sheep, mentioned above, showed hints of oscillations. Frank (1960) found marked oscillations in *Daphnia* populations, as did Slobodkin (1954). Terao and Tanaka (1928) found the same thing in another waterflea, *Moina macrocopa*. These oscillations appeared under quite constant laboratory conditions (except in the case of the sheep) and therefore were not due to environmentally induced fluctuations in $r_0$ or $K$. Why should such cycles occur?

### 9.1.2 Corrections and Time Lags

Before attempting directly to answer the question of oscillations, let's examine the assumptions of linearity and simultaneity made in the derivation of the logistic equation.

    First, there are few good measures of fecundity and mortality as they vary with population density, but what information exists indicates nonlinear relations. Indeed there is no particular reason to believe the relations to be linear and good reasons to believe them not to be. Consider some hypothetical creature in which $\bar{b}$ falls linearly and $\bar{d}$ rises linearly with $n$ (Fig. 9.3). The values $\bar{b}$ and $\bar{d}$ are taken to be the mean values of $b$ and $d$ over time. Mean equilibrium population size, $\bar{n}$, occurs when $\bar{b} - \bar{d} = 0$, that is when the lines $\bar{b}$ and $\bar{d}$ cross. Since the population will be fluctuating about this point with environmentally induced changes in the $\bar{b}$ and $\bar{d}$ lines, the frequency with which any value of $n$ occurs will fall off in some manner with the distance between $n$ and $\bar{n}$. The most frequent value of $n$ will be $\bar{n}$. Now, natural selection will act to modify physiology, behavior, etc., in such a way as to maximize $\bar{b}$ and minimize $\bar{d}$. But those characteristics which are adaptive at one population density may be disadvantageous at another. Since the population is most often near $\bar{n}$, it is in relation to population densities near $\bar{n}$ that natural selection operates most strongly. Thus $\bar{b}$ should rise and $\bar{d}$ should fall, but primarily at values of $n$ close to $\bar{n}$. For very high or very low $n$, those characteristics beneficial at $\bar{n}$ may prove detrimental. Selection will change the shape of the curves in Fig. 9.3 to more closely resemble those in Fig. 9.4. (An example of a birth rate curve closely resembling this predicted form is given by Frank, Boll, and Kelly (1957) for *Daphnia pulex*). When populations are very sparse it may be difficult to find a mate, there will be less social stimulation, and $\bar{b}$ may be lower than at some higher population level. Selection may perfect cooperative hunting techniques, such as feeding flock behavior; these techniques are efficient at some population levels, but not when too many animals gather together and sweep an area clean as they move. Thus, as populations get too large, food intake per individual may fall off, followed by birth rate. Aggressive behavior in mice is presumably advantageous

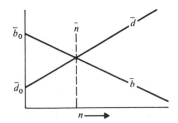

**Fig. 9.3.** Birth and death rates as functions of population size.

at normal population levels but at high levels, where probability of contact is much increased, may be lethal. Slash-and-burn agriculture is efficient strategy when there are enough individuals to perform the necessary work and so long as there are few enough people that they can move on when the land gives out. Such practices by today's South American populations have proved disastrous. An economy of waste is efficient so long as enough individuals are present to keep it operating through their consumption, but in today's overpopulated world it has already begun to sow the seeds of death through destruction of the environment.

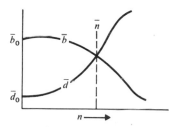

Fig. 9.4. Birth and death rates as functions of population size.

Because of the nonlinear relation between $\bar{b}$ or $\bar{d}$ and $n$, it has been popular to "improve" the accuracy of the logistic equation by adding higher-order terms. Thus, we write

$$\frac{d\ln n}{dt} = r_0(1 - \gamma_1 n - \gamma_2 n^2 - \cdots).$$

If, in fact, we could treat all individuals in a population as identical—no age or sex differences—then, by adding enough terms, we could theoretically generate an accurate expression. The biological meaning of the higher order terms, however, is obscure.

Even after correction for nonlinearity the previously mentioned oscillations are not accounted for. To explain these we must examine the assumption of *simultaneous* change in $b$, $d$ with changing $n$. In fact, such simultaneity does not occur; there are a number of kinds of time lag (Slobodkin, 1959). For example, the birth rate in one year may reflect shortages in food occurring over the previous year. These shortages may be due to high population levels in that previous year or even the year before that. It is easy to see how lags of this sort might generate oscillations in population size. Another kind of time lag relates to the period of immaturity. The number of individuals entering a *breeding* population, if it requires one year to mature, is proportional to the size of the breeding population one year ago, while the number leaving the breeding population is proportional to the present population size. Combining these two types of lag, we can write

$$\frac{dn(t)}{dt} = b_0\left[1 - \frac{n(t - \delta_1)}{K'}\right]n(t - \delta_2) - d_0 n(t),$$

where $\delta_1$, $\delta_2$ are the appropriate time lags, and $n(t)$ is the size of the breeding population at time $t$. This equation has been studied by Wangersky and Cunningham (1957) and shown to give rise to oscillations. The solution is difficult and will not be given here.

Frank (1960) tried numerical solutions to similar sorts of equations in an effort to explain the fluctuation patterns of his laboratory *Daphnia* (waterflea) populations, and discovered that population size appeared to be controlled by biovolume of the population five days previously. Population oscillations will be discussed further in 9.3.

## 9.2 LIFE TABLES

As was made abundantly clear in Chapter 6, most animals show seasonal reproductive peaks or breed only over one short period in a year. In such cases population fluctuates seasonally and an instantaneous rate of population growth, $r$, is inconstant. Also, different age groups in a population possess different reproductive capabilities and suffer different mortality rates. To incorporate these facts into an equation like the logistic one would be difficult and un-rewarding. However, two approaches exist whereby these complications may be dealt with. The first is the *life tables* approach which derives from the treatment of human actuarial tables. The second is a more complete approach developed by Leslie (see 9.3).

### 9.2.1 Basic Equations

We let $n_x(t)$ be the number of individuals reaching age $x$ at time $t$, $l_x$ be the probability that an individual survives to age $x$ (*survivorship*), and $m_x$ be the number of young born per individual at age $x$ (*fecundity*). Then $n_0(t)$ is the number of offspring born of all individuals of all ages at time $t$:

$$n_0(t) = \sum_x n_x(t)m_x.$$

But if mortality rates are invariant in time, $n_x(t) = n_0(t - x)l_x$, so

$$n_0(t) = \sum_x n_0(t - x)l_x m_x.$$

Now suppose that a stable age distribution exists; that is, suppose that the relative number of individuals of different ages remains constant. Then, if $R$ is the ratio of individuals in the population at one time to the same measure in the previous time unit,

$$n_x(t) = n_x(t - y)R^y,$$

so that

$$n_x(t - y) = n_x(t)R^{-y}$$

for all $y$. Thus

$$n_0(t - x) = n_0(t)R^{-x},$$

and
$$n_0(t) = \sum_x n_0(t) R^{-x} l_x m_x = n_0(t) \sum_x R^{-x} l_x m_x,$$

and dividing both sides by $n_0(t)$

$$1 = \sum_x R^{-x} l_x m_x. \tag{9.5}$$

This is perhaps the most important basic equation in ecology and will be used a number of times later in this book. The reader should be thoroughly familiar with its form and derivation. The reader should also realize that the equation is accurate only when a stable age distribution exists. It can be shown (Appendix 4) that in a constant environment population fluctuations will always be damped (if more than one age class reproduces) and that, concurrent with this, a stable age distribution will come about. It is thus reasonable to apply Eq. (9.5) (though with caution) to a variety of populations both for purposes of calculating population growth ($R$) and for genetic considerations (Discussion III) (but see part 9.5).

Notice when a stable age distribution exists that

$$n_x(t) = n_0(t - x) l_x = n_0(t) R^{-x} l_x,$$

so that

$$\frac{n_x(t)}{\sum_x n_x(t)} = \frac{n_0(t) R^{-x} l_x}{\sum_x n_0(t) R^{-x} l_x} = \frac{R^{-x} l_x}{\sum_x R^{-x} l_x} = c_x. \tag{9.6}$$

The value $c_x$ is the proportion of the population made up of individuals of age $x$. Notice when $R = 1$ (the population is unchanging) that $c_x$ is proportional to $l_x$. This fact may be extremely useful in estimating population age structures.

An important feature of Eq. (9.5) is this. Note that it can be written in the form

$$R^z - a_1 R^{z-1} - a_2 R^{z-2} - \cdots - a_z = 0.$$

We shall write this $F^{(z)} = 0$. Suppose $\lambda$ is a real, positive root of $F^{(z)}$. Then

$$\frac{dF^{(z)}}{d\lambda} = z\lambda^{z-1} - a_1(z-1)\lambda^{z-2} - \cdots - a_{z-1}$$

$$= \frac{z}{\lambda}\left[\lambda^z - a_1 \frac{z-1}{z}\lambda^{z-1} - a_2 \frac{z-2}{z}\lambda^{z-2} - \cdots - \frac{1}{z}a_{z-1}\lambda\right]$$

$$> \left\{\frac{z}{\lambda}\left[\lambda^z - a_1\lambda^{z-1} - a_2\lambda^{z-2} - \cdots - a_{z-1}\lambda - a_z\right] = 0\right\}.$$

Thus $F^{(z)}$ is always increasing with $R$ when $R = \lambda$. Consider now the case for $z = 2$. $F^{(2)} = R^2 - a_1 R - a_2 = 0$. The turning points (values of $R$ at which the slope of $F^{(2)}$ changes sign) are given by that value of $R$ for which $dF^{(2)}/dR = 2R - a_1 = 0$. Thus there is one positive, real turning point. But when $R = 0$, $F^{(2)} < 0$. Thus $F^{(2)}$ must curve downward and cross the $R$ axis to the left of the origin, reach a turning point to the right of the origin, and rise, crossing the $R$ axis at exactly one real, positive point. The equation $F^{(2)} = 0$, therefore, has one and only one real, positive root. We have now shown that $F^{(k)}$ (for $k = 2$) has exactly one real, positive root. Consider the case for $F^{(k+1)}$. We know that the slope of $F^{(k+1)}$ is positive for that value of $R$ for which $F^{(k+1)} = 0$. The turning point of $F^{(k+1)}$ is given by that value of $R$ such that

$$\frac{d}{dR} F^{(k+1)} = 0.$$

But this expression is equivalent to the general form $F^{(k)} = 0$ which we have just shown to have one real, positive root. Thus there is only one real turning point for $F^{(k+1)}$ to the right of the origin. Furthermore $F^{(k+1)} < 0$ for $R = 0$. Thus, as with $F^{(k)}$, the slope of $F^{(k+1)}$ drops, crosses the $R$ axis to the left of the origin, reaches a turning point to the right of the origin, and crosses the $R$ axis at exactly one real, positive point. Having proved the case of $k$, and that it therefore follows for $k + 1$, we have, by induction, proved the case for all $z$. The equation

$$1 = \sum_x R^{-x} l_x m_x$$

has exactly one real, positive root.

### 9.2.2 Life Tables

Suppose we designate the proportion of a population dying in the age interval $x$ to $x + 1$ with the symbol $d_x$. Now clearly the survivorship at age 0 ($l_0$) is 1.0. Also, clearly, the survivorship at age one ($l_1$) is $1 - d_0$, the survivorship at age two ($l_2$) is $1 - (d_0 + d_1)$, and so on. We write

$$l_x = 1 - \sum_{y=0}^{x-1} d_y, \quad \text{or} \quad l_x = l_{x-1} - d_{x-1}, \quad \text{or} \quad d_x = l_x - l_{x+1}. \quad (9.7)$$

We now define $q_x$ to be the *age-specific mortality*; that is, the probability that an individual reaching age $x$ will die before it reaches age $x + 1$. Clearly $d_x$, being the probability that an animal dies in the interval $x$ to $x + 1$, is the probability that an animal dies in that interval provided it reaches it ($q_x$) times the probability it reaches it ($l_x$). Thus $d_x = q_x l_x$, and

$$q_x = \frac{d_x}{l_x} = \frac{l_x - l_{x+1}}{l_x}. \quad (9.8)$$

We now introduce a well-known paper containing data to which we may apply Eqs. (9.7) and (9.8). In a very ingenious study, Edmondson (1945) took water plants on which the sessile rotifer *Floscularia conifera* lives and dusted them with carmine powder and then, after 24 hours, charcoal. Those rotifers present on the plants ingested the powders and deposited them in two rings in the tubes which this species constructs and in which its members live. The distance between the pink and the black ring gave the size-specific growth rate and allowed Edmondson to describe the relation between age and size. The fraction of all individuals dying in the 24-hour interval (those whose tubes possessed pink but not black rings) gave the mortality, $d_x$. The calculated values of $l_x$, $q_x$, along with the observed $d_x$ values for *Floscularia conifera* are given in Table 9.1.

**Table 9.1.** Life table data for the rotifer *Floscularia conifera* (from Edmondson, 1945)

| Age in days | Observed $x$ | $d_x$ | Calculated $l_x = 1 - \sum_{y=0}^{x-1} d_y$ | $q_x = \dfrac{d_x}{l_x}$ |
|---|---|---|---|---|
| 0–1 | 0 | 0.020 | 1.000 | 0.020 |
| 1–2 | 1 | 0.200 | 0.980 | 0.204 |
| 2–3 | 2 | 0.060 | 0.780 | 0.077 |
| 3–4 | 3 | 0.000 | 0.720 | 0.000 |
| 4–5 | 4 | 0.300 | 0.720 | 0.416 |
| 5–6 | 5 | 0.140 | 0.420 | 0.333 |
| 6–7 | 6 | 0.060 | 0.280 | 0.214 |
| 7–8 | 7 | 0.140 | 0.220 | 0.636 |
| 8–9 | 8 | 0.040 | 0.080 | 0.500 |
| 9–10 | 9 | 0.020 | 0.040 | 0.500 |
| 10–11 | 10 | 0.020 | 0.020 | 1.000 |

Suppose that this rotifer population had possessed a stable age distribution and that $R = 1$. Then $c_x$ (Eq. 9.6) is given by $l_x/\sum_x l_x$, which, for $x = 0, 1, 2, \ldots$, is 0.190, 0.186, 0.148, 0.137, 0.137, 0.080, 0.053, 0.042, 0.015, 0.008, 0.004.

Let's look at another example. Murie (1944) collected and aged skulls of the Dall mountain sheep, *Ovis dalli*. Since the collection consisted of skulls of individuals living over a great period of time in the past, and since there was no reason to believe the population to have been steadily increasing or decreasing over that time, $R$ was taken to be very close to 1.0, and a stable age distribution ($c_x$'s were assumed to have fluctuated very closely about some mean) was assumed. This being so, the proportion of skulls dead at age $x$ to $x + 1$ can be taken as a good estimate of $d_x$. Table 9.2 gives Murie's $d_x$ values as well as calculated $l_x$ (Eq. 9.7), $q_x$ (Eq. 9.8) and $c_x$ (Eq. 9.6) values. Further discussion and examples of life tables are given in Deevey (1947). We summarize here by re-listing the pertinent equations

$$c_x = R^{-x} l_x \Big/ \sum_x R^{-x} l_x \tag{9.6}$$

$$l_x = 1 - \sum_{y=0}^{x-1} d_y = l_{x-1} - d_{x-1} \tag{9.7}$$

$$q_x = d_x / l_x. \tag{9.8}$$

**Table 9.2.** Life table data for the Dall mountain sheep (from Murie, 1944)

| Age (years) | $x$ | $d_x$ | $l_x$ | $q_x$ | $c_x$ |
|---|---|---|---|---|---|
| 0–1 | 0 | 0.199 | 1.000 | 0.199 | 0.132 |
| 1–2 | 1 | 0.012 | 0.801 | 0.015 | 0.106 |
| 2–3 | 2 | 0.013 | 0.789 | 0.0165 | 0.104 |
| 3–4 | 3 | 0.012 | 0.776 | 0.0155 | 0.102 |
| 4–5 | 4 | 0.030 | 0.764 | 0.0393 | 0.101 |
| 5–6 | 5 | 0.046 | 0.734 | 0.0626 | 0.097 |
| 6–7 | 6 | 0.048 | 0.688 | 0.0699 | 0.091 |
| 7–8 | 7 | 0.069 | 0.640 | 0.108 | 0.084 |
| 8–9 | 8 | 0.132 | 0.571 | 0.231 | 0.076 |
| 9–10 | 9 | 0.187 | 0.439 | 0.426 | 0.058 |
| 10–11 | 10 | 0.156 | 0.252 | 0.619 | 0.033 |
| 11–12 | 11 | 0.090 | 0.096 | 0.937 | 0.013 |
| 12–13 | 12 | 0.003 | 0.006 | 0.500 | 0.008 |
| 13–14 | 13 | 0.003 | 0.003 | 1.000 | 0.004 |
| >14 | ⩾14 | — | 0.000 | — | 0.000 |

## 9.2.3 The Calculation of Growth Rates

We turn now to a discussion of the use of life tables in calculating $R$ (or $r$, bearing in mind that this $r$ is a discrete, not a continuous rate of increase). Suppose we are given the data shown in Table 9.3. We now apply Eq. (9.5) and obtain

$$1 = \sum_x R^{-x} l_x m_x = R^{-0} \times 1.00 \times 0 + R^{-1} \times 0.50 \times 1 + R^{-2} \times 0.30 \times 1$$

$$+ R^{-3} \times 0.10 \times 2 = 0.5R^{-1} + 0.3R^{-2} + 0.2R^{-3}.$$

Clearly $R = 1$ is a solution. Furthermore, as proved above, the equation can have only one real, positive root; in this case $R = 1$. When the population depicted reaches a stable age distribution, it ceases to change in size.

**Table 9.3**

| Age | $x$ | $l_x$ | $m_x$ |
|---|---|---|---|
| 0–1 | 0 | 1.00 | 0 |
| 1–2 | 1 | 0.50 | 1 |
| 2–3 | 2 | 0.30 | 1 |
| 3–4 | 3 | 0.10 | 2 |

Suppose that the data in Table 9.3 were as shown except that $m_2 = 2$. Then we would write (after Eq. 9.5)

$$1 = 0.5R^{-1} + 0.6R^{-2} + 0.2R^{-3}.$$

Clearly $R$ is now greater than 1.0. Its value, found by iteration, is $R = 1.17$.

Before applying this technique to real data, we present a problem often posed by nature. Animals don't breed continuously at a steady rate; they also do not reproduce all at once over an infinitesimally short period. Thus to assume that all reproduction occurs at age $x$ rather than over the interval $x$ to $x + 1$ is unrealistic. When production of young is highly seasonal this assumption does not lead to serious errors, but if a population breeds over long portions of the year it will. One not terribly satisfactory way around this problem is to use some corrected fecundity value, $F_x$, and an appropriately corrected survivorship value, $L_x$. The simplest method is to estimate reproduction over the $x$ to $x + 1$ age interval ($F_x$) as occurring, on average, at age $x + \frac{1}{2}$. The appropriate survivorship values are given by

$$L_x = \frac{l_x + l_{x+1}}{2},$$

Thus a life table may take the form

| Age | $x$ | $l_x$ | $L_x$ | $F_x$ |
|-----|-----|-------|-------|-------|
| 0–1 | $\frac{1}{2}$ | 1.000 | 0.750 | 1 |
| 1–2 | $1\frac{1}{2}$ | 0.500 | 0.300 | 2 |
| 2–3 | $2\frac{1}{2}$ | 0.100 | 0.050 | 0 |

$R$ is now found by setting (after Eq. 9.5)

$$1 = (R^{-\frac{1}{2}} \times 0.750 \times 1) + (R^{-1\frac{1}{2}} \times 0.300 \times 2) + (R^{-2\frac{1}{2}} \times 0.050 \times 0)$$
$$= 0.75R^{-\frac{1}{2}} + 0.60R^{-1\frac{1}{2}} : R = 1.36.$$

Where some more realistic *pivotal age* for fecundity exists (perhaps reproduction occurs, on average, at age $x + \frac{1}{3}$), $L_x$ can be corrected appropriately. Leslie and Ranson (1940) give the following data for a laboratory population of the vole *Microtus agrestis* (Table 9.4).

Setting $1 = (R^{-8} \times 0.83349 \times 0.6504) + \cdots$, it can be determined, with much bother, that $R = 1.09$. As noted in 9.1 we may now calculate $r$ ($=\ln R$), but the biological meaning of $r$ in a population breeding at discrete intervals (and in which $r$ may vary with $n$) is not clear.

The reader should develop his understanding of the techniques in Section 9.2 of this chapter by finding $q_x$, $c_x$, and recalculating $R$ for the data in Table 9.4.

**Table 9.4.** Life table data for a laboratory population of *Microtus agrestis* (from Leslie and Ranson, 1940)

| Age | Pivotal age (in weeks) | $L_x$ | $F_x$ |
|---|---|---|---|
| 0–12 | 8 | 0.83349 | 0.6504 |
| 12–20 | 16 | 0.73132 | 2.3939 |
| 20–28 | 24 | 0.58809 | 2.9727 |
| 28–36 | 32 | 0.43343 | 2.4662 |
| 36–44 | 40 | 0.29277 | 1.7043 |
| 44–52 | 48 | 0.18126 | 1.0815 |
| 52–60 | 56 | 0.10285 | 0.6683 |
| 60–65 | 64 | 0.05348 | 0.4286 |
| 68–76 | 72 | 0.02549 | 0.3000 |

## 9.3 THE LESLIE MATRIX

While life tables and the accompanying equations tell us much about a population's age structure, they do not provide a simple means by which temporal changes in $n_x$ can be calculated. To make such calculations, ecologists generally make use of a technique devised by Leslie (1945).

Because it involves the use of matrices, a subject alien to many ecologists in the past, the Leslie approach has generally been deleted from courses and books, and ignored even by many active population ecologists. Its importance, however, makes its inclusion mandatory here. For the student unfamiliar with matrix algebra, some simple explanations and rules follow.

### 9.3.1 Matrices

In the simplest case a matrix is merely an array of numbers. It is usually designated by a capital letter or a small letter in brackets:

$$A = \{a_{ij}\} = \begin{pmatrix} a_{11} & a_{12} & a_{13}\cdots \\ a_{21} & a_{22} & a_{23} \\ a_{31} & a_{32} & a_{33} \\ \vdots & & \end{pmatrix}$$

A matrix may have any number of rows or columns but in this chapter we are concerned only with square matrices (number of rows = number of columns) and row or column matrices, often denoted by a capital letter and an arrow:

$$\vec{N} = (n_1 n_2 \ldots), \quad \text{or} \quad \vec{N} = \begin{pmatrix} n_1 \\ n_2 \\ n_3 \\ \vdots \end{pmatrix}.$$

*Addition of matrices* is done element by element.

$$A + B = \begin{pmatrix} a_{11} & a_{12} & a_{13} \ldots \\ a_{21} & a_{22} \\ a_{31} & a_{32} \\ \vdots \end{pmatrix} + \begin{pmatrix} b_{11} & b_{12} \ldots \\ b_{21} & b_{22} \\ b_{31} & b_{32} \\ \vdots \end{pmatrix}$$

$$= \begin{pmatrix} a_{11}+b_{11} & a_{12}+b_{12} \ldots \\ a_{21}+b_{21} & a_{22}+b_{22} \\ c_{31}+b_{31} & a_{32} \mid b_{32} \\ \vdots \end{pmatrix}.$$

Note that only matrices with the same number of rows and columns can be added. Several examples follow.

$$\begin{pmatrix} 1 & 0 & 2 \\ 3 & 3 & 1 \\ 0 & 0 & 1 \end{pmatrix} + \begin{pmatrix} 2 & 2 & 1 \\ 0 & 0 & 1 \\ 2 & 2 & 2 \end{pmatrix} = \begin{pmatrix} 1+2 & 0+2 & 2+1 \\ 3+0 & 3+0 & 1+1 \\ 0+2 & 0+2 & 1+2 \end{pmatrix} = \begin{pmatrix} 3 & 2 & 3 \\ 3 & 3 & 2 \\ 2 & 2 & 3 \end{pmatrix},$$

$$\begin{pmatrix} 1 & 0 \\ 0 & 2 \end{pmatrix} + \begin{pmatrix} 5 & 2 \\ 2 & 1 \end{pmatrix} = \begin{pmatrix} 1+5 & 0+2 \\ 0+2 & 2+1 \end{pmatrix} = \begin{pmatrix} 6 & 2 \\ 2 & 3 \end{pmatrix},$$

$$(2 \quad 1) + (3 \quad 3) = (2+3 \quad 1+3) = (5 \quad 4),$$

$$\begin{pmatrix} 5 \\ 5 \\ 1 \end{pmatrix} + \begin{pmatrix} 2 \\ 0 \\ 7 \end{pmatrix} = \begin{pmatrix} 5+2 \\ 5+0 \\ 1+7 \end{pmatrix} = \begin{pmatrix} 7 \\ 5 \\ 8 \end{pmatrix},$$

$$\begin{pmatrix} 2 & 0 & 5 \\ 3 & 1 & 1 \end{pmatrix} + \begin{pmatrix} 2 & 0 \\ 1 & 1 \end{pmatrix} = \text{undefined.}$$

*Scalar multiplication* occurs when a matrix is multiplied by a constant (a matrix of one row and one column).

$$kA = k \begin{pmatrix} a_{11} & a_{12} \ldots \\ a_{21} & a_{22} \\ \vdots \end{pmatrix} = \begin{pmatrix} ka_{11} & ka_{12} \ldots \\ ka_{21} & ka_{22} \\ \vdots \end{pmatrix}.$$

**Examples:**

$$3 \times \begin{pmatrix} 1 & 0 \\ 2 & 0 \end{pmatrix} = \begin{pmatrix} 3 \times 1 & 3 \times 0 \\ 3 \times 2 & 3 \times 0 \end{pmatrix} = \begin{pmatrix} 3 & 0 \\ 6 & 0 \end{pmatrix}$$

$$5 \times \begin{pmatrix} 2 \\ 2 \\ 1 \\ 0 \end{pmatrix} = \begin{pmatrix} 5 \times 2 \\ 5 \times 2 \\ 5 \times 1 \\ 5 \times 0 \end{pmatrix} = \begin{pmatrix} 10 \\ 10 \\ 5 \\ 0 \end{pmatrix}$$

$$\lambda^2 \begin{pmatrix} 1 & 0 & 0 & 2 \\ 2 & 3 & 3 & 1 \\ 1 & 1 & 1 & 2 \\ 0 & 2 & 2 & 1 \end{pmatrix} = \begin{pmatrix} \lambda^2 & 0 & 0 & 2\lambda^2 \\ 2\lambda^2 & 3\lambda^2 & 3\lambda^2 & \lambda^2 \\ \lambda^2 & \lambda^2 & \lambda^2 & 2\lambda^2 \\ 0 & 2\lambda^2 & 2\lambda^2 & \lambda^2 \end{pmatrix}.$$

The real usefulness of matrices lies in their properties in multiplication. We define *matrix multiplication* in the following manner: The value of the element in the $i$th row, $j$th column of the product matrix, $C = A \times B$, is given by the sum of the successive products of the elements of the $i$th row of the $A$ matrix and the $j$th column of the $B$ matrix. Consider the following matrices:

$$A = \begin{pmatrix} a_{11} & a_{12} & a_{13} \\ a_{21} & a_{22} & a_{23} \\ a_{31} & a_{32} & a_{33} \end{pmatrix} \qquad B = \begin{pmatrix} b_{11} & b_{12} & b_{13} \\ b_{21} & b_{22} & b_{23} \\ b_{31} & b_{32} & b_{33} \end{pmatrix} \qquad C = \{c_{ij}\} = A \times B.$$

To find the element of $C$ in the (say) second row, third column, $c_{23}$, we take all elements of the second row of $A$ and multiply them, in sequence, with the elements of the third column of $B$, and add

$$\begin{pmatrix} \cdot & \cdot & \cdot \\ \cdot & & c_{23} \\ \cdot & & \end{pmatrix} = \begin{pmatrix} a_{11} & a_{12} & a_{13} \\ a_{21} & a_{22} & a_{23} \\ a_{31} & a_{32} & a_{33} \end{pmatrix} \begin{pmatrix} b_{11} & b_{12} & b_{13} \\ b_{21} & b_{22} & b_{23} \\ b_{31} & b_{32} & b_{33} \end{pmatrix}$$

$$= \begin{pmatrix} & \cdot & \\ \cdots & a_{21}b_{13} + a_{22}b_{23} + a_{23}b_{33} \\ & \cdot & \end{pmatrix}.$$

Similarly

$$\begin{pmatrix} c_{11} \cdots \\ \vdots \end{pmatrix} = \begin{pmatrix} a_{11} & a_{12} & a_{13} \\ a_{21} & a_{22} & a_{23} \\ a_{31} & a_{32} & a_{33} \end{pmatrix} \begin{pmatrix} b_{11} & b_{12} & b_{13} \\ b_{21} & b_{22} & b_{23} \\ b_{31} & b_{32} & b_{33} \end{pmatrix}$$

$$= \begin{pmatrix} a_{11}b_{11} + a_{12}b_{21} + a_{13}b_{31} \cdots \\ \vdots \end{pmatrix}$$

Some numerical examples follow.

$$\begin{pmatrix} 1 & 2 & 5 \\ 0 & 1 & 1 \\ 2 & 0 & 1 \end{pmatrix} \begin{pmatrix} 3 & 2 & 2 \\ 1 & 1 & 2 \\ 2 & 2 & 2 \end{pmatrix}$$

$$= \begin{pmatrix} 1\times3 + 2\times1 + 5\times2 & 1\times2 + 2\times1 + 5\times2 & 1\times2 + 2\times2 + 5\times2 \\ 0\times3 + 1\times1 + 1\times2 & 0\times2 + 1\times1 + 1\times2 & 0\times2 + 1\times2 + 1\times2 \\ 2\times3 + 0\times1 + 1\times2 & 2\times2 + 0\times1 + 1\times2 & 2\times2 + 0\times2 + 1\times2 \end{pmatrix}$$

$$= \begin{pmatrix} 3+2+10 & 2+2+10 & 2+4+10 \\ 0+1+2 & 0+1+2 & 0+2+2 \\ 6+0+2 & 4+0+2 & 4+0+2 \end{pmatrix} = \begin{pmatrix} 15 & 14 & 16 \\ 3 & 3 & 4 \\ 8 & 6 & 6 \end{pmatrix},$$

$$\begin{pmatrix} 5 & 1 \\ 1 & 2 \end{pmatrix} \begin{pmatrix} 0 & 1 \\ 2 & 1 \end{pmatrix} = \begin{pmatrix} 5\times0 + 1\times2 & 5\times1 + 1\times1 \\ 1\times0 + 2\times2 & 1\times1 + 2\times1 \end{pmatrix} = \begin{pmatrix} 0+2 & 5+1 \\ 0+4 & 1+2 \end{pmatrix} = \begin{pmatrix} 2 & 6 \\ 4 & 3 \end{pmatrix},$$

$$(2 \quad 1 \quad 1) \begin{pmatrix} 3 & 0 & 3 \\ 1 & 1 & 2 \\ 2 & 0 & 0 \end{pmatrix}$$

$$= (2\times3 + 1\times1 + 1\times2 \quad 2\times0 + 1\times1 + 1\times0 \quad 2\times3 + 1\times2 + 1\times0)$$

$$= (6+1+2 \quad 0+1+0 \quad 6+2+0) = (9 \quad 1 \quad 8)$$

$$(3 \quad 0 \quad 1) \begin{pmatrix} 2 \\ 2 \\ 1 \end{pmatrix} = (3\times2 + 0\times2 + 1\times1) = 6 + 0 + 1 = (7)$$

$$\begin{pmatrix} 2 & 5 & 5 \\ 1 & 0 & 2 \end{pmatrix} \begin{pmatrix} 1 \\ 1 \\ 2 \end{pmatrix} = \begin{pmatrix} 2\times1 + 5\times1 + 5\times2 \\ 1\times1 + 0\times1 + 2\times2 \end{pmatrix} = \begin{pmatrix} 2 + 5 + 10 \\ 1 + 0 + 4 \end{pmatrix} = \begin{pmatrix} 17 \\ 5 \end{pmatrix}$$

$$\begin{pmatrix} 1 & 1 & 3 \\ 2 & 0 & 1 \\ 2 & 2 & 2 \end{pmatrix} \begin{pmatrix} 5 \\ 1 \end{pmatrix} = \text{undefined.}$$

The reader should verify to his own satisfaction that matrix multiplication is associative $(A(BC) = (AB)C)$, but not commutative (that is, $A \times B$ does not necessarily equal $B \times A$). It will also be helpful if the reader can get a feeling for the notation usually used in matrix algebra. For example, the formal definition of matrix multiplication:

$$C = A \times B$$

is written:

$$\{c_{ij}\} = \left( \sum_k a_{ik} b_{kj} \right).$$

Suppose we denote the diagonal, square matrix

$$\begin{pmatrix} 1 & 0 & 0 \dots \\ 0 & 1 & 0 \\ 0 & 0 & 1 \\ 0 & 0 & 0 \\ \vdots \end{pmatrix}$$

by the letter $I$. The product of $A \times I$, then, is given by

$$\{c_{ij}\} = \left( \sum_k a_{ik} i_{kj} \right).$$

But $i_{kj}$ is zero except when $k = j$. Thus $\{c_{ij}\}$ is zero for all values except $\{a_{ij} i_{jj}\} = \{a_{ij} \cdot 1\} = \{a_{ij}\}$. That is, $A \times I = \{a_{ij}\} = A$. Similarly, $I \times A = A$. $I$ is called the *identity matrix*.

**Examples:**

$$\begin{pmatrix} a_{11} & a_{12} & a_{13} \\ a_{21} & a_{22} & a_{23} \\ a_{31} & a_{32} & a_{33} \end{pmatrix} \begin{pmatrix} 1 & 0 & 0 \\ 0 & 1 & 0 \\ 0 & 0 & 1 \end{pmatrix}$$

$$= \begin{pmatrix} a_{11} \times 1 + a_{12} \times 0 + a_{13} \times 0 & a_{11} \times 0 + a_{12} \times 0 + a_{13} \times 1 \\ a_{21} \times 1 + a_{22} \times 0 + a_{23} \times 0 & a_{21} \times 0 + a_{22} \times 0 + a_{23} \times 1 \\ a_{31} \times 1 + a_{32} \times 0 + a_{33} \times 0 & a_{31} \times 0 + a_{32} \times 0 + a_{33} \times 1 \end{pmatrix}$$

$$= \begin{pmatrix} a_{11} & a_{12} & a_{13} \\ a_{21} & a_{22} & a_{23} \\ a_{31} & a_{32} & a_{33} \end{pmatrix},$$

$$\begin{pmatrix} 1 & 0 & 0 \\ 0 & 1 & 0 \\ 0 & 0 & 1 \end{pmatrix} \begin{pmatrix} 2 \\ 3 \\ 3 \end{pmatrix} = \begin{pmatrix} 1 \times 2 + 0 \times 3 + 0 \times 3 \\ 0 \times 2 + 1 \times 3 + 0 \times 3 \\ 0 \times 2 + 0 \times 3 + 1 \times 3 \end{pmatrix} = \begin{pmatrix} 2+0+0 \\ 0+3+0 \\ 0+0+3 \end{pmatrix} = \begin{pmatrix} 2 \\ 3 \\ 3 \end{pmatrix},$$

$$\begin{pmatrix} 2 \\ 3 \\ 3 \end{pmatrix} \begin{pmatrix} 1 & 0 & 0 \\ 0 & 1 & 0 \\ 0 & 0 & 1 \end{pmatrix} = \text{undefined.}$$

As a review of these basic operations, the following examples are presented.

$$3 \left[ \begin{pmatrix} 1 & 2 \\ 3 & 0 \end{pmatrix} + \begin{pmatrix} 2 & 1 \\ 1 & 2 \end{pmatrix} \right] \begin{pmatrix} 5 & 0 \\ 1 & 1 \end{pmatrix} + \begin{pmatrix} 0 & 1 \\ 1 & 1 \end{pmatrix} I$$

$$= 3 \begin{pmatrix} 1+2 & 2+1 \\ 3+1 & 0+2 \end{pmatrix} \begin{pmatrix} 5 & 0 \\ 1 & 1 \end{pmatrix} + \begin{pmatrix} 0 & 1 \\ 1 & 1 \end{pmatrix} \begin{pmatrix} 1 & 0 \\ 0 & 1 \end{pmatrix}$$

$$= 3\begin{pmatrix} 3 & 3 \\ 4 & 2 \end{pmatrix}\begin{pmatrix} 5 & 0 \\ 1 & 1 \end{pmatrix} + \begin{pmatrix} 0\times 1 + 1\times 0 & 0\times 0 + 1\times 1 \\ 1\times 1 + 1\times 0 & 1\times 0 + 1\times 1 \end{pmatrix}$$

$$= 3\begin{pmatrix} 3\times 5 + 3\times 1 & 3\times 0 + 3\times 1 \\ 4\times 5 + 2\times 1 & 4\times 0 + 2\times 1 \end{pmatrix} + \begin{pmatrix} 0 & 1 \\ 1 & 1 \end{pmatrix}$$

$$= 3\begin{pmatrix} 18 & 3 \\ 22 & 2 \end{pmatrix} + \begin{pmatrix} 0 & 1 \\ 1 & 1 \end{pmatrix} = \begin{pmatrix} 54 & 9 \\ 66 & 6 \end{pmatrix} + \begin{pmatrix} 0 & 1 \\ 1 & 1 \end{pmatrix} = \begin{pmatrix} 54 & 10 \\ 67 & 7 \end{pmatrix}.$$

$$\begin{pmatrix} 2 & 1 & 1 \\ 5 & 0 & 2 \\ 2 & 2 & 2 \end{pmatrix} - kI = \begin{pmatrix} 2 & 1 & 1 \\ 5 & 0 & 2 \\ 2 & 2 & 2 \end{pmatrix} - k\begin{pmatrix} 1 & 0 & 0 \\ 0 & 1 & 0 \\ 0 & 0 & 1 \end{pmatrix}$$

$$= \begin{pmatrix} 2 & 1 & 1 \\ 5 & 0 & 2 \\ 2 & 2 & 2 \end{pmatrix} - \begin{pmatrix} k & 0 & 0 \\ 0 & k & 0 \\ 0 & 0 & k \end{pmatrix} = \begin{pmatrix} 2-k & 1 & 1 \\ 5 & 1-k & 2 \\ 2 & 2 & 2-k \end{pmatrix}.$$

### 9.3.2 The Leslie Matrix

We now consider Leslie's fundamental ideas on population growth. We define the square matrix:

$$L = \begin{pmatrix} F_0 & F_1 & F_2 \cdots & \\ P_0 & 0 & 0 & \\ 0 & P_1 & 0 & \\ 0 & 0 & P_2 & \\ \vdots & & & \end{pmatrix}.$$

As before, $F_x$ is the fecundity in the age interval $x$ to $x + 1$ and $P_x$ is the probability of surviving through the same interval. To simplify understanding and maintain consistency, we note that if all reproduction occurs at one, well-defined season, the number of young at time $t + 1$ arising from individuals age $x - 1$ at time $t$, is

$$n_x(t + 1)m_x = n_{x-1}(t)(1 - q_{x-1})m_x.$$

Also $P_x = 1 - q_x$. Bearing this in mind the above matrix can be written in the form

$$L = \begin{pmatrix} (1-q_0)m_1 & (1-q_1)m_2 & (1-q_2)m_3 & . & . & 0 \\ (1-q_0) & 0 & 0 & 0 & . & 0 \\ 0 & (1-q_1) & 0 & 0 & . & 0 \\ 0 & 0 & (1-q_2) & 0 & . & 0 \\ . & . & . & & . & . & 0 \end{pmatrix}$$

We denote the number of individuals of all ages in a population with the column matrix

$$\vec{N}^{(t)} = \begin{pmatrix} n_0(t) \\ n_1(t) \\ n_2(t) \\ \vdots \end{pmatrix},$$

where $n_x(t)$ designates, as before, the number of individuals of age $x$ at time $t$. Note now that

$$L\vec{N}^{(t)} = \begin{pmatrix} (1-q_0)m_1 & (1-q_1)m_2 & \cdot & \cdot \\ (1-q_0) & 0 & 0 & \cdot \\ 0 & (1-q_1) & 0 & \cdot \\ \cdot & \cdot & \cdot & \cdot \end{pmatrix} \begin{pmatrix} n_0(t) \\ n_1(t) \\ n_2(t) \\ \cdot \end{pmatrix}$$

$$= \begin{pmatrix} \sum_x (1-q_{x-1})m_x n_{x-1}(t) \\ (1-q_0)n_0(t) \\ (1-q_1)n_1(t) \\ \cdot \end{pmatrix}.$$

But

$$\sum_x (1-q_{x-1})m_x n_{x-1}(t) = \sum_x n_x(t+1)m_x$$

is the number of young entering the population at time $t+1$, or $n_0(t+1)$ (see above). Also, $(1-q_x)n_{x-1}(t)$ is the number of individuals of age $x-1$ at time $t$ surviving to age $x$ at time $t+1 = n_x(t+1)$. Thus:

$$L\vec{N}^{(t)} = \begin{pmatrix} n_0(t+1) \\ n_1(t+1) \\ n_2(t+1) \\ \vdots \end{pmatrix} = \vec{N}^{(t+1)},$$

and the general expression is

$$\vec{N}^{(t)} = L\vec{N}^{(t-1)} = L(L\vec{N}^{(t-2)}) = \cdots = L^t\vec{N}^{(0)}. \tag{9.9}$$

Note that this equation takes into consideration the lag between the time of birth of an individual and the time it enters the breeding population, and that it permits the simultaneous calculation of $n_x(t)$ for all $x$ at any time. Note also that $c_x(t) = n_x(t)/\sum_x n_x(t)$ can be calculated by this method, whereas the method given in Eq. (9.6) applies only to the *stable* age distribution. Using this approach we may define our time intervals in weeks or days (if we have a computer handy for the calculations) rather than years and thus also account for seasonal periodicity.

Let us apply the Leslie matrix approach to some hypothetical examples.

**Example 1**

Suppose we are given the life table data:

| Age | $x$ | $q_x$ | $m_x$ |
|-----|-----|-------|-------|
| 0–1 | 0 | 0.6 | 0 |
| 1–2 | 1 | 0.6 | 1.25 |
| 2–3 | 2 | 0.5 | 3.75 |
| 3–4 | 3 | 1.0 | 0 |

Then

$$L = \begin{pmatrix} (0.4)(1.25) & (0.4)(3.75) & (0.5)(0) & 0 \\ (0.4) & 0 & 0 & 0 \\ 0 & (0.4) & 0 & 0 \\ 0 & 0 & (0.5) & 0 \end{pmatrix} = \begin{pmatrix} 0.5 & 1.5 & 0 & 0 \\ 0.4 & 0 & 0 & 0 \\ 0 & 0.4 & 0 & 0 \\ 0 & 0 & 0.5 & 0 \end{pmatrix}.$$

Let $\vec{N}^{(0)}$ be $\begin{pmatrix} 0 \\ 100 \\ 0 \\ 0 \end{pmatrix}$.

Then

$$\vec{N}^{(0)} = \begin{pmatrix} 0 \\ 100 \\ 0 \\ 0 \end{pmatrix}, \quad n_{\text{total}} = 100$$

$$\vec{N}^{(1)} = \begin{pmatrix} 0.5 & 1.5 & 0 & 0 \\ 0.4 & 0 & 0 & 0 \\ 0 & 0.4 & 0 & 0 \\ 0 & 0 & 0.5 & 0 \end{pmatrix} \begin{pmatrix} 0 \\ 100 \\ 0 \\ 0 \end{pmatrix} = \begin{pmatrix} 150 \\ 0 \\ 40 \\ 0 \end{pmatrix}, \quad n_{\text{total}} = 190$$

$$\vec{N}^{(2)} = \begin{pmatrix} 0.5 & 1.5 & 0 & 0 \\ 0.4 & 0 & 0 & 0 \\ 0 & 0.4 & 0 & 0 \\ 0 & 0 & 0.5 & 0 \end{pmatrix} \begin{pmatrix} 150 \\ 0 \\ 40 \\ 0 \end{pmatrix} = \begin{pmatrix} 75 \\ 60 \\ 0 \\ 20 \end{pmatrix}, \quad n_{\text{total}} = 155$$

$$\vec{N}^{(3)} = \begin{pmatrix} 0.5 & 1.5 & 0 & 0 \\ 0.4 & 0 & 0 & 0 \\ 0 & 0.4 & 0 & 0 \\ 0 & 0 & 0.5 & 0 \end{pmatrix} \begin{pmatrix} 75 \\ 60 \\ 0 \\ 20 \end{pmatrix} = \begin{pmatrix} 127.5 \\ 30 \\ 24 \\ 0 \end{pmatrix}, \quad n_{\text{total}} = 181.5$$

If we continue the process and plot $n_{total}$ against time, we obtain Fig. 9.5.

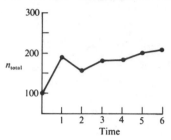

**Figure 9.5**

**Example 2.** Let

$$L = \begin{pmatrix} 0 & 10 & 0 \\ 0.2 & 0 & 0 \\ 0 & 0.1 & 0 \end{pmatrix}, \qquad \vec{N}^{(0)} = \begin{pmatrix} 100 \\ 0 \\ 0 \end{pmatrix}.$$

Then

$$\vec{N}^{(0)} = \begin{pmatrix} 100 \\ 0 \\ 0 \end{pmatrix},$$

$$\vec{N}^{(1)} = \begin{pmatrix} 0 & 10 & 0 \\ 0.2 & 0 & 0 \\ 0 & 0.1 & 0 \end{pmatrix} \begin{pmatrix} 100 \\ 0 \\ 0 \end{pmatrix} = \begin{pmatrix} 0 \\ 20 \\ 0 \end{pmatrix},$$

$$\vec{N}^{(2)} = \begin{pmatrix} 0 & 10 & 0 \\ 0.2 & 0 & 0 \\ 0 & 0.1 & 0 \end{pmatrix} \begin{pmatrix} 0 \\ 20 \\ 0 \end{pmatrix} = \begin{pmatrix} 200 \\ 0 \\ 2 \end{pmatrix},$$

$$\vec{N}^{(3)} = \begin{pmatrix} 0 & 10 & 0 \\ 0.2 & 0 & 0 \\ 0 & 0.1 & 0 \end{pmatrix} \begin{pmatrix} 200 \\ 0 \\ 2 \end{pmatrix} = \begin{pmatrix} 0 \\ 40 \\ 0 \end{pmatrix},$$

$$\vec{N}^{(4)} = \begin{pmatrix} 0 & 10 & 0 \\ 0.2 & 0 & 0 \\ 0 & 0.1 & 0 \end{pmatrix} \begin{pmatrix} 20 \\ 40 \\ 0 \end{pmatrix} = \begin{pmatrix} 400 \\ 0 \\ 4 \end{pmatrix}.$$

Continuing the process we obtain Fig. 9.6.

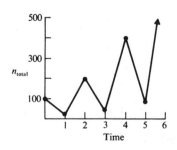

**Figure 9.6**

These hypothetical populations show the same kind of cycling observed in laboratory populations.

Continuing our discussion of the Leslie matrix we note that when stable age distribution comes about, each age group is multiplying at the same rate as every other age group. That is

$$n_x(t + 1) = Rn_x(t), \qquad \text{for all } x.$$

Written in matrix form this becomes

$$\vec{N}^{(t+1)} = R\vec{N}^{(t)},$$

or

$$\vec{N}^{(t+1)} = R\vec{N}^{(t)} = R(I\vec{N}^{(t)}) = (RI)\vec{N}^{(t)}.$$

But since $\vec{N}^{(t+1)} = L\vec{N}^{(t)}$, we can write

$$L\vec{N}^{(t)} = (RI)\vec{N}^{(t)}.$$

Hence

$$(L - RI)\vec{N}^{(t)} = \begin{pmatrix} 0 \\ 0 \\ 0 \\ \vdots \end{pmatrix}. \qquad (9.10)$$

We now note (Appendix 3) that because $(L - RI)$ is a square matrix with no all-zero rows, its inverse can be written

$$(L - RI)^{-1} = \frac{\text{Adj}\,(L - RI)}{|L - RI|}.$$

Suppose that the inverse exists—that is, suppose $|L - RI| \neq 0$. Then

$$(L - RI)^{-1}(L - RI)\vec{N}^{(t)} = (L - RI)^{-1}\begin{pmatrix} 0 \\ 0 \\ 0 \end{pmatrix} = \begin{pmatrix} 0 \\ 0 \\ 0 \end{pmatrix}$$

$$(L - RI)^{-1}(L - RI)\vec{N}^{(t)} = (I)\vec{N}^{(t)} \neq \begin{pmatrix} 0 \\ 0 \\ 0 \end{pmatrix}.$$

We thus have a contradiction, and must conclude that

$$|L - RI| = 0. \qquad (9.11)$$

What this means is shown by the following. Consider a population with three age classes:

$$L = \begin{pmatrix} (1-q_0)m_1 & (1-q_1)m_2 & 0 \\ 1-q_0 & 0 & 0 \\ 0 & 1-q_1 & 0 \end{pmatrix}.$$

Then (see Appendix 3)

$$|L - RI| = \begin{vmatrix} \begin{pmatrix} (1-q_0)m_1 & (1-q_1)m_2 & 0 \\ 1-q_0 & 0 & 0 \\ 0 & 1-q_1 & 0 \end{pmatrix} - R\begin{pmatrix} 1 & 0 & 0 \\ 0 & 1 & 0 \\ 0 & 0 & 1 \end{pmatrix} \end{vmatrix}$$

$$= \begin{vmatrix} \begin{pmatrix} (1-q_0)m_1 & (1-q_1)m_2 & 0 \\ 1-q_0 & 0 & 0 \\ 0 & 1-q_1 & 0 \end{pmatrix} - \begin{pmatrix} R & 0 & 0 \\ 0 & R & 0 \\ 0 & 0 & R \end{pmatrix} \end{vmatrix}$$

$$= \begin{vmatrix} (1-q_0)m_1 - R & (1-q_1)m_2 & 0 \\ (1-q_0) & -R & 0 \\ 0 & (1-q_1) & -R \end{vmatrix}$$

$$= \begin{pmatrix} (1-q_0)m_1 - R & (1-q_1)m_2 & 0 & (1-q_0)m_1 - R & (1-q_1)m_2 \\ (1-q_0) & -R & 0 & (1-q_0) & -R \\ 0 & (1-q_1) & -R & 0 & (1-q_1) \end{pmatrix}$$

$$= (R^2(1-q_0)m_1 - R^3 + 0 + 0) - (0 + 0 - R(1-q_0)(1-q_1)m_2)$$

$$= -R^3 + R^2(1-q_0)m_1 + (1-q_0)(1-q_1)m_2$$

$$= -R^3(1 - R^{-1}(1-q_0)m_1 - R^{-2}(1-q_0)(1-q_1)m_2).$$

But at stable age distribution, this must be zero. Thus

$$1 - R^{-1}(1-q_0)m_1 - R^{-2}(1-q_0)(1-q_1)m_2 = 0,$$

and

$$1 = R^{-1}(1-q_0)m_1 + R^{-2}(1-q_0)(1-q_1)m_2.$$

Since $1-q_0$ is the probability of living to age 1, and $(1-q_0)(1-q_1)$ is the probability of surviving to age 2, this expression finally becomes

$$1 = R^{-1}l_1 m_1 + R^{-2}l_2 m_2,$$

or, for the general case of several age groups,

$$1 = R^{-1}l_1 m_1 + R^{-2}l_2 m_2 + \cdots = \sum_x R^{-x}l_x m_x.$$

This equation should have a familiar look (see Eq. 9.5). For further discussion, see Appendix 4.

### 9.3.3 Limited Environments

Until now we have treated $m_x$ and $q_x$ as if they were invariant with population size. In dealing with population changes, however, this assumption of invariance can lead us wildly astray unless $n$ is well below the carrying capacity. There are a number of ways in which corrections can be made. One is given below. Suppose, for example, that all $m_x$, $1 - q_x$ values are depressed to a fraction $c(n) = 1 - n/K'$ of their maximum values. Then the effective value of $L$ is, in fact, $c(n)L$, and

$$\vec{N}^{(t+1)} = c(n) L \vec{N}^{(t)} = \left(1 - \frac{n(t)}{K'}\right) L \vec{N}^{(t)}.$$

If we let $K' = 400$ and re-use Example 1 from part 9.3.2 (Fig. 9.5), we obtain

$$\vec{N}^{(0)} = \begin{pmatrix} 0 \\ 100 \\ 0 \\ 0 \end{pmatrix} \ldots n = 100$$

$$\vec{N}^{(1)} = (1 - 100/400) \begin{pmatrix} 0.5 & 1.5 & 0 & 0 \\ 0.4 & 0 & 0 & 0 \\ 0 & 0.4 & 0 & 0 \\ 0 & 0 & 0.5 & 0 \end{pmatrix} \begin{pmatrix} 0 \\ 100 \\ 0 \\ 0 \end{pmatrix}$$

$$= (0.75) \begin{pmatrix} 150 \\ 0 \\ 40 \\ 0 \end{pmatrix} = \begin{pmatrix} 112.5 \\ 0 \\ 30 \\ 0 \end{pmatrix}, \quad n = 142.5$$

$$\vec{N}^{(2)} = (1 - 142.5/400) \begin{pmatrix} 0.5 & 1.5 & 0 & 0 \\ 0.4 & 0 & 0 & 0 \\ 0 & 0.4 & 0 & 0 \\ 0 & 0 & 0.5 & 0 \end{pmatrix} \begin{pmatrix} 112.5 \\ 0 \\ 30 \\ 0 \end{pmatrix}$$

$$= (0.644) \begin{pmatrix} 56.25 \\ 45 \\ 0 \\ 15 \end{pmatrix} = \begin{pmatrix} 36.23 \\ 29 \\ 0 \\ 9.66 \end{pmatrix}, \quad n = 74.9.$$

The results (dashed line) are superimposed in Fig. 9.7 on the maximum growth curve as shown in Fig. 9.5 (solid line). The depressing effects of population density are obvious.

To find the equilibrium population we first note that in the absence of population-depressing effects, the value of $R$ would take on the value $R_0$ such that $|L - R_0 I| = 0$. Next we note that since the effective value of $L$ is really

$cL$, the true value of $R$ must be such that $|cL - RI| = 0$. But, at equilibrium, $R = 1$. Thus $|cL - I| = 0$, so $|L - 1/c\,I| = 0$. But $|L - R_0 I| = 0$. Thus $1/c = R_0$. But $c$ has been defined as $(1 - n/K')$. Thus, at equilibrium,

$$R_0 = \frac{1}{1 - n/K'} = \frac{K'}{K' - n},$$

and

$$n = K'(R_0 - 1).$$

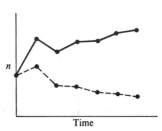

$$n \qquad \text{Time}$$

**Figure 9.7**

In the particular example used here,

$$|L - R_0 I| = \begin{vmatrix} 0.5 - R_0 & 1.5 & 0 & 0 \\ 0.4 & -R_0 & 0 & 0 \\ 0 & 0.4 & -R_0 & 0 \\ 0 & 0 & 0.5 & -R_0 \end{vmatrix} = R_0{}^4 - 0.5R_0{}^3 - 0.6R_0{}^2 = 0,$$

so

$$R_0 = 1.06,$$

and

$$\hat{n} \text{ (or } K) = (400)(1.06 - 1.00) = (400)(0.06) = 24.$$

As in the case with no density-dependent depression of the population (simple Leslie matrix approach), numbers tend to oscillate.

It is almost certainly unrealistic to suppose that both $1 - q_x$ and $m_x$ vary linearly with $n$ (see discussion of $\bar{b}$ and $\bar{d}$ in part 9.1.2). Nevertheless the approach used in this section must be considered superior to the logistic equation in that it accounts for age differences and the birth-to-breeding time lag. By writing

$$\vec{N}^{(t+1)} = \left(1 - \frac{n(t - \delta)}{K'}\right) L\vec{N}^{(t)}$$

rather than

$$\vec{N}^{(t+1)} = \left(1 - \frac{n(t)}{K'}\right) L\vec{N}^{(t)},$$

the method also allows for time lags in the effect of density on growth.

Note when stable age distribution occurs that population birth, $b$, and death, $d$, rates can be defined. In fact, where depression of survival and birth rate are as shown above, (and writing $m_x(\text{max}) = m_x^*$, $\dot{q}_x(\text{min}) = q_x^*$):

$$b = \sum_x m_x c_x = \sum_x m_x^* \left(1 - \frac{n}{K'}\right) c_x = \left(1 - \frac{n}{K'}\right) \sum_x m_x^* c_x,$$

which we write $(1 - n/K')b_0$, and

$$d = \sum_x q_x c_x = \sum_x \left[1 - (1 - q_x^*)\left(1 - \frac{n}{K'}\right)\right] c_x$$

$$= \sum_x [c_x - (1 - n/K')c_x + (1 - n/K')c_x q_x^*]$$

$$= n/K' + (1 - n/K') \sum_x c_x q_x^*,$$

which we write $n/K' + (1 - n/K')d_0$. Combining these expressions, we have

$$b - d = b_0(1 - n/K') - d_0(1 - n/K') - n/K'.$$

Now define

$$K'' = K' \frac{b_0 - d_0}{1 + b_0 - d_0}.$$

Substituting for $K'$, we obtain

$$b - d = (b_0 - d_0)(1 - n/K''),$$

or

$$r = r_0(1 - n/K'').$$

At stable age distribution, the technique defined in 9.3.3 is essentially identical to that described by the logistic equation. The important difference is that in the case of logistic growth, $b$, $d$, and $r$ are continuous variables ($r = d\ln n/dt = \ln R$), whereas above they are discrete variables.

Before discussing the evolution of birth and death rates, we must note that in many species—perhaps most species other than birds and mammals—mortality and fecundity may depend more on size (or some other character) than age. It is possible to modify the expressions of sections 9.2 and 9.3 so that all expressions remain valid. The symbol $m_x$ now refers to fecundity at size $x$, and $q_x$ is the probability of surviving the transition from size $x$ to $x + 1$. Since newborn individuals fit in the size class 0 (or minimum $x$), we can write (where $t_x$ is the time required to grow to size $x$):

$$n_0(t) = \sum_x n_x(t)m_x = \sum_x n_0(t - t_x)l_x m_x = \sum_x n_0(t)R^{-t_x}l_x m_x,$$

so that

$$1 = \sum_x R^{-t_x}l_x m_x. \tag{9.12}$$

The implicit assumptions are that $t_x$ is constant for a given $x$ (that growth rates for given sized individuals do not vary over time), and that the size distribution of individuals is constant.  When these two criteria are met, clearly there also exists a stable age distribution.  Stable age distribution is thus a necessary but not sufficient condition for the validity of Eq. (9.12).

## 9.4 THE EVOLUTION OF BIRTH AND DEATH RATES

One section of Chapter 6 was devoted to the subject of age of reproduction.  It is clear from what information we have, including the data of Leslie and Ranson given previously in this chapter, that fecundity varies with age.  In fact the relation between $m_x$ (or $F_x$) and age ($x$) follows a rather consistent pattern in all species, rising at first and then falling as the animal gets older.  It has also been noted that the age-specific mortality seems to fall, reaching a minimum in most species sometime prior to the age of earliest reproductive capability, and then rising again (see Caughley, 1966, for data on mammals).  Why these patterns occur, and the selective forces molding them, are of considerable interest to the population ecologist.

The most significant of the early attempts to examine selection for age specificity in mortality and fecundity is reported by Medawar (1957). Hamilton (1966) later made corrections in Medawar's approach and Emlen (1970), following Hamilton's method, extended the subject considerably.

Because it is necessary to make the genetic calculations tractable, we begin with these assumptions:

1.  The population is panmictic and the changes in gene frequency over one generation resulting from selection for $m_x$ or $q_x$ are very small.

2.  There is no epistasis.

3.  Those alleles affecting the trait in question ($q_x$ or $m_x$ for some $x$) do not affect mortality or fecundity at other ages.  That is, there is no pleiotropy.

4.  A stable age distribution exists.

Because there is a stable age distribution, each age group increases at the same rate over time by a factor $R$.  Thus if we take the unit of time over which $R$ is measured to be the time between successive breeding seasons, $R$ changes only very slightly (due to genetic change) from time unit to time unit, and we can use $R$ in place of $\overline{W}$ as a measure of fitness (see also discussion, Section III).  Let $p_k$ be the frequency of one of two alleles at the $k$th locus.  Then, after Chapter 1 (ignoring epistasis)

$$\Delta p_k = \frac{p_k(1-p_k)}{2R}\frac{\partial R}{\partial p_k}.$$

The change in $R$ due to the selected change in $p_k$ is given by the Taylor series

$$\Delta_{p_k}R = \frac{\partial R}{\partial p_k}\Delta p_k + \frac{1}{2}\frac{\partial^2 R}{\partial p_k^2}(\Delta p_k)^2 + \cdots$$

which, for $\Delta p_k$ very small (assumption above) becomes

$$\Delta_{p_k} R \rightarrow \frac{\partial R}{\partial p_k} \Delta p_k = \frac{p_k(1-p_k)}{2R} \left(\frac{\partial R}{\partial p_k}\right)^2.$$

If we denote by $Q_x$ the set of all loci affecting $q_x$, then the change in $R$ due to selected change of all $p_k$ in $Q_x$—that is the change in $R$ due to selected change in $q_x$—is given by

$$\Delta_{q_x} R = \sum_{p_k \in Q_x} \frac{\partial R}{\partial p_k} \Delta p_k = \sum_{p_k \in Q_x} \frac{p_k(1-p_k)}{2R} \left(\frac{\partial R}{\partial p_k}\right)^2.$$

If there are no pleiotropy effects,

$$\frac{\partial R}{\partial p_k} = \frac{\partial R}{\partial q_x} \frac{\partial q_x}{\partial p_k},$$

so that

$$\Delta_{q_x} R = \sum_{p_k \in Q_x} \frac{p_k(1-p_k)}{2} \left(\frac{\partial q_x}{\partial p_k}\right)^2 \frac{1}{R} \left(\frac{\partial R}{\partial q_x}\right)^2 = \Phi_q(x) \frac{1}{R} \left(\frac{\partial R}{\partial q_x}\right)^2. \tag{9.13}$$

The value on the left of this equation can also be written

$$\Delta_{q_x} R = \frac{\partial R}{\partial q_x} \Delta q_x + \frac{1}{2} \frac{\partial^2 R}{\partial q_x^2} (\Delta q_x)^2 + \cdots$$

which, for $\Delta q_x$ very small (assumption again), approaches

$$\Delta_{q_x} R = \frac{\partial R}{\partial q_x} \Delta q_x. \tag{9.14}$$

Setting the $\Delta q_x R$ value of Eqs. (9.13) and (9.14) equal, we obtain

$$\Delta q_x = \Phi_q(x) \frac{1}{R} \frac{\partial R}{\partial q_x}. \tag{9.15}$$

The same argument can be followed with respect to fecundity, in which case:

$$\Delta m_x = \Phi_m(x) \frac{1}{R} \frac{\partial R}{\partial m_x}. \tag{9.16}$$

In Eqs. (9.16) and (9.15), $\Phi$ is clearly a function of the frequency of alleles in the genome and the effect a change in allele frequency has on $q_x$ or $m_x$. It is thus a measure of the ability of the population to respond to selection. When only two alleles occur at each locus, $\Phi_q(x) = V_A(q_x)$ (see 2.4). The terms $\Delta q_x$ and $\Delta m_x$ are the changes in mortality and fecundity affected by selection and $|(1/R)(\partial R/\partial q_x)|$, $|(1/R)(\partial R/\partial m_x)|$ are the selection pressures acting to bring about those changes. The calculation of these selection intensity equations follows from Eq. (9.5):

$$1 = \sum_y R^{-y} l_y m_y,$$

where

$$l_y = (1 - q_0)(1 - q_1) \cdots (1 - q_{y-1}).$$

Differentiate both sides of the equation by $q_x$. Then

$$0 = \sum_y - yR^{-y-1} \frac{\partial R}{\partial q_x} l_y m_y + \sum_y R^{-y} \frac{\partial l_y}{\partial q_x} m_y.$$

But

$$\frac{\partial l_y}{\partial q_x} = \frac{\partial}{\partial q_x} [(1 - q_0)(1 - q_1) \cdots (1 - q_{y-1})]$$

is zero for $y < x + 1$, and

$$(1 - q_0)(1 - q_1) \cdots (1 - q_{x-1})(-1)(1 - q_{x+1}) \cdots (1 - q_{y-1})$$

$$= -\frac{(-q_0)(1 - q_1) \cdots (1 - q_x) \cdots (1 - q_{y-1})}{(1 - q_x)} = -\frac{l_y}{1 - q_x}$$

for $y \geqslant x + 1$. Thus

$$0 = \sum_{y \geqslant 0} - yR^{-y-1} l_y m_y + \sum_{y \geqslant x+1} R^{-y} l_y m_y \cdot \frac{1}{1 - q_x},$$

so that

$$\frac{1}{R} \frac{\partial R}{\partial q_x} = -\frac{\sum_{y \geqslant x+1} R^{-y} l_y m_y}{\sum_{y \geqslant 0} yR^{-y} l_y m_y} \cdot \frac{1}{1 - q_x}. \tag{9.17}$$

Since, prior to the age of first reproduction, selection pressure on $q_x$ varies inversely with $1/(1 - q_x)$, unless the capability of the genetic machinery to respond, $\Phi_q(x)$, changes with age, an equilibrium should be reached during this stage in which $q_x$ and $(1/R)(\partial R/\partial q_x)$ remain constant.

Differentiating both sides of Eq. (9.5) by $m_x$, we have

$$0 = \sum_y - yR^{-y-1} - \frac{R}{m_x} l_y m_y + R^{-x} l_x,$$

so that

$$\frac{1}{R} \frac{\partial R}{\partial m_x} = \frac{R^{-x} l_x}{\sum_{y \geqslant 0} yR^{-y} l_y m_y}. \tag{9.18}$$

The general forms of the selection intensity versus age curves, as generated from the above equations, with $R = 1$, are shown in Figs. 9.8(a) and (b). Alpha denotes first age of reproduction.

Referring to Fig. 9.8(a), consider a deleterious trait which affects survival and which crops up variably at ages $a$ or $b$. We assume that some of the variability in the age of onset of this trait is additive–genetic (Chapter 2). Then, since selection intensity against the trait is greater at age $a$ than at age $b$, the trait will be more suppressed at the earlier age and the mean age of its onset will be pushed back toward old age. The same argument can be made for deleterious traits affecting fecundity. Deleterious characters will tend, under natural selection, to drift toward old age. Beneficial traits clearly will be pushed toward—and perhaps beyond—age of first reproduction. $[(1/R)\,(\partial R/\partial m_x)]$ values do not apply prior to this age since no reproduction occurs at all here].

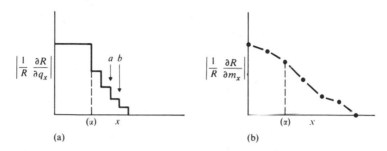

**Fig. 9.8.** Selection intensity as a function of age : (a) selection intensity on mortality rate, (b) selection intensity on fecundity.

We note now that the final distribution of beneficial and deleterious traits will result from an equilibrium of selection "forces" acting as described above and opposing selection "forces" or mechanical limitations. Young individuals grow in size and, despite our earlier second assumption, mortality and fecundity at different ages are not independent; therefore the general result of pushing beneficial traits up to age $\alpha$, and bad traits toward old age, is to produce a $q_x$ curve which drops from birth to some age at or prior to $\alpha$, and then rises throughout life. As stated previously, this is the usual pattern observed in nature. An oyster or a fish which grows throughout life and becomes more and more immune, therefore, from physical—and perhaps predatory—danger, represents a case in which mechanical considerations may advance the age of minimum mortality. In species such as this the $q_x$ curve continues to fall until very late in life and offers a quite different appearance. When viewed closely, however, the general pattern is the same.

If maximum efficiency in reproduction occurs at age $k$, then clearly the nature of development dictates that $k$ should not precede $\alpha$, but probably follow it. Thus fecundity should first rise with age (after $\alpha$), and then fall. In spite of the fact that selection intensity for successful reproduction may occur at (say) age one, increasing size or experience in finding and attracting a mate and caring for young, may extend the age of maximum fecundity beyond age one. Thus such factors as growth and learning affect the details of the age-specific mortality and fecundity curves, but their general shape remains the same (Figs. 9.9a and b). Figures 9.10(a) and (b) below show the $F_x$ and $q_x$ curves for man—the first is taken from Keyfitz (1968) and represents the U.S. population in 1964, the second, from Pearl (1940), represents the U.S. population from 1929 to 1931.

The effects of environmental perturbations and population fluctuations on the evolution of $q_x$ and $m_x$ curves are beyond the scope of this book. The interested reader is referred to Emlen (1970).

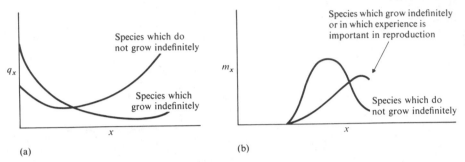

Fig. 9.9. Mortality rate as a function of age.

Note that in species which suffer low mortality, by virtue of better buffering against the physical environment—large animals—or by virtue of being predators rather than prey, the curves in both Figs. 9.9 and 9.10 are extended to the right, favoring increased fecundity and survivorship at older ages. Opportunistic and small species should display shorter life spans even when protected from harm than large species; in the former, there has simply been little selection pressure to keep the homeostatic properties viable at advanced ages.

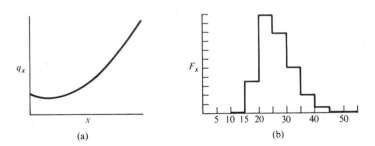

Fig. 9.10. Age-specific mortality and fecundity in man.

## 9.5 APPARENT EXCEPTIONS

While the arguments presented in this chapter are sound, their application to the real world may often be less so. It has been noted that the size of a new year class, $n_0$, should be proportional to the number of breeding animals:

$$n_0 = bn = n \sum_x c_x m_x.$$

But if environmental fluctuations are great relative to a species' homeostatic capabilities, $m_x$ may vary so wildly from year to year that the relation between $n_0$ and $n$ becomes little more than a scatter. In such situations it is highly unlikely that anything even approaching a stable age distribution could come about. Such difficulties are not great when we deal with large homeotherms. They are not insignificant in the case of small homeotherms, however, and in most plants and poikilotherms they may render the arguments made in this chapter almost meaningless. Species which are so affected by environmental fluctuations are generally opportunistic (6.3.1) and by virtue of the helplessness of their many small seeds, eggs, and/or larvae, most opportunistic species fit the arguments of this chapter only in a very rough manner. Such species, through the tremendous mortality of their young, are in addition likely to show great genotypic differences (perhaps reflecting themselves in $m_x$, $q_x$ differences) from place to place and year class to year class, which accentuates the difficulty.

There are also a large number of species which proliferate clonally for long periods of time between occasional bouts of sexual reproduction. In the case of such species the expressions given in parts 9.1 through 9.4 are clearly over-simplified. The expressions also do not account for the rather common practice of sex changing (serial hermaphroditism).

Except in the case of commercially important fish, expressions adequately describing population growth in such seemingly unbehaved species are practically nonexistent. Rules devised by fisheries people tend to be rough and empirical. Clearly there is need for research in this field. Meanwhile, bear in mind that the discussions in this chapter (augmented where necessary to account for sex changes, bursts of sexual or asexual reproduction, etc.) are generally applicable, if very rough. The difficulties are largely statistical and not substantive.

# 10
# Population Regulation

As Lack (1954) points out, the degree to which populations fluctuate in size falls far short of the amplitudes of which they are capable. Gunn and Symmons (1959) report that the red locust, *Nomadaeris septemfasciata*, increases as much as 750 times over several years and 140 times over two seasons of outbreak, and some insect species may vary as much as 10,000-fold over several years. But even this is less than what might occur given the growth potential of these small species. Suppose, for example, a birth rate of two million eggs per female per year and a sex ratio of $1 : 1$. Then each individual produces on average $10^6$ young per year and the population adds $10^6 n$ new members. At equilibrium, $10^6 n$ also die each year. But a variation in fecundity or number of deaths of only one percent results in a new population of size

$$n' = n + 10^6 n - 10^6 n \pm (0.01 \times 10^6 n) = n \pm 10^4 n \qquad \text{(really } n + 10^4 n \text{ or } 0\text{)}.$$

Thus either $R = n'/n \approx 10^4$, or the population becomes extinct.

Terrestrial vertebrates fluctuate much less in number, about the most variable being the voles which may vary as much as 100-fold over three or four years, but usually much less. Birds seldom fluctuate over more than an order of magnitude and larger mammals still less. Yet the deer-mouse, *Peromyscus maniculatus* (Blair, 1940) has an average litter size of four, young females produce on average two litters in mid-summer and fall, and older females two in the spring as well. One old pair, then, produces $2 \times 2$ litters $\times 4$ young $= 16$ young per year, and the first eight of these produce $8/2$ pairs $\times 2$ litters $\times 4$ young $= 32$ more. The old pair thus produces $16 + 32 = 48$ descendants for an $R$ value of $48/2 = 24$. Note the potential of such an $R$ value. In two years, the population will multiply $24 \times 24 = 576$-fold, and in three, $24^3 = 13,824$ times. If the population is to remain constant, exactly 22 of every 24 young must die before reaching reproductive age. (Actually this figure may be lowered slightly to allow more reproduction the first year, since there will be some die-off before the second).

In this case the proportion surviving is $2/24 = e^{-\delta}$, where $\delta$, the instantaneous death rate, is 2.485. Note that if the death rate is doubled, the proportion surviving to reproduce is $e^{-2(2.485)} = e^{-4.97} = 1/144$, so that each parental pair leaves $24 \times 1/144 = 1/6$ of an offspring on average—clearly the population would become extinct quite rapidly $(R = (1/6)/2 = 1/12)$. On the other hand, if $\delta$ were halved, the number of young surviving would be $24 \times e^{-1/2(2.485)} = 6.9$, so that $R = 6.9/2 = 3.45$, and the population would increase rapidly. If the population is to remain within narrow limits, the death rate must average very close to 2.485 and never stray far from the birth rate. The probability that this should occur fortuitously in species after species seems vanishingly small, an observation which has led many ecologists to believe that populations are regulated.

## 10.1 DENSITY REGULATION

Those who believe that populations are regulated have suggested two means by which regulation might come about.

Wynne-Edwards (1962) feels that population regulation mechanisms have evolved because they benefit a population: Individual members avoid running short of food or shelter and are less likely to attract predators. Voluntary depression of $R$ becomes the mechanism for a relatively hunger-free and predation-free existence. Unfortunately the evolution of population regulation to this end would require group selection to override natural-selective pressures acting to increase $R$. This explanation of population regulation, therefore, will not be seriously considered in this book.

The other suggestion is that regulation is simply an unavoidable fact of life: Populations which grow too large run out of some resource, or attract predators, and thereby limit their own size. Populations which decline over long periods of time simply die out and are, or may be, replaced by individuals from more successful, adjacent populations. If this explanation is correct, then populations are limited at an upper bound but not at a lower bound. Such populations, although often referred to as regulated are, more accurately, limited. The factor(s) which limit the population, the *limiting factors*, clearly must be inversely related to population size or density (number per unit area) and are labeled *density-dependent*. Factors totally unrelated to population size are called *density-independent*.

### 10.1.1 Historical Perspective

In spite of the narrow ranges within which many populations fluctuate, a large number of workers, primarily insect ecologists, have insisted that some populations fluctuate with no indication of density-dependent influences. These workers note that insects suffer high mortality in bad weather and may also cut back on reproduction as such times, often by entering diapause. In good weather, on the other hand, insect populations show rapid expansion. In other words, growth seems to be affected primarily by the physical environment so that control, if it exists, must be due to density-independent factors. Simple diffusion, or perhaps some behavioral response to population density, tends to buffer fluctuations by providing net migration from high-density to low-density areas. Andrewartha (1957) notes that the grasshopper *Austroicetes cruciata* eats grass, the supply of which is never short. However, in dry weather, the green grass turns brown and becomes inedible and the sparsity of green grass leads to an energy loss for most individuals. Thus the grasshoppers—or many of them—starve. The amount of wet and dry weather is clearly independent of grasshopper numbers. Davidson and Andrewartha (1948), working with the thrip, *Thrips imaginis*, discovered that changes in population density were highly correlated with changes in rainfall, evaporation rates, and temperature. Other workers have compiled considerable data also supporting the notion that almost all insect populations fluctuate with changes in weather and not with density-dependent factors. In fish, Watt (1968) points out that 94% of the variance from year to year in recruitment of four-year-old small-mouth bass, *Micropterus dolomieui*, is accounted for by the temperatures between June and October in the year when the recruits are spawned, and Beverton and Holt (1957, p. 72) state: "We know of no definite evidence showing the dependence on density of the natural mortality rate in adult fish populations." Further discussion of this viewpoint can be found in Andrewartha and Birch (1954).

In answer to the insect ecologists, the vertebrate ecologists have likened unregulated population growth to a random walk in which the size of the steps is undetermined. One might think of such unregulated populations in terms of the drunk on the subway platform, lurching about randomly. Eventually he will either find himself in the pit and extinct, or an infinite distance away. Only if he has a tendency to move away from the pit and only if his outward progress is stopped by the walls of the station (and these are both position or density-dependent factors) will he, in the long run, remain on the platform. Unregulated populations must eventually random-walk either to extinction or to very large sizes and, ultimately, infinite sizes. In addition, the vertebrate ecologists have pointed out that weather, the factor which Andrewartha and Birch claim as the prime determinant of insect population size, is really not density-independent. Thus in low-density populations most individuals may find adequate shelter from heavy rains or cold temperature, but fewer do so in high-density populations. In such a case, mortality is obviously higher in denser populations and weather must be considered density-related. At high enough population densities it may be population limiting. Note that a shortage of shelter does not imply an absolute shortage in terms of limited numbers of shelter areas; it may simply be that they are hard to locate in a hurry so that not all individuals find them when they are needed. Smith (1961) illustrated rather nicely that the effects of weather must be density-dependent (assuming thrips really are affected almost entirely by changes in weather) when he found from the thrips data of Davidson and Andrewartha that the changes in population were related to population density. In 14 years of data, he found significant negative correlations between population density and subsequent change in population density.

MacArthur (1958) took a somewhat different approach in his study of warblers. Looking at population records on a year-to-year basis he listed, in order, the years of increase, $I$, and decrease, $D$, and noted the number of runs (consecutive series of $I$'s and $D$'s). In the case of the black-throated green warbler, *Dendroica virens*, he found 27 runs in 28 consecutive changes (29 years). The blackburnian warbler, *D. fusca*, showed 19 runs in 23 changes. In both cases, the number of runs is significantly higher (and thus the length of the runs significantly lower) than would be expected in a random sequence of *I*'s and *D*'s, implying that an *I* was more likely to be followed by a *D* than another *I*, and vice versa. This suggests density regulation of the warbler populations.

The effect of density-dependent factors on bird reproduction is neatly shown by the data of Lack (1966). The mean clutch size of great tits, *Parus major*, clearly declines as the number of breeding pairs in an area increases (Table 10.1).

**Table 10.1.** Clutch size and population density

| Number of pairs/area | Average clutch size | Number of years of observation |
|:---:|:---:|:---:|
| 1–10 | 11.1 | 1 |
| 11–20 | 9.9 | 1 |
| 21–30 | 9.8 | 5 |
| 31–40 | 9.4 | 3 |
| 41–50 | 8.9 | 3 |
| 51–60 | 8.7 | 2 |
| 61–70 | — | — |
| 71–80 | — | — |
| 81–90 | 8.0 | 1 |

At higher densities fish grow more slowly, and smaller fish usually lay fewer eggs (part 6.4). Fruit flies lay fewer eggs when crowded, there is a lower percentage of individuals pupating, and the emerging adults are smaller so that, in turn, fewer eggs are laid. In *Daphnia pulex*, $q_x$ for young individuals increases with population density (Frank, Boll, and Kelly, 1957). More information of this sort is available in Lack (1954).

## 10.1.2  A Simple Synthesis

It seems clear that all populations must be limited by density-related factors, and yet—at least at most insect population levels—weather, the primary influence on population change, is claimed by some to be density-independent.

For years insect and vertebrate ecologists have disagreed on this matter and some continue to disagree. The problem has been neatly resolved by Horn (1968).

First, divide all population-influencing factors into two categories, density-dependent and density-independent. If we consider only the first, it is clear that $dn/dt$, where $n$ is population density, increases with $n$ at first and then declines. A guide to density-mediated population change, though it is only a very rough approximation, is given by the logistic equation, $dn/dt = r_0 n(1 - n/K)$ (Fig. 10.1a). We let this curve be that which obtains under optimal conditions with respect to density-independent factors ($r_0$ takes on its maximum value). Now consider the effect on $dn/dt$ of density-independent factors—that is, those factors which affect $r_0$ and thus affect $dn/dt$ proportionately to $n$. The plot of $dn/dt$ against $n$ is a straight line (Fig. 10.1b). We suppose that the effects of population-regulating and density-independent factors are additive and note that since Fig. 10.1(a) assumes optimal conditions, the density-independent effect can only be negative (Fig. 10.1b). We now turn Fig. 10.1(b) around and superimpose both curves (Fig. 10.2), which we shall designate $D$, density-related, and $I$, density-independent. Equilibrium, $\hat{n}$, occurs when $D - I = 0$.

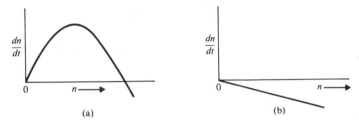

**Fig. 10.1.** Rate of population change as a function of population size : (a) only density-dependent factors are operating, (b) only density-independent factors are operating.

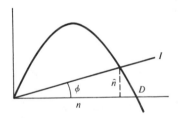

**Fig. 10.2.** (See text for explanation.)

Now consider two situations, one in which density-independent effects are very great ($\phi$ is large, Fig. 10.3a) and one in which they are slight ($\phi$ is small, Fig. 10.3b). In both cases, if $I$ varies between positions 1 and 2, the value of $\hat{n}$ will vary between $s$ and $u$ or $t$ and $v$. If $D$ varies between positions 1 and 2, $\hat{n}$ will vary between $s$ and $t$ or $u$ and $v$. In the case of Fig. 10.3(a), the distances $su$ and $tv$ are less than $st$ and $uv$. Thus most variation in $n$ is due to changes in $I$. In Fig. 10.3(b), $su$ and $tv$ are considerably greater than $st$ and $uv$, meaning that most variation in $n$ is due to $D$. In both cases populations are limited (by population-limiting, density-dependent factors) but where density-independent factors have a strong effect on mortality and fecundity it is possible that virtually all population variation can be accounted for by factors unrelated to population density. In view of the fact that insect populations fluctuate with much greater amplitude than most vertebrate populations (fish are exceptions), and in view of the fact that insect populations are more prone to the effects of changes in density-independent factors (being smaller and thus less buffered against the physical environment) and thus correlate highly with them, it is not surprising that the insect ecologists and vertebrate ecologists have staked out opposing positions. It is unfortunate that these extreme positions are still held in some quarters.

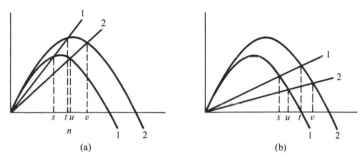

**Fig. 10.3.** (See text for explanation.)

A few more words need to be said about the role of migration in population limitation. It is widely observed that animals tend to leave population-dense areas, thus buffering changes in density and, over large areas, appearing to regulate local populations (Calhoun and Webb, 1953). In the grasshopper *Phaulacridium vittatum*, density-related mass migrations are commonplace. Before the migration takes place, the high densities result in food shortage and consequent cannibalism and corpse eating (Clark *et al.*, 1967). After migration these habits disappear. Lidicker (1962) has suggested that some locusts may have evolved a tendency to migrate even before food becomes short. Of course, this assumes that over the majority of recent evolutionary time the locusts, when local densities became high, had somewhere better to go; otherwise, selection would have favored those remaining. When local outbreaks are not highly correlated with each other in time, Lidicker suggests migration might be expected to evolve and regulate local populations. The population over an entire collection of local populations in certain situations may also be regulated by migration. When the unusual event occurs of simultaneous outbreaks over most of the area, the migratory habit will still manifest itself even though it can no longer relieve population pressure. Since most local areas are densely populated, many of the migrants must settle in submarginal areas where they die or fail to reproduce. The average fitness for the entire population thus declines and further population growth is damped or even stopped.

Even in the absence of mass migrations, peripheral populations (see 4.4) may be held at levels where density-dependent factors never operate. Here local populations may have fitnesses lower than one, their numbers being held above zero by immigration from more beneficent environments. The number of individuals per unit area is directly dependent upon local, density-independent factors and indirectly dependent on density-related factors in the population from which the immigrants came (Nicholson, 1957).

It is possible, of course, that while survival is largely under the influence of density-independent factors, fecundity is primarily density-related. By killing off large numbers of individuals of such a species, then, one relieves the breeding population of depressing effects with the result that next year's crop of this species is not diminished but greatly increased. This is one of the hazards connected with the use of chemical pesticides.

It has also been argued that populations of one species are limited by densities of other species and that density-dependent factors are not of critical importance or, perhaps, not even of negligible impact. Two examples are cited in Andrewartha and Browning (1961): The tsetse fly, *Glossina morsitans*, lays its eggs on antelopes which appear to be unaffected by that fact. Thus the fly population cannot be regulated through its ill effects on its hosts but rather simply by the ease with which it finds a host, a consideration directly dependent on antelope population density. The sheep tick *Ixodes ricinus* waits in grass for a passing sheep onto which it leaps. To complete its life cycle, the tick must find a sheep, feed, drop off, and repeat the performance two or more times. The probability of surviving to reproduce would seem strictly independent of tick population density, but directly related to sheep population density.

## 10.2 POPULATION-LIMITING FACTORS

In a classic paper, Hairston, Smith, and Slobodkin (1960) proposed some general rules for the factors limiting terrestrial populations. The world is green—that is, not all plant food is devoured, which implies that herbivores do not run short of food. But if herbivores do not run short of food then their mortality and fecundity should not be affected by food shortage, so that some other factor must limit their populations. The most likely alternative is predation. Furthermore, since the plants are not eaten down—in fact the major part of plant matter dies and decays —then the plants cannot be herbivore-limited but must be resource-limited. But if carnivores hold their prey populations down then they are, in fact, limiting their own food supply and so must be food-limited. While to a very rough degree these generalizations seem true, it would be a mistake to accept their derivation at face value. In spite of the authors' words, some of their logic can, in fact, be disputed; and there are other objections. Two papers in particular have leveled criticisms, some valid and many not, at the reasoning of Hairston, Smith, and Slobodkin (Murdoch, 1966; Ehrlich and Birch, 1967). Criticisms, some taken from these two papers, are presented below.

It is not true that a green world implies herbivores not to be food-limited. The important point is not whether food is available but whether it is easily enough harvested and sufficiently rich in energy and nutrient to result in a *net* gain to the herbivores. Deer would starve in a forest if no green, or only a small amount of green, remained below six or eight feet above the ground. On the island of Rottnest off Western Australia the quokka, *Seton brachyurus*, a small marsupial, dies of starvation in dry times in the midst of plenty. The food is present but its protein content is not high enough to sustain the animals (Main *et al.*, 1959). In the tropics, many phytophagous (plant-eating) insects eat shoot tips but ignore older leaves. It may be reasonable to assume that older leaf tissue does not contain the materials essential to the insects' survival or reproduction. Food need not be sparse in an absolute sense but only in the sense relative to the cost of its acquisition and to its nutrient and caloric yield. Limitation by a factor which is apparently not in any short supply has been termed limitation through *relative shortage* of a resource (Andrewartha and Birch, 1954). Note carefully, though, that relative shortages will limit populations only if the cost of acquisition is density-related.

Predators tend to catch only very young, post-reproductive, or sick prey. Thus, though predators may be limited by their prey (a relative shortage of prey), they may, in turn, affect populations of that prey only slightly. This fact does not affect Hairston, Smith, and Slobodkin's argument with respect to herbivores and carnivores, but does affect their logic regarding plants and herbivores; that is, "predators being food-limited does not imply their food is predator-limited," is equivalent to "plants are not herbivore-limited does *not* imply herbivores are not food-limited."

In general the hypotheses of Hairston, Smith, and Slobodkin seem true for large, terrestrial organisms, but less true elsewhere.

### 10.2.1 Changes in the Environment

We now discuss in more detail some of the factors which may play a role in population limitation.

Populations may be limited by self-induced changes in the environment which are detrimental to their further growth. In Gause's (1934) experiments with protozoans, population growth of competing species was inhibited by the production of waste products. *Tribolium* populations may be inhibited by the release of gases by their members. Sparse populations will exist under low concentrations of waste products since biodegradation balances waste production at very low levels. Dense populations will have to put up with much higher concentrations which may prove detrimental to further growth. This should be especially true for populations well above their normal density and the waste product concentrations they have evolved to cope with or even take advantage of (Ludwig and Boost, 1939).

### 10.2.2 Resources

In animals, mineral nutrients may be limiting secondarily, through their effects on plant abundance or ion concentrations in plants. Nikolskii (1963) notes that catches of pelagic fish vary with the phosphorus content of water. But in general when we speak of resources as limiting factors in animals we are speaking of the food supply in terms of total biomass (weight of organic material) or number of prey items per unit area.

Slobodkin (1954) fed *Daphnia obtusa* on algal food in the laboratory and found an almost linear relation between equilibrium population density, $\hat{n}$, and food level for $(50 \leqslant \hat{n} \leqslant 250)/50$ c.c. of pond water. Clearly, in this laboratory situation, *Daphnia* was food-limited. Eisenberg (1966) placed freshwater snails, *Lymnaea elodes*, in small enclosures in their natural environment and discovered that increasing or decreasing the number of snails or the amount of predation had virtually no effect on the population density the next year. Nor did mortality change with a change in population density. But when additional food, in the form of spinach, was added, fecundity increased and the next year's population of young individuals was up four to nine-fold. In the whirligig beetle, *Dineutes*, at least in the laboratory, food shortage leads to cannibalism (Istock, 1966) and thus, presumably, would act as a population-limiting factor. Ward (1965) has shown that the African quelea, *Quelea quelea*, fights for food in the lean season and suffers higher mortality when food supply drops, and that high food density is correlated with population density. There seems little doubt that the quelea is (at least partially) food-limited.

Resource limitation in plants occurs in two basic ways. First, where water is scarce it may be competed for, thus limiting plant growth and reproduction. In desert communities, survival may depend on a plant's ability to put out a massive network of shallow roots to trap water quickly over as great an area as possible. Such strategy precludes close spacing of plants and results in bare areas of ground. Of course, this strategy is beneficial only so long as the water acquired can be held, and only so long as the sparse precipitation is fairly regular. In areas where there may be occasional very long droughts, plants are more likely to sink long tap roots to the water table. In this situation also the plant whose roots grow fastest survives. Interestingly, in extreme deserts such as those found on the Peruvian coast, where rains may come only once in several years, water may not be the limiting factor. Here, neither the wide, shallow, nor deep tap root strategy benefits the plant, for water is too scarce and the water table too low. These plants usually possess no adaptations for dry climate except the ability of their seeds to remain dormant for long, hot, dry periods of time. When the rain comes, the plants germinate, grow rapidly, set seed, and die. The limiting factor in this case may be soil nutrients.

Mineral nutrients, especially in the form of phosphates and nitrates, play an extremely important role in plant population limitation. This is made clear when we note that the addition of fertilizer to soil results in increased growth rates under a tremendous variety of conditions. Different species have different soil chemistry requirements and tolerances, and the addition of fertilizers may cause changes in soil chemistry and pH; these facts mean that, in still other cases, the added nutrients may fail to show beneficial results only because of secondary effects. Addition of organic material—high in $PO_4^{\equiv}$ and/or $NO_3^-$—to aquatic systems increases productivity and, if in sufficient amounts, allows such rapid growth of the most opportunistic species that all other species are crowded out. Algal blooms result, coming about with such suddenness and persistence that herbivore populations cannot keep pace. Hence the algae dies, rots, and depresses the oxygen content of the water, further inhibiting animal growth. Nutrient additions thus allow greater productivity, but with occasionally dramatic changes in the whole biological community.

### 10.2.3 Predation

The next chapter is devoted to predation, and so only one example is offered here. It has been estimated that in the period around 1918 there was enough food on the Kaibab plateau to support an estimated 30,000 deer. There were at the time only about 4000 deer. Subsequent to that year, man eliminated the puma, wolf, and coyote, and by 1925 the deer population had risen to 100,000. These vast numbers clearly represented a growth which overshot its food supply, and in two winters of severe starvation the population dropped to 60,000. Later declines brought the deer to 10,000, roughly one-third the number that originally might have been supported by the food supply. But the plants have not been able to fully recover because of overgrazing, with the result that the deer are still starving. It seems clear that the deer are now food-limited, but at one time were predator-limited (Leopold, 1943). The data cited above have recently been questioned (Caughley, 1970) and many wildlife biologists now feel that deer are naturally limited by shortage of winter forage rather than predation. Nevertheless, even if incorrect, the above figures still make an excellent hypothetical example.

The discussion of the Kaibab deer raises an important point. Animals in which food plays a major role in population limitation have evolved means of maximizing fitness in the face of famine. Producing so many young that some run real risks of starvation is clearly wasteful of time and energy, and so such species have evolved the ability to maximize fitness through reduction of brood size (see 6.3.1). Such species seldom face starvation even though food is in short supply, unless the effects of food shortage in population limitation are felt outside the breeding season.

### 10.2.4 Space and Spacing

It is often stated that territoriality acts as a population-limiting factor. As the density of a bird population increases, more individuals squeeze into the nesting area, but the number successfully obtaining territories does not keep pace with the population density. Thus population increase means proportionately fewer individuals producing offspring. But a question of consistency arises. If territoriality has evolved in response to food needs, for example, (part 8.5) then the real limiting factor is food. Furthermore, the territorial habit acts to maximize the food available to each breeding bird. Thus territoriality maximizes the population level maintainable under a given food situation (Brown, 1969). There is nothing inaccurate about speaking of territory-limited populations so long as one keeps the proper perspective and does not lose sight of the fact that territoriality exists in response to more basic needs.

Space becomes a very important limiting factor in marine intertidal areas, at least for plants and filter feeders. The abundance of nutrients, both organic and inorganic, appears sufficient for such species to grow profusely, until limited by a shortage of space to which holdfasts may attach, or larvae settle. This subject is discussed more fully in 15.3.3. Space is also critically tied to water as a limiting factor in dry climate plants. Here the acquisition of water depends on the area covered by a plant's root system. Toward the apparent end of providing individual plants with more space, many species have evolved growth inhibitors or germination inhibitors which serve to discourage crowding. The evolution of symbiotic relationships with certain insects serves the same function (12.4.2).

Forest plants are generally thought of as light-limited; those genotypes which can grow fast enough to reach open sunlight ahead of others, and spread a canopy maximizing their own light acquisition and minimizing that of their neighbors, maximize their relative fitness but thereby limit the number of individuals surviving and reproducing. As in the case of water, light and space are inseparable, for if a plant can inhibit growth around it for a sufficient radius, it need not compete for light.

## 10.2.5 Social Factors

Social factors may limit populations in an indirect way. For example, aggressive interactions between groups may result in buffer areas which are uninhabited. As local populations grow, only the fortunate cross the buffer and adjacent areas to successfully settle elsewhere. Thus social factors may limit, indirectly, the resources and space available and thus the entire population. The aggressive behavior of small mammals may serve to limit their populations in this way, or at least to retard growth.

Social factors also may limit populations through increased fighting and deleterious physiological effects associated with crowding. As discussed in 9.1.2, individuals should be adapted to the most common range in density of their populations. When populations become abnormally high, normal behavior may become maladaptive and novel stresses may upset behavioral or physiological homeostasis. Calhoun (1952) has observed that rats, *Rattus norvegicus*, tend to form groups with aggressively maintained buffer zones between them. He also has noted that as population density rises in this species, the number of fights, and hence wounds, increases. This leads to more infections and a higher mortality rate.

Let us consider the ultimately simple system in which neither groups nor buffers form and no abnormal behavior occurs. We assume that the incidence of fights is roughly proportional to the chance that two individuals in appropriately ugly or potentially ugly moods find themselves in the same place at the same time ($\propto n^2$). Then

$$\frac{dn}{dt} = b_0 n - d_0 n - \gamma n^2,$$

so that

$$n = \frac{n_0 e^{(b_0 - d_0)t}(b_0 - d_0)}{n_0 \gamma (e^{(b_0 - d_0)t} - 1) + (b_0 - d_0)}. \tag{10.1}$$

Equilibrium is reached, when $t \to \infty$, at

$$\hat{n} \to \frac{b_0 - d_0}{\gamma}. \tag{10.2}$$

Such a population is clearly self-limiting. Of course, the assumption that fights occur with frequency proportional to $n^2$ may be incorrect, but some data exist to indicate this to be a reasonable approximation in at least some cases. Hazlett (1968) kept varying numbers of the hermit crab, *Pagurus bernhardus*, in a tank, and counted the number of fights which took place over ten-minute intervals. The results are shown in Table 10.2. Under the mass action hypothesis (number of fights $\propto n^2$), the best fit is given when the proportionality constant is 0.655, and the agreement with observed data is good.

Table 10.2. Aggression as a function of population size

| Number of animals | Number of observed fights/10 minutes | Predicted number of fights |
|:---:|:---:|:---:|
| 2 | 5 | 2.6 |
| 3 | 8 | 5.9 |
| 4 | 11 | 10.5 |
| 5 | 17 | 16.4 |
| 6 | 24 | 23.6 |
| 7 | 33 | 32.1 |
| 8 | 37 | 41.9 |
| 9 | 51 | 53.1 |

The above, however, may be a special case since, at least in animals with some degree of social behavior, we would expect the number of fights to increase with increasing abnormality of behavior under the stress of an unusual social situation (abnormally high population density). The situation is not so simple, however. Southwick (1955) kept six 25 ft × 6 ft pens of house mice, *Mus musculus*, with unlimited food and water. All populations grew and eventually leveled off. The number of fights did increase faster than $n^2$ but this was not all. The males in some instances ran amuck, tearing up nests and eating young. In all cases, increased numbers of wounds and ectoparasites were observed. The results of these experiments, though, do not help with quantifying changes in behavior; individual differences between pens were very large. Lloyd and Christian (1967) have run very similar experiments with slightly different results. Fighting did not necessarily rise faster than $n^2$. As with Southwick's experiments, however, aggression varied greatly between populations. Infant mortality fluctuated considerably but showed no definite trend with population density. Birth rate dropped as density increased.

Krebs (1970) studying fluctuating populations of the voles *Microtus ochrogaster* and *M. pennsylvanicus* in the field, found that aggression (measured in number of approaches and attacks between wild mice brought into the laboratory) reached its peak in populations approaching or at their high points.

Studying wild rats in Baltimore, Christian and Davis (1956) found that populations in different city blocks showed little interblock migration and were, therefore, essentially isolates. They examined 21 of these isolates and found a relation between the state of growth and size of population and the adrenal weight of its members (Table 10.3). A simple interpretation is that high population densities induce an increase in the need for adrenalin with corresponding increases in adrenal weight. Thus populations in the process of increase may not yet suffer this consequence of high density, but adrenal weights should be high in peak and declining populations. In male rats, change in pituitary weights roughly parallels change in adrenal weights (Selye, 1956). Christian (1956) finds the same phenomenon in house mice—as populations peak, not only do birth rate and infant survival both decline, but the adrenal weights of males are 25% higher than in populations of fewer individuals. Females showed the same tendencies but to a smaller extent. Here, increased adrenal weight was correlated with hypertrophy of the cortical zona fasciculata and delayed onset of puberty. Delayed puberty was also found in males of crowded populations. In 1961, Christian reported a decrease in the size of the seminal vesicles in male house mice from large populations, especially in those individuals low in the dominance hierarchy, and also reduced thymus weight.

**Table 10.3.** Physiological correlates of population change

| State of population | Adrenal weights | |
|---|---|---|
| | males | females |
| low, stationary | above average | slightly below average |
| low, increasing | below average | below average |
| high, increasing | average | average |
| high, stationary | above average* | above average* |
| decreasing | above average | above average |
| | *highly variable | |

A 1964 review by Christian and Davis details the effects of such changes: slowed growth, delayed or inhibited sexual maturity, increased intra-uterine mortality, and inadequate lactation. Endocrine changes with, presumably, similar consequences also occur in deer mice, *Peromyscus*; voles, *Microtus*; rats, *Rattus*; woodchucks, *Marmota*; rabbits, *Sylvilagus*; and Japanese deer, *Cervus*. It is not difficult to envision these effects acting as a negative feedback mechanism to limit population growth. In addition, these physiological changes seem to leave individuals less resistant to further stress. If deprived of water, individuals of the vole *Microtus pennsylvanicus* from large populations are more likely to die than individuals from uncrowded populations (Warnock, 1965).

*Endocrine strain* (resulting from *population stress*) may have prolonged effects. That is, an increased death rate may persist even after a population is thinned. There may also be maternal effects: Young of stressed mothers have characteristics of stressed animals even when they grow up in sparse populations (Chitty, 1957). This, where true, will introduce a lag effect into population density limitation and may result in population cycling. However, the data which lead one to the above conclusion could also be accounted for if the effects of stress were not passed on to offspring via the uterine environment, but were genetically transmitted. Genetic considerations are examined in part 10.4.

Lest the reader depart armed with enthusiasm for endocrine strain as a population-limiting mechanism it should be pointed out that such strain is not ubiquitous, even within species. Strecker and Emlen (1953), working with *Mus* in pens, found no increase in adrenal weight of individuals in crowded populations with unlimited food, but the birth rate did drop and the death rate rose. Clough (1965) kept *Microtus pennsylvanicus* in large outdoor pens in their natural habitat, and artificially manipulated population rise. He found no correlation between population size and adrenal, thymus, or spleen weight, and no secondarily induced susceptibility to other stress (survival time in cold water was not reduced in individuals from crowded populations). In both of the above cases it is possible that populations were limited before reaching levels at which endocrine strain became significant, but how crowded must individuals be before the strain becomes significant, and how often do natural populations reach this size? As Chitty (1960) puts it: "While there is no difficulty at all in interfering experimentally with reproductive processes, the difficulty comes in producing an effect which corresponds with anything going on in nature."

Finally, we note, with Murton (1967), that endocrine strain is neither unexpected nor, as some have suggested, a mechanism evolved to control population size. Social responses are geared to population levels under which most individuals spend the most time. At abnormal levels, social responses may become maladaptive.

## 10.2.6 The Interdependence of Limiting Factors

In spite of the comments of many authors, no single factor can be held responsible for limiting a given population. A simple illustration follows. Predators kill those individuals they can most easily capture—that is the weakest as well as the very young and very old. But an animal is usually weak because of lack of food, or disease, or failure to consistently find adequate shelter. Furthermore, food shortage will necessitate longer periods foraging in the open where individuals are more easily preyed upon. If animals face heavy predation they may withdraw to shelter and thereby find less time to feed. Can we, then, say whether a population is food-limited or predator-limited?  Social relations are affected by hunger—and thus, perhaps, indirectly by predation—and social factors may result in guarded food supplies or buffer zones.  Social factors and food (and predation) cannot be considered independently.

Another sort of food-social interdependence is suggested by the work of Aumann and Emlen (1965).  Fifty-five populations of *Microtus pennsylvanicus* showed a strong correlation between density and the concentration of soil sodium.  Several simple experiments were set up to investigate this correlation.  It was found that, in the laboratory, mouse populations offered dilute salt water as well as distilled water experienced higher birth rates and maintained greater densities than populations to which only distilled water was offered.  Further, populations with more males used more salt water.  Finally, individuals showed stronger preference for salt water in dense than in sparse populations.  Since it is known that adrenalectomized animals show increased appetite for salt, the above results suggest that dense populations experience adrenal strain which is alleviated by increased sodium intake.  The degree to which social factors influence a population's size, then, may be partly dependent on the amount of sodium available to the population.  Again, food (this time in the form of one basic nutrient) and social factors cannot be separated in their role of population limitation.  Warnock (1965) has shown that the amount of cover greatly affects the amount of wounding or mortality induced by crowding.  Thus the feedback effects of overpopulation cannot be considered independently of considerations of cover.

The gecko *Gehyra variegata* of Australia seems to be population-limited by a combination of space and social behavior.  In this species the males "defend" territories beneath the loose bark of dead trees, one male to a tree.  Females live within each male's territory, but are themselves aggressive, a fact which denies to the males more than two or three mates.  During the winter tolerance is high, but in the summer extra lizards are chased off and die.  Thus the breeding population is limited by a combination of the number of dead trees and bad tempers (Bustard, 1970).

Clearly, plants will be limited by a combination of space, water, nutrients, and light, the interrelations and relative importance of the various factors changing from place to place.

In spite of the interdependence of limiting factors, however, we will continue to speak of $x$-limitation throughout the rest of this book, meaning that a partial regression of population change on $x$ shows a significantly larger coefficient than a partial regression of population change on any other potentially limiting factor.

### 10.2.7 "Density"

We are now in a position to consider the meaning of population "density".  As it is formally used, of course, it is simply a number of individuals per unit area.  But in terms of effective density to a predator (5.3.3, 5.4), or the mediation of social interaction, is this measure of density really appropriate?  Warnock (1965) found that in *Microtus pennsylvanicus* the important factor was not number per area, but total number of different individuals contacted. Thus large populations over large areas suffered roughly the same stress (measured by hypertrophy of spleens and female adrenals) as did large populations in small areas. Thus the nature of the spacing system, as well as density, affects birth rates and death rates.

It is difficult, and perhaps impossible, to devise a measure which will replace density as a tool in examining population size and limitation. For food limitation, simple density may be quite adequate. For predator limitation the situation is quite different (the reader is referred to Chapters 5 and 7). It seems not too difficult to devise at least a roughly meaningful measure for social interactions. Lloyd (1964) has devised such a measure, which he calls *mean crowding*. Following the thesis that population stress varies with the number of different individuals encountered, we first define an *ambit* as that area over which an animal confines its wanderings. If we assume that all of a home range is covered (see 8.6), home ranges and ambits are the same. Suppose now that we divide an area into a number of adjoining areas the size of ambits and measure the number of individuals found in each—call it $n_i$ for the $i$th area (we are clearly disregarding the fact that ambits vary in size). The number of different individuals encountered by a hypothetical animal wandering over the $i$th ambit is $n_i - 1$. Thus a measure of crowding is given by the average of $n_i - 1$, weighted according to the number of individuals in each area, or

$$m^* \text{ (mean crowding)} = \frac{\sum n_i(n_i - 1)}{\sum n_i}.$$

Ideally, of course, one would measure actual ambits rather than areas the size of a mean ambit, but this is clearly impractical. Note that where population density is $\rho$, $\sigma^2$ gives the variance in $n$ over all areas, and $Q$ is the number of areas measured:

$$m^* = \frac{\sum n_i(n_i - 1)}{\sum n_i} = \frac{\sum n_i^2}{\sum n_i} - 1 = \frac{\sum n_i^2 - (\sum n_i)^2/Q + (\sum n_i)^2/Q}{\sum n_i} - 1$$

$$= \frac{\sum n_i^2 - (\sum n_i)^2/Q}{Q} \cdot \frac{Q}{\sum n_i} + \frac{(\sum n_i)^2}{(\sum n_i)Q} - 1$$

$$= \frac{\sigma^2}{\rho} + \rho - 1 = \rho + \left(\frac{\sigma^2}{\rho} - 1\right).$$

## 10.3 POPULATION FLUCTUATIONS

It has already been mentioned that endocrine strain, with time lags, might produce fluctuations in populations by allowing for overcompensation for high population density. Here we consider other density-dependent factors affecting—producing or damping—fluctuations.

It is obvious that where birth rates and death rates are not constant there will be seasonal fluctuations in population size. This simple observation raises two important points:

1. Density-related deaths in one season may be due to one factor primarily (say food), while in another season most density-related deaths are due to predation. This complicates the picture of what limits populations. If density-related deaths due to food in one season are considerable, and density-related deaths due to something else in another season are rare, however, we may say both that the population is food-limited, and that it is limited in that season of food shortage.

2. If a population is limited by resource shortage in one season, it will exist amid plenty during the rest of the year.

Fluctuations measured over long periods of time, from one season to the same season in successive years, may arise (theoretically) in populations which respond to density-dependent factors suddenly. That is, instead of $r$ (or $R$) declining gradually as $n$ increases, it suddenly drops when $n$ reaches a critical point. The population crashes, then slowly grows back. Nicholson (1954a, b) calls this a *relaxation oscillation*.

If adults but not young individuals contribute to population density feedback, then $n$ will continue to grow during the maturation period of young, even when the adult population is approaching the carrying capacity (if resource-limited). Thus the population will overshoot $K$ and be forced back, because of the maturation lag again, below $K$. Oscillations produced for this reason Nicholson has called *lag oscillations*. Nicholson (1948) also describes a case of *scramble* competition (see 12.1.3) in which larvae of the blowfly, *Lucilia cuprina*, "scrambled" for 1 gram of beef brain in such a way that if their number was high, all got something to eat, but few got enough. Thus the next generation was much less numerous. The relation between the number of larvae feeding on the brain and the subsequent number of adults emerging is given in Fig. 10.4.

Suppose each adult produces 25 larvae and that a population starts with 80 larvae. According to Fig. 10.4, these 80 larvae will result in about 5

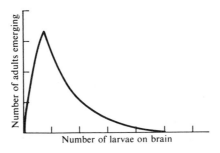

**Fig. 10.4.** The relation between larval density and emerging adults: Blowflies on a gram of sheep brain (from Nicholson, 1948).

adults which will give rise to about 125 larvae. These 125 larvae, in turn, will generate about 2 adults and subsequently about 50 larvae. Carrying these simple calculations farther it becomes clear that the blowfly population will oscillate. If each adult gives rise to only 4 larvae, the oscillations are somewhat damped. The two cases are reproduced in Figs. 10.5.

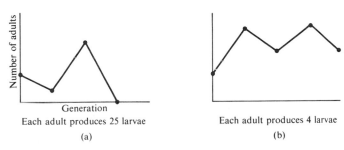

Each adult produces 25 larvae

(a)

Each adult produces 4 larvae

(b)

**Fig. 10.5.** Population oscillations in the blowfly (see text).

In part 9.3 we discussed the nature of the population oscillations accompanying that population's approach to stable age distribution. Differences in age-specific mortality and fecundity are responsible for these oscillations which eventually damp out. However, it is possible that social interactions continually change $q_x$ and $m_x$ values so as to promote continuing oscillations. For example, when population densities are high, rodent males often tend to attack young animals and drive them into submarginal areas where they suffer high mortality. As a result, the number of reproductives in the next generation is greatly reduced and the population may drop. Recovery in the following generation leads to cycling. Another means by which age differences may lead to oscillation—discussed very briefly by Slobodkin (1954)—is inter-age class competition. The simplest case is that in which the presence of old individuals depresses reproduction or increases death rate in younger animals (this needn't involve submarginal habitat as in the above case). So long as old individuals are abundant, $R$ is low. But these individuals are only very slowly replaced (low birth rates or high mortality among young), and so they eventually decline in number, competitive interference drops, and $R$ increases. Eventually individuals of the large age classes produced during these latter years also become old, and inhibition of growth begins again.

The above cycles depend upon time lags and/or age differences. We now ask whether population-limiting factors can lead to cycling independent of these considerations. We return to Fig. 10.3 and enlarge that portion where the lines $I$ and $D$ cross, recalling that $dn/dt$—or in a seasonally breeding species, $\Delta n$—is given by $D-I$ (Fig. 10.6).

We consider that range of $n$ close enough to $\hat{n}$ that the lines $I$ and $D$ may be considered straight. Now suppose that $n$ has the value $A$. Recalling that the subsequent change in $n$, $\Delta n$, is given by $D - I$, we conclude that $n$ will change by an amount

$$P_1 - P_2 = (\hat{n} - A)\tan\theta_1 - (\hat{n} - A)\tan\theta_2 = (\hat{n} - A)(\tan\theta_1 - \tan\theta_2).$$

Similarly, if $n$ takes the value $B$, its subsequent change will be to the left an amount $(B - \hat{n})(\tan\theta_1 - \tan\theta_2)$. Thus if $\tan\theta_1 - \tan\theta_2 > 1$, $n$ will always swing back past $\hat{n}$, resulting in population oscillations. If $\tan\theta_1 - \tan\theta_2 > 2$, the oscillations will increase in amplitude.

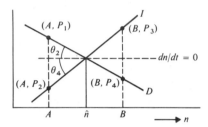

**Figure 10.6**

To put these inequalities into more conventional terms, note that $\tan\theta_1$ is a measure of the severity of the density-independent factors. We shall call this term $r_0 c$, and observe that $c$ is the proportional drop in $r$ which would occur were density-dependent factors absent. $\tan\theta_2$ is merely $d(\Delta n)/dn$ which, to the extent the logistic equation is applicable, is given by

$$d(\Delta n)/dn = \frac{d}{dn}[r_0 n(1 - n/K)] = r_0(1 - 2n/K).$$

Thus $\tan\theta_1 + \tan\theta_2 = r_0(1 + c - 2n/K)$, and the above conditions become

$2 > r_0(c - 1 + 2n/K) > 1$      Population oscillates, but damps      (10.3a)

$r_0(c - 1 + 2n/K) > 2$      Population oscillates with increasing amplitude.

(10.3b)

Fluctuations and instability may be inherent properties of a system even in a constant environment (as assumed above) when $r_0$ or $c$ is large, both conditions which are most likely met in populations affected primarily by density-independent factors. The fact that, with increasingly large displacements, the value of $\Delta n$ relative to $n$ declines (due to the fact that $D$ is not really straight) means that population fluctuations can increase in amplitude only within limits. Thus if $r_0(c - 1 + 2n/K) > 2$, oscillations may be sustained without damping, but will never increase beyond some (theoretically) definable point.

## 10.4 GENETIC CONSIDERATIONS

There are at least three ways in which the genetic structure of a population may affect its size and its fluctuations. Picture a population at equilibrium. Suppose now that the physical environment suddenly becomes more benign—that is, the value of $c$ in Eqs. (10.3) suddenly decreases. The population then increases and two things happen. First, released from the normal level of genetic deaths, abnormal gene combinations increase in frequency (Chapter 1, Discussion I). Second, the population is less restricted by the physical, more restricted by the biotic (density-related—food, predation, social) factors so that selection favors greater adaptation to the latter at the expense of the former (4.2.2). Now suppose that the physical environment reverts to its original state. The population is no longer adapted to that original state. Furthermore, it has accumulated new gene combinations that have survived due to relaxed selection pressures from the physical environment. Thus the reversion of the physical environment certainly brings about a drop in the population to levels below that at which it was once in equilibrium. A new, lower equilibrium arises and we may assume that selection subsequently favors a slow rise in $n$. Fluctuation in the physical environment thus keeps individuals somewhat less adapted to one specific set of circumstances— it favors *eurytopy* (generalization) with respect to the physical environment. It also allows, through the accumulation of new gene combinations, for faster evolutionary change (see 2.6). Carson (1968) has discussed this kind of population-genetic interaction in some detail. Pimentel (1961) feels that such genetic feedback may act to check population growth.

Evidence for another hypothesis involving genetic feedback control on populations has been, so far, largely indirect, although some direct evidence is now being sought (Tamarin and Krebs, 1969). The hypothesis can be attributed to Chitty (1960): Genetic dimorphism (or polymorphism) exists so that one form predominates during population increase and peak, another during decline. Several conceivable variations exist but the following seems most plausible. The first morph is highly aggressive and mates frequently as a result. Thus, as the population grows, this morph displaces the less aggressive, second morph. When the population reaches high levels, though, high levels of aggression become maladaptive and the population experiences social changes which lead to its decline. The aggressive first morph dies off more rapidly than the second because of greater numbers of wounds or higher endocrine strain. This scheme seems to provide for genetic change with population change and assures repeated population change as a result of the genetic change; the population experiences self-induced cycles.

The first step in supporting this hypothesis was to demonstrate clearly that gene frequencies do, in fact, change with the stage of a cycling population. Chitty (1960) and DeLong (1967) had already noted that *Microtus agrestis* and feral *Mus musculus* populations often declined during the breeding season, even in the presence of abundant food and cover and in the absence of high predation, but these declines could have been due to lags in endocrine strain. The fact that individual size varied with the cycle and that large parents produced large offspring suggested genetic correlations with population size (Chitty and Phipps, 1966) but were not clear evidence of that possibility. Studies aimed specifically at finding genetic correlations with population cycle stages have been carried out by Krebs and his co-workers with voles (Tamarin and Krebs, 1969). Transferrin, an iron-binding $\beta$-globulin in vertebrate serum, occurs in various forms, each controlled by a single allele at one locus. *Microtus pennsylvanicus* possesses the alleles $T_F^E$ and $T_F^C$, *M. ochrogaster* the alleles $T_F^E$ and $T_F^F$. Blood samples were taken with toe clips from several populations at different parts of their cycles. Calling $p$ the frequency of one allele, these workers plotted $\Delta p$ against $p$ and found $\Delta p$ to change direction at a critical value of $p$. This, of course, implies heterosis (see 1.5.3). However, no excess of heterozygotes was found over that implied by the Hardy–Weinberg law. Thus the allele must be maintained either through pleiotropy or frequency-dependent selection (the contribution of an allele to fitness declines as its frequency increases). Pleiotropic maintenance seems doubtful over any extended period of time and frequency-dependent selection, of course, is what we need to support Chitty's hypothesis. Following the frequencies in the transferrin alleles, it was found in *Microtus pennsylvanicus* that $T_F^C$ increased during population increase and $T_F^E$ increased during population decline. $T_F^C$ mice had generally higher survival, and at high population densities higher growth rates. In *M. ochrogaster*, $T_F^E$ increased with the population and $T_F^F$ increased during population decline. $T_F^E$ mice generally had higher survival, were larger, and were (males and perhaps females) in breeding condition more of the time. These data are highly suggestive but not yet at all conclusive in support of Chitty's hypothesis.

A third manner in which genetic change affects populations is that explored by Wellington (1957, 1960, 1964) in the western tent caterpillar, *Malacosoma pluviale*. In this case population density does not necessarily affect genetic change, but genetic change certainly affects population growth. Wellington found he could divide tent caterpillars into two basic categories: active and independent animals, and others. The others were followers, but ranged from active to so sluggish that some never freed themselves from their eggs and so died without ever feeding. The more active caterpillars tend to give rise to active adults (although the correlation is not perfect) so that activity seems to be at least partly genetically determined. The moths lay eggs primarily in areas suitable for nesting, and in normal years these areas might be quite restricted. The active moths, however, range farther than their relatives and thus waste more eggs in unsuitable areas. As a result the genes favoring high activity drop in frequency and the population becomes more sluggish. This occurs in spite of the fact that the sluggish caterpillars are more disease prone and more likely to starve because of their unwillingness to move about. Thus the population declines. When a good year comes along, however, more areas are suitable for egg laying, and the remaining active moths disperse and reproduce with great success. The population soars. The decline in activity and fitness begins again the next year.

# 11
# Predator-Prey
# Interactions

The effect of prey on predator populations has already been reviewed directly in Chapter 10 and indirectly as it affects reproduction in Chapter 6. We therefore start here with a discussion of how predation affects prey populations.

It is clear that predation does not act as a population limiter in the same sense as resource shortage: For a given number of predators, more prey does not imply higher mortality from predation. In fact, prey mortality is likely to be lower since with more prey, the number of predations per prey individual is lower. Thus if predator populations remained constant, mortality would vary inversely with prey density and population instability would be encouraged. Predator populations, though, do react to increases in prey density through immigration, increased reproduction, and, perhaps, lowered mortality, and thus increase in number. If a predator population rises proportionately more than the prey population to which it is responding, then predation clearly may act as a population-limiting factor on the prey.

But the true situation is more complicated than this. First, as will be seen from Eq. (11.1) and following calculations, the prey–predator system is inherently oscillatory (Fig. 11.1). The oscillations so produced we shall call *equilibrium oscillations*. Oscillations also arise from time lags and changes in age structure. For example, the increased birth rate of predators responding to high food concentration needs time to work its effect on the rate of predation: young predators eat less than adults and require time to grow up. Thus the predator response lags the prey stimulus. As a result, when the prey population levels

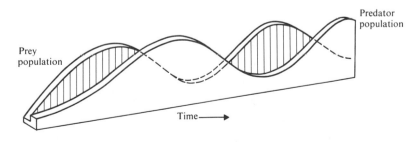

**Fig. 11.1.** Predator and prey oscillations, showing the time lag.

off, the predators may produce that number of young which, in the presence of the current adult population, can find plenty to eat but which, when the young mature, will find itself short of food. The predators thus overeat their prey, which decline, and, now having much less to eat, either suffer higher mortality, lower fecundity, or both, and so also decline.

Prey–predator cycling probably plays the primary role in a number of naturally occurring population cycles. Between the Arctic tundra and the Arctic conifer belt, for example, voles show a fairly regular four-year cycle (Elton, 1942). The varying hare, *Lepus Americanus*, shows a nine to eleven-year cycle along with its predators the lynx, *Lynx Canadensis*, the goshawk, *Accipiter gentilis*, and the great horned owl, *Bubo Virginianus*, (see Chitty, 1948), and the cycles of lemmings, *Dicrostonyx groenlandicus* and snowy owl, *Nyctea Scandiaca*, in the Arctic, appear to be related (Shelford, 1945). This list is by no means exhaustive. While there is good evidence linking these cycles to the simple hypothesis above, however, there are also some discrepancies and a good deal of controversy over the matter. This chapter is an attempt to review the conflicting evidence and explore the theoretical arguments pertaining to prey–predator interactions.

## 11.1 PREY-PREDATOR CYCLES

### 11.1.1 The Intensity of Predation

If predation is to act as a prey population-limiting factor, then it is clear that it must account for a reasonable proportion of prey mortality. In prey species with $R_0 = 2$, predators which, on the average, take 10% a year will not be adequate to the job.

There are an unfortunately small number of prey–predator sets for which both $R$ (prey) and intensity of predation are known. But occasionally a death rate due to predation has been ascertained and this may serve as an initial guide to the impact of predation. *Pisaster ochraceus*, the west coast sea star, has been reported to take very roughly 25% of available *Tegula funebralis*, the black turban snail. This amounts to just under 90% of the total mortality of adults (Paine, 1969). If the turban snail population is limited through factors influencing adult survival, then predation is very important among those factors. Pearson (1964) estimates 88% of the *Microtus californicus* population (and 33% of the *Reithrodontomys* population) in a field near San Francisco to be eaten each year by predators. For a species with an $R_0$ value of perhaps 24 (see introduction to Chapter 10), predation of this magnitude may indeed be very significant in population limitation. Krebs (1966), although he feels 22% to be more accurate than 88%, agrees that predation is probably very important in limiting California vole populations. Darling (1937) estimated 50% of the mortality of young red deer during their first year of life to be due to predation—mostly by fox, eagle, and wildcat—and H. N. Southern (cited in Lack, 1954) believes that one-fourth to one-half of all mortality in the wood mouse, *Apodemus sylvaticus*, in Oxford, can be accounted for by tawny owls, *Strix aluco*, and most of the rest by weasels. Slobodkin, Smith, and Hairston (1967) have considered these sorts of data—and given more examples—and conclude that predation is indeed very often heavy enough for predators to limit herbivores (but see 10.2).

### 11.1.2 The Theory of Prey-Predator Cycles

Let us accept the notion that, at least in some cases, prey may be predator limited. We now consider the simple case in which prey are limited strictly by predation by one species and predators strictly by the number of one prey species. Beginning with the rough approach of Lotka and Volterra, ignoring time lags, we let $n_1$ be the prey population, $n_2$ the predator population and write

$$dn_1/dt = b_1 n_2 - \delta n_1 n_2$$
$$dn_2/dt = \beta n_1 n_2 - d_2 n_2 \tag{11.1}$$

These are known as the Lotka–Volterra equations.

If we define $x = \beta n_1/d_2$, $y = b_1 n_2/\delta$, these equations become:

$$\begin{cases} dx/dt = \dfrac{\beta}{d_2}(dn_1/dt) = \dfrac{\beta}{d_2}\left(b_1 \dfrac{d_2 x}{\beta} - \delta \dfrac{d_2 x}{\beta}\dfrac{b_1 y}{\delta}\right) = b_1 x - b_1 xy \\[3mm] dy/dt = \dfrac{b_1}{\delta}(dn_2/dt) = \dfrac{b_1}{\delta}\left(\beta \dfrac{d_2 x}{\beta}\dfrac{\delta y}{b_1} - d_2 \dfrac{\delta y}{b_1}\right) = d_2 xy - d_2 y. \end{cases}$$

Dividing, we obtain

$$dy/dx = \frac{d_2 xy - d_2 y}{b_1 x - b_1 xy},$$

so that

$$(b_1 x - b_1 xy)\, dy/dt + (d_2 y - d_2 xy)\, dx/dt = 0, \qquad b_1 \frac{1-y}{y}\, dy + d_2 \frac{1-x}{x}\, dx = 0,$$

and

$$b_1 \ln y - b_1 y + d_2 \ln x - d_2 x = \ln C,$$

or

$$\frac{n_1^{d_2}}{e^{\beta n_1}} \cdot \frac{n_2^{b_1}}{e^{\delta n_2}} = C.$$

The relation is graphed in Fig. 11.2, and is a closed loop. This simple model leads to regular prey–predator cycles of unchanging amplitude.

The graph (Fig. 11.2) represents a center—that is, it depicts the equilibrium oscillations. The additional oscillations due to time lags and age structure changes may or may not damp, and may be maintained or driven by the equilibrium oscillations.

In making all of the assumptions necessary to set up the Lotka–Volterra equations—no age-specific variation, no time lags, amount of interaction propor-

tional to the product of prey and predator numbers—we have clearly oversimplified nature. Since simple models of this sort may possibly give erroneous results
it is useful to look at other simple models with somewhat different sets of assumptions to see if the predictions arising from both are the same. An alternative
approach, merely qualitative, but nonetheless more amenable to the relaxation of
assumptions, follows. It is taken loosely from the arguments of Rosenzweig and
MacArthur (1963), Rosenzweig (1969), and Maly (1969). Age differences are not
taken into consideration.

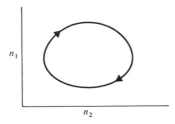

**Fig. 11.2.** The dynamic relation of predator and prey populations.

We note that as the prey, $n_1$, becomes more numerous, the number of predators, $n_2$, which are necessary to block further growth of $n_1$ increases, very
roughly, proportionately. (This keeps number of predators per individual prey
at a constant level). At very high values of $n_1$, however, density-regulating factors
in the form of resource shortages come into play and the number of predators
needed to block prey population growth drops. The opposite may be true for
very small $n_1$. Thus we obtain a curve of the sort pictured in Fig. 11.3, where
the arrows show the direction of change in the prey population. Rosenzweig
(1969) has calculated this curve from the data of Huffaker (1958) on predatory and
herbivorous mites, noting the numbers of each and the subsequent change in the
prey population, and Maly (1969) has done the same for *Paramecium* preyed upon
by the rotifer *Asplanchna*. A similar curve can be constructed for the predator.
Since the number of prey per predator must be, roughly, some constant value to
ensure predator population growth, regardless of $n_2$, the appropriate figure is of

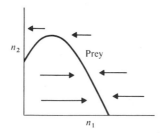

**Fig. 11.3.** The conditions for increase in the prey population (see text for explanation).

the general sort shown in Fig. 11.4. Where, as we are assuming here, the number of predators born is dependent solely on the food available, we can write:

$$\frac{dn_2}{dt} = \beta n_1 - d_2 n_2,$$

so that at equilibrium,

$$n_2 = \frac{\beta}{d_2} n_1.$$

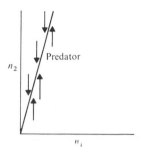

**Fig. 11.4.** The conditions for increase in the predator population (see text for explanation).

Superimposing Figs. 11.3 and 11.4, we can construct the three situations depicted in Fig. 11.5. Bearing in mind the fact that $\Delta n_1$ must be zero when the point $n_1$, $n_2$ falls on the prey curve, and that $\Delta n_2$ must be zero when $n_1$, $n_2$ lies on the predator curve, we can see that the path of $n_1$, $n_2$ over time roughly follows that marked with the dashed line.

This approach, though far from precise, tells us a good deal. If the predator is highly efficient at capturing prey, it needs fewer prey to maintain itself and the predator curve (Fig. 11.4) is steeper (Fig. 11.5a). The prey–predator cycle in this case is unstable and leads to the extinction of the prey (and subsequently the predator). A slightly less efficient predator will interact in the manner of Fig. 11.5b, showing damped oscillations. A predator which is inefficient (Fig. 11.5c) will reach equilibrium with its prey and not cycle.

If the predator population is affected by shortage of resources other than the prey, or is, itself, affected by predation, Fig. 11.5b may take either of the following two forms (Fig. 11.6). In the first figure, other limiting factors come to bear on the predator population only when it has reached high levels (i.e. they are unimportant relative to the prey as limiting factors) and they have no significant effect on the prey–predator cycles. Where these other factors are major forces in population limitation, however, the cycles are damped to nothing and a stable, noncyclic equilibrium between prey and predator obtains. This assumes, of course, an environment static with respect to all factors other than the prey and predator populations.

Suppose the environment provides refuges for a number of prey individuals, areas in which they cannot be reached by predators. Then we may redraw Fig. 11.5(a) as in Fig. 11.7. It is clear that such refuges prevent prey extinction and keep the prey–predator cycles within bounds.

If the prey presents a relative rather than an absolute food shortage (part 10.2) the predator gains less food per unit prey at low prey densities, but becomes more

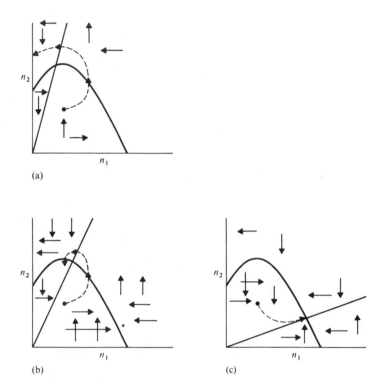

(a)

(b)                                (c)

**Fig. 11.5.** Diagrams showing the dynamic interaction of prey and predator populations.

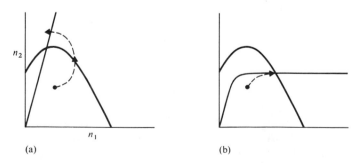

(a)                                (b)

**Fig. 11.6.** Diagrams showing the dynamic interaction of prey and predator populations.

efficient, owing to greater opportunity for choice, at high prey densities (Fig. 11.8). The effect of prey being in relatively short supply is to stabilize the prey–predator cycles.

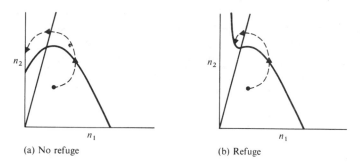

(a) No refuge                        (b) Refuge

**Fig. 11.7.** Diagrams showing the dynamic interaction of prey and predator populations.

### 11.1.3 Statistical "Cycles"

The above theory, although nonquantitative, is fairly straightforward. We should expect, at least occasionally, to find prey–predator cycles in nature. Where such cycles naturally damp we should expect perturbations in the physical environment to restart them. The next question is, do we really find such cycles? The opening section of this chapter outlined apparent cases of cycles. Showing them first to be more than statistical artifacts, and second, to be due to predator–prey interactions, is not a simple task.

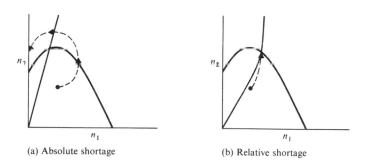

(a) Absolute shortage                  (b) Relative shortage

**Fig. 11.8.** Diagrams showing the dynamic interaction of prey and predator populations.

Cole (1954) suggested that the "cycles" observed were really spurious. Consider $N$ samples in time and define a population peak by any three samples, the first and third of which are lower than the second. Any three samples, $a$, $b$, $c$ can be arranged in six ways, of which two place the highest in the center. Thus the probability of finding that any given series of three samples defines a peak is

$2/6 = 1/3$. In a series of $N$ samples taken from a randomly fluctuating population, there are $N - 2$ groups of three consecutive samples. The expected number of peaks is thus $1/3 \times (N - 2) = (N - 2)/3$ and the number of cycles is $[(N - 2)/3] - 1 = (N - 5)/3$. The mean cycle length is $N/[(N - 5)/3] = 3N/(N - 5)$ which, as $N$ becomes large, approaches three. Thus, in a randomly fluctuating population sampled each year, there will be spurious cycles averaging three years or slightly longer in length. The compound peaks (the peaks themselves will show peaks) will average $3 \times 3 = 9$ years or slightly longer. When it is recalled that most natural "cycles" average either three or four or nine or ten years from peak to peak, it indeed begins to look as if these "cycles" are not regular, biologically caused events. (See Fig. 11.9). The cycle of the lemmings in Alaska is about three years (Pitelka, 1957) and small rodents and foxes in Arctic and Alpine tundra cycle roughly every three or four years (Keith, 1963). Red squirrels also fluctuate with periodicity of about three years although Kemp and Keith (1970) have shown the period to be significantly less than that predicted by Cole. Grouse appear to show both three or four-year cycles and nine or ten-year cycles (Keith, 1963), and mink, lynx, snowshoe hare, fox, fisher, muskrat, goshawks, and great horned owls, in Canadian forests, all fluctuate with a quite regular periodicity of nine or ten years (MacLulich, 1937; Keith, 1963). Coyotes cycle with an average period of 8.6 years over Saskatchewan, Manitoba, and Alberta (Keith, 1963). Whether or not there are biological causes behind these cycles, it seems apparent that statistical artifacts may arise. Random population changes might serve to make otherwise regular cycles appear rather irregular.

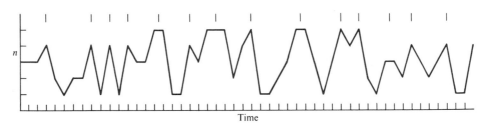

**Fig. 11.9.** Random population fluctuations and "cycle" length.

### 11.1.4 The Biological Evidence

One difficulty in defending the existence of prey–predator cycles is that, because of complicating biological factors, most of which cannot be quantitatively assessed, such cycles, if they exist, would not be perfectly regular. Thus the biologist is forced into a definition of "cycle" which is considerably less precise than might be desired. According to Davis (1957), "In ecology, use of the term 'cycle' refers to a phenomenon that recurs at intervals. The intervals are variable in length, but it is implied that their variability is less than one would expect by chance and that reasonably accurate predictions can be made." The words "implied" and "reasonably accurate" do not evoke ecstasies of confidence in the rigor of the approach, Another, better, and more often accepted definition is Cole's (1954) "—a sequence of events repeating in a definite order but without any necessary implication of strict regularity in time." Sophisticated techniques can distinguish regular cycles, such as harmonic cycles, from background noise. These techniques can also distinguish cycles of the sort Davis defines if one interprets his "implied" as meaning "necessary." Cole's "cycles" cannot be distinguished from random noise in a population regulated within limits. We therefore make no attempt to defend cycles as biologically induced patterns on the basis of their regularity, and pass on to the evidence offered by the correlations and phase relations between prey and predator population changes.

Most of the evidence comes from Arctic or northern forest regions in Canada. This is where the suspected cycles occur, and fur-trading records allow reasonable estimates of a number of species over long periods of time. Records from the Hudson's Bay Company go back 206 years for the lynx, *Lynx canadensis*, and are available by sub-regions of Canada between 1921 and 1934. These records show violent but rather regular fluctuations, with a fair degree of synchrony over different parts of Canada, for a period of more than 100 years (Elton and Nicholson, 1942). Similar data, from pelts, for the varying hare, *Lepus americanus*, show its numbers to rise and fall in confluence with those of the lynx (MacLulich, 1937). The correlation coefficient $= 0.55 \pm 0.05$ The period of the lynx averages 9.7 years, that of the hare, 9.6. MacLulich believed the two "cycles" to be related. Keith (1963) concurs, noting that the changes in the lynx population lag those in the hare population, as theory dictates. Butler (1953) examined these "10-year cycles", and attempted to smooth over any statistical variation by averaging overlapping triplets of censuses (numbers 1, 2, 3; numbers 2, 3, 4, etc.). If every peak in the lynx population fluctuations is counted, Cole's statistical peaks may be seen as bumps, spaced at 3.08 year intervals, on the "10-year cycles."

Figure 11.10, taken from Butler (1953) shows the temporal relation between the peak years of the rabbit population and those of its predators. The first predator to respond is the lynx. As pointed out by Watt (1968), this fact is significant. Lynx breed in their second year and bear young in the same year. Fisher, whose population response comes after a greater time lag, display delayed implantation. They breed at age one or two, but bear their young a year later.

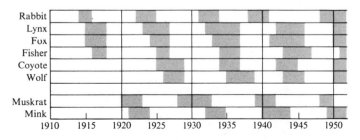

**Fig. 11.10.** The temporal relation between peak years in the snowshoe hare and its predators.

The best study of species with three to four-year "cycles" is probably that of Pitelka (1957), near Point Barrow, Alaska. Early in the summer the lemming (*Lemmus trimucronatus*) population declines somewhat because of heavy predation by nesting jaegers and owls. The population may recover in late summer. Suppose the lemming population is low in one summer. The vegetation is thus eaten but lightly and much remains through the following winter under the snow. This provides a good food supply. Furthermore, the low numbers of lemmings provide little food for their nesting predators whose numbers fall or, at least, fail to increase. The winter, then, provides ideal conditions for population growth and the next spring finds a somewhat enlarged lemming population. The number of predators is low, so although they now experience a reasonably rich nesting season they have little effect on the lemmings, which continue to increase in number during the second winter. But in the second spring the lemming population is large enough to decimate its food supply, and its predators are now numerous. Lemming numbers decline from predation over the spring and summer, and of starvation over the winter. The decline continues through the next winter and the following spring. Four years after the observations began, the lemming population has returned to its original level. The "prey" and "predator" here appear to be the plants and the lemmings. The effect of predation is less significant than that of starvation, and Pitelka believes that the predators may act to force an early lemming decline and thus prevent mass starvation leading to more violent oscillations.

Pearson (1966) estimated population density in the California vole, *Microtus californicus*, by trapping, and judged predator populations by the number of scats of feral cats, raccoons, *Procyon lotor*, gray foxes, *Urocyon cinereoargenteus*, and striped and spotted skunks, *Mephitis mephitis* and *Spilogale putorius*. The number of mice eaten was ascertained by the appearance of body parts in the scats. During what appeared to be $1\frac{1}{2}$ full population cycles, he found very close agreement between the number of mice and the number of mice eaten, the latter lagging the former.

Pearson (1966) also notes that when *Microtus* invaded predator-free Brooks Island in San Francisco Bay and replaced *Mus*, its population grew and showed seasonal fluctuations but no cycling, while the mainland population continued to oscillate as much as 100-fold.

The examples above are strongly indicative that prey–predator cycles exist. However, there is some counter evidence. If such cycles occur in Canada, one would expect them also to exist in climatically similar areas in Europe and Russia. While there is some possible indication of cycling in Russia, Greenland, and Iceland, cycles seem to be entirely lacking in Europe (Keith, 1963). Another somewhat disturbing fact is that muskrat, *Ondatra zibethica*, and mink, *Mustela vison*, which do not eat hares, "cycle" in phase with those predators which do (Moran, 1953a). Hares introduced onto lynx-free Anticosti Island in Canada "cycled" in phase with the mainland population (Elton and Nicholson, 1942). Rowan (1950) notes that even some fish populations "cycle" in unison with the lynx. All this supports the notion that cycling involves more than statistical fluctuations, and suggests a cause of cycling other than the interaction of a few prey and predator species. Moran (1953b), considering this conclusion, points out that "cycles" are roughly in phase over all of Canada and, to some degree, correlated with weather changes. He suggests that the "cycles" are indeed maintained by prey–predator interaction, but are periodically triggered and thus kept roughly in phase over wide areas by weather factors. Leslie (1959) has shown that age-class-caused cycling of out-of-phase populations can be brought into phase by random perturbations common to the environments of the different populations. It seems reasonable that prey–predator oscillations could also be brought into phase in this manner. Watt (1968) also believes the "cycle" to be of inherent rhythm (prey–predator) although driven by random forces of another sort.

No one appears anxious to give up the idea of prey–predator cycles. We concur, viewing the bulk of evidence and the rather inescapable theoretical arguments.

## 11.2 COMPLICATIONS IN THE PREY-PREDATOR INTERACTIONS

### 11.2.1 Trophic Factors

The effect of the addition of a top predator to a simple prey–predator system is not easy to assess. Let us suppose first that the second species of a three-level system is primarily food-limited. This is what Pitelka believes to be the case for his Arctic lemmings. This implies the top predators either to be somewhat inefficient or to use alternative foods, in which case the relation between predator and top predator, ignoring changes in the prey population, should be more stable (Figs. 11.5 and 11.6) than that between the lower predator and its prey. Since the interaction of prime importance is between a species and that factor with the most profound influence on its population, the dominant interaction here is between prey and lower predator—the less stable interaction. Addition of a top predator level in this case should, as Pitelka suggests, stabilize the cycling of the lower predator and its prey. In the case of a top predator being so efficient that the lower predator becomes primarily predator-limited (this is Hairston, Smith, and Slobodkin's (1960) argument), then the primary interaction is that of an efficient top predator and its prey, less stable than the less important interaction between lower predator and prey. In this case, addition of a top predator should reduce the stability of the cycle between the prey and lower predator.

But this is a simplification. It is conceivable that during cycling the second species is alternately under the greatest influence from predators and food. What of maturation time lags? Generally, as we have seen, these tend to reduce stability. But in three-species systems the plot thickens. A system otherwise stabilized by addition of a top predator may be made unstable by such an addition if the time lag between birth and maturation is sufficiently greater in the top species than in the second species. The predator–top predator cycle, normally more stable than the prey–predator cycle, is now made considerably less stable. By the same token, a top predator addition otherwise reducing stability may actually increase stability because of maturation time lags. Of course the fact that the top predator may eat not only the lower predator, but the prey too, also complicates the picture.

In addition to the problems posed by more than two trophic levels, predator and prey, there is also the matter of alternative sources of prey. Ignoring maturation time lags, a simple graphical approximation becomes possible. The existence of alternative foods means that a predator will feed less intensely on the prey. The predator curve (Fig. 11.5) is thus less steep. It will vary as relative abundances of its prey species change, but should always remain less steep than it would be if only one prey were available. In addition, the predator might (normally) be expected to shift its preferences toward the more common prey species (see 7.5). Thus, when a prey becomes scarce, since all prey will not vary in phase, it will in general be *relatively* scarce also (and vice versa) and the prey curve will be higher at the left and drop more rapidly to the right (Fig. 11.11). Both considerations lead to the conclusion that alternative foods stabilize prey–predator cycles. Holling (1965) says, "Because of the marked buffering effects of alternative foods . . . numerical responses of euryphagous predators to changes in the density of prey species are unlikely to be as great as might otherwise be expected."

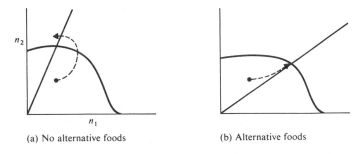

(a) No alternative foods          (b) Alternative foods

**Fig. 11.11.** Diagrams showing the dynamic interaction of prey and predator populations.

But again, reality is not so simple. As relative food abundances change, preferences change, but not instantaneously. Predators often must learn to efficiently capture and utilize new prey. It may even be, as in the case of *Thais lapillus*, cited in Chapter 7, that the dietary change is very sudden, and follows a long lag during which food abundances are changing. It may be argued that where such lags occur, predator populations continue to decimate a declining prey population while sustaining themselves on occasional meals on their alternative prey. In such cases the alternative prey serves to allow further exploitation of a prey than would occur in its absence, and cycles are exaggerated. Thus Pearson (1966) argues that *Microtus californicus* cycling is made less stable by the presence of other species suitable as prey to the feral cats, raccoons, gray foxes, and striped and spotted skunks. We may add that when a predator is food-limited, it is likely to show only slight food preferences (part 7.2). Thus dietary preference changes will be small and alternative prey are very unlikely to lend instability to a prey–predator system. This is more likely when a higher predator exists so that food becomes less important as a limiting factor on the lower predator.

It is known that at least some predators take those prey individuals which are most easily caught and that these, other than sick animals, are primarily the very young and very old. This has been shown for the mountain lion, *Felis concolor*, preying on mule deer, *Odocoileus hemionus*, and elk, *Cervus canadensis* (Hornocker, 1970). Pimlott (1967), examining the prey of wolves, found that fawns and old animals were killed more often than young adults, and Crisler (1956) noted that wolves had great difficulty in killing healthy adult caribou and seldom tried. She estimated the potential prey at about 1.8% of the caribou stragglers and thus at well under 1.8% of the total population. Such selective hunting, although poorly documented, should be the rule. But very young, old, and sick animals are those with the least reproductive importance. Thus selective hunting of this sort buffers prey–predator cycling. In fact Errington (1946) writes that predators, by taking (in a sense) only "excess" individuals, may have very little effect indeed on prey populations. This conclusion seems particularly apt in the case of the muskrats he was studying. In these animals, predation within the area of territories is very low while "excess" individuals driven from the area suffer a very high mortality. In the case of these muskrats, predation actually acts to stabilize populations and prevent cycles.

## 11.2.2 Dispersal

The above discussions all assume a closed system. In nature this is almost never found. If $R$ for both prey and predator were exactly 1.0, and if access to and from an area occurred in response to food and predation level changes, then time lags in dispersal and maturation could result in prey–predator cycling. But in fact the dispersal lag may be very small so that, while local outbreaks of prey are quickly taken advantage of, the predators leave before their decimation of the food supply feeds back on their own numbers. Thus dispersal can also—and probably usually does—buffer prey–predator cycling.

The interrelation between dispersal ability, patchiness of the environment, and food preference is rather untidy. Looking at predation by tawny owls on 1992 marked mice, Southern and Lowe (1968) recovered the remains of 160 in owl pellets. Analysis of the contents and location of the pellets indicated that bank voles, *Clethrionomys glareolus*, were caught in rough proportion to their numbers in both heavy cover and on bare ground, while woodmice, *Apodemus sylvaticus*, were more likely to be caught, relative to their numbers, in brambles and not heavy cover. Thus when the common food is woodmice, owls might best place their territories in open country with brambles, while when voles are more important foods, the owls might best nest elsewhere. If relative mouse abundances were suddenly to change midway through the breeding season, preference change would suffer a considerable lag because of the impossibility of changing nest site.

The result, without the inevitable further complications, would be decreased stability in mouse–owl fluctuations.   Change in feeding habitat with changes in the relative abundances of different prey has been observed in another rodent–owl situation by Glue (1967).   Barn owls, *Tyto alba*, tend to concentrate feeding efforts in grass and scrubland when voles, which are there, are common, but switch to corn ricks and hedgerows to catch rats when the vole population is low.

In the simplified environment of the laboratory, most of those factors buffering prey–predator cycles in nature are absent, and almost always, prey–predator systems very quickly go extinct.   An exception is the experiment of Gause *et al.* (1936) in which the yeasts *Saccharomyces exiguus* or *S. pombe* acted as prey, and *Paramecium bursaria* as predator.   The stability of this system is probably attributable to sedimentation which provides a refuge (see Fig. 11.7) for the yeast.

Classic laboratory experiments in prey–predator relations are those of Huffaker (1958), who used as prey the mite *Eotetranychus sexmaculatus* and as predator *Typhlodromus occidentalis*, another mite.   The prey were fed oranges, and distributed in various ways over variously complex "universes."   The various experiments and their results are listed below in skeletal form.   For more details the reader is referred to Huffaker's original paper.

1.   Four oranges, half covered with wax, were regularly spaced on a tray several inches apart.

2.   Eight oranges were distributed as above.

3.   Six oranges, not wax covered, were distributed as above.

4.   Four oranges, each half covered with wax, were distributed at random among 36 rubber balls of the same size.

5.   Eight oranges, each half covered with wax, were distributed at random among 32 rubber balls of the same size.

In all five of the above experiments, the prey population rose, followed by the predator population which ate all the prey and then itself went extinct.

6.   Twenty oranges, each 9/10-covered with wax, were regularly interspersed with rubber balls.   In this case the prey fell after an initial rise, and rose a second time before the crash.

7.   Forty oranges, each 19/20-covered with wax, were regularly interspersed with 40 balls. Vaseline barriers partly isolated three areas in the tray. The cycles still were hopelessly unstable.

8.   One hundred and twenty oranges, 19/20-covered with wax, were interspersed with rubber balls over three trays narrowly connected, and each divided into 10 partly isolated areas by vaseline barriers. Prey were initially placed on two oranges, predators on the same two oranges 11 days later. The populations crashed.

9.   The above experiment was repeated, this time starting with more prey scat-

tered over all oranges, and with wooden pegs to which the prey might attach silk threads so as to swing out over the barriers. A fan provided the necessary air currents and predators were released over 27 oranges, five days after the release of the prey. This time the system made three complete oscillations before it crashed.

We conclude that stability in a simple prey–predator system is dependent on complexity sufficient to provide the prey with at least temporary refuge.

A natural system, somewhat similar to that of Huffaker, in which persistence of the prey depends on its dispersing one step ahead of the predator, is that of the prickly pear, *Opuntia sp.*, and its herbaceous moth, *Cactoblastis sp.* In this case the moth decimates local populations, but not before their descendants have sprung up elsewhere. Locally there exists a highly unstable, cycling system, but over a wide area both populations are patchy but quite stable (Dodd, 1940, 1959; Nicholson, 1947).

The dispersion pattern of the prey population tends to profoundly affect the prey–predator interaction. Where individuals of the prey species in question are scattered about among individuals of other species, the time and energy expended between food items increases for the predator. This is due to the added travel between items, and perhaps also the time used by the predator in search and recognition (decision-making). Because the cost (time and energy) of food acquisition per item climbs, the net return to the predator is lowered. A mixed species community of prey will support a smaller biomass of predators than a mono-species community of the same productivity (part 13.1). For this reason mono-culture crops often suffer disastrous predation (or infection) unless treated with pesticides. Unfortunately most pesticides also kill the insect pest's predators, thus encouraging the pest's return in even greater numbers in subsequent years. Furthermore, steady use of pesticides encourages the evolution of resistant pest strains. Farmers are thus caught in a need for applying more and more pesticides, a spiral which in the end must be self defeating yet, if broken, would almost certainly prove disastrous to crops. The need for a balanced pest control program utilizing biological controls as well as (and perhaps eventually to the exclusion of) chemical pesticides should be painfully obvious so long as monoculture farming exists.

### 11.2.3 Opportunistic Species

In some areas, prey–predator cycling should be expected, in others not. Areas of high species diversity permit more choice of diet, meaning more types of alternative prey, and thus, probably, more stable, less-pronounced cycles. In addition, such areas, through their biological complexity, may also be physically complex: One can picture a simple community of a few grasses and shrubs and contrast it with a complex community of grasses, shrubs, trees, vines, bromeliads, etc., and conclude intuitively that the latter should possess a greater variety of refuges for prey. On this basis also, cycling should be less pronounced in more diverse environments. Prey–predator cycles of the sort proposed in the Arctic and north temperate zones (relatively simple communities) are, in fact, not observed in the more complex tropics.

The existence of prey–predator cycling also depends on the prey and predator species in question. The discussions in part 11.1 and so far in this part (11.2) are strongly biased toward terrestrial vertebrates, species whose populations fluctuate little under the influence of density-independent factors. Where insects, for example, are involved—either as prey or predator—the situation is less predictable. Opportunistic species, with high fecundity and mortality rates, do not live in populations approaching their carrying capacities; their populations seldom approach equilibrium, and Figs. 11.3 through 11.8 therefore do not apply. Furthermore, high mortality in one year with a consequently small breeding class does not necessarily lead to a subsequently low population. A large adult population does not necessarily follow from a large larval crop (which may succumb in massive numbers to a late frost or a hailstorm, for example), nor does a large larval crop follow from a large population of adults (whose eggs may be destroyed by drought or who laid few eggs because of rain, etc.) Thus, while the number of such species may greatly affect that of their predators, the opposite is not necessarily true. We do not read of damselfly–blackbird cycles.

With respect to opportunistic species as predators, the same general conclusions can be drawn. Here, outbreaks of insects, for example, may cause havoc with plant foods, but except in extreme cases where the vast majority of the food population is destroyed, the reciprocal effect may be slight.

## 11.3 THE HOST-PARASITE INTERACTION

In many instances parasites kill their hosts and thus affect host populations in much the same way as predators. The two primary differences are that they take longer to kill, and leave more for scavengers and decomposers. In most cases, though, parasites do not kill their hosts directly, but may weaken them to other stresses. Parasites also may sterilize their hosts. Trematodes of the clam *Transenella*, for example, eat away their hosts' ovaries, often completely (Obrebski, 1968). This may not cut host numbers down immediately, but obviously inhibits their replacement. Can host–parasite cycles occur? Since the relation is not essentially different from the prey–predator relation except in quantitative aspects, there seems no reason why they should not. In fact, what are apparently such cycles have been found in the system of the aphid *Acrythosiphon pisum* and its parasite *Aphidius smithi* (Vanden Bosch *et al.*, 1967), and elsewhere. As in the prey–predator system, the host–parasite system also displays time lags. For example, if one plots the percentage of black-headed budworm (*Acleris variana*) pupae parasitized against the larval density (number per 100 square feet of foliage), no discernible pattern is found. But if the percentage of pupae parasitized is plotted against larval density in the previous generation, a direct relation can be seen (Morris, 1959). This implies that high larval density leads to more parasites and thus higher rates of parasitism leading, in turn, to lower larval densities, and vice versa. In examining the possible cycling of parasite and host, correlations in host density and rates of parasitism should be made between different generations. Within-generation correlations will be lacking or, perhaps, seriously misleading (Miller, 1963; Solomon, 1964).

A simple model of the host–parasite interaction has been constructed by Nicholson and Bailey (1935).

Let $H_n$ = number of hosts in the $n$th generation,

$P_n$ = number of parasites in the $n$th generation,

$a$ = "searching power" of the parasites.

Then the rate at which hosts of one generation are decimated by parasites (assume death follows from infestation) is

$$\frac{1}{H_n}\frac{dH_n}{dt} = -aP_n,$$

so that the proportion of hosts remaining to reproduce after one generation is

$$\frac{H_n^{(t+1)}}{H_n^{(t)}} = e^{-aP_n}.$$

Then, where $R_0$ is the value of $K$ for the host, in the absence of infestation

$$H_{n+1} = R_0 H_n e^{-aP_n}.$$

The number of hosts destroyed is

$$R_0 H_n - H_{n+1} = R_0 H_n (1 - e^{-aP_n}),$$

and if we assume that, on average, each host destroyed gives rise to one adult parasite, then

$$P_{n+1} = R_0 H_n (1 - e^{-aP_n}).$$

Trends in the host–parasite system can be generated by taking a given $H_0$ and $P_0$ and calculating $H_1$, $P_1$, then $H_2$, $P_2$, etc. Using the house fly, *Musca domestica*, and its pupal parasite, the wasp *Mormoniella vitripennis*, DeBach and Smith (1941) tested the above model. They first experimented until they found a combination of fly, parasite, and fly food in which each parasite found, on average, one house fly every 24 hours (= one generation). This combination consisted of 18 wasps, 36 flies, and 3 quarts of barley, and was used to start the experiment. This relationship meant that, by the next generation, one-half of the flies would be destroyed, so that for a constant population, $R_0$ must be 2.0, and $a = -\ln(H_{n+1}/H_n)/P_n$. Through calculations at various times during the course of the experiment a mean value for $a$ of 0.045 was found. The values of $P_0 = 18$, $H_0 = 36$, $a = 0.045$ were then used to generate $P_n$, $H_n$, and the results were compared with the observed populations (Table 11.1). The observed numbers of replicate means are rounded to the nearest digit. The fit is good over most of one cycle, which means that

$a$ = constant can be estimated from the data such that the equations are reasonably good predictors of population changes.

**Table 11.1.** Parasite and host populations over time (from DeBach and Smith, 1941)

|        |    |    |    | $n$ |    |    |    |    |           |
|--------|----|----|----|-----|----|----|----|----|-----------|
|        | 0  | 1  | 2  | 3   | 4  | 5  | 6  | 7  |           |
| $P_n$  | 18 | 21 | 18 | 15  | 11 | 9  | 11 | 14 | predicted |
|        | 18 | 20 | 19 | 15  | 11 | 9  | 9  | 11 | observed  |
| $H_n$  | 36 | 31 | 26 | 22  | 23 | 29 | 37 | 47 | predicted |
|        | 36 | 32 | 26 | 22  | 22 | 26 | 34 | 45 | observed  |

## 11.4 GENETIC CONSIDERATIONS

When a very efficient predator has begun decimating a prey population to what seems certain extinction, it is hard to believe that selection for better defense by the prey may effect changes rapidly enough to save that prey.   Nevertheless, exactly this may occur.  In 1915, disease very nearly wiped out the plentiful oyster population on Prince Edward Island.  The oyster fisheries were destroyed.  But in 1929 the oysters increased again, and reached normal (pre-blight) levels in 1930. However, when oysters from elsewhere were brought in, they died of the same disease.  The local population had apparently evolved a new resistance to the disease (Needler and Logie, 1947).   African bovids have apparently evolved resistance to the trypanosomes carried by the tsetse fly; introduced cattle have not.  The same situation holds for the American chestnut blight.  European and Asian chestnuts which have long evolved with the fungus causing the blight are effectively immune from its deadly effects.

We have noted that a very efficient predator may destroy its food source and thus cause its own demise.  A "prudent predator" will curb its appetite in the interests of posterity.  Just how prudence might evolve is not clear—natural selection acting for efficiency should certainly overbalance group selection acting for inefficiency.  But, as Slobodkin (1968) has pointed out, the prey have evolved patterns—very young, old, and sick animals are most vulnerable—which effectively render the predator prudent.  In the case of parasite–host interactions, however, group selection will be considerably stronger.  Whole, isolated parasite populations may be totally wiped out by the death of their host, and competition between genotypes within the prey need not involve who can eat the most or most thoroughly destroy host homeostasis.  In addition the prey may evolve resistance to predation and hosts may evolve resistance to parasitism.  In both prey–predator and host–parasite systems one or both species are affected by natural selection in such a way that predation becomes effectively less efficient and cycles are stabilized.  An example follows.

The European rabbit, *Oryctolagus cuniculus*, was purposely introduced into Australia several times. It took hold from 24 pairs in 1859 and grew to several hundred millions before the myxomatosis virus was released as a control agent in 1950. During the first outbreak of the disease, a tremendous portion of the rabbit population died. A characteristic sample from Lake Urana shows the result of inoculating 100 rabbits. The population declined from 5000 in September 1950 to about 50 within six weeks (99% mortality). In the second year, after some population build-up, the disease struck again, killing this time about 90%. Over all of Australia, the first-year epidemic killed 97% to 99%, the second-year epidemic 85% to 95%, and the third-year epidemic only 40% to 60%. Laboratory tests showed that the rabbits were indeed more resistant (which is hardly surprising, considering the immense selection pressure), and also that the virus had become less virulent. The mechanism by which the virus had lost its former potency is straightforward; rabbits taking longer to die were, for a longer period of time, available to mosquitos which fed on the rabbits and passed the disease (Fenner et al., 1953; Fenner, 1965).

Pimentel (1961, 1966) and Pimentel et al. (1963) have extended the above genetic arguments somewhat and claim that as a prey (or host) is declining from predation (or parasitism), selection for resistance is high but selection for predation efficiency is low. Thus the predator becomes lax and the prey more likely to escape, and the prey now begins to increase in number. But at the population low of the prey and early in its subsequent increase, the predators are suddenly hard put to find food. Thus the selective tables are turned, the prey becomes lax and, as it declines, the predator becomes more efficient. This scheme may, theoretically, lead to oscillations. In attempts to demonstrate this Pimentel et al. (1963) and Pimentel and Stone (1968) have shown that *Musca domestica* becomes increasingly resistant to parastism by *Nasonia vitripennis*, that the parasite becomes less virulent, and that cycles becomes less violent after several generations of coevolution in the laboratory. Unfortunately, the critical question of whether resistance and virulence change significantly with the stage of the cycle could not be ascertained. The mathematical model devised by Pimentel in conjunction with the above hypothesis has recently been criticized by Lomnicki (1971). Neither author has considered age-specific effects or large rates of change in gene frequency, both factors that must be considered important in the question of genetic-population interactions.

# 12
# Interspecific Competition and Mutualism

Everybody has a feeling for what "competition" must mean, but the feelings differ. Non-ecologists tend to view competition as an active process by which one party actively chases or intimidates the other for the purpose of keeping something for itself. Ecologists in general think of competition as a passive process in which one species, being more efficient than another, gets the bulk of the contested something. Any kind of interspecific interference may be thought of as competition. So also may competition be reflected in the relative ability of species to escape predation. Intraspecific competition clearly includes all the above and is the process through which natural selection acts on gene frequencies. Interspecific competition, as the term is usually used, does not include the ability to escape predators, except in the sense that species may compete for areas of shelter which make them less likely prey.

If we are to treat interspecific competition in any kind of rigorous way, we must have a formal definition. Two such definitions follow; they differ primarily in their wording, but there are other nuances of difference as well.

Milne (1961) says: "Competition is the endeavor of two (or more) animals to gain the same particular thing, or to gain the measure each wants from the supply of a thing when that supply is not sufficient for both (or all)." For those objecting to "endeavor" and "want," Birch (1957a) says: "Competition occurs when a number of animals (of the same or different species) utilize common resources the supply of which is short; or if the resources are not in short supply, competition occurs when the animals seeking that resource nevertheless harm one another in the process." Both definitions are used by different ecologists.

In this book we use a slight variant of Birch's definition. This difference does not represent fierce individuality on the part of the author nor a splitter spirit, but an attempt to avoid circularity. Populations which vie for a resource in short supply cannot be recognized unless the shortage somehow "harms" them. But what we must mean by harm is that $r$ (or $R$) and therefore $r_0$ or $K$ are lowered from what they would be in the absence of (interspecific) competition. Thus, in the second part of Birch's definition, we cannot tell whether competition is occurring unless we know that one or more species are "harmed" as a result of competition. But we cannot know whether such harm has occurred unless we assume the existence of competition; in fact there is no way of knowing whether the harm is or is not due to competition, for the effects of food and predator-limitation are interdependent (10.2.6). The only workable, noncircular definition of (interspecific) competition which retains the spirit of Birch's definition is: (interspecific) competition occurs when two or more species experience depressed fitness ($r_0$ or $K$) attributable to their mutual presence in an area. This definition is less likely to lead one to circular reasoning and is theoretically workable. Of course, proving that competition exists in any given natural situation is still a difficult exercise.

## 12.1 TYPES OF COMPETITION

### 12.1.1 Active Competition

Active competition is the result of passive competition. That is, when two populations mutually depress their respective $r_0$ or $K$ values it may be advantageous for each to exclude the other from its area of habitation so as to simulate those conditions which would obtain in the other's absence. Both $r_0$ and $K$ are components of fitness (Section III, Discussion), so the fitness of any genotype is increased by excluding individuals of the other species. Of course, active exclusion will not always occur, for it may be that the advantage of ousting a competitor is more than offset by the cost of ousting him.

One form of active competition is interspecific territoriality (8.5.5). Redwing (*Agelaius phoeniceus*) and yellow-headed (*Xanthocephalus xanthocephalus*) blackbirds exclude each other from individual (actually family) territories, as do the hummingbirds *Calypte anna* and *Selasphorus sasin* (Pitelka, 1951), and the eastern (*Sturnella magna*) and western (*Sturnella neglecta*) meadowlarks (Lanyon, 1956). Division of an area into interspecific territories ensures an increased resource supply to those individuals fortunate enough to possess territories, and thereby increases the fitness of their genotypes, but does not result in lessened interspecific competition, as measured in depressed $r_0$ or $K$ for the population as a whole because the total amount of resource per population remains unchanged. To alleviate the effects of competition (on $r_0$ and $K$) there must be widespread separation of populations—that is active competition must bring about a significant degree of allopatry. This may occasionally happen among ants where, for example, Brian (1952) has observed species of *Formica* in cut-over pine woods attack and destroy colonies of *Myrmica*. Of course the effects of the competition cannot be totally negated by this means, for the cost of exclusion itself affects $r_0$ and/or $K$.

### 12.1.2 Passive Competition

Suppose two populations compete for some resource and do not actively exclude each other from broad areas. A given amount of resource in the form of shelter, nest-holes, or other space-factors, will support a given number of individuals of both species. If the population of one increases, then that of the other must decrease. A given amount of nutrient resource supports only so much total metabolism by both species together. As the metabolism of one species increases due to a growing population, that of the other must decrease, resulting eventually in population decline.   An often-used phrase is that the more efficient species displaces the less efficient.

The word "efficient" deserves some comment.   The most "efficient" species is that which turns the limiting resource most rapidly into individuals.   Thus "efficiency," when both populations are well below the joint carrying capacity, is simply $r_0$ (which as we have seen in Chapter 6 is really more a function of energy allocation than efficiency, in the true sense of the word), and when both populations are at joint-$K$ (where $R$ would equal 1 for both species were there no competition), is the degree to which the depression of $K$ by the other species can be avoided.   Thus "efficiency" is related to efficiency only by analogy and might best be dropped from common usage.

Occasionally, using resources in different ways or at different times eliminates or alleviates competition.   For example, the barnacles *Balanus balanoides* and *Chthamalus stellatus* on the English coast overlap in their intertidal distribution and compete for space.   When *Balanus* is absent, *Chthamalus* grows very well at low tidal levels, but *Balanus*, when present, grows faster and larger, and undercuts or overgrows its competitor, thus killing it.   At high tidal levels, *Balanus* cannot adapt to the physical environment and *Chthamalus* flourishes (Connell, 1961). Because the contested resource, space, is used in different ways—there is only a slight overlap in tidal level used—competition over the whole population is slight.   In Ceylon, the crow *Corvus splendens* and the flying fox *Pteropus giganteus* roost in the same trees.   However, the crows roost at night, and the bats during the day.   Thus competition for roosting space is avoided (Willey, 1904).   Pryer (1884) gives a similar account for swiftlets, *Collocalia sp.*, and free-tailed bats, *Myctinomus plicatus*, in northern Borneo.

The evolution of time differences in utilization of the same resource has often been cited as a means of alleviating competition.   This explanation may be true for space-resources such as roosting trees, but cannot be true in the case of food. What matters in the case of food is whether or not it is available.   Whether it is eaten at night or during the day does not matter, because in either case it ceases to be available.

### 12.1.3 "Contest" and "Scramble" Competition

Nicholson (1957) divides competition in another way.   His uses were meant to apply to intraspecific competition but can also be applied to competition between species.   In *contest competition*, each individual possesses its own supply of resources.   Thus while some individuals may not get enough to survive, some always have sufficient.   Competition between raptor nestlings (6.3.1) is an example of contest competition, as is food competition between individuals on territories "defended" for their food supply.   Interspecific territoriality leads to contest competition between individuals of different species; area exclusion and consequent habitat separation, as in the cases of the barnacles or ants given above, results in contest competition between populations of different species.   *Scramble competition*, as the name suggests, is a scramble for resources with no one individual necessarily obtaining a sufficient amount. The example of the sheep blow-flies and the gram of brain (Nicholson, 1948) given in part 10.3 is a case of scramble competition. Populations of competing species usually do not "defend" adequate resources as a group, and so most interspecific competition must be said to be scramble competition, although unlike the case of intraspecific scramble competition, instability does not normally result.

### 12.1.4 Interference

It seems reasonable to assume that, when the behavior patterns of one species depress the fitness of another, active competition is being displayed. Often, however, the resource in short supply, or the resulting advantage in defense from predation—or any advantage at all—is not obvious, and it may be that the depression of the other species is merely a fortuitous by-product of normal social behavior. Competitive effects of this sort are called *interference*.

In two open-field enclosures, DeLong (1966) placed either *Mus musculus* or both *Mus* and *Microtus californicus*. In the two pens, *Mus* showed similar fecundity and body growth, but the recruitment of young *Mus* into the population was much lower in the presence of *Microtus*. Furthermore, the number of juvenile *Mus* per lactating female declined over several years as the *Microtus* population increased. Laboratory experiments supported the conclusion that *Microtus* were destroying *Mus* nests. In another enclosure experiment, Caldwell (1964) found *Mus* and *Peromyscus polinotus* each to withdraw into opposite halves of a common enclosure. There were no apparent differences in vegetation or physiography to explain this separation which was therefore probably socially induced. As the *Mus* population declined, *Peromyscus* extended its range in the enclosure. These species do not "defend" large group territories, and so it is doubtful that active competition for some shared resource such as food was the cause of the separation. It seems inescapable that separation resulted from interspecific avoidance at the individual level. The reason for such avoidance is not clear, but it is quite clear that where thwarted, in areas of overlap, the tendency for interspecific avoidance might lead to aggression or endocrine strain, perhaps, and thus interference.

## 12.2 THE THEORY OF COMPETITION

### 12.2.1 General Arguments

We use the logistic equation as an approximation to reality, and note that, according to the equation, the addition of one individual of a species to a population depresses its population's carrying capacity by a fraction $1/K$.

$$\frac{dn}{dt} = rn\left(1 - \frac{1}{K}n\right).$$

We let the relative amounts by which species 1 depresses species 2 and species 2 depresses species 1 be, respectively, $\alpha_{12}$, $\alpha_{21}$. Thus

$$dn_1/dt = (r_1)_0 n_1\left[1 - \frac{n_1}{K_1} - \alpha_{21}\frac{1}{K_1}n_2\right] = (r_1)_0 n_1\left(1 - \frac{n_1 + \alpha_{21}n_2}{K_1}\right) \quad (12.1)$$

$$dn_2/dt = (r_2)_0 n_2 \left[ 1 - \frac{n_2}{K_2} - \frac{\alpha_{12} n_1}{K_2} \right] = (r_2)_0 n_2 \left( 1 - \frac{\alpha_{12} n_1 + n_2}{K_2} \right). \qquad (12.2)$$

These are the Lotka–Volterra competition equations.

At equilibrium,

$$\hat{n}_1 = K_1 - \alpha_{21} \hat{n}_2 \qquad (12.3)$$

$$\hat{n}_2 = K_2 - \alpha_{12} \hat{n}_1. \qquad (12.4)$$

That is, when $n_1 < K_1 - \alpha_{21} n_2$, $n_1$ increases, and vice versa. When $n_2 < K_2 - \alpha_{12} n_1$, $n_2$ increases, and vice versa. Plotting the lines 12.3, 12.4, we obtain four possible configurations (Fig. 12.1). The arrows show the direction in which the populations move: Below the solid line, $n_1$ increases; below the dashed line, $n_2$ increases. Now suppose that the points $P$, $Q$, $R$, represent initial values of $n_1$ and $n_2$. Where Figs. 12.2(a), (b), (c), (d) correspond to 12.1(a), (b), (c), (d), the arrows show the joint trend in population growth. In the first two cases, either one species or the other wins, the loser becoming extinct. In the third case there exists an unstable equilibrium, and the winner depends upon the initial conditions (and relative $r_0$ values). Only in the last case is coexistence possible.

In the configuration leading to coexistence (Fig. 12.2d), the following inequalities hold:

$$K_2/\alpha_{12} > K_1$$

$$K_1/\alpha_{21} > K_2.$$

Rearranged, these become

$$1/K_1 > \alpha_{12}/K_2$$

$$1/K_2 > \alpha_{21}/K_1.$$

But $1/K_1$ is the fraction by which an individual of species 1 depresses its own growth rate, $r_1$, while $\alpha_{12}/K_2$ is the fraction by which the same individual depresses the growth rate, $r_2$, of its competitor. The conditions for coexistence are thus that each species must depress its own value $r$ more than it depresses that of its competitor. Equilibrium is given by that point at which the lines

$$\hat{n}_1 = K_1 - \alpha_{21} \hat{n}_2 \qquad \text{and} \qquad \hat{n}_2 = K_2 - \alpha_{12} \hat{n}_1 \qquad \text{cross,}$$

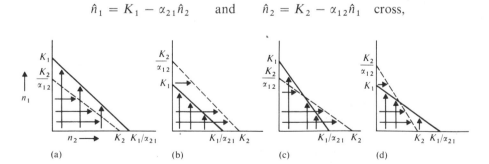

Fig. 12.1. Conditions for increase in populations of competing species: four possible configurations.

or

$$\hat{n}_1 = K_1 - \alpha_{21}\hat{n}_2 = \frac{K_2 - \hat{n}_2}{\alpha_{12}},$$

or

$$\hat{n}_1/\hat{n}_2 = \frac{K_1 - \alpha_{21}K_2}{K_2 - \alpha_{12}K_1}. \tag{12.5}$$

If species are quite similar (their niches are broadly overlapping) with respect to those factors limiting their populations, alpha values may be quite large and the inequalities necessary for coexistence may not hold. If species are quite dissimilar, they affect each other less and less—alphas decline toward zero—and coexistence becomes more likely. The above approach is taken from Gause and Witt (1935).

Consider the following approach. Let the frequencies of $n_1$, $n_2$ be $p$, $1 - p$. Then we can write

|  | Species 1 | Species 2 |
|---|:---:|:---:|
| Initial population | $n_1$ | $n_2$ |
| Initial frequency | $p$ | $1 - p$ |
| $R$ value | $R_1$ | $R_2$ |
| Population in next time unit | $n_1 R_1$ | $n_2 R_2$ |
| Frequency in next time unit | $p' = \dfrac{n_1 R_1}{n_1 R_1 + n_2 R_2}$ | $(1 - p') = \dfrac{n_2 R_2}{n_1 R_1 + n_2 R_2}$ |

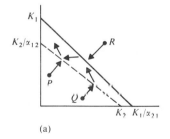

(a)

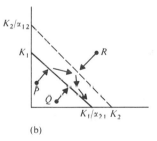

(b)

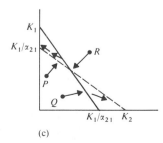

(c)

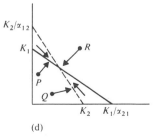

(d)

**Fig. 12.2.** The dynamics of competitive interaction.

Thus
$$\Delta p = p' - p = \frac{n_1 R_1}{n_1 R_1 + n_2 R_2} - p = \frac{p R_1 - p^2 R_1 - p(1-p) R_2}{p R_1 + (1-p) R_2},$$

or
$$\Delta p = \frac{p(1-p)(R_1 - R_2)}{p R_1 + (1-p) R_2}. \tag{12.6}$$

Equilibrium occurs when $p = 0$, $p = 1$, or when $R_1 = R_2$. If, as in the Gause model,

$$R_1 = e^{r_1} = \exp\left[ (r_1)_0 \left( 1 - \frac{n_1 + \alpha_{21} n_2}{K_1} \right) \right]$$

(and the same form for $R_2$), and $R_1 = R_2 = 1$ at equilibrium, then

$$\ln R_1 = \ln R_2 = 0 = (r_1)_0 \left( 1 - \frac{\hat{n}_1 + \alpha_{21} \hat{n}_2}{K_1} \right) = (r_2)_0 \left( 1 - \frac{\hat{n}_2 + \alpha_{12} \hat{n}_1}{K_2} \right),$$

so that
$$\frac{\hat{n}_1 + \alpha_{21} \hat{n}_2}{K_1} = \frac{\hat{n}_2 + \alpha_{12} \hat{n}_1}{K_2},$$

and, as in Eq. (12.5),
$$\frac{\hat{n}_1}{\hat{n}_2} = \frac{K_1 - \alpha_{21} K_2}{K_2 - \alpha_{12} K_1}.$$

In experiments with *Paramecium caudatum* and *P. aurelia*, variety three and four, Hairston *et al.* (1968) found that the first and third of these groups depressed the second more when together than did either one alone. The $\alpha$ value of two species with respect to another is neither the sum nor the mean of the two alphas, necessarily: Also alphas may change over different patches in an environment and if species change their pattern of patch use as their populations vary, the direction of competition may change. Alphas may also change with relative and absolute values of $n_1$ and $n_2$, and changes in alpha may show time lags. None of these difficulties are handled by the Gause–Witt model and Eq. (12.6) becomes quickly cumbersome while still making no claims of accuracy. For the moment, we note that populations near equilibrium can be treated with the Gause–Witt diagrams and that while density-related changes in $\alpha$ may provide for the co-existence of competing species in some instances, coexistence is expected commonly only when niches overlap little or when there are extenuating circumstances. The question of how little the overlap must be is covered in Chapter 15, and the matter of extenuating circumstances is discussed below.

## 12.2.2 Competitive Exclusion

Except for the extenuating circumstances, competition between ecologically similar species should result in the local extinction of all but one. Let's look at some experimental evidence. Gause (1934) grew two species of yeast, *Saccharomyces sp.* and *Schizosaccharomyces sp.* together and alone on an extract of brewer's yeast with water and sugar in anaerobic cultures. From the single-species cultures he measured $K_1/K_2$ at 2.241. Eqs. (12.1) and (12.2) were then rearranged to give

$$\alpha_{12} = \left( K_2 - \frac{K_2 dn_2/dt}{(r_2)_0 n_2} - n_2 \right) \bigg/ n_1$$

$$\alpha_{21} = \left( K_1 - \frac{K_1 dn_1/dt}{(r_1)_0 n_1} - n_1 \right) \bigg/ n_2,$$

and from the mixed cultures, the alpha values calculated:

$$\alpha_{12} = 3.15$$

$$\alpha_{21} = 0.439.$$

Thus $1/K_1 < \alpha_{12}/K_2$ and $1/K_2 > \alpha_{21}/K_1$ and species 1 should be expected to win (Fig. 12.2a)—which it did. Species 1 inhibited species 2 by the production of alcohol, but the effect of species 2 on species 1 was not apparent.

Gause (1935) also experimented with *Paramecium aurelia* and *P. caudatum*. In this case the accumulation of waste by-products hurt *P. caudatum* more than *P. aurelia*, and the latter emerged from the contest victorious. The winner, however, could be changed by changing the medium frequently and thus minimizing the waste products. Vandermeer (1969) repeated Gause's experiments with another variety of *Paramecium aurelia*, also studying competition of this species with *Blepharisma sp.* and *P. bursaria*, and competition between these last two species. He found that alpha values varied during the course of the experiment (this has also been observed repeatedly in class laboratory exercises) but was able to determine values which satisfactorily predicted the outcomes observed. In general, but not in every case, one species won and one became extinct.

Slobodkin (1964), working with Hydrida, found that the large brown *Hydra littoralis* and small green *Chlorohydra viridissima* coexisted if grown in the dark: in the light, however, the green hydra excluded the brown. Both experiments were begun with equal numbers of the two species.

In competition experiments with *Drosophila melanogaster* and *D. simulans*, Tantawy and Soliman (1967) found that *D. melanogaster* eliminated its competitor in 80 days at 25°C. At 15°C, however, *D. simulans* was the winner after 150 days. The experiments, in both cases, were begun with 122 4-day old, inseminated female *D. melanogaster* and 600 4-day old, inseminated female *D. simulans*.

There are many other such experiments, including those using the beetles *Rhizopertha* and *Calandra* by Birch (1953a), and many using the flour beetles, *Tribolium sp.*, by T. Park, D. Mertz, P. Dawson, and others. In nearly all cases, exclusion of one of the experimental species was the result, although in many cases the winner could be changed by changing the experimental conditions.

The strong tendency for exclusion in virtually all laboratory experiments, and the theoretical basis for expecting exclusion as the result of competition between ecologically similar species, led Hardin (1960) to suggest *competitive exclusion* as a basic ecological law. It is difficult to accept as a basic law a statement that says that competition between two species will inevitably lead to the demise of one provided the two species aren't too different. One would like to know what "too different" meant, providing it was not circular. There are also other difficulties with a "law" of competitive exclusion. Cole (1960), in a negative reply to Hardin, pointed out a number of the extenuating circumstances. Some of these are discussed below. (See also Ayala, 1969, 1970; Gause, 1970.)

### 12.2.3 Relative Shortages

A population is limited by resources in relatively short supply only if the cost of acquisition of the resource is density related. Thus competition exists for such resources in the same way it exists for resources in absolute short supply. However, there exists a possibly common case, exemplified by Andrewartha's grasshoppers which died of starvation because the grass turned brown and lost its protein value, and by the quokkas on the island of Rottnest (10.2). This is the situation where reproduction is held at sufficiently low levels, or mortality at sufficiently high levels by an abundant resource of poor quality, so that secondary factors which otherwise might be insignificant in population limitation become the major limiting factors. Whether the resource is considered limiting or not now becomes a matter of semantics. But since it is not directly limiting, two species cannot be said to be competing for it, and competitive exclusion will not occur.

### 12.2.4 Dispersal

Which of two species excludes the other in competition depends on environmental conditions. We may thus view a gradient of some sort at one end of which one species wins in competition and where at the other end the other species wins. The dividing line along the gradient at which the winner changes may be very well defined, but to the extent that the species can move about, dispersal will act to maintain both along a certain portion of the environment (Fig. 12.3).

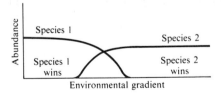

**Figure 12.3**

Since any gradient will undoubtedly display several competitive species patterns of this sort it is probably often impossible to determine, in the field, whether a given species coexists in an area where it is found, or is merely a straggler from an adjacent area or patch type.

## 12.2.5 Statistical Considerations

When two species' realized niches (as defined in Section II) overlap little, there is little competitive effect; exclusion does not occur, and the alphas, being small in value, have little effect on the equilibrium population sizes. When niche overlap (with respect to limiting factors) is great, the competitive effect may be to eliminate one species, and will certainly affect the population sizes of both species. However, very broadly overlapping niches usually imply close phylogenetic affinities and hence similar $R$, $K$, and $\alpha$ values. The more broadly overlapping the niches, the more the two competing species tend to resemble a single species. But when $R$, $K$, and $\alpha$ values are very similar, the exclusion process becomes less and less rapid and less and less clear-cut in the presence of stochastic variation. Recall that when the two species have identical (competitive sub-) niches, exclusion is merely a random-walk process. Thus the direction of competitive exclusion between two sibling species, for example, may not be highly predictable, and the process may take a very long time so that the two species give every appearance of coexistence. Thus Miller (1964), says: "Where competition exists solely of the differential exploitation of a common resource, and such forms of interference as territoriality and aggression are lacking, there is seldom any evidence that competition has a limiting effect. In fact, it has been demonstrated recently that sibling species of *Drosophila* larvae, which do not have strict space requirements and compete almost entirely by exploitation, can exist in equilibrium."

## 12.2.6 Predation

In his experiments with Hydrida, Slobodkin (1964) found that in lighted conditions the green hydra eliminated the brown. If both species were equally predated (removal of fixed proportion of the population increments since the last census), however, exclusion took a longer time, and at predation levels of 90%, the two species coexisted. In his 1961 book, Slobodkin explains at some length the reasons for this, using logistic equations with the competitive term (Eqs. 12.1, 12.2) and an additional term depicting losses from predation. This technique is used in further discussion in Chapter 15, but for now we use a more straightforward approach. Note that if $R$ takes the value $R^*$ in the absence of predation, and that if $m_1$, $m_2$ denote the proportion of species 1, 2 taken per predator per time unit, then $R_1 = R_1^* - m_1 n_p$, $R_2 = R_2^* - m_2 n_p$, and Eq. (12.6) becomes

$$\Delta p = \frac{p(1-p)[(R_1^* - m_1 n_p) - (R_2^* - m_2 n_p)]}{[p(R_1^* - m_1 n_p) + (1-p)(R_2^* - m_2 n_p)]}, \tag{12.7}$$

where $n_p$ is the predator population. In the absence of predation, species 1 wins when $R_1 > R_2$. But, at the same time, predator food preference for species 1 may result in $R_1^* - m_1 n_p < R_2^* - m_2 n_p$, so that the outcome of competition is changed. It is also possible that when $R_1^* = R_2^*$, $R_1 - m_1 n_p \neq R_2^* - m_2 n_p$, and vice versa. Thus predation may either favor coexistence where it would not otherwise occur or destroy coexistence. The "law" of competitive exclusion may be significantly affected by predation.

## 12.3 INTERSPECIFIC COMPETITION AND EVOLUTION

### 12.3.1 Niche Separation

In 1946, Elton reported on the number of animal species and genera in 55 communities, widely differing in physical characteristics, all over the world. The ratio of the number of species to number of genera was similar in all cases and about 86% of the genera had only one species. Combining a number of these communities, however, he found that the ratio of species to genera greatly increased. He concluded that competition resulted in a strong tendency for closely related species to be found in separate habitats. Closely related species are usually also ecologically quite similar.

Lack (1945) comments that the cormorant, *Phalacrocorax carbo*, and the shag, *P. aristotelis*, are very similar in appearance and habits, that they are distinguished in the field only with great difficulty, and that they occur in the same areas around Britain. They avoid competition in subtle ways, the cormorant nesting on the flat, broad cliff ledges, the shag in caves, hollows, or on narrow ledges. Of the cormorant's diet, 59% consists of flatfish, prawns, and shrimp, foods which comprise only 3% of the shag's diet. The shag concentrates (82%) on sand eels and sardines, while the cormorant gains only 1% of its food from these delicacies. These birds show clear niche separation in using both different foods and different nest sites.

MacArthur (1958), watching five species of wood warblers in spruce forests in the eastern United States, discovered that these birds used different parts of the trees and thereby overlapped only slightly in foraging area. If foraging areas are considered resources this again is a case of niche separation between related species. The Cape May warbler, *Dendroica tigrina*, spends its time foraging in the top 40% of the trees, mostly on the outer parts of branches. The blackburnian warbler, *D. fusca*, uses the top 60%, primarily the top 20%, of the trees on the outer and mid parts of the branches. The black-throated green warbler, *D. virens*, forages in the mid 20% of the trees from the trunk to the outer parts of the branches. The bay-breasted warbler, *D. castanea*, forages at mid levels and low in the trees, and the myrtle warbler, *D. coronata*, forages low. The foraging movements also differ, the Cape May moving mostly up and down and the blackburnian radially. The black-throated green moves tangentially as does the myrtle, while the bay-breasted forages radially. Where foraging height broadly overlaps, movements differ; where movements are similar, foraging heights are separate. These birds appear to avoid (or minimize) interspecific competition by spatially and behaviorally dividing their environment.

Habitat (or niche) separation of the sort found by MacArthur, though much less marked, has also been found in English titmice by Hartley (1953).

A good example of niche separation has been reported by Carleton (1965) for the least chipmunk, *Eutamias minimus*, and the golden-mantled ground squirrel, *Citellus lateralis*, in Colorado. Of 57 plant and 5 insect species common in the area in summer, only one (the dandelion, *Taraxacum*) is highly edible for both species.  Of 15 less-used foods eaten by the ground squirrel, 8 or more are also used by the chipmunks.  Of 9 less-used foods of the chipmunk, 8 are also eaten by the ground squirrel; 80% of the diet of both animals is the dandelion.  The two species, which we presume to be food-limited—at least to a strong degree—appear to be very similar indeed in their diets.  But when one examines their utilization of the one really important food species it becomes clear that competition may be ameliorated by niche separation.  The chipmunk shows a marked preference for dandelion flowers and seeds, the ground squirrel for dandelion stems.

The black-capped (*Parus atricapillus*) and chestnut-sided (*P. rufescens*) chicadees are sympatric in British Columbia and the northern Pacific states. They may forage in mixed flocks in the winter, but black-caps are sighted $3\frac{1}{2}$ times as often in deciduous trees than in conifers and chestnut-sided chicadees are seen 5 times as often in conifers as deciduous trees.  Also the black-cap usually forages below 5 ft from the ground and none is ever seen above 70 ft.  The chestnut-sided chicadee forages mostly between 45 and 50 ft and 12% occur above 70 ft (Smith, 1967).

A slightly different approach to niche separation was taken by Schoener (1968) in his study of four *Anolis* lizards on Bimini.  Schoener measured the height, color, and size of the perches used by the lizards, whether or not the lizards were found among leaves and whether they perched on branches or on the ground.  If $p_{ix}$ represents the frequencies of species $x$ in the $i$th situation, then a measure of the difference (niche separation) between two species $x$ and $y$ can be given by

$$1 - \tfrac{1}{2} \sum_i |p_{xi} - p_{yi}|.$$

Schoener calculated these values for all lizard species pairs and also calculated differences in diet.  He found that if overlap in habitat was great, difference in diet was also great, and that if diets were similar, the species were quite separated in their use of habitat.

## 12.3.2 The Role of Intraspecific Competition

In 1949, Svärdson suggested that intraspecific competition for resources would act to broaden niches—species would become more generalized in diet, etc., and that interspecific competition would tend to make niches narrower (fewer dimensions and, along any environmental gradient a smaller range of responses). Since intraspecific competition for resources is a result of resource shortage (by definition), the first part of his statement follows immediately (part 7.2).  But since, for a given number of competitors and a given range of resources, niche broadening must lead to greater niche overlap, it also follows that interspecific competition will oppose niche broadening.

Svärdson's observation leads to two important points.  First, the breadth and overlap of niches tend toward equilibria between the "forces" of intra- and interspecific competition.  Second, and of more immediate interest, a sudden bloom of food will alleviate, momentarily, both kinds of competition and result in the greater use of those foods of greatest "value" (see 7.2.1) for the individual species. Niches will momentarily narrow. Overlap, however, may either decrease or increase. The usual conclusion that less intense competition permits greater niche overlap is true (by the same token, decreased competition intensity permits less niche overlap) but is not equivalent to the statement that less intense competition leads to greater niche overlap, or more intense competition leads to less overlap. This latter statement has no obvious basis in theory. There is no clear reason why increased competition should necessarily result in greater niche separation. Some data indicate that it does (Hartley, 1953; Newton, 1967), which is not surprising, Other data undoubtedly show that, in other cases, it doesn't.

Kohn (1966) found that species of vermivorous cone snails, *Conus sp.*, are more specialized in diet when sympatric than when allopatric. This supports a generalization which follows from Svärdson's observations. Since niche breadth represents an equilibrium between intra- and interspecific "forces," realized niches will be broader when no competitors are present, narrower when interspecific competition occurs.

### 12.3.3 Character Displacement

It is not always clear whether niche separation has resulted from competition (selection favors divergence) or whether coexistence occurs because separation existed in the first place. Most cases of niche separation between coexisting species probably involve a little of both paths. It is often possible, though, to demonstrate cases where competition has clearly resulted in selection pressures successfully forcing niche separation. This situation has been called *character displacement* by Brown and Wilson (1956) who first described the phenomenon in some detail. Slight ecological differences, often manifesting themselves in morphological or other differences, between two species which are allopatric (do not occur together) become exaggerated in zones of sympatry (where they occur together). For example, trout and char, when together, react quickly to the competitive situation and separate in space, the char moving to deeper, the trout to shallower water (Chapman, 1966). An example from Lack (1947) involving Galapagos finches is given in Table 12.1. In both species, beak depth differences occurring between allopatric populations are exaggerated in areas where the species are sympatric.

**Table 12.1.** Character displacement in Galapagos finches (from Lack, 1947)

| Island | Finch species present | Beak depth (mm) range | mean |
|--------|----------------------|----------------------|------|
| Crossman | *Geospiza fuligonosa* | 8.0–11.0 | 9.5 |
| Daphne | *G. fortis* | 8.0–11.5 | 10.0 |
| Albemarl, | ( *G. fuligonosa* | 7.5–9.5 | 8.5 |
| Indefatigable | ( *G. fortis* | 10.5–15.5 | 13.0 |
| Abington, Bindloe, | ( *G. fuligonosa* | 6.5–9.0 | 8.0 |
| James, Jervis | ( *G. fortis* | 10.5–13.0 | 12.0 |

### 12.3.4 Ecotones and Competition

An ecotone is a steep ecological gradient. We note that over an ecological gradient there will be a range over which one species can survive (average fitness over time = 1) and another, overlapping range over which some slightly ecologically divergent competitor can survive. If we assume—as must clearly always be the case—that a species is most likely to be successful in competitively excluding another where its fitness in the absence of competition (for a given population density) is highest, then the result of competition is as shown in Fig. 12.4.

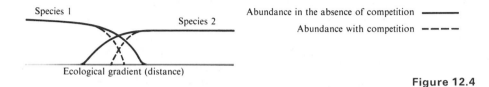

Figure 12.4

The distance over which one species replaces the other as most common is shortened. Over steep gradients this distance tends to be particularly small, and selection pressures against dispersal of adults into submarginal parts of the gradient may be very strong. We thus often find well-defined zones characterized by particular species; sudden transitions in the biological community occur at certain points along the gradient. Particularly good examples occur along altitudinal gradients where lowland, lower montane, upper montane, subalpine, and alpine tundra are often very clearly demarcated. The vole *Microtus pennsylvanicus* is common across the prairie to the base of the Colorado foothills, where it is abruptly replaced by *M. montanus*. Intertidal areas are also good ecotones resulting in sharp transition, with tidal level, from one group of species to another.

## 12.4 COMMENSALISM AND MUTUALISM

### 12.4.1 Commensalism

We now turn to a situation opposite to that of competition—where two species benefit from, or at least do not harm, each other ($\alpha_{ij} \leqslant 0$).

*Commensalism* is the relationship between two organisms when one is benefited and the other unaffected (that is, $\alpha_{ij} < 0$, $\alpha_{ji} = 0$). Orb-spinning spiders are often associated with small commensal spiders which live on their webs, neither helping nor harming their hosts, but living off the food remains left by the host. The commensal species is usually smaller than that minimum size which the host bothers to eat and is thus safe from attack. In other cases this is not true. *Conopistha sp.*, which occurs with *Nephila sp.*, is eaten if tossed into the latter's web. It probably avoids death by avoiding the host and the sticky strands of the host's web (Kaston, 1965). Water trapped in bromeliads, pitcher plants, and other plants, provides a habitat often for whole communities, and the yellow and pink primrose moth is often found hidden within the flower of the evening primrose, *Oenothera sp*. In such cases the plants serve as hosts for a variety of commensal species.

## 12.4.2 Plant-Animal Mutualism

"Mutualism" describes a relationship between two or more species which benefits all ($\alpha_{ij} < 1$ for all $i$, $j$). Everyone is familiar with the mutual relation between some insects and plants in which the insect, while getting food from the plant, aids in pollination. Plants have often evolved characteristics which attract the appropriate pollinating species. Orchids may mimic female insects in copulatory posture, carrion flowers smell like rotten meat and attract flies. The colors of blossoms may attract—or at least be an indicator of—the kind of organism which pollinates a plant species. Some flowers bloom white, at dusk, to attract bats while hummingbird-pollinated flowers are tubular (accommodating the long bill) and usually red (at least in California). The color red seems to serve as a symbol of appropriate flowers (releaser) to the hummingbird and probably represents a case of mutual mimicry by widely divergent plant taxa (Grant, 1966).

Burrs are an annoying example of a plant commensalism in which the plant gets its seeds dispersed, but in many cases seed-dispersal mechanisms are mutualistic. Fruits may have evolved their often rich, sweet taste to attract birds which spread the seeds in their droppings. In fact plants often avoid competition between offspring and parent by ensuring such dispersal; the seeds of some species will not germinate until they have passed through the digestive tract of an animal.

Green algal *symbionts* (the term for species benefiting each other) occur commonly in protozoans, sponges, coelenterates, flatworms, aschelminthes, and mollusks. The former often are dependent (*obligate* symbionts) on the latter for a place to live, and the latter may somehow be able to use the energy fixed from the sun by the algae. Recall that in Slobodkin's Hydrida experiments, the green (with algal symbionts) *Hydra* excluded the brown *Hydra* only in lighted conditions.

Also, intestinal flora—bacteria or protozoans—are common in the guts of many animals. Without them, cows and cockroaches would not digest the cellulose in the cell walls of their food and would starve.

One type of plant–animal mutualism deserves special mention because of the magnificent work by Dan Janzen. It has been known for some time that plants attract insects with the use of special nectaries on their leaves (see, for example, Bailey, 1924). Certain plants, *Myrmecophytes*, are regularly equipped with ant-adapted *domatia*, extrafloral nectaries, and other food-rich structures. These structures seem to serve no other purpose than to attract and feed ants (Brown, 1960). Janzen (1966, 1967) has examined in detail the relation of ants, *Pseudomyrmex ferruginea*, *P. nigrocincta*, and *P. spinicola*, to the swollen-thorn acacias, *Acacia sp.*, in Central America. The base of the large thorns of these acacias is swollen and is hollowed out by the ants to use for a nest. Foliar nectaries (sugar) and *beltian bodies* (protein), food-rich structures on the leaf tips, supply the ants with food. The ants, in turn, patrol the host plant and drive off or kill all other insects or plant material on the acacia and from an area around the plant, on the ground. The ant-inhabited acacias thus grow in cylindrical spaces free of all other organisms save the ants. Without ants, the nectaries and beltian bodies still form, but the plants are quickly defoliated by insects or rodents, shaded by other faster growing plants, and destroyed. The ants, by clearing an area around the acacias, also protect them from the fires which commonly sweep the acacia habitat during the dry season.

Some species of ants of the genus *Azteca* are obligate occupants of the *Cecropia*, a large-leafed tree of the tropics. The ants eat and kill tendrils of vines where they touch the tree and thus destroy one source of shading. This allows *Cecropia* to grow to the forest canopy. The tree in turn produces large, hollow internodes where the ants live. Regularly spaced thin spots in the internode walls allow the ants to drill entrances. In addition, *mullerian bodies*, rich in protein and carbo-hydrate, grow from the modified petiole bases and provide the main food source for the ants (Janzen, 1968).

### 12.4.3 Animal-Animal Mutualism

We begin with a familiar case of mutualism involving the cowbird, *Molothrus ater*, or the cattle egret, *Bubulcus ibis*, and cattle. It seems obvious that cowbirds benefit from the insects stirred up by or attracted to cattle. Heatwole (1965) has shown that cattle egrets require fewer steps per time to catch equivalent numbers of insects when following cattle than when feeding away from cattle (they also do well following farm machinery for which they may forsake the cows). It is not certain that the cattle benefit from the relationship but it seems likely that the birds, by taking insects off their backs, may keep down the incidence of disease and the level of annoyance. The birds may also act as sentinels for wild bovids. Much the same situation exists with respect to bee-eaters, *Merops nubicus*, which ride on the backs of African bustards and ostriches, and oxpeckers, *Buphagus erythrorhynchus*, which follow buffaloes, rhinos, giraffes, and large antelopes (Welty, 1962).

The Egyptian plover, *Pluvianus egyptius*, gains its food by picking insects off the backs, legs and even gums of crocodiles, benefiting their hosts by removing leeches and other ectoparasites.

The greater honeyguide, *Indicator indicator*, is an African bird which is a symbiont of the honey badger or ratel, *Mellivora capensis*, and man (and perhaps baboons). These birds, after discovering a beehive, seek out a ratel or man in the proper frame of mind, and with considerable racket, lead their symbiont to the hive. The ratel or man thereupon opens the hive for the honey and leaves the bird to feed upon the beeswax, the mainstay of its diet (Skead, 1951). Honeyguides are also symbionts to a bacterium, *Micrococcus cerolyticus*, and a yeast, *Candida albicans*, which aid them in digesting the wax. Even the bacterium and the yeast are mutualistic, being more effective in breaking down the wax—to their own benefits—when together than when apart (Friedmann, 1967).

The ants *Crematogaster parabiotica* and *Camponotus femoratus* jointly inhabit a large ball of earth and epiphytes around a tree branch. *Crematogaster* is small and lives near the surface of the ball, *Camponotus* is larger and lives deeper in the ball. The galleries of both species open into one another and the two species may forage in common files. Their broods, however, are kept separate. A slight disturbance to this integrated community brings *Crematogaster* out to deal with it and thus saves *Camponotus* the bother. A larger disturbance, with which *Crematogaster* can't cope, rouses *Camponotus*. Both species thus defend their ball to the other's benefit (Wheeler, 1928).

Ants of many species commonly keep aphids or other plant lice. The ants eat a metabolic by-product of these animals known as *honeydew* which is apparently harmful to their producers if not removed: in the absence of ants, aphids take great pains to get rid of it (Das, 1959: Bombosch, 1962, cited in Sudd, 1967). The ants also protect their symbionts from predation. To check this latter assertion, Wellenstein (1952) greased bands on trees to keep ants away. The isolated plant lice declined in numbers and the number of predators increased. Some ants may even construct shelters for their aphids. The ant *Crematogaster striatulus* covers its symbiont scale, *Pseudococcus njalensis*, with little shelters complete with doors, through which *Crematogaster* workers but no other larger ants or predatory lady-bug larvae can enter (Strickland, 1951, cited in Sudd, 1967). Another ant, *Acropyga paramaribensis*, keeps scales, and disturbed workers pick them up for safety along with their own brood. Females in mating flight carry along immature scales, presumably to use in stocking a new colony (Bunzli, 1935, cited in Sudd, 1967).

The fish *Nomeus sp.* lives among the tentacles of the Portuguese man-o'-war, *Physalia sp.*; other fish, *Amphiprion percula*, *A. bicinctus*, *A. clarkii*, *Tetradrachmum trimoculatus*, live among the tentacles of the east Indian anemones *Discosoma giganteum* and *Actinia hemprichi*. These fish somehow escape harm from the nematocysts which normally paralyze or kill, and from these nematocysts gain a measure of protection (Moser, 1931; Gohar, 1948; Davenport, 1955). The fish, in turn, are believed to bring food items to their hosts. For recent information and discussion the reader is referred to Mariscal, 1970.

One of the most interesting cases of mutualism is the *cleaning symbiosis* of tropical marine habitats. Here, small fish such as *Elacatinus oceanops*, *Thalassoma bifasciatum*, *Bodianus rufus*, *Gramma hemichrysos*, or shrimp such as *Periclimensis pedersoni* or *Hippolysmata californica*, clean ectoparasites from the bodies, mouth, and gill cavities of larger fish. Even such predatory fish as the moray eel, *Gymnothorax mordax*, may be cleaned. Cleaner-fish or shrimps often remain in particular *cleaning stations*, learned by local customers. They are identified by their bright, characteristically striped coloration and their behavior. Cleaning-shrimp, for example, whip their antennae and sway back and forth. Large fish, such as groupers, also have evolved certain postures, which tell the cleaner when to enter or when to leave. In the former case they may swim very close and present the gill region (Eibl-Eibesfeldt, 1955; Feder, 1966). The advantage to both cleaner and cleaned is obvious. The advantage to the cleaned was dramatically demonstrated by Limbaugh (in Feder, 1966), who removed all known cleaners from two small, isolated reefs in the Bahamas. The number of fish declined "drastically" in only a few days and most, except for territorial fish, were gone within two weeks. Those remaining were covered with blotches, sores and swellings.

As one might expect in any stable relationship such as the cleaning symbiosis, evolution is bound to produce an exploiter of the system. In this case it is the blenny, *Aspidontus taeniatus*, which mimics the appearance, swimming pattern, and behavior of the cleaner-fish *Labroides dimidiatus*, the most common cleaner in the Indo-Pacific. The large fish to be cleaned swims past this blenny, presents its gill cavity, and the blenny enters, proceeding to feed not on ectoparasites, but the fish itself (Eibl-Eibesfeldt, 1959).

## 12.5 THE EVOLUTION OF COMPETITIVE RELATIONSHIPS

We return now to competitive relationships and examine various genetic and evolutionary considerations which, we shall see, may be of great importance.

Each species is composed of individuals of different genotypes; different genotypes possess different fitnesses and, presumably, different abilities to withstand active competition and interference from competing species. Therefore, it seems reasonable to assume that the "competitive viability" (the performance under competition) of a population is, in part, a function of sampling—that is, influenced by the founder principle (1.2.2). If this is true, laboratory competition experiments should be more variable in outcome when begun with small populations. Dawson and Lerner (1966) raised *Tribolium castaneum* and *T. confusum* together on whole-wheat flour and yeast medium at 29°C, 70% relative humidity, and found considerably greater variation in outcome between replicates when the experiments were started with two pairs of each than when they were started with ten pairs of each. Lerner and Dempster (1962), using the same species and conditions, found the results to be more consistent when starting the experiments with inbred —and thus more genetically homogeneous—strains.

A species in competition suffers a drop in fitness due to the presence of its competitor. Thus selection acts within competing species to lessen the effect of the competitor on them. This selection should be strongest on that species suffering the greatest depression of fitness—that is, the losing competitor (see Discussion, Section III). It seems possible then, that the stronger selection acting on the losing species might, before extinction could occur, turn the tables and change the direction of selection. This hypothesis was suggested by Park and Lloyd (1955) who raised *Tribolium castaneum* and *T. confusum* together for several generations. Without replacement of *T. confusum*, *T. castaneum* always won the competition under the laboratory conditions. It was reasoned, however, that a *T. confusum* strain raised for several generations with *T. castaneum* might do better when pitted against stock *T. castaneum*. This expectation was not borne out and Park and Lloyd rejected the hypothesis. The hypothesis has been revived, however, by Pimentel *et al.* (1965) and Ayala (1966). Pimentel calls the hypothetical change in competitive dominance a "genetic feedback" theory and claims it can lead to the coexistence of two competing species.  To test this hypothesis, Pimentel *et al.* (1965) constructed 16 small cages joined by tubes, into which they placed house flies and blowflies, along with sugar for food.  Eggs were laid in a medium of dried milk, yeast, agar, water, and liver.  The populations were allowed to vary with (almost) no meddling and by the 55th week it appeared that the blowflies had been all but eliminated.  At this point, however, the blowfly population suddenly began recovering, and eventually excluded the house flies.  Experimental flies were taken from the above populations at the 38th week, at which time the blowflies were still in decline and placed in competition against other flies in single cages.  The results are shown in Table 12.2.  Although the statistics are not overwhelming, it seems quite clear that the experimental blowflies were better competitors (under the laboratory conditions) than wild blowflies.  It thus appears that they had experienced strong selection for competitive viability.  The house flies, which were not losing in competition, improved in competitive viability less than the blowflies, if at all.

**Table 12.2.** Competition in flies (see text for explanation)

|  | Number of runs won by: | |
|  | house flies | blowflies |
| --- | --- | --- |
| Wild house flies and wild blowflies | 2 | 2 |
| Experimental house flies and wild blowflies | 3 | 2 |
| Wild house flies and experimental blowflies | 0 | 5 |
| Experimental house flies and experimental blowflies | 0 | 5 |

Roughly similar experiments have been carried out by Seaton and Antonovics (1967).  These workers reared larval wild-type and genotype "dumpy" *Drosophila melanogaster* in milk bottle cultures, then removed the virgin females and mated them with their own genotype.  Thus there was no interbreeding between the two competing "species."  Each generation was started with six wild-type and six dumpy fertilized females from the preceding generation which were allowed four days to lay eggs and then removed.  Limited food was provided for the larvae.  After four generations of larval coevolution, twenty inseminated females, different relative numbers of the two strains, were placed together, allowed to lay eggs, and the number of larvae emerging noted.  This same procedure was carried out with stock flies also.  The authors found that with stock flies, serious competitive inhibition occurred.  The total yield for a mixture of 10 : 10 females was considerably lower than that for all wild-type or all dumpy.  But after four generations of selection under competitive conditions, the competitive inhibition was less, and the combined yield for the 10 : 10 mixture was often greater than for either species alone.

On the basis of the above experiments, some simple concepts of competition may need to be changed. The evolutionary process may be quite rapid, and may be capable of altering the outcome of a competitive situation which, at its outset, seems clearly predictable. It may provide for coexistence of two competing species which, at any given time, display all the necessary requisites for a straight-forward case of competitive exclusion.

# Discussion III

## III.1 SELECTION AND ITEROPARITY

The derivation (Chapter 1) of the equation

$$\Delta p_k = \frac{p_k(1 - p_k)}{2\overline{W}} \frac{\partial \overline{W}}{\partial p_k},$$

in addition to random mating (nonrandom mating can be corrected for), no epistasis (epistasis can also be accounted for), and very small differences between the fecundities and mating propensities of different genotypes, assumes that the species under consideration are *semelparous*—individuals breed only once in their lifetimes. Most species, at least vertebrates, are *iteroparous*, breeding several times throughout their lives. Iteroparous species may experience changes in age structure. Different age classes display different fecundities and suffer from different mortality rates; therefore a changing age structure results in changing, perhaps drastically changing, relative fitness values over time. Strictly speaking, then, all the selection arguments used so far in this book, including that used in Chapter 9 with reference to the evolution of birth and death rates, have been, at best, nonrigorous. Simple expressions for selection in iteroparous populations not at stable age distribution have never been worked out, although computer analyses of simple models employing the Leslie matrix have been published (Anderson and King, 1970; King and Anderson, 1971). Among other things, these analyses show fluctuations in $p$ with fluctuating population size (as might be expected). When the age distribution is stable, however, simple equations (expressing $\Delta p$ for both discretely and continuously breeding populations) can be derived—and recall that populations approach a stable age distribution over time unless disturbed.

Assume a very large, randomly breeding population and, as in Chapter 1, no epistasis. Where $\Delta p$ is very small so the population as a whole is in Hardy–Weinberg equilibrium, we can write

$$n_{ijx}(t + 1) = R_{ijx}n_{ijx}(t).$$

Here $n_{ijx}(t)$ is the number of individuals of genotype $ij$, age $x$, at time $t$ and $R_{ijx}$ is the ratio of $n_{ijx}(t + 1)$ to $n_{ijx}(t)$. But since all age groups change in size proportionately, $R_{ijx} = R_{ij}$, and

$$n_{ij}(t + 1) = R_{ij}n_{ij}(t).$$

From the arguments of Section I, it follows easily that

$$\Delta p_k = \frac{p_k(1-p_k)}{2R} \frac{\partial R}{\partial p_k} = \frac{p_k(1-p_k)}{2} \frac{\partial \ln R}{\partial p_k}, \tag{III.1}$$

where $R$ is the mean value of all $R_i$ and $\Delta p_k$ is the change in $p_k$ over one time unit (rather than one generation). We note that, by the arguments in Chapter 1, selection acts to increase $\ln R$. But in a semelparous species, $W_{ij} = (R^T)_{ij}$ where $T$ is generation time. Hence $\ln R = \ln(\overline{W^{1/T}})$ (see 1.5.1).

From the above calculations it should be clear that when stable age distributions obtain the selection arguments made throughout this book are equally valid for iteroparous and semelparous species. For populations undergoing great fluctuations as a result of susceptibility to environmental change, stable age distributions will not occur and the selection arguments apply much more loosely.

## 3.2 *K*-SELECTION, $r_0$-SELECTION AND $\alpha$-SELECTION

Having related the concepts of fitness and selected change directly with the population parameter, $R$, we proceed to break $R$ into its component parts to see how each responds to selection.

The selection intensity on a character of value $x$ over the whole population is estimated by

$$\left| \frac{1}{R} \frac{\partial R}{\partial x} \right| \quad \text{(see part 9.4)}. \tag{III.2}$$

Writing $R = e^r$, or $r = \ln R$, this becomes

$$\left| \frac{1}{R} \frac{\partial R}{\partial x} \right| = \left| \frac{\partial \ln R}{\partial x} \right| = \left| \frac{\partial r}{\partial x} \right|.$$

We now recall that, at least to a rough approximation, we can write

$$r = r_0(1 - n/K - \alpha m/K) - C,$$

where $C$ is the density-independent effect, and $\alpha/K$ is the degree to which each individual of another species, of population size $m$, depresses $r$. Selection pressures on $r_0$, $K$, $\alpha$, now become

$$\left| \frac{\partial r}{\partial r_0} \right| = 1 - \frac{n}{K} - \frac{\alpha m}{K},$$

$$\left| \frac{\partial r}{\partial K} \right| = \frac{r_0(n + \alpha m)}{K^2}, \tag{III.3}$$

$$\left| \frac{\partial r}{\partial \alpha} \right| = \frac{r_0 m}{K}.$$

From these simple equations it is clear that

**1.** When $n \ll K$, the value of $\partial r / \partial r_0$ is large. Thus selection acts strongly to increase $r_0$. Populations held by inclement weather or predation to levels well below their carrying capacities experience primarily $r_0$-selection and may be expected to be opportunistic. Opportunistic species may be expected to be highly efficient at producing zygotes.

**2.** From the above it is clear that species from which $n \ll K$ possess large $r_0$ values. Thus it is not clear how $r_0(n + \alpha m)/K^2$ varies with $n/K$. Over long periods of time

$$r_0 \left[ 1 - \left( \frac{n + \alpha m}{K} \right) \right] - C$$

must average zero, so that

$$\frac{r_0(n + \alpha m)}{K} \approx r_0 - C.$$

Thus $\overline{|\partial r / \partial K|} \approx (r_0 - C)/K$ and there seems no immediate guide as to the relative strengths of selection on $K$ for populations approaching as opposed to populations normally far below their carrying capacities. However, selected change in $K$ is slowed by strong selection pressure on $r_0$ (4.2.2), and so $K$-selection may be expected to be strongest when $n \approx K$.

**3.** Selection for avoiding competitive inhibition acts most strongly when the competitor is numerous relative to the carrying capacity (i.e. when the species in question is relatively scarce—while in the process of being excluded), and in opportunistic species (large $r_0$).

Interestingly enough, the degree to which the species in question depresses its competitor's carrying capacity does not enter the equations. This is in keeping with the findings of Seaton and Antonovics (1967) that there is a general lessening of competitive inhibition as a result of selection. Winning by virtue of hurting one's competitor may evolve by group selection if the disadvantage to individuals of possessing the necessary mechanism is very slight.

$K$ and $r_0$-selection are concepts which have been around for some time. They have recently been given voice in these terms by Cody (1966), MacArthur and Wilson (1967), and King and Anderson (1971). These authors have argued, as above, that unstable environments may keep populations below their carrying capacities and thus favor larger $r_0$ values. Conversely, in stable areas such as the tropics, $K$-selection will be relatively stronger (see Cody's discussion of clutch size in birds, part 6.3). The relation between $r_0$, $K$, and $\alpha$-selection as they pertain to Park and Lloyd, Ayala and Pimentel's thesis of the reversal of competitive dominance (12.5) is obvious.

## 3.3 THE INTERDEPENDENCE OF $r_0$ AND $K$

In the logistic equation, $r_0$ and $K$ are defined as independent variables. Throughout the latter part of Chapter 9 and subsequent chapters in Section III, however, we have gradually developed the notion of carrying capacity as the number of individuals which an environment can support. This $K$ is somewhat more broadly defined than that in the logistic equation, and is not necessarily independent of $r_0$. For example, suppose that both $b$ and $d$ vary linearly with $n$:

$$b = b_0(1 - \delta_1 n), \qquad d = d_0(1 + \delta_2 n).$$

Then $\hat{n} = K$ when $b = d$, or $b_0(1 - \delta_1 \hat{n}) = d_0(1 + \delta_2 \hat{n})$, and

$$K = \hat{n} = \frac{b_0 - d_0}{\delta_2 d_0 + \delta_1 b_0} = \frac{r_0}{\delta_1 r_0 + (\delta_1 + \delta_2)d_0}$$

(for experimental data, see Birch, 1953a and b).

From the above expression it is clear that $\hat{n}$ may depend on $r_0$ as well as on other factors impinging on $K$ (food supply, etc., as they affect $\delta_1, \delta_2$). Thus we note that population size may be limited either by the population-limiting factors discussed in Chapter 10 (affecting $K$ via $\delta_1, \delta_2$), or by factors affecting $r_0$, and that these factors may not be the same.

## 3.4 FREQUENCY AND DENSITY-DEPENDENT SELECTION

It has been observed that scarce morphs may have a mating advantage in *Drosophila* (Section I). In Chapter 5 it was noted that mimetic morphs become more fit as their proportion in a prey population decreases. Allard and Adams (1969) report that in barley, wheat, and a composite, cross wheat, higher productivity is shown by populations of mixed varieties. Such observations indicate that the fitness of a genotype may be dependent on its relative frequency in a population.

The fitness of a genotype may be dependent also on its absolute frequency— that is, its population density. A simple model of frequency-dependent selection in which density is also considered is given below.

We examine selection for an allele exhibiting complete dominance in a panmictic population and, as before, ignore epistasis. Where $n_1$ is the number of individuals of genotype $A_1$—, $n_2$ the corresponding number for the $A_2 A_2$ genotype, and $\alpha_{ij}$ is the competitive effect of genotype $i$ on $j$, we write, after the Lotka–Volterra competition equations,

$$\frac{1}{n_1}\frac{dn_1}{dt} = r_1(\text{max})\left(1 - \frac{n_1 + \alpha_{21}n_2}{K_1}\right) \propto 1 - \frac{(1 - q^2) + \alpha_{21}q^2}{K_1}n,$$

$$\frac{1}{n_2}\frac{dn_2}{dt} = r_2(\text{max})\left(1 - \frac{n_2 + \alpha_{12}n_1}{K_2}\right) \propto 1 - \frac{q^2 + \alpha_{12}(1 - q^2)}{K_2}n.$$

At population–genetic equilibrium both expressions equal zero. Thus

$$K_1 = [(1 - q^2) + \alpha_{21}q^2]n,$$

so that

$$q^2 = \frac{K_1 - n}{(\alpha_{21} - 1)n}. \qquad \text{(III.4a)}$$

Similarly,

$$q^2 = \frac{K_2 - n\alpha_{12}}{(1 - \alpha_{12})n}. \qquad \text{(III.4b)}$$

Setting the two expressions equal to one another,

$$n = 0, \quad \frac{K_1(1 - \alpha_{12}) + K_2(1 - \alpha_{21})}{1 - \alpha_{12}\alpha_{21}}. \qquad \text{(III.5)}$$

For the non-trivial case, substituting back into (III.4),

$$\hat{q} = \sqrt{\frac{K_1 - \alpha_{21}K_2}{K_1(1 - \alpha_{12}) + K_2(1 - \alpha_{21})}}.$$

The value of $\hat{q}$ falls between zero and one if and only if

$$K_1 > \alpha_{21}K_2, \qquad K_1\alpha_{12} < K_2.$$

That is, when

$$\frac{1}{K_2} > \frac{\alpha_{21}}{K_1}, \qquad \frac{1}{K_1} > \frac{\alpha_{12}}{K_2}.$$

Not surprisingly these are the same conditions required for the coexistence of two competing species. It is possible that where $K$ and $\alpha$ values indicate fixation for one allele, sufficient differences in $r$ may reverse the outcome, depending on the initial state of the system.

Notice that at equilibrium, $r_1 = r_2 = \bar{r}$, so that genetic (segregational) load is zero.

For alternative derivations of the above concepts the reader is referred to Charlesworth (1971), King and Anderson (1971), Roughgarden (1971) and Clarke (in press).

## BIBLIOGRAPHY, SECTION III

Allard, R. W. and J. Adams, 1969, Population studies in predominantly self-pollinating species. XIII. Intergenotypic competition and population structure in barley and wheat, *Amer. Natur.* **103**: 621–645.

Anderson, W. W., and C. E. King, 1970, Age-specific selection, *Proc. Nat. Acad. Sci.* **66**: 780–788.

Andrewartha, H. G., 1957, The use of conceptual models in population ecology, *Cold Spr. Harbor Symp. Quant. Biol.* **22**: 219–236.

Andrewartha, H. G., and L. C. Birch, 1954, *The Distribution and Abundance of Animals*, University of Chicago Press, Chicago.

Andrewartha, H. G., and T. O. Browning, 1961, An analysis of the idea of "resources" in animal ecology, *J. Theoret. Biol.* **1**: 83–97.

Aumann, G. D., and J. T. Emlen, 1965, Relation of population density to sodium availability and sodium selection by microtine rodents, *Nature* **208**: 198–199.

Ayala, F. J., 1966, Reversal of dominance in competing species of *Drosophila, Amer. Natur.* **100**: 81–83.

Ayala, F. J., 1969, Experimental invalidation of the principle of competitive exclusion, *Nature* **224**: 1076.

Ayala, F. J., 1970, Invalidation of competitive exclusion defended, *Nature* **227**: 89–90.

Bailey, I. W., 1924, The anatomy of certain plants from the Belgian Congo, with special reference to myrmecophytism, *Bull. Amer. Mus. Nat. Hist.* **45**: 585–620.

Beverton, J. H., and S. J. Holt, 1957, *On the Dynamics of Exploited Fish Populations*, Ministry of Agriculture, Fishery Investigations, Great Britain.

Birch, L. C., 1953a, Experimental background to the study of the distribution and abundance of insects, *Evol.* **7**: 136–144.

Birch, L. C., 1953b, Experimental background to the study of the distribution of insects, II, The relation between innate capacity for increase in numbers and the abundance of three grain beetles in experimental populations, *Ecol.* **34**: 712–726.

Birch, L. C., 1957a, The meanings of competition, *Amer. Natur.* **91**: 5–18.

Birch, L. C., 1957b, The role of weather in determining the distribution and abundance of animals, *Cold Spr. Harbor Symp. Quant. Biol.* **22**: 203–218.

Blair, W. F., 1940, A study of prairie deer-mouse populations in southern Michigan, *Amer. Midl. Natur.* **24**: 273–305.

Bombosch, V. S., 1962, Untersuchung über die Auslösung der Eiablage bei *Syrphus Corollae* Fabr. (Dipt. Syrphidae) *Z. Angew. Entom.* **50**: 81–88.

Brian, M. V., 1952, The structure of a dense natural ant population, *J. Anim. Ecol.* **21**: 12–24.

Brown, J. L., 1969, The buffer effect and productivity in tit populations, *Amer. Natur.* **103**: 347–354.

Brown, W. L., 1960, Ants, acacias, and browsing mammals, *Ecol* **41**: 587–592.

Brown, W. L., and E. O. Wilson, 1956, Character displacement, *Syst. Zool.* **5**: 49–64.

Bunzli, G. H., 1935, Untersuchungen über coccidophile Ameisen aus den Kaffeefeldern von Surinam, *Mitt. Schweiz. Entom. Ges.* **16**: 453–593.

Bustard, H. R., 1970, The role of behavior in the natural regulation of numbers in the gekkonid lizard, *Gehyra variegata, Ecol.* **51**: 724–728.

Butler, L., 1953, The nature of cycles in populations of Canadian mammals, *Can. J. Zool.* **31**: 242–262.

Caldwell, L. D., 1964, An investigation of competition in natural populations of mice, *J. Mammal.* **45**: 12–30.

Calhoun, J. B., 1952, The social aspects of population dynamics, *J. Mammal.* **33**: 139–159.

Calhoun, J. B., and W. L. Webb, 1953, Induced emigrations among small mammals, *Science* **117**: 358–360.

Carleton, W. M., 1965, Food habits of two sympatric Colorado sciurids, *J. Mammal.* **47**: 91–103.

Carson, H. L., 1968, The population flush and its genetic consequences, in R. C. Lewontin (Ed.), *Population Biology and Evolution*, Syracuse University Press.

Caughley, G., 1966, Mortality patterns in mammals, *Ecol.* **47**: 906–918.

Caughley, G., 1970, Eruption of ungulate populations, with emphasis on Himalayan thar in New Zealand, *Ecol.* **51**: 53–72.

Chapman, D. W., 1966, Food and space as regulators of salmonid populations in streams, *Amer. Natur.* **100**: 345–357.

Charlesworth, B., 1971, Selection in density-regulated populations, *Ecol.* **52**: 469–474.

Chitty, D., 1957, Self regulation of numbers through changes in viability, *Cold Spr. Harbor Symp. Quant. Biol.* **22**: 277–280.

Chitty, D., 1960, Population processes in the vole and their relevance to general theory, *Can. J. Zool.* **38**: 99–113.

Chitty, D., and E. Phipps, 1966, Seasonal changes in survival in a mixed population of two species of vole, *J. Anim. Ecol.* **35:** 313–331.

Chitty, H., 1948, The snowshoe rabbit inquiry, 1943–46, *J. Anim. Ecol.* **17:** 39–44.

Christian, J. J., 1956, Endocrine responses to population size in mice, *Ecol.* **37:** 258–273.

Christian, J. J., 1961, Phenomena associated with population density, *Proc. Nat. Acad. Sci.* **47:** 428–499.

Christian, J. J., and D. E. Davis, 1956, The relationship between adrenal weight and population status of urban Norway rats, *J. Mammal.* **37:** 475–486.

Christian, J. J., and D. E. Davis, 1964, Endocrines, behavior, and population, *Science* **146:** 1550–1560.

Clark, L. R., P. W. Geier, R. D. Hughes, and R. F. Morris, 1967, *The Ecology of Insect Populations in Theory and Practice*, Methuen, London.

Clarke, B., Density-dependent selection, *Amer. Natur.*, in press.

Clough, G. C., 1965, Variability of wild meadow voles under various conditions of population, density, season, and reproductive activity, *Ecol.* **46:** 119–134.

Cody, M. S., 1966, A general theory of clutch size, *Evol.* **20:** 174–184.

Cody, M. S., 1968, On the methods of resource division in grassland bird communities, *Amer. Natur.* **102:** 107–147.

Cole, L. C., 1954, Some features of random population cycles, *J. Wildl. Mgt.* **18:** 2–24.

Cole, L. C., 1960, Competitive exclusion, *Science* **132:** 348–349.

Connell, J. H., 1961, The influence of interspecific competition and other factors on the distribution of the barnacle, *Chthamalus stellatus, Ecol.* **42:** 710–723.

Cooper, W. S., 1913, The Climax Forest of Isle Royale, Lake Superior, and its development. *Bot. Gaz.* **55:** 1–44, 115–140, 189–235.

Crisler, L., 1956, Observations of wolves hunting caribou, *J. Mammal.* **37:** 337–346.

Darling, F., 1937, *A Herd of Red Deer,* Natural History Library (64), New York.

Das, G. M., 1959, Observations on the association of ants with coccids of tea, *Bull. Entom. Res.* **50:** 437–448.

Davenport, D., 1955, Specificity and behavior in symbiosis, *Quart. Rev. Biol.* **30:** 29–46.

Davidson, J., 1938a, On the ecology of the growth of the sheep population in South Australia, *Tr. Roy. Soc. S. Austr.* **62:** 141–148.

Davidson, J., 1938b, On the growth of the sheep population in Tasmania, *Tr. Roy. Soc. S. Austr.* **62:** 342–346.

Davidson, J., and H. G. Andrewartha, 1948, The influence of rainfall, evaporation, and atmospheric temperature on fluctuations in the size of a natural population of *Thrips imaginis* (Thysanura), *J. Anim. Ecol.* **17:** 200–222.

Davis, D. E., 1957, The existence of cycles, *Ecol.* **38:** 163–164.

Dawson, P. S., 1966, Development rate and competition ability in *Tribolium, Evol.* **20:** 104–116.

Dawson, P. S., and I. M. Lerner, 1966, The founder principle and competitive viability of *Tribolium, Proc. Nat. Acad. Sci.* **55:** 1114–1117.

DeBach, P., 1949, Population studies of the long-tailed mealybug and its natural enemies on citrus trees in southern California, 1946, *Ecol.* **30:** 14–25.

DeBach, P., and H. S. Smith, 1941, Are population oscillations inherent in the host-parasite relation? *Ecol.* **22:** 363–369.

Deevey, E. S., Jr., 1947, Life tables for natural populations of animals, *Quart. Rev. Biol.* **22:** 283–314.

DeLong, K. T., 1966, Population ecology of feral house mice: interference by *Microtus, Ecol.* **47:** 481–484.

DeLong, K. T., 1967, Population ecology of feral house mice, *Ecol.* **48:** 611–634.

Dodd, A. P., 1940, The biological campaign against prickly pear, *Comm. Prickly Pear Bd.,* Brisbane, Queensland.

Dodd, A. P., 1959, The biological control of prickly pear in Australia, *Monog. Biol.* **8:** 565–577.

Edmondson, W. T., 1945, Ecological studies of sessile rotatoria, II, Dynamics of populations and social structure, *Ecol. Monog.* **15:** 141–172.

Ehrlich, P. R., and L. C. Birch, 1967, The "Balance of nature" and "population control", *Amer. Natur.* **101:** 97–107.

Eibl-Eibesfeldt, I., 1955, Uber Symbiosen Parasitismus und Andere Besondere Zwischenartliche Beziehungen Tropischer Meerfische, *Z. tierpsychol.* **12:** 203–219.

Eibl-Eibesfeldt, I., 1959, Der Fisch *Aspidontus taeniatus* als Nachahmer des Putzers *Labroides dimidiatus, Z. tierpsychol.* **16:** 19–25.

Einarsen, A. S., 1945a, Some factors affecting ring-neck pheasant population density, *Murrelet,* **26:** 2–9.

Einarsen, A. S., 1945b, Some factors affecting ring-neck pheasant population density, *Murrelet* **26:** 39–44.

Eisenberg, R. M., 1966, The regulation of density in a natural population of the pond snail *Lymnaea elodes, Ecol.* **47:** 889–905.

Elton, C. S., 1942, *Voles, Mice and Lemmings: Problems in Population Dynamics,* Oxford University Press, Oxford.

Elton, C. S., 1946, Competition and the structure of ecological communities, *J. Anim. Ecol.* **15:** 54–68.

Elton, C. S., and M. Nicholson, 1942, The ten-year cycle in numbers of the lynx in Canada, *J. Anim. Ecol.* **11:** 215–244.

Emlen, J. M., 1970, Age specificity and ecological theory, *Ecol.* **51:** 588–601.

Errington, P. L., 1946, Predation and vertebrate populations, *Quart. Rev. Biol.* **21:** 144–177, 221–245.

Feder, H. M., 1966, Cleaning symbiosis in the marine environment, in S. M. Henry (Ed.), *Symbiosis,* Vol. I, Academic Press, New York.

Fenner, F., 1965, Myxoma virus and *Oryctolagus cuniculus*: two colonizing species, in H. G. Baker and G. I. Stebbins (Eds.), *The Genetics of Colonizing Species,* Academic Press, New York.

Fenner, F., I. D. Marshall, and C. M. Woodroffe, 1953, Studies on epidemiology of infectious myxomatosis in rabbits, I, Recovery of Australian wild rabbits (*Oryctolagus cuniculus*) myxomatosis under field conditions, *J. Hyg.* **51:** 225–244.

Fitch, H. S., F. Swenson, and D. F. Tillotson, 1946, Behavior and food habits of the red-tailed hawk, *Condor* **48:** 205–237.

Frank, P. W., 1960, Prediction of population growth form in *Daphnia pulex* cultures, *Amer. Natur.* **94:** 357–372.

Frank, P. W., C. D. Boll, and R. W. Kelly, 1957, Vital statistics of laboratory cultures of *Daphnia pulex* (DeGeer), as related to density, *Physiol. Zool.* **30:** 287–305.

Friedmann, H., 1967, Avian symbiosis, in S. M. Henry (Ed.), *Symbiosis,* Vol. II, Academic Press, New York.

Gause, G. F., 1931, The influence of ecological factors on the size of populations, *Amer. Natur.* **65:** 70–76.

Gause, G. F., 1934, *The Struggle for Existence,* Williams and Wilkins, Baltimore.

Gause, G. F., 1935, *La Théorie mathématique de la lutte pour la vie,* Hermann, Paris.

Gause, G. F., 1970, Criticism of invalidation of principle of competitive exclusion, *Nature* **227:** 89.

Gause, G. F., N. P. Smaragdova, and A. A. Witt, 1936, Further studies of interaction between predator and prey, *J. Anim. Ecol.* **5:** 1–18.

Gause, G. F., and A. A. Witt, 1935, Behavior of mixed populations and the problem of natural selection, *Amer. Natur.* **69:** 596–609.

Glue, D. E., 1967, Prey taken by the barn owl in England and Wales, *Bird Study* **14:** 169–183.

Gohar, H. A. F., 1948, Commensalism between fish and anemone (with a description of the eggs of *Amphiprion bicinctus* Rüppell), *Publ. Mar. Biol. Stat. Ghardaqa, Egypt,* No. 6 pp. 35–44.

Grant, K. A., 1966, A hypothesis concerning the prevalence of red coloration in California hummingbird flowers, *Amer. Natur.* **100:** 85–97.

Gunn, D. L., and P. M. Symmons, 1959, Forecasting locust outbreaks, *Nature* **184:** 1425.

Hairston, N. G., J. Dallan, R. K. Colwell, D. J. Futuyma, J. Howell, M. D. Lubin, J. Mathias, and J. H. Vandermeer, 1968, The relationship between diversity and stability: an experimental approach with protozoa and bacteria, *Ecol.* **49:** 1091–1101.

Hairston, N. G., F. Smith, and L. B. Slobodkin, 1960, Community structure, population control, and competition, *Amer. Natur.* **94:** 421–425.

Hamilton, W. D., 1966, The moulding of senescence by natural selection, *J. Theoret. Biol.* **12:** 12–45.

Hardin, G., 1960, The competitive exclusion principle, *Science* **131:** 1292–1297.

Hartley, P. H. T., 1953, An ecological study of the feeding habits of the English titmice, *J. Anim. Ecol.* **22:** 261–288.

Hazlett, B. A., 1968, Effects of crowding on the agonistic behavior of the hermit crab *Pagurus bernhardus, Ecol.* **49:** 573–575.

Heatwole, H., 1965, Some aspects of the association of cattle egrets with cattle, *Anim. Beh.* **13:** 79–83.

Holling, C. S., 1965, The functional response of predators to prey density and its role in mimicry and population regulation, *Mem. Entom. Soc. Canada* number **45.**

Horn, H. S., 1968, Regulation of animal numbers: a model counter example, *Ecol.* **49:** 776–778.

Huffaker, C. B., 1958, Experimental studies on predation: dispersion factors and predator-prey oscillations, *Hilgardia* **27:** 343–383.

Hutchinson, G. E., 1951, Copepodology for the ornithologist, *Ecol.* **32:** 571–577.

Hutchinson, G. E., 1957, Concluding Remarks, *Cold Spr. Harbor Symp. Quant. Biol.* **22:** 415–427.

Istock, C. A., 1966, Distribution, coexistence, and competition of whirligig beetles, *Evol.* **20:** 211–234.

Janzen, D. H., 1966, Coevolution of mutualism between ants and acacias in Central America, *Evol.* **20:** 249–275.

Janzen, D. H., 1967, Fire, vegetation structure, and the ant–acacia interaction in Central America, *Ecol.* **48:** 26–35.

Janzen, D. H., 1968, Allelopathy by myrmecophytes: the ant *Azteca* as an allelopathic agent of *Cecropia, Ecol.* **50:** 147–153.

Kaston, B. J., 1965, Some little-known aspects of spider behavior, *Amer. Midl. Natur.* **73:** 336–356.

Keith, L. B., 1963, *Wildlife's Ten-Year Cycle,* University of Wisconsin Press, Madison.

Kemp, G. A. and L. B. Keith, 1970, Dynamics and regulation of red squirrel (*Tamiasciurus hudsonicus*) populations, *Ecol.* **51:** 763–779.

Keyfitz, N., 1968, *Introduction to the Mathematics of Population,* Addison-Wesley, Reading, Mass.

King, C. E. and W. W. Anderson, 1971, Age-specifiic selection II, the interaction between *r* and *K* during population growth, *Amer. Natur.* **105:** 137–156.

Kohn, A. J., 1966, Food specialization in *Conus* in Hawaii and California, *Ecol.* **47:** 1041–1043.

Krebs, C. J., 1966, Demographic changes in fluctuating populations of *Microtus californicus, Ecol. Monog.* **36:** 239–273.

Krebs, C. J., 1970, *Microtus* population biology: behavioral changes associated with the population cycle in *Microtus ochrogaster* and *Microtus pennsylvanicus, Ecol.* **51:** 34–52.

Lack, D. 1945, The ecology of closely related species with special reference to the cormorant (*Phalacrocorax carbo*) and shag (*Phalacrocorax aristotelis*), *J. Anim. Ecol.* **14**: 12–16.

Lack, D., 1947, *Darwin's Finches,* Cambridge University Press, Cambridge.

Lack, D., 1954, *The Natural Regulation of Animal Numbers,* Oxford University Press, London.

Lack, D., 1966, *Population Studies of Birds,* Clarendon Press, Oxford.

Lanyon, W. E., 1956, Territoriality in the meadowlark, genus *Sturnella, Ibis* **98**: 485–489.

Leopold, A., 1943, *Game Management,* Scribner, New York.

Lerner, I. M., and E. R. Dempster, 1962, Indeterminism in interspecific competition, *Proc. Nat. Acad. Sci.* **48**: 821–826.

Leslie, P. H., 1945, On the use of matrices in certain population mathematics, *Biometrika* **33**: 183–212.

Leslie, P. H., 1959, The properties of a certain lag type of population growth and the influence of an external random factor on a number of such species, *Physiol. Zool.* **32**: 151–159.

Leslie, P. H. and R. M. Ranson, 1940, The mortality, fertility rate of natural increase of the vole (*Microtus agrestis*) as observed in the laboratory, *J. Anim. Ecol.* **9**: 27–52.

Levins, R., 1968, *Evolution in Changing Environments,* Princeton University Press, Princeton, New Jersey.

Lidicker, W. Z., Jr., 1962, Emigration as a possible mechanism permitting the regulation of population density below carrying capacity, *Amer. Natur.* **96**: 29 33.

Lloyd, J. A., and J. J. Christian, 1967, Relationship of activity and aggression to density in two confined populations of house mouse (*Mus musculus*), *J. Mammal.* **48**: 262–269.

Lloyd, M., 1964, Mean crowding, *J. Anim. Ecol.* **36**: 1–30.

Lomnicki, A., 1971, Animal population regulation by the genetic feedback mechanism: a critique of the theoretical model, *Amer. Natur.* **105**: 413–421.

Ludwig, W., and C. Boost, 1939, Uber das Wachstum von Protisten-populationen und den Allelokatalytischen Effekt, *Archiv Protist* **92**: 453–484.

MacArthur, R. H., 1958, Population ecology of some warblers of northeastern coniferous forests, *Ecol.* **39**: 599 619.

MacArthur, R. H., and E. O. Wilson, 1967, *The Theory of Island Biogeography,* Princeton University Press, Princeton, New Jersey.

MacLulich, D. A., 1937, Fluctuations in the numbers of the varying hare (*Lepus americanus*), *University Toronto Studies, Biol. ser.,* No. 43.

Main, A. R., J. W. Shield, and H. Waring, 1959, Recent studies on marsupial ecology, *Monog. Biol.* **8**: 315–331.

Maly, E. J., 1969, A laboratory study of the interaction between the predatory rotifer *Asplanchna* and *Paramecium, Ecol.* **50**: 59–73.

Mariscal, R. N., 1970, A field and laboratory study of the symbiotic behavior of fishes and sea anemones from the tropical Indo-pacific, *University California Publ. Zool.,* Vol. 91.

Medawar, P. M., 1957, *The Uniqueness of the Individual,* Methuen, London.

Miller, C. A., 1963, Parasites and the spruce budworm, in R. F. Morris (Ed.), *The Dynamics of Epidemic Spruce–Budworm Populations, Mem. Ent. Soc. Can.* No. 31.

Miller, R. S., 1964, The ecology and distribution of pocket gophers in Colorado, *Ecol.* **45**: 256–272.

Milne, A., 1961, Definitions of competition among animals, in *Symp. of the Society for Exper. Biol.* No. XV, Mechanisms in Biological Competition.

Moran, P. A. P., 1953a, The statistical analysis of the Canadian lynx cycle, I, Structure and prediction, *Austr. J. Zool.* **1**: 163–173.

Moran, P. A. P., 1953b, The statistical analysis of the Canadian lynx cycle, II, Synchrony and meteorology, *Austr. J. Zool.* **1**: 291–298.

Morris, R. F., 1959, Single-factor analysis in population dynamics, *Ecol.* **40**: 580–588.

Moser, J., 1931, Beobachtungen über die Symbiose von *Amphiprion percula* (Lacepede) mit Aktinien. *Sitzungsberichte der gesellschaft Naturforschender*, Freunder von 16 Juni.

Murdoch, W. W., 1966, "Community structure, population control, and competition"— a critique, *Amer. Natur.* **100**: 219–226.

Murie, A., 1944, *The Wolves of Mount McKinley*, U.S. Dept. Inter. (Faunal series) No. 5.

Murton, R. K., 1967, The significance of endocrine stress in population control, *Ibis.* **109**: 622–623.

Needler, A. W. H., and R. R. Logie, 1947, Serious mortalities in Price Edward Island oysters caused by a contagious disease, *Trans. Roy. Soc. Canad.*, 3rd ser., Sect. V, **41**: 73–94.

Newton, I., 1967, The adaptive radiation and feeding ecology of some British finches, *Ibis* **109**: 33–98.

Nicholson, A. J., 1947, Fluctuation of animal populations, *Rept. 26th meeting ANZAAS*, Perth, W. Australia.

Nicholson, A. J., 1948, Competition for food among *Lucilia cuprina* larvae, *Proc. 8th Intern. Cong. Entom.* pp. 277–281.

Nicholson, A. J., 1954a, Compensatory reactions of populations to stresses and their evolutionary significance, *Austr. J. Zool.* **2**: 1–8.

Nicholson, A. J., 1954b, An outline of the dynamics of animal populations, *Austr. J. Zool.* **2**: 9–65.

Nicholson, A. J., 1957, Self-adjustment of populations to change, *Cold Spr. Harbor Symp. Quant. Biol.* **22**: 153–173.

Nicholson, A. J., and V. A. Bailey, 1935, The balance of animal populations, Part I., *Proc. Zool. Soc. Lond.* 551–598.

Nikolskii, G. V., 1963, *The Ecology of Fishes*, Academic Press, New York and London.

Obrebski, S., 1968, On the population ecology of two intertidal invertebrates and the palaeoecological significance of size–frequency distributions of living and dead shells of the bivalve *Transenella tantilla*, Ph.D. thesis, University Chicago.

Paine, R. T., 1969, The *Pisaster–Tegula* interaction: prey patches, predator food preference, and intertidal community structure, *Ecol.* **50**: 950–961.

Park, T., and M. Lloyd, 1955, Natural selection and the outcome of competition, *Amer. Natur.* **96**: 235–240.

Pearl, R., 1930, *The Biology of Population Growth*, Knopf, New York.

Pearl, R., 1940, *Introduction to Medical Biometry and Statistics*, Saunders, Philadelphia and London.

Pearson, O. P., 1964, Mouse predation: an example of its intensity and bioenergetics, *J. Mammal.* **45**: 177–188.

Pearson, O. P., 1966, The prey of carnivores during one cycle of mouse abundance, *J. Anim. Ecol.* **35**: 217–233.

Pimentel, D., 1961, Animal population regulation by the genetic feedback mechanism, *Amer. Natur.* **95**: 65–79.

Pimentel, D., 1966, Complexity of ecological systems and problems in their study and management, in K. E. F. Watt (Ed.), *Systems Analysis in Ecology*, Academic Press, New York.

Pimentel, D., E. H. Feinberg, P. W. Wood, and J. T. Hayes, 1965, Selection, spatial distribution, and the coexistence of competing fly species, *Amer. Natur.* **99**: 97–109.

Pimentel, D., W. P. Nagel, and J. L. Madden, 1963, Space–time structure of the environment and the survival of parasite–host systems, *Amer. Natur.* **97**: 141–167.

Pimentel, D., and F. A. Stone, 1968, Evolution and population ecology of parasite–host systems, *Canad. Ent.* **100**: 655–662.

Pimlott, D. H., 1967, Wolf predation and ungulate populations, *Amer. Zool.* **7**: 267–278.

Pitelka, F. A., 1951, Ecological overlap and interspecific strife in breeding populations of hummingbirds, *Ecol.* **32**: 641–661.

Pitelka, F. A., 1957, Some aspects of population structure in the short-term cycle of the brown lemming in northern Alaska, *Cold Spr. Harbor. Symp. Quant. Biol.* **22**: 237–251.

Pryer, H., 1884, An account of a visit to the birds-nest caves of British North Borneo, *Proc. Zool. Soc. Lond.* pp. 532–538.

Rosenzweig, M. L., 1969, Why the prey curve has a hump, *Amer. Natur.* **103**: 81–87.

Rosenzweig, M. L., and R. H. MacArthur, 1963, Graphical representation and stability conditions of predator-prey interactions, *Amer. Natur.* **97**: 209–223.

Roughgarden, J., 1971, Density-independent natural selection, *Ecol.* **52**: 453–468.

Rowan, W., 1950, Canada's premier problem of animal conservation, *New Biol.* **9**: 38–57.

Schoener, T. W., 1968, The anolis lizards of Bimini: resource partitioning in a complex Fauna, *Ecol.* **49**: 704–726.

Seaton, A. P. C., and J. Antonovics, 1967, Population interrelationships, I, Evolution in mixtures of *Drosophila* mutants, *Hered.* **22**: 19–33.

Sclyc, II., 1956, *The Stress of Life,* McGraw-Hill, New York.

Shelford, V. E., 1945, The relation of snowy owl migration to the abundance of the collared lemming, *Auk* **62**: 592–596.

Skead, C. J., 1951, Notes on honeyguides in southeastern Cape Province, South Africa, *Auk* **68**: 52–62.

Slobodkin, L. B., 1954, Population dynamics in *Daphnia obtusa* (Kurz), *Ecol. Monog.* **24**: 69–88.

Slobodkin, L. B., 1959, Energetics in *Daphnia pulex* populations, *Ecol.* **40**: 232–243.

Slobodkin, L. B., 1961, *The Growth and Regulation of Animal Numbers,* Holt, Rinehart, and Winston, New York.

Slobodkin, L. B., 1964, Experimental populations of Hydrida, *J. Anim. Ecol.* **33** (suppl): 131–148.

Slobodkin, L. B., 1968, How to be a predator, *Amer. Zool.* **8**: 43–51.

Slobodkin, L. B., F. Smith, and N. Hairston, 1967, Regulation in terrestrial ecosystems and the implied balance of nature, *Amer. Natur.* **101**: 109–124.

Smith, F. E., 1961, Density dependence in the Australian thrips, *Ecol.* **42**: 403–407.

Smith, S. M., 1967, An ecological study of winter flocks of black-capped and chestnut-backed chickadees, *Wilson Bull.* **79**: 200–207.

Solomon, M. E., 1964, Analysis of processes involved in the natural control of insects, in Cragg, J. B. (Ed.), *Advances in Ecological Research* Vol. II, Academic Press, London and New York.

Southern, H. N., and V. P. W. Lowe, 1968, The pattern of distribution of prey and predation in tawny owl territories, *J. Anim. Ecol.* **37**: 75–97.

Southwick, C. H., 1955, Regulatory mechanisms of house mouse populations: social behavior affecting litter survival, *Ecol.* **36**: 627–634.

Strecker, R. L., and J. T. Emlen, 1953, Regulatory mechanisms in house mouse populations: the effect of limited food supply on a confined population, *Ecol.* **34**: 375–385.

Strickland, A. H., 1951, The entomology of swollen shoot of Cacao, I, The insect species involved, with notes on their biology, *Bull. Entom. Res.* **41**: 725–748.

Sudd, J. H., 1967, *An Introduction to the Behavior of Ants,* St. Martins, New York.

Svärdson, G., 1949, Competition and habitat selection in birds, *Oikos* **1**: 157–174.

Tamarin, R. H. and C. J. Krebs, 1969, Microtus population biology, II, Genetic changes at the transferrin locus in fluctuating populations of two vole species, *Evol.* **23**: 187–211.

Tantawy, A. O., and M. H. Soliman, 1967, Studies on natural populations of *Drosophila,* VI, Competition between *Drosophila melanogaster* and *Drosophila simulans, Evol.* **21**: 34–40.

Terao, A., and T. Tanaka, 1928, Population growth of the water-flea, *Moina macrocopa* (Strauss), *Proc. Imper. Acad.* (Japan) **4**: 550–552.

Vanden Bosch, R., C. F. Lagace, and V. M. Stern, 1967, The interrelationship of the aphid *Acyrthosiphon pisum* and its parasite *Aphidius smithi* in a stable environment, *Ecol.* **48**: 993–1000.

Vandermeer, J. H., 1969, The competitive structure of communities: an experimental approach with protozoa, *Ecol.* **50**: 362–371.

Wangersky, P. J., and W. J. Cunningham, 1957, Time lags in population models, *Cold Spr. Harbor Symp. Quant. Biol.* **22**: 329–338.

Ward, P., 1965, Feeding ecology of the black-faced diock, *Quelea quelea*, in Nigeria, *Ibis* **107**: 173–214.

Warnock, J. E., 1965, The effects of crowding on the survival of meadow voles (*Microtus pennsylvanicus*) deprived of cover and water, *Ecol.* **46**: 649–664.

Watt, K. E. F., 1968, *Ecology and Resource Management* (*A Quantitative Approach*), McGraw-Hill, New York.

Wellenstein, V. G., 1952, Zur Ernährungsbiologie der Roten Waldameise (*Formica rufa L.*), *Z. Planzenkrankh.* **59**: 430–451.

Wellington, W. G., 1957, Individual differences as a factor in population dynamics: the development of a problem, *Can. J. Zool.* **35**: 293–323.

Wellington, W. G., 1960, Qualitative changes in natural populations during changes in abundance, *Can. J. Zool.* **38**: 289–314.

Wellington, W. G., 1964, Qualitative changes in populations in unstable environments, *Can. Ent.* **96**: 436–451.

Welty, J. C., 1962, *The Life of Birds,* Saunders, Philadelphia and London.

Wheeler, W. M., 1928, *The Social Insects,* Harcourt Brace, New York.

Willey, A. 1904, Crows and flying foxes at Barberyn, *Spolia Zeylanica* **2**: 50–51.

Wynne-Edwards, V. C., 1962, *Animal Dispersion in Relation to Social Behaviour,* Oliver and Boyd, Edinburgh.

# Section IV
# The Ecology of Communities

# 13
# Community Structure and Energetics

To the evolutionist, community ecology, particularly community energetics, is a partially explored wilderness. The subjects covered under these headings are clearly related to the evolution of organisms, but at the same time sufficiently far removed from the basic evolutionary laws that it is difficult for the evolutionist even to define his role in studying them. Most of the knowledge of communities comes from classical ecologists, chemists, and systems-analysts, people who describe, correlate, and compile information. This knowledge offers a fair amount of prediction potential, but nevertheless lacks the ultimate (as opposed to proximate) sorts of understanding that makes evolutionists comfortable. The following two chapters paint a brief picture of empirical ecology, with an occasional sally into evolutionary theory. The last two chapters are largely theoretical.

## 13.1 TROPHIC STRUCTURE

A biological *community* is a collection of organisms in their environment. There have been numerous valiant attempts at more precise definitions, but since each worker tends to think of "community" in his own way—at one instant in time, over long time periods, organisms must be interacting, organisms need not be interacting, etc.—it is pointless to follow these attempts here. The term *ecosystem* has also been variously defined and, by some, considered to have a different meaning from community. We shall consider an *ecosystem* to consist of a community along with its physical-environmental setting. Other words, such as biocoenoses, biotopes, are in the author's opinion best left unused.

It is clear that all life, to grow, reproduce and do work, must have an energy supply. This energy comes ultimately, in almost all cases, from the sun. Solar radiant energy is converted into chemical bond energy in green plants, the so-called *primary producers*, and is passed on through a food chain, or *trophic chain* to herbivores, the primary carnivores, perhaps secondary carnivores, and at all points along the line, to decomposers. Of course, trophic chains are seldom—or never—so simple as one primary producer → one herbivore, etc. A partial diagram of energy flow, expanded from that of Paine (1966) for the rocky intertidal community of the Pacific northwest coast is given in Fig. 13.1. The phytoplankton and benthic algae are the primary producers, and starfish, birds, and predatory fish are top predators, but in between the trophic *levels* break down: are barnacles and mussels herbivores or primary carnivores or both?

In reality we are far more often concerned with *trophic webs*, although as an approximation for simple systems, the trophic chain is not an altogether invalid concept. It is often useful in showing the relative amounts of energy passed between different (perhaps compound) trophic levels.

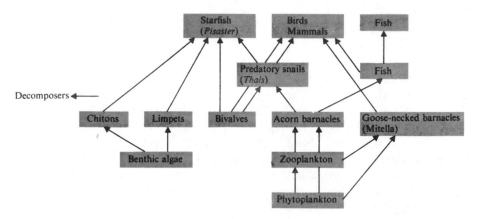

Figure 13.1. Trophic web for the Pacific northwest coast rocky intertidal.

A certain amount of the energy assimilated (ingested *and retained*) by organisms in a given trophic level goes to heat production. Much of the energy is used for digestion and work and almost always less than 40% characteristically goes into growth and reproduction (see Fig. 13.3). The amount of energy—often measured in terms of organic material, or *biomass*—produced at a given trophic level is called the *net productivity* of the trophic level. Examples of energy assimilated and net productivity in a simple aquatic community in Silver Springs, Florida, are in Table 13.1, taken from the data of Odum (1957).

Table 13.1

|  | Energy assimilated | Net productivity |
|---|---|---|
| primary producers | 20,810 | 8833 |
| herbivores | 2390 | 1478 |
| primary carnivores | 383 | 67 |
| top carnivores | 21 | 6 |
| decomposers | 5060 | 460 |
| (Energy in kg cal/m$^2$/yr $\times 10^{-3}$) | | |

Energy is lost to a system by way of emigration of organisms, and the blowing or washing away of plants, animals, and organic litter. It is gained from adjacent communities in the same manner. Not all energy passes up the food chain. Much goes directly to decomposers. Saprophagous organisms and copraphagous animals may eat the remains or fecal material of their predators. In deciduous forests up to one-third or even one-half of the annually *fixed* (photosynthesized) carbon goes to the forest floor as litter, usually in the form of leaves. Suppose we let $x$ be the amount of such litter, $L$ the annual production, $c$ the proportion going directly into litter, and $k$ the rate of decomposition. Then

$$\frac{dx}{dt} = cL - kx,$$

so that, at equilibrium,

$$k = cL/x, \quad \text{or} \quad x = cL/k.$$

For one cohort of leaves, we can also write

$$\frac{dx}{dt} = -kx, \quad \text{so that} \quad \frac{x}{x_0} = e^{-kt}.$$

Measuring $x$ and knowing $x_0$, or knowing $c$ and $L$ will enable us to calculate $k$. In this manner, $k$ was estimated by Olson (1963) to vary from about 4.0 in the South American and African tropics, to roughly 0.25 in southeastern United States pine forests, to 0.0625 in Minnesota pine forests. Although $L$ is lower in northern than in tropical forests, the change in $L$ with latitude is less than the change in $k$. Furthermore, $c$ tends to be lower in the tropics than in temperate zones. The measures thus offer an empirical explanation for the deep litter layers found in northern forests and the lack of litter in tropical forests. The ultimate cause of lower decomposer activity in colder climates in not known. If there is so much food about, what limits decomposer populations?

Net primary productivity (=gross productivity − respiration) is highest in the tropics. It ranges from less than 200 grams of organic material per square meter per year in dry deserts to between 200 and 300 g/m$^2$/yr in desert grasslands, up to about 1000 g/m$^2$/yr in moist shrub and grasslands. Temperate forests show 1200–1500 g/m$^2$/yr productivity, while for tropical forests (and marshlands) the figure is 1500–3000 (Woodwell and Whittaker, 1968). Between 10 and 25 g/m$^2$/day is characteristic of estuaries, coral reefs, alluvial plains, and sugar cane fields (figures from MacArthur and Connell, 1966). Specific figures include 1124 g/m$^2$/yr = 3.08 g/m$^2$/day in scrub oak forest on Long Island, New York (Woodwell and Whittaker, 1968), the figure given earlier of 8833 kcal/m$^2$/yr = 24.2 cal/m$^2$/day in a spring in Florida (Odum, 1957), and 8205/365 = 22.5 cal/m$^2$/day in a Georgia salt marsh (Teal, 1962).

The comparison of grams and calories is justifiable on the following basis. Organic material generally varies little in its protein, fat, and carbohydrate proportions between animals or between plants. Therefore, to an approximation, calories and grams are interconvertible with the use of a simple multiplier. Note, however, that a gram of dry organic material may possess a caloric value anywhere between four and six. To compare with the figures cited from Odum and Teal above, one would multiply that of Woodwell and Whittaker by something in the neighborhood of four to six.

A fairly detailed description of energy flow in a Georgia salt marsh is given in Fig. 13.2 modified from a diagram in Teal (1962). (See also Table 13.2.)

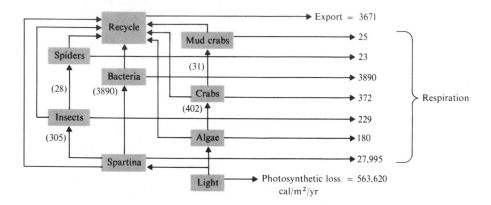

**Fig. 13.2.** Energy flow in a Georgia salt marsh (modified from Teal, 1962).

**Table 13.2.** Energy balance sheet for a Georgia salt marsh (from Teal, 1962)

| | | |
|---|---:|---|
| Import | 600,000 | |
| Loss in photosynthesis | 563,620 | |
| Gross productivity | 36,380 | |
| Respiration by producers | 28,175 | |
| Net productivity | 8,205 | |
| Bacterial respiration | 3,890 | |
| Primary consumer respiration | 596 | (insects, crabs) |
| Secondary consumer respiration | 48 | (spiders, mud crabs) |
| Total energy dissipated by consumers | 4,534 | |
| Net export | 3,671 | |

## 13.2 EFFICIENCY

The field of community energetics makes use of several measures of efficiency (for a review, see Lindeman, 1942). We mentioned assimilation efficiencies in Chapter 7. Now let us look at how the assimilated energy is used. Part, clearly, goes for respiration—entropy changes, kinetics, and homeostasis (maintenance)—and is lost as heat. The remainder goes into metabolic products (hair, feathers, saliva, etc.), growth, and reproduction—net productivity. Let's look at a very simple quantitative model: fitness, denoted by $R$, is the number, per parent, of offspring surviving to reproduce. To a rough approximation, $R$ is proportional to the amount of energy put into reproduction. The proportion of the assimilated energy left for homeostasis can be written

$$(1 - p - a)$$

where $p$ is the proportion going into reproduction and growth (ultimately reflecting itself in reproduction) and $a$ is the proportion used in acquiring food. The higher this value the lower the instantaneous mortality rate which may thus, to an approximation, be written

$$1 - k(1 - p - a).$$

The value, $k$, here, is a constant denoting the value of homeostasis in survival in relation to the cost of homeostasis. But if the above expression describes mortality, then at reproductive age ($=1$ time unit), the proportion of offspring surviving is

$$e^{k(1 - p - a) - 1}$$

and $R$ becomes

$$\propto p e^{k(1 - p - a) - 1}.$$

We assume that natural selection favors those values of $p, k$ maximizing $R$, and accordingly write

$$0 = \frac{\partial R}{\partial p} = e^{k(1 - p - a) - 1}[1 - pk], \qquad \text{so} \qquad p = 1/k,$$

where $k$ is $k$(optimal). The optimal value of $k$ is determined by biochemical factors and, except for two considerations, might be expected to be similar for all animals. Homeotherms have higher homeostatic value—else homeothermy would not have evolved. Thus $k$ should be higher and therefore $p$ lower in homeotherms than in poikilotherms. As the importance of food as a limiting factor increases, the time spent in foraging activities (and thus $a$) increases. But the cost of homeostasis does not necessarily decrease. Thus, on average, there is less energy remaining for growth and reproduction and $p$ should decline. Table 13.3 gives $p$ values, normally referred to as *growth efficiencies*, gathered by Golley (1968).

   In Chapter 7, assimilation efficiencies for a number of organisms were given. Suppose we consider invertebrates, which assimilate roughly 40% of what they eat and turn roughly 25% of that into growth—that is, into food for their predators. Such invertebrates are turning their food into food for their predators (assuming the vast bulk of them are eaten and that they eat the vast bulk of food available to them) with an efficiency $(0.40 \times 0.25) = 0.10$. This figure is known as their *ecological efficiency*. The ecological efficiency of vertebrates, in spite of their high assimilation efficiencies, is quite low because only a small proportion of the energy they assimilate goes into production (growth and reproduction). (See Table 13.3; Turner, 1970.) The variability between species in both assimilation and growth efficiencies means that ecological efficiencies also vary widely. However, at least in aquatic communities, it has been found that when all species on a trophic level are considered together, the ecological efficiency of the levels varies surprisingly little, either within or between communities. Looking at Table 13.1, we may calculate ecological efficiencies to be

$$1478/8833 = 17\% \text{ for herbivores}$$

$$67/1478 = 5\% \text{ for primary carnivores}$$

$$6/67 = 9\% \text{ for top carnivores}$$

Slobodkin (1968) noted that ecological efficiencies (over trophic levels) varied from between 4% and 5% to between 20% and 25% and averaged 10% to 15%, with no apparent relation to trophic level. His experimental *Daphnia* varied, in the laboratory, between 8% and 12% at stable population levels.

Let us take 10% as an approximation of ecological efficiency at all trophic levels and look at a consequence of this figure. Only $(1/10)^n$ of the primary

**Table 13.3.** Growth efficiencies for homeothermic and poikilothermic animals (after Golley, 1968)

|  | Species | Growth efficiency |
|---|---|---|
| Homeothermic species for which food may play an important role in population limitation | marsh wren | 0.5 |
|  | savanna sparrow | 1.1 |
|  | other sparrows | 2.1 |
|  | deer mice | 1.6 |
|  | old field mouse | 1.8 |
|  | ground squirrels | 2.9 |
|  | meadow vole | 3.0 |
|  | weasel | 2.4 |
|  | elephant | 1.5 |
| Poikilothermic species for which role of food in population limitation is questionable | harvester ant | 0.3 |
| Poikilothermic species for which food is of little importance (relative to weather, predation, or environmental instability) in population limitation | spittlebug | 9.1 |
|  | grasshoppers | 37.2 |
|  | grasshoppers | 36.6 |
|  | Orthoptera | 15.6 |
|  | plant hoppers | 25.5 |
|  | aribatid mites | 21.4 |
|  | isopod | 17.9 |
|  | mussel | 29.8 |
|  | nematodes | 24.7 |
|  | snail | 14.0 |

production is available as food $n$ levels removed from the primary producers. But a predator at this level needs to be of some minimum size to acquire its food, and thus has minimum food needs. It also needs to maintain a population dense enough to allow mating opportunities adequate to its population's survival. The number of trophic levels is thus limited. It should be highest when productivity is consistently high such as in the moist tropics, and lowest when productivity is low and erratic such as on deserts or in the Arctic.

The question is often asked: If there is more food available lower in the food chain, why do species become top carnivores? The answer is that fitness depends not on the amount of food available, but on the food available per predator. Top carnivores have less food and correspondingly fewer competitors for that food. It is a basic tenet of population limitation that if more food exists at one place—or at one trophic level—the predator population will expand to use it. A hypothetical extreme generalist would find it a toss-up at which trophic level he chose to live.

If individuals of some genetic line within a population could move down one trophic level without finding increased numbers of competitors, it would certainly pay them, in terms of a temporary increase in $R$, to do so. If the population to which this genetic line belonged were food-limited, its carrying capacity would increase about ten fold. Such a shift was made by man during the neolithic revolution when he learned to cultivate and use plant food, and led the way to the first great period of human population growth.

A consequence of the low value characterizing growth efficiency concerns the fate of assimilable but not metabolizable materials. Only a fraction of assimilated material is retained, the rest is metabolized. Thus while about 75% to 95% (see Table 13.3) of all food assimilated by an individual ultimately leaves the body as water, $CO_2$, and other waste products, nonmetabolized material is often stored. Its concentration rises perhaps $100/(100 - 75) = 4$ to $100/(100 - 95) = 20$ times that in the food level. This concentration of nonmetabolites continues up the food chain so that top carnivores may possess concentrations several orders of magnitude above those of herbivores. Such a fact may prove useful to animals at times, but in our day of rampant pollution it is proving disastrous. DDT and other chlorinated hydrocarbons, for example, are stored in fat and only very slowly metabolized. They are so-called persistent pesticides which occur throughout the world in almost undetectably low amounts in water, but have been concentrated many times in food chains so that concentrations in top carnivores are often at acute or even lethal levels. Whether use of these pesticides is the primary cause of the drastically depleted populations of such species as the peregrine falcon, *Falco peregrinus*, cannot be known, but its effects are certainly great in this respect and will almost certainly cause extinction of these species if continued.

## 13.3 PRODUCTIVITY AND BIOMASS

*Biomass* or *standing crop* is the amount of organic matter. It may be measured over an entire community, a trophic level, or a species population.

We have commented above that productivity is often measured in mass equivalents. Therefore, where $P'$ = net productivity, $B$ = biomass, and $M$ is the rate at which biomass is lost (through death to a population, through decomposition to a community), we can write

$$dB/dt = P' - M.$$

Clearly, at equilibrium, $P' = M$. We now look at each term for the case where $B$, $P'$, $M$ are measured over a single species population.

The number of individuals dying between age $x$ and $x + dx$ is given by

$$n_x \mu_x \, dx = \frac{n_{\text{total}} R^{-x} l_x \mu_x \, dx}{\int R^{-x} l_x \, dx}$$

($l_x$ is survivorship to age $x$, $\mu_x$ is the instantaneous rate of mortality at age $x$, and $R$ carries its usual meaning; for derivation of the discrete case see Chapter 9.) If $w_x$ is the average mass of an individual aged $x$, then

$$M = \frac{n_{\text{total}} \int R^{-x} l_x \mu_x w_x \, dx}{\int R^{-x} l_x \, dx}.$$

But

$$B = \int n_x w_p \, dx = \frac{n_{\text{total}} \int R^{-x} l_x w_p \, dx}{\int R^{-x} l_x \, dx}$$

so that at equilibrium:

$$M/B = P'/B = \frac{\int R^{-x} l_x \mu_x w_x \, dx}{\int R^{-x} l_x w_x \, dx} = \frac{\int l_x \mu_x w_x \, dx}{\int l_x w_x \, dx}. \tag{13.1}$$

We return to this equation later. In the meantime, for the sake of simplicity in illustrating a point, note that if $\mu_x = $ constant, then

$$P'/B = \frac{\mu \int l_x w_x \, dx}{\int l_x w_x \, dx} = \mu.$$

Note also that when $\mu$ is constant,

$$l_x = e^{-\mu x}$$

so that average life expectancy ($\bar{x}$ such that $e^{-\mu \bar{x}} = 0.5$) is

$$\bar{x} = \ln 2/\mu.$$

Thus

$$P'/B = \ln 2/\bar{x} \tag{13.2a}$$

and

$$B = P'\bar{x}/\ln 2. \tag{13.2b}$$

For several species at one trophic level, this becomes

$$B = \sum B = \frac{\sum_i P_i' \bar{x}_i}{\ln 2}.$$

Now at equilibrium $P'$ decreases to roughly 10% of its value for every trophic level removed from the primary producers. But since predators are usually larger than their prey, and since larger animals usually have longer life expectancies (more environmental buffering, see also part 9.5), the value of $B$ should not drop as greatly as $P'$. This reasoning applies to animals (excluding parasites and decomposers); the opposite reasoning would generally apply to the change in $P'/B$ between the primary producer and herbivore levels, at least in terrestrial communities. Thus $P'/B$ should be highest for herbivores (or perhaps primary producers) and should fall at each successive trophic level above the herbivores. Table 13.4 gives the pertinent figures for Odum's (1957) Silver Spring.

**Table 13.4**

|        | Producers | Herbivores | Primary carnivores | Top carnivores | Decomposers |
|--------|-----------|------------|--------------------|----------------|-------------|
| $P'$   | 8833      | 1478       | 67                 | 6              | 460 kcal/m$^2$/yr |
| $B$    | 809       | 37         | 11                 | 1.5            | — g/m$^2$   |
| $P'/B$ | 10.92     | 39.95      | 6.09               | 4.00           | —           |

The declining levels of $P'$ or $B$ with trophic level are often illustrated with drawings of pyramids, the broad (large $P'$, $B$ value) base representing the primary producers and the narrow (small $P'$, $B$) apex indicating the top carnivores (decomposers spoil the pyramidal shape and are usually ignored in this illustrative scheme). The productivity pyramid must always look something like a pyramid. Note, however, that if $\bar{x}$ rises faster than $P'$ falls, for all species together, with a change to a higher trophic level, $B$ may actually *increase*. Another way of viewing this: A level composed of populations with very rapid turnover rates can display very high productivity with a very small biomass. A higher trophic level with lower productivity may, by virtue of a much slower turnover rate, have a large biomass. According to Riley (1956), the volume under each square meter of surface water in Long Island Sound contains 16 g of phytoplankton, which support 32 g of zooplankton and bottom fauna. Harvey (1950) reports corresponding figures of 4 g and 21 g for the English Channel. Biomass pyramids may be inverted.

## 13.4 YIELD

At equilibrium, individuals of one trophic level are devouring individuals of the level below them at such a rate that the birth rate of any given species in the lower level equals the death rate. But there is more than one rate of predation which allows for such an equilibrium. If predation rate goes up, prey numbers may decline. But if prey numbers drop the prey population becomes less inhibited by density-related factors and displays an increasing birth rate or a declining nonpredatory death rate. A new equilibrium may be possible.

Experiments with *Tribolium confusum* show that as higher and higher proportions of the adult population are removed per unit time, productivity of the remaining adults increases (Watt, 1955). The same has been found by Slobodkin (1959, 1961) to be true for *Daphnia* populations, and Hayne and Ball (1956) found benthic productivity in small ponds in Michigan to be 17 times higher in the presence of a normal predatory fish population than in the absence of fish.

A question of practical interest to fishermen, foresters, and game managers is just what rate of predation is sustainable by a prey population and what rate of predation results in the maximum sustainable yield. A simple way to explore this problem is to make use of the logistic equation:

$$dn/dt = r_0 n(1 - n/K) - Mn_p n,$$

where $M$ is the predation rate and $n_p$ the number of predators (see part 12.2.6, Eq. 12.8; part 15.3.3, Eq. 15.9). The yield $Mn_p n = Y$. At equilibrium, of course,

$$Y = Mn_p n = r_0 n - \frac{r_0 n^2}{K}.$$

Sustainable yield is maximum when $dY/dn = 0$, so that $n = K/2$ and

$$Y(\text{max}) = r_0\left(K/2 - \frac{(K/2)^2}{K}\right) = r_0 K/4,$$

$$M(\text{opt})n_p = \frac{r_0 K/4}{K/2} = r_0/2.$$

Note that since $Mn_p n = r_0 n(1 - n/K)$, we can write

$$Mn_p = r_0 - \frac{r_0 n}{K},$$

so that

$$n = K(1 - Mn_p/r_0),$$

so that

$$Y = Mn_p n = KMn_p(1 - Mn_p/r_0).$$

It would be possible to raise $Mn_p$ to $r_0$, but this would drive the prey population close to extinction and would result in a very low yield. This can be seen in Fig. 13.3 which plots yield against predation intensity ($Mn_p$). Predation at an intensity above $r_0$ will result in extinction of the prey.

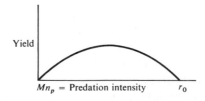

**Fig. 13.3.** Yield as a function of predation intensity.

While the above calculations serve as an illustration of natural processes, they are based on the not very accurate logistic equation and should therefore be considered valid only in their qualitative trends. An obvious factor, ignored above, is age. Different age classes in a population are usually exploited by predators at different rates, and different age classes may contribute to and respond to population density stress in different degrees. Watt (1955), for example, found that maximum sustainable yield in *Tribolium confusum* was much different from that predicted by the above equations, due to age factors. When exploitation occurred at a rate of 5.4 beetles per day, $n$ was about equal to $K/2$. This yield, then, should have been the maximum sustainable yield. But, when about 10 beetles per day were removed, the population dropped to $K/10$ and stabilized. The number of adults in the two populations were not significantly different, while the young animals were clearly less numerous in the second population. Increased "predation" had alleviated the inhibiting effects of high population density without significantly reducing the number of reproductives, and thus allowed for higher productivity than that predicted by the logistic model, and a higher sustained yield. In vertebrate populations predation on the old and sick may keep the mean age low and close to that of maximum fecundity. As a consequence, productivity may increase. It is also possible that where predation stabilizes prey–food cycling, food quantity is less often pushed to very low points and average prey number increases over that obtainable in the absence of predation. Howard (1965) feels that this is the case for North American small rodents, and points also to the positive correlation between liberal hunting regulations and deer herd size in Wisconsin (see Dahlberg and Guettinger, 1956).

For further information on age effects, see Watt's (1955) paper, or MacArthur (1960).

## 13.5 SUCCESSION

In any community there is a species or group of species which, by virtue of their size and/or abundance, affect the physical environment profoundly and appear to determine the nature of the whole community. These species are called *dominants*. (Of course, these species depend on as well as determine the nature of other organisms in the environment.) In forests, the dominants are clearly trees, in grassland, the larger and more numerous grass species. In the rocky marine intertidal, the dominant life forms may consist of mussels and barnacles as well as various plant species. For areas of similar climate and geology, the dominants which appear when the community is in equilibrium are essentially the same species. The process by which a community reaches this equilibrium, often passing through many vegetational stages, each with different dominants, is called *succession*.

When a community is seriously disturbed—a forest is slash-cut, for example— *pioneer* species are the first to recolonize. These, eventually, are replaced by other dominants which are themselves replaced, and so on until the dominant flora and fauna characteristic of the stable (equilibrium) community in that region, the *climax*, is reached. Depending on the disturbance and whether the disturbed area is wetter or drier than the more mesic (intermediate) climax, different pioneer species may become established. Thus different successional pathways to the climax may occur. Such pathways are called *seres*. Occasionally a pioneer dominant will also be a climax dominant so that succession never occurs. This is true, for example, of lodgepole pine, *Pinus contorta*, in drier, upper montane (see 14.5) areas in the front range of the Colorado Rockies, but is an usual phenomenon in terrestrial environments. Dry seres (those that begin under drier conditions than finally exist because of buffering of the environment, transpiration of dominants, etc., in the climax community) are known as *xeric*, or *xerarch*. *Hydrarch* succession begins with more moist conditions.

Succession that begins with a disturbance is known as *secondary succession*. Succession that begins with bare rock or some other dry, lifeless substrate (xerarch) or water (hydrarch) is *primary succession*.

Primary hydrarch succession begins with open water. As organisms die and settle, the body of water very gradually fills in. As the water becomes shallower, the first emergent plants appear. These die, and their bulk of nondecayed material hurries the fill-in process. Eventually, around the edges of the water appear grasses and sedges, and then herbs and shrubs. A transect from the center of an old lake past the shoreline shows a gradient of successional stages all the way from open water to the terrestrial climax of the region. Details of the process, of course, depend on the temperature, geological chemistry, etc. of the region. As indicated in Fig. 13.4, primary xerarch succession may begin in a number of ways. We will briefly describe bare rock surface succession. The first organisms to colonize are lichens, which eventually die, forming a very thin layer of soil. In this soil may grow other lichens or mosses, and as the soil builds in thickness, grasses grow. Acidic secretions may eat away parts of the rock surface which is then incorporated into the soil, and roots may grow into small rock crevices, splitting off parts of the rock. Soil thus accumulates and larger plants are able to successfully colonize the area, bringing with them animals characteristic of the successional stage. Eventually, whether the first step is open water or bare rock, the same climax community is established.

The fact that, except for local, micro-climatic or edaphic variations, a given area always displays the same climax dominants, allows us to characterize geographic areas by their dominants. Thus South Dakota is said to be largely *Andropogon–Agropyron*, long-grass prairie, and the Newfoundland, Nova Scotia, and New Brunswick area, in spite of Longfellow's claims, is characterized by spruce, *Picea*, and balsam fir, *Abies*. Figure 13.4 gives a detailed description of succession to a spruce–fir–birch climax on Isle Royale in Lake Superior (Oosting, 1958, after Cooper, 1913).

"Climax" is usually defined as meaning stable or unchanging. Unfortunately, while this definition is often convenient, it is an oversimplification. The rate at which different species replace one another in succession slows down considerably at some point in time and it is here that we speak of climax. Further change does occur, however. More will be said about such *senescent* communities later.

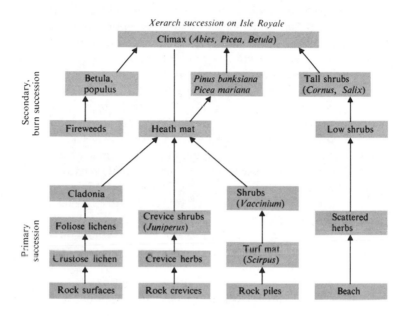

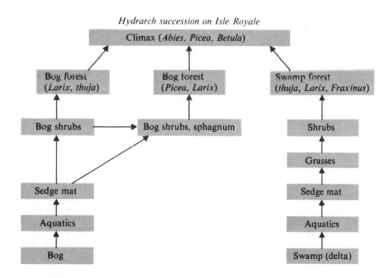

**Fig. 13.4.** Succession on Isle Royale (after Cooper, 1913).

## 13.5.1 A Competitive Approach

The discussion in this section refers primarily to forest communities, because it is on such communities that the most work has been done. The general argument, however, should be applicable to other communities as well. We assume that all organisms have evolved the ability to reproduce under the physical and chemical conditions of their environment. These conditions will certainly change, but we presume that the organisms will retain the ability to reproduce themselves. A catch in this assumption is that replacement of one's population may occur only after a number of individuals have died and made space for offspring. During the sterile period between the establishment of a population and the time when room is first made for a new generation, other species not requiring the extra room may colonize and come into competition with the original population. If, in the pool of species from which new colonizers are constantly arriving, there exists a species which can competitively exclude the original population, then that population will be replaced. We may picture succession as the constant arrival of competitive colonizers and the exclusion of previously successful colonizers. Which species successfully establish themselves and which are competitively excluded and when depends on the physical and chemical conditions prevailing at different times in the region. These conditions are clearly affected by the dominant species of the moment and of the past. One may thus characterize a successional, as opposed to climax, dominant as a species which alters its environment in a way which allows another species to reproduce more successfully than itself, thus bringing on its own demise. The process of competitive replacement continues until a species or group of species becomes established which cannot be out-competed by any of the species available for colonization. At this point, assuming that this species can replace itself, relative stability (still subject to climatic, geological, and genetic changes) is attained. This point represents the climax community.

The kinds of interactions which change the physical and chemical environment are myriad; a few are listed here. Legumes, with their nodules of nitrogen-fixing bacteria, add nitrates to the soil. Most other plants remove them. Soil levels build, and soil profiles (affected by the varying pH values of different leaf types or the differing degrees of buffering against rainfall, temperature changes, and evaporation) change over time. Soil acidity changes with the type and decomposition rates (affected by buffering by dominants) of litter. Soil microarthropods and annelids affect the texture and chemistry of the soil. Different plant species encourage the existence of different animals, each of which eats different foods, packs, digs up, or burrows in the ground, and defecates in its own manner. The factor of greatest importance in forest succession is light. When the reproduction of one group of dominants in their own shade is less successful than that of a group of colonizing dominants, there will be competitive exclusion of the former group. Species which, when mature, throw considerably more shade than existed at the time they colonized the area may be replaced by more *shade-tolerant* species. We would expect those species which are early successional dominants to be adapted to bright light conditions—and thus not shade-tolerant—while later successional and climax species should be adapted to shade conditions—and thus less sun-tolerant. This is, in fact, the case. It is unlikely that a species can adapt to both bright and shady conditions so that it is a superior competitor to all other potential colonizers. Thus it is unusual to find the same species as both pioneer and climax dominants.

Some particularly interesting observations on light competition have been made by Horn (1970). Horn notes that when leaves are scattered over several vertical layers (multilayer situation) each layer acts, to an approximation, as a neutral filter. Thus lower layers receive essentially the same quality light but less of it. Since photosynthetic efficiency falls off significantly only when light intensity drops below about 20% of direct sunlight, each layer in a multilayer situation operates at full efficiency down to that point. Thus canopy species will do well to be multilayered, but species growing under shady conditions should have fewer layers. Early successional species of trees are usually multilayered. Later successional species, in their smaller growth stages, and subcanopy species in climax forests, are generally monolayered, spreading their leaves very densely but such that (ideally) no leaves are shaded by others. In eastern and midwestern deciduous forests, dogwoods, *Cornus*, are monolayered, as are the seedlings and young saplings of maples. Adult maples, reaching the canopy, are multilayered. The naturalist looking for multi and mono-layers should note that the leaves of a monolayer plant needn't all lie in one plane. The density of the "neutral filter," i.e., the amount of shadow cast by leaves in a multilayer, is related to the smallest circle which can be inscribed in a leaf. Thus the shadow can be minimized with least diminishing of the photosynthetic surface by making the leaves lobed. Early successional and canopy multilayered species generally possess lobed leaves (maple, oak, for example) while subcanopy mono-layered species, which do best by maximizing leaf surface and do not suffer from shadows cast, such as dogwood, show unlobed leaves.

Certain classes of organisms establish themselves only after a previous class has matured and somewhat altered the environment, which suggests that suc-cession may be a discrete rather than continuous process. In this connection Odum (1960) has found that productivity changes with succession in jumps from one steady state to another, each associated with a set of dominant life forms.

Species which are both pioneer and climax dominants exist because of rather interesting adaptations. In the case of the lodgepole pine mentioned earlier, the cones open when heated by fire. Thus when a fire burns out a lodgepole forest or an area close enough to such a forest as to have cones scattered about, the pine seeds are the first there, lodgepoles become established and, since the trees throw dense shade, prevent other species from colonizing. Lodgepole pine stands in the Colorado front range are characterized by sparse undergrowth and trees all of nearly the same size, punctuated here and there by shrubs or aspen, *Populus tremuloides*, where a fallen tree has opened up new ground for colonization without fire.

## 13.5.2 A Thermodynamic Approach

Margalef (1958, 1963, 1968) views communities as being in different stages of "maturity." His stages clearly relate to successional stages and his definition of maturity reflects the changes which occur with succession. More "mature" communities are distinguished from less mature by low productivity to biomass ratio ($P/B$), more complex food webs, a higher proportion of animals to plants, and greater efficiency of food use (note that $P$ here denotes gross productivity. Net productivity, $P'$, referred to earlier in this chapter, is $P - R$, where $R$ is respiration. Also $P$ and $R$ here refer to the entire community). Other authors (for example Odum, 1969) have noted the same trends with succession. Let's explore these trends.

Early in succession, primary or secondary, biomass is low in the community as a whole. Thus there is little competition for light, water, etc., and productivity per unit biomass should be high. But as biomass accumulates (see above discussion of primary succession), there is less light and water per unit biomass, so productivity is increasingly inhibited. Productivity, though, depends on the total leaf surface area which increases with biomass. Thus productivity may increase with succession, but less rapidly than biomass, and $P/B$ should decline. This theoretical argument is not rigorous, but the conclusion is borne out by data.

As biomass builds it is more likely to subdivide into microhabitats or patch types and consequently a more diverse fauna (see 15.3.1) with more complex interactions.

Early successional species are more likely to be successful if they possess the ability to disperse into the area, establish themselves, and multiply. Competition occurs in a growing community and competitive success thus depends primarily on relative rates of population increase. In later successional stages, growth in biomass is much slower and competitive success depends on the ability to maintain a viable population. Thus $r_0$-selection is stronger in early succession, $K$-selection in later stages (see Discussion, Section III). $K$-selection results in fine tuning in which organisms increase their efficiency of use of their resources. Greater efficiency in food use means that there can be a greater ratio of animal to plant biomass. Again, the theoretical argument is not rigorous, but the conclusions accord with observation.

If nutrients are suddenly added to a community, $P/B$ will clearly increase. Population equilibria will thus be upset and competitive situations which were stable may become unstable, with that species possessing the highest $r_0$ excluding others. Diversity and trophic complexity decline. According to Margalef, the system is forced into a more immature state. Margalef (1963) postulates an interesting corollary: Suppose two communities, one more mature than the other, are linked, with energy flow between them. Since $P/B$ is higher in the more immature system, energy, bound up in biomass, shows a net flow into the more mature system. The increase in biomass in the more mature system enhances the successional process there, but the loss in biomass from the other system retards succession. The difference in maturity between the two communities is thus exaggerated. Boundaries between subregions of differing maturity should tend to become more sharply defined with time. This appears to be true at least for marine algae (Margalef, 1958).

With succession, $P/B$ declines and $B$ grows. But as $B$ increases, respiration should increase also. Thus $P/R$ should fall with succession. Of course, if $B$ does not continually increase with succession—and there is no reason to believe it should in late stages—or if the increased efficiency of organisms late in succession extends to basal metabolic needs, $R$ may actually decrease and lead to an increasing $P/R$ value. If our climax is, in fact, a stable community—or if we define it as such—then at climax $P = R$ and there is no further accumulation of biomass. Table 13.5, which is part of a table (slightly modified) appearing in Odum (1969), summarizes the changes occurring during succession.

**Table 13.5.** Changes in communities undergoing succession (modified from Odum, 1969).

| Community attribute | Early succession | Late succession |
|---|---|---|
| 1. Gross productivity/community respiration ($P/R$) | $\gtrsim 1$ | $\rightarrow 1$ |
| 2. Gross productivity/biomass ($P/B$) | high | low |
| 3. Biomass supported/unit energy flow (efficiency of food utilization) | low | high |
| 4. Net community productivity ($P'$) | high | low |
| 5. Food chains | simple | complex webs |
| 6. Total biomass ($B$) | small | large |
| 7. Selection pressures | $r_0$-selection | $K$-selection |

It is possible, of course, returning for a moment to the competitive approach, that a climax dominant will be unable to reproduce in its own shade. In this case $P/B$ continues to drop and $P/R$ drops below 1.0. The community is said to be senescent, and gradually disintegrates until dying dominants fall (in a forest) making way for new colonizers.

### 13.5.3 Disclimaxes

Succession does not always go to completion. In climatically unstable areas, for example, periodic disturbances disrupt community organization, perhaps kill off some species, and may expose new soil surface for exploitation. This disorganization may be taken advantage of by opportunistic species characteristic of earlier successional stages, and the community may retain the appearance of an immature stage. Another way to view the situation: In 15.3.4 it is shown that in very unstable environments there is less competitive exclusion. Thus the competitive process by which succession proceeds is alleviated and succession slows or stops. Finally, periodic disturbances may disrupt reproduction thus lowering $B$ and consequently maintaining $P/B$ at a high level. Thus succession is held up at an immature stage. When this happens, the community is said to maintain a *disclimax*.

One common cause of disclimax is fire. In the early 1800's and before, prairie fires, often intentionally set by Indians, prevented the easternmost parts of the American long grass prairie from becoming forest. Only the occasional burr oak with its fire resistant bark was able to establish itself. Now these fires no longer burn and the forest is encroaching on the prairie. The old oaks which were the first trees to survive are sometimes still seen. Fire had resulted in an extension of the prairie into regions which would otherwise have been forested. Frost heaving in Arctic and alpine tundra regions is another factor causing disclimaxes. Lumbering is a source of disturbance. If the lumber yield is to be climax now and in the future it is best to disturb the community as little as possible by careful thinning. If the yield is to be in a pioneer species tree, clear cutting and perhaps burning are desirable. This throws the community back into the pre-pioneer stage of secondary succession.

Heavy grazing by animals keeps $P/B$ values high and may result in disclimaxes. Hope-Simpson (1940) describes the vegetation on the chalk downs in England to be quite different on areas grazed by rabbits and those not, for example. Pastureland may appear the same year after year, apparently being in a stable state, but differ greatly between ungrazed, lightly grazed, and heavily grazed areas. Animals may have evolved the habit of preferentially eating colonizing species and thus halting succession in a stage beneficial or necessary to their continued local survival. Such evolution would have to have occurred through kin selection or group selection, but as Hershkowitz (1962) points out, many small mammals occur in semi-isolated pocket populations. For small mammals, at least, such evolution may be possible. It would be difficult, and perhaps of only academic interest, to prove that feeding habits conducive to disclimax had evolved to this purpose since natural selection at the individual level may have favored food preferences with, fortuitously, the same result. For example, it seems likely that plant species often evolve animal attract-ants in response to selection pressures for dispersal. Such selection pressures might be expected to be stronger in early successional plants, which are generally more opportunistic. Also, a plant which is opportunistic must put energy into production at the expense of defense. Both considerations would lead to food preferences in earlier successional plant species.

### 13.5.4 Micro-succession

Regardless of the successional status of a community, small subunits of that community are experiencing succession on a much smaller, much more rapid scale. Every time a tree dies and rots, *micro-succession* occurs.

Dead pine logs in North Carolina first attract fungi and beetles which eat the phloem and loosen the bark. Loosening of the bark allows other species to enter, including scavengers and fungus feeders. A few predators such as centipedes follow, with ants, using the surface of the log. The action of the fungi and beetles puts holes in the bark through which water enters. By the second year most of the phloem is gone and with it the phloem-feeding beetles. Sapwood-eating insects such as termites now appear, as do frass-feeders. Since there is now more water entering and evaporating from the log, decay occurs more rapidly and softwood feeders become more numerous.   The emergence holes of insect larvae are now numerous and some are used by solitary wasps.   By the third year there is little bark left and the sapwood is quite rotten and soft.   The final stages of decay set in and the log is incorporated into the humus (Savely, 1939).

A classic study is that of Winston (1956) who looked at micro-succession in acorns in Illinois.   Since several alternative successional pathways exist, verbal description is difficult.   The events are diagramed in Fig. 13.5.

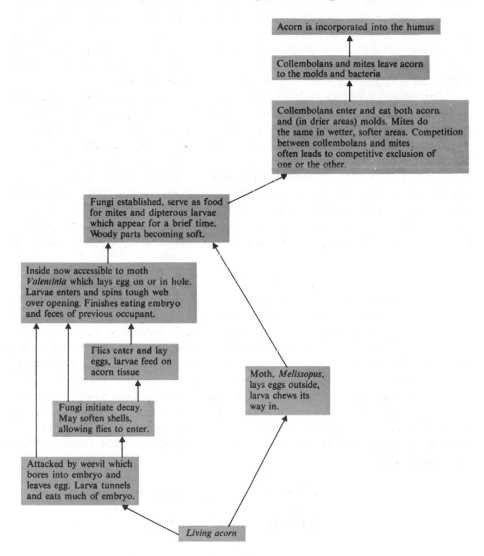

**Fig. 13.5.** Micro-succession in an acorn (from Winston, 1956).

## 13.6 STABILITY

The word stability has been used in a variety of ways and it often appears with no definition.  By stability do we mean constancy of productivity or numbers of individuals?  Do we mean predictability—and, if so, of what?  Do we refer to the ability or tendency of a community—or part thereof—to return to its original state, or some other state, when perturbed?  Can communities in successional stages, or communities oscillating with respect to some parameter, be considered stable?  Lewontin (1969) has approached this problem, but his solutions strike this author as more of mathematical interest than biological usefulness.  For the purposes of the moment we define stability, with Leigh (1965), as the inverse of the frequency with which a population reaches some given size (say $n = 0$—extinction).  If the definition is to be extended from one species to the whole community it clearly needs amendment.  For now we consider one species at a time.

With approach of climax during succession, species tend to persist longer and, at climax, theoretically persist indefinitely.  It is therefore tempting to try and find a connection between stability and the (other) characteristics of a mature community.  Margalev (1963) considers biomass, $B$, as the "keeper of organization," so that $B/P$ is a measure of the information per unit energy flow.  Since $B/P$ increases with succession, the amount of information per unit energy flow increases with succession and mature communities contain more information per unit energy flow with which to maintain their integrity.  They are thus more stable.  In addition to vagueness as to the meaning of "stable," this argument is terribly sloppy.  Some of it, however, may be salvaged: For any given population at equilibrium, Eq. (13.2) is valid

$$P'/B = \ln 2/\bar{x},$$

where $\bar{x}$ is the life expectancy.  As $P'/B$ declines with succession (assuming $P'/B$ to vary with $P/B$), $\bar{x}$ increases.  When $P'/B$ is high, life expectancy is short.  Short life expectancy is associated with small and, usually, opportunistic species.  We have already noted that early successional species are apt to be opportunistic.  But opportunism is associated with great population fluctuations and thus, presumably, greater chance for local extinction—hence instability.  Leigh's (1965) argument is more rigorous and leads to the same conclusion.  We let $N_i$ be the biomass of species $i$, $e_i$ be the population growth rate of species $i$ in the absence of other species, $a_{ij}$ be the interaction (prey–predator) coefficient of species $i$ and $j$, and make the simplifying assumption that $a_{ij} = -a_{ji}$.  Then, making use of the Lotka–Volterra equations, we have

$$dN_i/dt = e_i N_i + \sum_{j=1}^{n} a_{ij} N_i N_j$$

$$P \text{ (productivity)} = \tfrac{1}{2} \sum_{i,j} |a_{ij}| N_i N_j$$

$$B \text{ (biomass)} = \sum_{j=1}^{n} N_j.$$

If $q_i$ represents the equilibrium value of $N_i$, then the frequency with which $\ln(N_i/q_i)$, in its fluctuations, crosses a value $c$ is given by

$$\text{frequency} = \frac{1}{\pi} \sqrt{\sum_j q_i q_j a_{ij}^2} \; e^{-q_i c^2/2\theta}$$

where $\theta$ is a constant (see Leigh, 1965, for this and following derivations). Now if we assume that selection acts on $a_{ij}$ to minimize this frequency for a given $P$ (group selection?), we can write (assuming $q_i \approx q_j$):

$$\frac{\partial}{\partial |a_{ij}|} \left[ \sum_{i,j} |a_{ij}|^2 \, q_i q_j - \lambda \left( \sum_{i,j} |a_{ij}| q_i q_j| \right) \right] = \sum_{i,j} (2|a_{ij}| q_i q_j - \lambda q_i q_j) = 0,$$

or

$$2\sum_{i,j} |a_{ij}| q_i q_j = \lambda \sum_{i,j} q_i q_j, \qquad 2(2P) = \lambda \sum_i q_i B = \lambda B^2,$$

so

$$\lambda = 4P/B^2, \qquad \text{and} \qquad |a_{ij}| = 2P/B^2.$$

Substituting back into the frequency equation, and noting that $q_i \approx B/n$, we obtain

$$\text{frequency} = \frac{2P}{\pi B \sqrt{n}} \, e^{Bc^2/2\theta n} \left( = \frac{1}{\text{stability}} \right).$$

Thus as $P$ increases stability declines. An increase in $B/P$ or $B$, or an increase in $n$, all result in greater stability. As $P/B$ declines with succession stability should increase.

Another connection between maturity and stability was first suggested by Nicholson (1933). If the food web is complex—the usual situation in mature communities—a species may change greatly in population density without greatly affecting its predators which have alternative foods. The effect upon its prey is also buffered by the fact that it is only one of several predators. If stability is defined as the magnitude in change of certain populations to perturbations in others, then communities with complex food webs are more stable than those with simple food chains. Note that this definition of stability is not the same as the one considered above, although the concepts are closely related. MacArthur (1955) formalized Nicholson's hypothesis by reasoning that alternative food pathways would be most effective in buffering against population changes if the energy flowing through each were equal. If one pathway carried nearly all the energy, there would be, effectively, only about one pathway. A good measure of this concept of stability should consider both the number and the equitability of energy flow through food pathways. In addition, consider the scheme depicted in Fig. 13.6. A good measure of stability should include the stability of the carnivore–herbivore interaction plus the average of the two herbivore–producer interactions. But the only measure of stability, $S$, which meets all these criteria is

$$S = -\Sigma q_i \log q_i,$$

where $q_i$ is the proportion of energy flowing along the $i$th pathway.  In the example (Fig. 13.6), the $q_i$'s would be

$$p_{15} \times p_{57} \qquad p_{25} \times p_{57} \qquad p_{36} \times p_{67} \qquad p_{46} \times p_{67}.$$

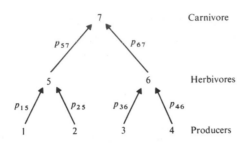

**Fig. 13.6.** A simplified trophic web (see text).

A word of caution is in order.  As given above, $S$ is a quantitative *definition* of a measure of stability.  Thus to conclude a system is stable on the basis of this measure is quite circular.  $S$ measures the degree of stability *given that this is the meaning one wishes to convey by the word "stability."*

There is undoubtedly more to stability than trophic complexity even if one wishes to equate stability directly to alternative food choice.  As pointed out in Chapter 11, time lags may determine whether the addition of alternative foods increases or decreases amplitude in population fluctuations.  Also, changes in population levels do not necessarily imply similar changes in energy flow.  The productivity of a species may increase as its numbers decline, or vice versa, depending on the cause of the population drop.  It is also possible that one species holds another below some critical level above which it would undergo massive population growth.  Those species which hold the trigger to population outbreaks in others, if they exist, have importance far out of proportion to their numbers or the energy flowing through them.  Paine (1969) has called these *keystone* species and supports his contention that they exist with his findings (1966) that when *Pisaster* (the keystone species) is removed from Washington state intertidal areas there follow striking changes in species composition.  Similar experiments have been carried out in New Zealand with the same results (Paine, personal communication).  Species may, of course, also affect species with whom they are not directly energetically linked.  Buffering effects of trees against temperature and humidity changes and the consequent effects on forest floor populations, or the use of woodpecker holes by other birds and mammals are two simple examples.

In conclusion, definitions and measures of stability are badly needed.  Both that of Leigh and that of MacArthur are useful, but neither necessarily measures what another person thinks of as community stability.

# 14

# The Origin and Maintenance of Communities

The geographical distribution of a species depends upon its ability to disperse and colonize new areas, as well as upon historical accident and its subsequent interaction with competitors and predators. This chapter discusses dispersal, colonization, and historical accident, and then relates these factors to community structure. The discussion borders on a number of broad fields of interest such as biogeography, but it is not our purpose to explore these fields in any detail.

## 14.1 DISPERSAL

### 14.1.1 General Considerations

Among vertebrates, most dispersal involves young animals. Where home ranges exist, competitive pressures and sometimes direct parental pressures drive the young away. In birds, retention of a juvenile appearance seems to allay aggression but once an adult pelage or plumage is attained the young animal must leave to seek its own range—or at least leave that of its parents—or face attack from larger, stronger, and more experienced competitors. For most mammals, a home range must eventually be established since continued wandering may be fatal (part 8.6.2). In the beaver, *Castor canadensis*, only one family occupies a given pond. When the young of one year arrive, the young of the previous year are driven out and forced to settle or provide themselves with a new pond (Bradt, 1938). Birds wander as young, perhaps because they can't often compete with older individuals and must travel far to find a place where they will not be chased from food or prevented from reproducing (Welty, 1962). First year herring gulls, *Larus argentatus*, captured, marked, and released on Kent Island, off New Brunswick, were recaptured one year later at an average distance of 1380 km from Kent Island. Second year birds averaged 695 km, third year birds, 465 km, and birds over three years old wandered an average distance of 467 km (Gross, 1940).

Many invertebrates and some vertebrates possess larval stages which may have evolved as a means of dispersal. A dispersal phase would often be advantageous in opportunistic and extremely important in sessile species. In locusts there may be two *phases*, the *gregaria* phase which swarms and migrates and the solitary, more sedentary *solitaria* phase. When population density rises, the locusts change from the solitaria to the gregaria form and move out from the area in tremendous numbers. The gregaria phase is adapted to dispersal over long distances by its reduced wing load (lower specific gravity) and the fact that it carries more fat and less water (Kennedy, 1956).

The advantages to an opportunistic species of an ability to disperse are obvious. For nonopportunistic species, dispersal is one way of avoiding situations of high population density or a deteriorating environment. Of course, the advantages of dispersing must be weighed against the hazards of settling unfamiliar areas with unfamiliar occupants (part 8.6).

To the advocate of group selection, dispersal (even if personally disadvantageous to the moving individuals) acts to increase genetic variance and thus leads to a greater ability to adapt to new situations and avoid competition. As pointed out in Chapter 3, group selection is probably a very weak force, though if unopposed by selection at the individual level it may be significant. Possibly the group advantage of increased genetic variance has played a role in the evolution of dispersal tendencies and mechanisms, though it has probably not been of prime importance.

Magnitudes of dispersal distances by individual organisms for a variety of species were published in a massive compendium of figures by Wolfenbarger (1946).

## 14.1.2 Long-distance Dispersal Mechanisms

The fact that organisms may move out from their places of birth, or from dense population situations, raises the possibility that long-distance dispersal might result in the colonization of previously unoccupied areas. We now look at the mechanisms by which animals disperse from place to place; in particular, how organisms cross such barriers as mountains and broad stretches of water.

One obvious means of dispersal is by active movement—the creature flies, swims, or walks. Long-distance dispersal is possible by such means if no barriers to movement exist. It would not be possible, however, for a warm water marine snail to disperse from the Caribbean to the Pacific Ocean (except through the Panama Canal), for it would either have to cross land, or traverse areas of cold water. Active dispersal between islands or between a mainland and islands may be accomplished by birds, bats, or strong-flying insects, but little else, and only when the climate between the areas is suitable for survival of the animal.

More important as mechanisms for dispersal are the various means by which organisms are passively carried. For example, Malone (1965) reports that two species of freshwater snails may be carried significant distances overland, attached to the feet of killdeer, *Charadrius vociferus*, and Revill *et al.* (1967) report the same for certain algae and protozoans on midges and crane-flies. Mud on the feet of birds may carry insects or insect eggs, snail eggs, and encysted organisms of many sorts. Agnew and Flux (1970) found 160 of 369 hares, *Lepus capensis*, examined in Kenya to be carrying, collectively, 810 plant burrs and seeds representing 17 species. Since these hares clean themselves at least once every day they may have considerable importance as dispersers of plant genes. Passive dispersal may also occur by air or water. Gressitt and his coworkers (cited in Carlquist,

1966) have captured many species of insects and spiders in aerial traps. In fact the species of trapped insects occurred in essentially the same proportions as they are found in the insect faunas of oceanic islands, suggesting aerial transport as the primary means of insect dispersal. Carlquist (1966) points out that, over islands, there are often centers of condensation which might act to direct settling of airborne organisms. Rotifers, tardigrades, and encysted creatures of many kinds are transported by wind, and spiders commonly spin sheets to aid them in catching the wind and getting airborne. One sometimes sees fields densely covered for miles with sheets of white, blowing and tumbling in the wind. Even strong fliers such as birds are carried passively, to some extent, in strong winds. After big storms we often find large numbers of *accidental* species of birds, carried from the American mainland as far as Hawaii.

Water flotation, of course, is an important vehicle for the transport of seeds and seed parasites. Rafts of debris may be large enough to transport such animals as mice and lizards quite long distances. Rodents (judging by the most nearly related mainland species) had to cross about 600 miles of ocean to reach the Galapagos Islands. *Rafting* seems the only plausible way by which they could have made the crossing. Rafting also includes the transport of organisms in ships by man. This is the way in which such species as the house mouse, *Mus*, rats, *Rattus* (unintentionally), and starlings, *Sturnus vulgaris* (intentionally), reached United States shores. It *may* account for a tremendous proportion of the recent dispersal of species between islands and continents.

There must be very few organisms larger than small insects which disperse several hundred miles or more and survive on landing. Cruden (1966) has pointed out that internal transport of seeds by birds is plausible only over fairly short distances since seeds pass through the intestinal tract in a few hours. Also preening makes the external long-distance transport of organisms by birds unlikely. Rafting, including chunks of debris large enough to carry mice or lizards, must be quite uncommon; and even when it occurs, how likely is it that a mouse or lizard is aboard, how likely that there are several—enough to start a new population  or that the debris will reach an ecologically benign shore before its passengers die. Long-distance dispersal is more likely to involve a series of stepping stones, a few individuals spreading a short distance to a nearby island, for example, building a population, and then spreading a bit farther, to the next island, in the same way. Evidence of such island hopping is found in the fact that the Hawaiian avifauna is primarily of Indo-Malayan ancestry even though America is the closest land mass. Between Hawaii and the Indo-Malayan area are many small islands, between Hawaii and America very few (Carlquist, 1966).

### 14.1.3  Habitat Selection

It is clearly advantageous for organisms to be able to pick out the best habitat in which to live and reproduce. Thus birds choose territory and nest sites in areas with familiar characteristics and barnacle larvae settle preferably on substrates of certain chemical characteristics, texture, and tidal level. *Peromyscus mani-*

*culatus bairdi,* a field dwelling subspecies, has an inherited preference for field to forest habitat which can be strengthened, and perhaps weakened, but not reversed by experience (Wecker, 1963). In damselflies, *Hetaerina americana, H. lugens, Ischnura demorsa,* and *I. damula,* on the other hand, the choice of a lotic or lentic environment by the adults depends largely on their position at the time of emergence (Johnson, 1966). Whatever the determinants of habitat preference in an animal, however, it almost invariably exists and, presumably, is adaptive. It is reasonable to assume that the distribution of an organism over the world depends, as well as on chance dispersal, on that organism's choice of habitat in the area where it arrives. If an appropriate habitat is lacking, a species may spend its time searching and fail to reproduce even though habitat adequate to successful reproduction is present. Such behavior, although maladaptive in the newly invaded area, was evolved elsewhere where it was not maladaptive. Lack (1933, 1938, 1940) has discussed the "psychological factor in bird distribution," pointing out that birds fail to colonize areas which, to our eyes, appear perfectly suitable. The rock pipit, *Anthus spinoletta,* for example is absent from areas which possess all the essentials for pipit survival and reproduction, but lack the superficial appearance of its usual habitat.

## 14.1.4 Migrations

Among large, mobile, or flying species capable of active, long-distance dispersal, it should not be surprising to find long-distance migration patterns. A reasonable (but untestable) explanation for the evolution of such patterns is the following. Where seasonal differences in food supply are great, for example, and it is possible that a very few individuals wander far enough in the lean season to find a richer area, selection should favor the genotypes possessed by those wanderers who move in the proper direction. When the original area provides more food than the new, selection favors those who wander in the homeward direction. Where orientation by the individuals is possible, regular migration patterns may appear. The patterns should reflect a balance between the advantages of wandering still farther for a better food supply and advantages of wandering less far and thus curtailing the hazards of a long excursion. In the case of birds, many fly just far enough to insure adequate food. On the other hand, some appear to overdo things. The Arctic tern, *Sterna paradisaea,* migrates from pole to pole, the sooty shearwater, *Puffinus griseus,* breeds off New Zealand, then scatters in all directions, often over vast distances. The golden plover, *Pluvialis dominica,* breeds either in Alaska, in which case it migrates to the Marquesas, off the southern coast of Chile, or in northeastern Canada, whence it migrates to southern Argentina.

Above, we have spoken as if food change were the important criterion in determining the existence of migration, a hypothesis elegantly and rather thoroughly supported by Lack (1954). But it is also possible that the impetus for migration is escape from temperature change. Of course it would be foolish to try to separate a temperature and a food hypothesis, for both are inextricably bound together. The masked weaver, *Ploceus cucullatus,* needs to eat 20% of its body weight per day at 18°C, 25% at 9°C, and 28% at 7°C (Schildmacher, 1929, cited in Welty, 1962). To reason as if food were an independent cause of migration, however, may prove useful in prediction. MacArthur (1959) has noted that most birds in the eastern North American deciduous forest, except in the far south, migrate—in the sense that they summer in the nearctic, winter in neotropical areas. On the west coast, however, fewer migrate and very few of those species nesting on the American great plains fly south for the winter. A sharp difference in the proportion of migratory and nonmigratory birds occurs at the eastern boundary of the great plains and the deciduous forest, near the Mississippi River. The reason, according to MacArthur, is probably that seasonal differences in food abundance are great in the eastern, but not southeastern forests, less on the temperate west coast and, in spite of the great temperature fluctuations, also less on the great plains. This is true because in prairie land seeds are plentiful and may be taken all winter in areas where snow has sublimed or been blown away by the wind. It should be possible to generalize these arguments to palearctic species as well.

Some examples of migratory animals follow. Among marine invertebrates it is very common for animals to move from deep water to the intertidal region, seasonally, for spawning. This is true of some crabs, sea urchins, prosobranch snails, and nudibranchs. Occasionally, one finds movement in the other direction as in the case of *Thais lapillus* (Pelseneer, 1935) or a convergence by reproducing animals on the lower intertidal from both directions as in *T. lamellosa* (Emlen, 1966a). Those cuttlefish which winter on the continental shelf south of England swim north in the spring to breed in the North Sea. Butterflies such as the monarch, *Danaus plexippus*, the red admiral, *Vanessa atalanta*, and the painted lady, *Vanessa cardui*, fly north in the spring, south in the fall, and the painted lady breeds at both ends of the journey. The desert locust, *Schistocerca gregaria*, flies north to Algeria and Morocco where it remains during March through June, the local rainy season, then returns south to the Sudan for the period from July to October, the rainy season in that area, breeding in both places (Donnelly, 1947). A number of fish such as the cod and the herring have weak-swimming or planktonic young which undergo a passive migration, carried by ocean currents. The eggs are laid in one place, grow up in another, and the adults return to where they were spawned to repeat the cycle. The larvae of the Arctic cod, *Gadus morhua*, for example, drift on the Spitsbergen current from Vest Fjord in Norway, where they hatch, to the southeastern Barents Sea. The adolescent fish later move to deeper water, join the adults, and in the autumn migrate back to Vest Fjord to spawn (Cushing, 1968). The American (*Anguilla rostrata*) and the European (*A. anguilla*) eels spawn in the Sargasso Sea where they overlap in distribution. Then the American species metamorphoses in its first year, the European eel in its second, and both cross the Atlantic, the former to the west, the latter to the east. They live separately in fresh water for several years before returning to the Sargasso Sea. Salmon of western Canada and the United States do the opposite, breeding in freshwater tributaries, the young, after variable periods of time, heading to the sea to spend usually two to four years.

Large proportions of temperate-breeding bird species migrate, usually between their nesting grounds in the spring and summer and more tropical areas, or regions in the opposite hemisphere during the rest of the year. A few move altitudinally, like the Clarkes nutcracker, *Nucifraga columbiana*. Mammals may migrate north and south, as in the case of the caribou, *Rangifer caribou*, which herds on the tundra in summer and moves south to timberline for the winter. Bison, too, used to migrate 200 to 400 miles north and south, and the gray whale, *Rachianectes glaucus*, swims south along the California coast in the winter, where it gives birth, and then returns to the Arctic in late spring (Bourliere, 1955). Vertical migration occurs in the elk, *Cervus canadensis*, and the mule deer, *Odocoileus hemionus*, which form winter and summer herds in areas perhaps only 10 to 60 miles apart, but quite different in climate.

## 14.2 COLONIZATION

### 14.2.1 Characteristics of an Easily Colonized Area

Consider a climax community. It is, by definition, quite stable, in almost any sense of the word. *K*-selection has been operating within similar habitats for a very long time and the resident species have become finely adaptively tuned to a myriad of interrelationships. Intraspecific competition has broadened niches, until an equilibrium point with interspecific competitive pressures has been reached (12.3.2). Species which are inferior competitors in this system have been ousted and are re-ousted as they re-invade. It may generally be assumed that if an uncontested food source or source of shelter existed at one time, something will have invaded and made use of it. (This last statement may be challenged; refer to 15.2.2.) A new species finely adapted to another community perhaps, but not specifically to this one, may be expected to have little chance of successfully competing with the indigenous species on their home ground. · A climax community is not the most likely spot to find successful colonization by new species.

Communities in early successional stages, unlike climax communities, are probably quite sensitive to perturbation (part 13.6). This suggests that a perturbation in the form of an invading species may disrupt the community so that normal competitive processes are altered. Replacement of one species by another, moreover, will be less a matter of who is most efficiently adapted (with respect to *K*-selection) and more a matter of who has the highest intrinsic rate of natural increase. For a highly fecund species, early successional stages should prove likely spots for colonization.

An *impoverished* community is one which, by definition, has unused niches and thus can admit more species with a minimum of niche shuffling due to competition. Impoverished areas are therefore susceptible to colonization. Impoverished areas may include islands, peninsulas and, in general, geologically very recent areas (see part 16.3). Many authors have commented on the low species diversity of almost any plant or animal taxon on islands (there are some exceptions) and on the tendency for the species present to expand their niches into those parts of the niche space (Discussion, Section II) used by competitors on the mainland. Van Valen (1965) suggests that there must be a selective advantage in increasing variation (niche breadth) in island species; the equilibrium point between intra and interspecific competition clearly seems, as should be expected, to have shifted.

The susceptibility of at least some islands to invasion is dramatically illustrated in the Hawaiian Islands which were impoverished not only with respect to birds but to bird parasites. Man introduced a number of avian species to these islands, and with them bird pox, bird malaria, and other diseases. The result has been massive extinction of endemic birds and replacement by new species. Mosquitoes, the vector of the two major bird diseases, can't survive above an altitude of about 600 m, and there only is the native avifauna still largely unchanged (Mayr, 1965; Warner, 1968).

Whether an area is impoverished cannot be determined until after it has been successfully colonized by a new species, without the subsequent elimination of some previously established species. The freshwater shrimp *Eucrangonyx gracilis* and the brackish water amphipod *Gammarus fasciatus*, both from North America, have spread, respectively, over parts of England and Ireland, without any detectable disturbance—so far—of the biological community. The same situation holds for the Caspian and Black Sea amphipods *Corophium curvispinum* and *Orchestia bottae* in England (Elton, 1958). With respect to amphipods, Britain, an island, was at least in the recent past impoverished.

## 14.2.2 Characteristics of Successful Colonizing Species

Since colonization is unlikely to occur in climax communities, we concentrate our attention on characteristics useful to survival in early successional stages and islands.

First we should expect species with high $r_0$ values (part 9.1) to be good colonizers. The reasons for this have already been given. It is difficult to state with any certainty, but it seems that the bulk of successfully introduced species have had high intrinsic rates of increase. Of all phytophagous insects imported to the United States, about 25% have become established (DeBach, 1965). The figure for weedy, annual plants may be even higher. But outside of gardens and protected areas, larger species with lower reproductive rates seem to have survived introduction much less consistently.

Another characteristic of a good colonizer is that it is able to shift its realized niche (Discussion II) if necessary, to avoid competition or make use of new opportunities. That is, the species should be a generalist. A specialist such as the fig, *Fiscus*, which is absolutely dependent for pollination on a particular agaonid wasp (Ramirez, 1969, 1970), will be unlikely to survive in a newly invaded area unless the wasp is present.

High $r_0$ and generalized habits are characteristic of fugitive species (Hutchinson, 1951; part 11.2.2) which are also highly mobile and thus likely to reach new areas often. As Wilson (1965) points out, the bulk of colonizers to new regions are fugitive species.

One more characteristic may be useful to a prospective colonizer. If individuals of a species travel in flocks there is a greater chance that enough will arrive in a new area together to make successful mating possible. Thus Mayr (1965) points out that such flock birds as the Tasmanian white eye, *Zosterops lateralis*; starlings; sparrows, and some thrushes, are very successful colonizers of isolated islands.

### 14.2.3 Human Disturbance

In his travels, man has introduced species to many new areas. Of his intentional introductions, many have been successful and, in some cases, have proved beneficial. The insect predators brought into the United States to control accidentally introduced insect pests of farm and orchard crops have, for the most part, done their work with rousing success. Few find fault with the introduction of pheasant to our shores. The introduction of a few nutria, *Myocastor coypus*, to the U.S. from South America looked good for the fur industry, and although populations of this animal began to multiply rapidly and might have become a serious problem, chemical controls were developed in time. The introduction of rabbits to Australia was only temporarily disastrous. Rabbits were brought under control in the 1950's with the use of *Myxomatosis* virus (see part 11.4).

Many successfully introduced species seem to have had little effect other than to further disrupt the few islands of natural wilderness left and to spoil large areas which might otherwise have proved good laboratories for the evolutionary ecologist. (These introductions have also served as natural experiments for the evolutionary ecologist). Species in this category include the monarch butterfly (into Australia), the cabbage butterfly, *Pieris rapae*, (into North America), and Canadian pond weed, *Potamogeton*, (into Europe). The starling, introduced into the United States in 1891 when about 80 birds were released in Central Park in New York City, is another species of this sort. It took the starling 25 years to cross the Allegheny Mountains, but once across it spread rapidly west. By 1922 starlings had appeared in Ohio, by 1926 in Indiana, and by 1954 over virtually the entire country. The first clear record of a starling in Alaska was in 1953 (Elton, 1958). The American muskrat, *Ondatra zibethica*, was introduced into Czechoslovakia in the form of five individuals kept by a landowner. Aided by further introductions, the escape of these muskrats has resulted in the spread of the species over all of Europe and northeastern Russia (Elton, 1958).

A number of introductions have proven extremely unfortunate. In bringing plant material from Europe, man has accidentally released a number of plant diseases in the United States. In Europe and Asia these diseases had coevolved with their hosts and formed a stable relationship. American plants similar enough to the Old World species to be attacked by these diseases, but without the resistance, include the chestnut, *Castanea dentata*, (chestnut blight) and elm, *Ulmus*, (Dutch elm disease).

The story of the Great Lakes lamprey, *Petromyzon marinus*, is one of the more dismal chapters in the story of ecological problems precipitated by man (see Fig. 14.1). Lampreys had long existed in the St. Lawrence River and Lake Ontario, but not in the other Great Lakes because of the barrier provided by Niagara Falls below Lake Erie. In 1830 a canal was opened for shipping between Lakes Erie and Ontario. The lamprey, preferring cool water, did not rapidly colonize the warm, shallow waters of Lake Erie, but its slow spread finally landed a critical number of individuals in deep, cold Lake Huron in 1939 and there a population explosion ensued. By 1953 Lake Huron was ruined for commercial fishing and the lampreys were increasing in Lakes Michigan and Superior. Control methods which involve treating tributaries individually with select poisons affecting only the ammocoete larvae of the lamprey have proved only partially successful.

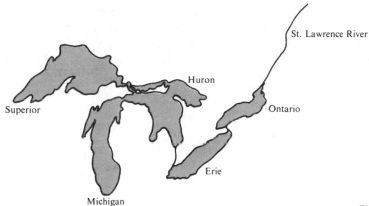

St. Lawrence River

Huron

Superior

Ontario

Erie

Michigan

**Figure 14.1**

Plant introductions have been amazingly successful. In much of the area once covered by long grass prairie, native species are in the minority. The long grass prairie with its miles of six to eight foot grass, that enormous forest of grass devoid of landmarks, proved an almost insurmountable barrier to early white explorers. It is gone now, its dominants scattered in small patches and clustered along old fencerows between farms and housing developments. The open prairie areas are now largely composed of such introduced species as corn, *Zea*; wheat, *Triticum*; butter-n-eggs, *Linaria vulgaris*; chicory, *Cichorium intybus*; poison hemlock, *Conium maculatum*; Queen Anne's lace, *Daucus carota*; clover, *Trifolium*; field daisies, *Chrysanthemum leucanthemum*; dandelion, *Taraxacum*; and in overgrazed areas, mullein, *Verbascum thapsus*. The presence of house sparrows, *Passer domesticus*; pigeons, *Columba livia*; house mice, *Mus musculus*; and rats, *Rattus*, seem only somewhat less overwhelming, and our bluebird, *Sialia sialis*, has been largely exterminated by starlings in some areas.

In view of the previous comments concerning the difficulty of colonizing climax areas, it seems legitimate to ask why so many new species have been so remarkably successful. Part of the answer lies in the coevolution of European and Asian species with man. Man was farming there for a long time, on a much larger scale than in America. Urbanization, also, had been proceeding for hundreds of years. Such species as the house sparrow and the pigeon became adapted to the kind of habitat provided by man. These species, then, were not invading a wild America, they were invading a human environment, something to which American species had been exposed for only a few generations. The strangers were the American, not the European and Asian species. It is not surprising that most introduced species in America are found in urban or suburban areas and around farms, and that their incidence drops precipitously in protected, wilder areas.

Of course, the human disturbance explanation is not the whole story, for there have been a fair number of successful colonizations of the Old World by American species. The American gray squirrel, *Sciurus carolinensis*, has been spectacularly successful over the midlands and much of the south of England where it has replaced the English gray squirrel, *S. vulgaris*. One reason for such great success of human introductions is that attempts to establish new species recur frequently and involve many individuals.

## 14.3 THE DISTRIBUTION OF PHYLA

Before man arrived, providing frequent opportunities for invasion of new territory by large numbers of species, it is likely that long-distance dispersal was a very slow process, and for large mammals and reptiles it was probably impossible to reach new islands or continents without the aid of periodically appearing land bridges. Under the above assumptions it is possible, from geological history and the fossil record, to establish the routes followed by taxa through their evolutionary history and to surmise their likely points of evolutionary origin.

Two things seem clear. First, at any time in the past, there were several large centers of adaptive radiation. Second, the connections between continental areas changed over time. The discussion below is taken largely from Elton (1958) and Kurten (1969).

During much of the Mesozoic, the world's land mass seems to have been divided into two supercontinents, Laurasia in the northern hemisphere and Gondwanaland in the southern hemisphere (Fig. 14.2). By the Cretaceous or perhaps the late Triassic these land areas had split into what were to become our present continents, but were differently related from today (Fig. 14.3). When mammals and birds first appeared, land connections within the two supercontinents were still well developed, but soon afterwards a number of splits developed and the parts of the continents began drifting away from each other. Europe and Asia (including the Far East) were cut in half and effectively isolated from one another by an arm of the Tethys Sea reaching to the Arctic Ocean in the north (Fig. 14.3). Europe and North America, on the other hand, remained quite close for a long time and a land bridge between North America and Asia existed, on and off, throughout the Cenozoic. To move from Europe to Asia, animals thus had to pass through North America. For this reason, North America in particular acted as a center of adaptive radiation for groups arising in the northern hemisphere. The insectivores, bats, primates, carnivores, artiodactyls, perissodactyls, rodents, lagomorphs, and pangolins are all of Laurasian origin. In Gondwanaland, the South American portion became divided into two parts by what is now the Amazon Basin. Edentates and oppossum rats as well as a number of extinct mammalian groups arose here. Africa became well separated from South America early in the Cretaceous and in the early Tertiary was divided into two or three islands. These islands were the areas which saw the first elephants, conies, aardvarks and, probably, sirenia.

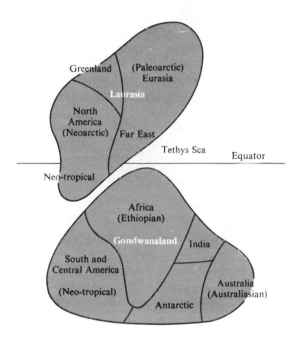

**Fig. 14.2.** Distribution of the world's land mass during the Mesozoic. (From "Continental Drift and Evolution," Bjorn Kurten. Copyright © 1969 by *Scientific American, Inc.* All rights reserved.

The early appearance in Africa of Laurasian species and the invasion of North America by elephants implies a land connection between these areas as early, perhaps, as the Eocene.

Little is known of the far eastern, Indian, and Australian centers although marsupials and monotremes presumably arose in Australia, an area isolated from the other continental areas for a very long time.

During the course of continental drift between the Cretaceous and the present, periodic land links developed. For example, after the middle Pliocene, there was a connection between North and South America. At this time tapirs, llamas, peccaries, foxes, deer, dogs, cats, otters, bears, raccoons, skunks, mastodons (originally from Africa), horses, antelopes, and voles crossed from the north to

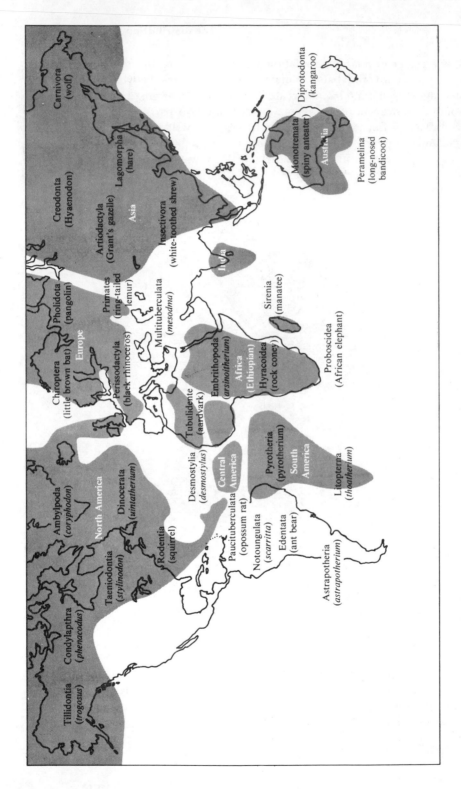

**Fig. 14.3.** Distribution of the world's land mass during the Cretaceous. (From "Continental Drift and Evolution," Bjorn Kurten. Copyright © 1969 by *Scientific American, Inc.* All rights reserved.)

the south. Porcupines spread north. Deer and elephants (one from North America and the other from Africa) re-invaded North America from Asia over the Bering Strait at various times in the Cenozoic. The Orient, isolated from Europe first by the Tethys Sea and then from Europe and the rest of Asia by the rise of high mountains, experienced little exchange with the rest of the world and as a result is, today, faunistically quite distinct. The separation of North Africa from Europe was bridged, and the early development of the Sahara Desert barrier resulted in a North African mammal fauna Eurasian rather than Ethiopian in character.

| Era | Period | Epoch | Approximate time from start of period to the present (millions of years) |
|---|---|---|---|
| Cenozoic | Quaternary | Recent | 0.01 |
| | | Pleistocene | 2 |
| | Tertiary | Pleiocene | 12 |
| | | Miocene | 25 |
| | | Oligocene | 31 |
| | | Eocene | 58 |
| | | Paleocene | 63 |
| Mesozoic | Cretaceous | | 135 |
| | Jurassic | | 181 |
| | Triassic | | 230 |
| Paleozoic | | | |

The Cenozoic history of mammals is one of invasions, extinctions, and re-invasions. The horse which originated in Laurasia, probably in North America, later became extinct in America, only to reappear with the Spaniards in the 1500's. The history of specific groups is often complex and cannot be covered in this book. At least one general conclusion seems clear, though. The early proximity of Europe and North America, and northern Asia and North America, and the later connection of Europe and Asia, has resulted in a great similarity in the mammalian species found in these areas. South America, Africa, the Indian subcontinent and South China and Malaysia, and Australia are characterized by quite different groups.

The history of bird evolution and dispersal follows a route similar to that of the mammals since it occurred over essentially the same period of time. Characteristic bird orders are found across North America, Europe, and northern Asia, and quite different orders occur in other parts of the world. Thus it makes sense to speak, in 14.1.2, of an "Indo-Malayan" avifauna in Hawaii.

Adaptive radiation occurs when new habitats are colonized as well as when barriers arise which subdivide populations.  Thus adaptive radiation spreads outwards from a center.  Since greater and greater distances from the center mean more and more different climates and ecological conditions, evolutionary change should be greatest in the youngest, most far-flung species.  Thus centers of adaptive radiation are characterized by ancestral (*primitive*) species.  Common ancestors, or creatures most nearly resembling them, should be indicators of geographic evolutionary centers.

An interesting sidelight on the dispersal, colonization, and history of species is often found in highly mobile groups such as birds.  One often finds a situation in which a species has dispersed progressively from one ancestral area, all the way around the world (or an ocean, large lake, etc.) until it meets its progenitors from the other direction.  Throughout the dispersal route the species has been gradually changing genetically and ecologically so that while at no point does it differ much from those populations on either side of it, by the time it meets itself from behind we see two quite distinct species.  The herring gull, *Larus argentatus*, occurs from England westward around the world, and gradually fades into a form known in Europe as the lesser black-backed gull, *L. fuscus*, which abuts the *L. argentatus* populations in England from the east.  In England, where the two exist together, interbreeding can take place, but this happens rarely.  Brown (1967) reports seeing one hybrid in four years of observation, and never a mixed pair.  Nevertheless the two "species" breed side by side in the same colonies.

## 14.4 THE DISTRIBUTION OF ECOSPECIES

When one speaks of different species or taxa, it is usually with reference to phyletic relationships, or at least what we believe to be phyletic relationships.  Suppose, however, that we wished to compare two areas which were faunistically distinct in terms of their phylogenetic histories.  We might find that the absolute and relative abundances of frass feeders—or any or all other classifications—ordered by rank in commonness, was essentially the same in the two areas.  We might find that these abundances could be accounted for by the same variables—say measures of temperature, precipitation, foliage density, etc.—in the same way.  Would we be justified in calling these two areas different on the basis of the different phylogenetic affinities of their inhabitants?  The answer, of course, depends on one's point of emphasis.  To a taxonomist the areas are different.  To an ecologist they are not.  The ecologist should concern himself with species in the *biological species* sense (Mayr, 1963) when he is dealing with sympatric populations, for he must know who is breeding with whom.  But in the comparison of separated areas, he needs to deal with *ecospecies*, that is, species which are *ecological equivalents*.  The term ecospecies is not to be confused with the term ecotype, which denotes geographically distinct populations of the same species, each adapted to its own specific habitat.  The smaller antelope of Africa and the pronghorn, *Antilocapra americana,* of North America exploit their environment in similar ways and may be considered ecologically equivalent. The African buffalo, *Syncerus caffer*, and the American bison, *Bison bison*, occupy much the same niche space (when food species are classified as ecospecies) and are thus ecospecies. The flying squirrel, *Glaucomys volans*, and the gliders *Acrobates*, *Petaurus*, and *Schoinobates* of Australia are ecospecies as are wolves, *Canis lupus*, of the northern hemisphere and hunting dogs, *Lycaeon picta*, of Africa. Ecospecies may bear only a remote phylogenetic relation to each other; the flying squirrel is a rodent, the gliders marsupials. The difficulty with the ecospecies concept, obviously, is that "equivalence" is hard to define and even harder to measure. It is best, at least at this stage of knowledge, to use *ecospecies* loosely, without any rigid definition, as meaning species which exploit their respective communities in a similar (defined as one wishes) manner.

Ecospecies arise from convergent or parallel evolution, and it is remarkable how similar such species can be in their habits and appearance. The African longclaw, *Macronyx croceus*, a pipit, is ecologically very similar to the American meadowlark, *Sturnella*. Both inhabit grasslands and forage by prodding the ground with their bills. Both are starling size and shape (the longclaw is slightly smaller), and both possess a yellow throat and belly, brown back, white outer tail feathers, a white line over the eye, and a black "V" on the chest. The red-shouldered wydah, *Coliuspasser axillaris*, an African ploceid (weaver finch), is a marsh breeder like the American redwing, *Agelaius phoeniceus*, and like that unrelated icterid, displays in the male a glossy black body with a red shoulder patch. In spite of differences in internal anatomy and different phylogenetic histories, these birds are clearly ecological equivalents.

Certain types of habitat seem to favor certain behavioral and physiological adaptations regardless of the history of the local fauna and flora, although convergence is not commonly as marked as in the examples just cited. Cold climates favor larger, stockier bodies (Bergmann's and Allen's rules) in poikilotherms as well as in homeotherms—even poikilotherms regulate temperature to some extent. Lindsey (1966) presents data showing a steady increase in the size of fish from tropical to temperate to cold waters. On deserts there seems to be an evolutionary convergence in the gait of small mammals. The *Dipodidae* of the great palearctic desert, the *Zapopidae* and *Heteromyidae* of the American deserts, the *Muridae* and *Dasyuridae* of Australian deserts, and the *Pedetidae* of southern African deserts all possess a jerboa-like form, with reduced front legs, enlarged hind legs and a jumping habit (Buxton, 1923). Forests, which for nonflying species offer poor visibility, encourage the development of acute hearing or sense of smell, while in open country vision is more often well developed. With vision of less importance, forest species are more often nocturnal than prairie species. Birds, which can rise above the forest and use their eyes, are nearly all diurnal except for those predators which make use of the food supply offered by nocturnal prey. Even on the prairies, small mammals to whom grass is a forest of sorts, are more nearly nocturnal than large mammals which can see over the grass. Because postures and movements are easily seen and localized, and because they allow for more detailed communication than sound or odor production, most highly social species are also visually oriented species. Forest mammals, which are usually nocturnal, are also usually solitary. Diurnal prairie mammals usually move in social groups. Nearly all simians are diurnal, with well-developed social structures within their groups. A few are nocturnal—and solitary. Most prosimians are both nocturnal and solitary.

It would be impractical for an ecologist to try to predict all species, their abundances and interactions, in any area, given only a few basic data. But it is perhaps not so impractical to hope that we may eventually be able to do this for ecospecies. First, of course, we need some kind of a classification of ecospecies and habitat types. There have been attempts to classify habitats but the value of these attempts cannot be assessed until ecospecies classifications are proposed. Also, unfortunately, most categorizing of habitat types involves "indicator species." Prediction of species present in such an area becomes, at least partly, circular. It would be interesting to know how the axes of the niche spaces of two areas might be placed in one to one correspondence, and whether the niches of ecospecies then occurred in the same characteristic portions of the niche spaces (were isomorphic).

## 14.5 LIFE ZONES

Because of convergent evolution, areas which possess similar climatic and physical aspects possess similar dominant and nondominant species. Thus we can subdivide the world into a number of *life zones* or *biomes*, each with broadly similar ecological communities, in terms of ecospecies. That portion of a land biome bounded by water is a *biotic province*. Thus the so-called boreal forest of Canada constitutes a biotic province, and the boreal forest of the Soviet Union constitutes another. Together they form a biome.

The general form of an ecological community—that is, to what biome it belongs—can be predicted from the shape of its *climatograph*. A climatograph is merely a month-by-month plot of precipitation against temperature and takes the forms illustrated in Fig. 14.4 (from Smith, 1940).

The boreal forest biome of Canada and Alaska (and the Soviet Union) extends into the United States in the extreme northeast, in a modified form in the northern midwest, and extends south along the Rocky Mountains, Olympics, and Cascades. The dominant species are spruce, *Picea*, and fir, *Abies*, with paper birch, *Betula papyrifera*, and *Vaccinium sp.* associated. The soil is acidic and the lakes are cold and acid. Sphagnum-tamarack (*Larix laricina*) bogs are common. Mammals include the moose, *Alces americana*; the lynx, *Lynx canadensis*; the pine marten, *Martes americana*; red squirrels, *Tamiasciurus hudsonicus*; and forest voles, *Clethrionomys gapperi*. This biome extends unbroken —except for water—all the way around the world from about 50°N latitude to the northern tree line. It is usually referred to by its Russian name, *taiga*.

The deciduous forest biome occurs over most of the eastern United States and much of Europe and southern USSR. It is characterized in America by maple, *Acer sp.*; beech, *Fagus grandifolia*; oak, *Quercus sp.*, and other trees. Undergrowth is highly variable. Major mammals include the white-tailed deer, *Odocoileus virginianus*, the gray squirrel, *Sciurus carolinensis*, the black bear, *Euarctes americanus*, and the opossum, *Didelphus marsupialis*.

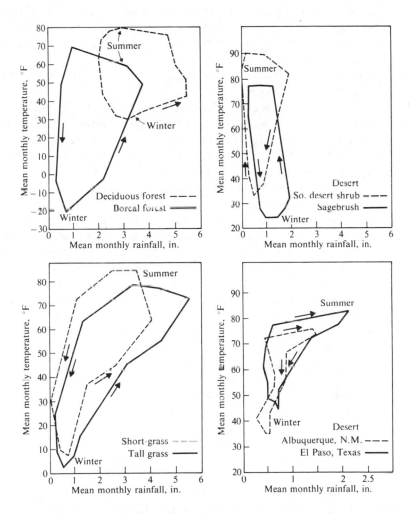

**Fig. 14.4.** Climatographs: plotting precipitation versus temperature on a month-by-month basis

The prairies of the United States and Canada are highly variable and grasslands are less often thought of as a worldwide biome. The long grass prairie is (was) characterized in the United States by *Andropogon scoparius, Agropyron repens*, and *Bouteloua sp.* in the highlands, and *Poa pratensis*, *Stipa spartea*, and others on slopes. In the lowlands, *Spartina pectinata* and *Panicum virgatum* are dominant. The short grass prairie, warmer, drier, and to the southwest is characterized by *Bouteloua gracilis*, *Buchloe dactyloides*, and other grass species. Important prairie mammals are (were) the bison, the pronghorn, and the coyote, *Canis latrans*. Deserts vary from short grass prairie region with open ground to sandy areas with very little plant life. Desert scrub such as creosote bush, *Larea sp.*, sagebrush, *Artemisia tridentata*, and various cactuses represent the dominant flora of these regions in the United States. Jack rabbits, *Lepus*, and various jumping mice, especially the kangaroo rats, *Heteromyidae*, are common. In some places, the coyote is also common.

The species mentioned above are, of course, characteristic of the climax community of the particular biome.

As air becomes less dense it cools. Therefore, an increase in altitude has the same effect on temperature as a change in latitude. The equivalence relationship is 360 feet of altitude for every degree of latitude. Thus as we rise from the plains into the Colorado Rockies we encounter, after a zone of variable vegetation, a coniferous forest very reminiscent of the taiga. This is one of the southerly arms of the boreal forest mentioned earlier. Eventually we reach tree limit and find a tundra similar to that north of the taiga. One cannot equate altitude and latitude quite so easily because, while heat by convection declines at high altitudes, solar radiation increases. At high latitudes it decreases. Soils may be better drained in mountainous areas, and wind conditions will be different. It is difficult to categorize the changes occurring with altitude, let alone compare them with latitudinal changes. North-facing slopes (in the northern hemisphere) receive little sun; between 5500ft and 8200ft in the Colorado front range the slopes are cool and moist much of the year and snow covered most of the winter. Douglas fir, *Pseudotsuga menzesii*, is the dominant species here. On south-facing slopes, temperatures are high, the air and soil are dry, and snow sublimes rapidly in winter. These slopes are characterized by scattered pine, *Pinus ponderosa*, grasses, composites, *Opuntia*, and *Verbascum*. Communities at these altitudes constitute the *montane* zone (Moir, 1969). The higher, *subalpine* forest representing the southern extension of the boreal forest is less variable with slope but is not quite equivalent to taiga. Whereas the boreal dominants to the north and east, *Picea mariana*, *P. alba*, and *Abies balsamea* are presumably available to colonize, the subalpine is dominated by the ecologically not quite equivalent *Picea engelmanni* and *Abies lasiocarpa* (Moir, 1969).

The "equivalence" between altitude and latitude also fails in another way. We would expect to find progressively lower timberlines to the north. This is indeed the case above about 30°N latitude; not so to the south. Timberline is at 6000ft in the northern Canadian Rockies, 11,000 to 14,000ft in the Central Rockies, but 10,500 to 13,000ft in the Sierra Nevada, and still lower in the tropics. The relation between altitude and latitude is in many ways illusory and should be handled with care.

## 14.6 LOCAL DYNAMICS IN DISTRIBUTION

Shelford (1945) observed that following lemming population crashes in Canada, snowy owls, *Nyctea scandiaca*, appeared in the United States in great numbers. He reasoned that the owls had extended their range because their populations had increased in response to high lemming densities and when that species disappeared had to emigrate to find food. Besides being eminently reasonable, this explanation hints at the possible existence of a very interesting phenomenon. Migration of a predator outward from some population center, progressively depressing prey populations along its path, with subsequent drop in predator numbers, might well trigger a wave of prey–predator cycles. During the increase phase in the predator population, densities will be highest toward the epicenter of the wave so that predator migration is always outward. In the decline phase migration will be inward so that, in areas of declining prey, predator populations will enhance the amplitude of the cycles. If the prey also migrate, the cycles in any one location will be buffered, and perhaps all but disappear. We assume for the moment, however, that since predators are, on the whole, considerably more mobile than their prey, such waves may in fact occur. Watt (1968) has discussed this phenomenon and calls it an *epizootic wave*.

If mobility of the predator is extremely high, one individual may traverse great distances in response to changing prey densities and the wave disappears— we see a single prey–predator cycle over a unified area. If the predators move very slowly, the wave pattern would theoretically resemble ripples spreading on a water surface. The pattern of waves would be extremely regular in a temporally and spatially homogeneous environment, only how they would get started in such a case is not clear. In reality, such cycles have been suggested to occur in the Canadian boreal forest, a spatially quite homogeneous but temporally quite unpredictable area, and the rate of migration of predators seems to be rapid enough that the waves, if they really exist, spread over the entire area in considerably less time than one cycle. Continued encouragement of cycling by back migration of predators during their decline seems virtually nonexistent. It is intriguing to think that such waves are self-perpetuating (although perhaps damped), but it is not likely that they are. Let us look at the evidence.

There are several apparent cases of continuing epizootic waves. Keith (1963) reports that hares, and maybe grouse, declined first in Nova Scotia and last in the Yukon and Alaska in the 1930's and 1940's. Recovery occurred first in the northern sections of the central provinces (Butler, 1953; Keith, 1963) and in the maritime provinces (Keith, 1963). Northern Manitoba and Alberta (two small centers which coalesce) seem to serve as epicenters for hares (Butler, 1953). In the nineteenth century, lynx peaks occurred first in the Athabasca Basin and spread outward. Since the 1920's the lynx center had been consistently in northern Manitoba or Saskatchewan (Keith, 1963). In 1943 there was a spruce budworm outbreak over an area of about 1000 square miles in northern Ontario. It spread to 15,000 square miles by 1948. The pattern was one of a wave; the percentage of trees defoliated at the epicenter declined from 1943 to 1950, and 25 miles away it rose from zero in 1943 to a peak in about 1947 and then declined. Fifty miles away, the peak was reached around 1949 or 1950 (Belyea, 1952). Another apparent case of an epizootic wave involves the Himalayan thar, *Hemitragus jemlahicus*, in New Zealand (Caughley, 1970).

It is entirely possible, of course, that the observed waves involve nothing more than the migration of prey and predator outward from the area of a population outbreak, and that these movements do not act to promote waves of prey–predator cycles at all. Where mass migration of prey species occurs, this seems likely, and there is some evidence of mass migrations by prey. Cox (1936) reports seeing tremendous numbers of snowshoe hares, *Lepus americana*, moving at night from high density tick-infested areas in Minnesota. Predators had converged to take advantage of the migrating animals and Cox comments on the surface of a frozen lake littered with carcasses. Keith (1963) comments that mass migrations are associated with high population densities in many—not necessarily predator—species, and that they probably occur in hares and have been observed in sharp-tailed grouse, *Pedioecetes phasianellus* and willow ptarmigan, *Lagopus lagopus*.

Whether waves of prey–predator cycles occur or whether epizootic waves merely represent wavelike migrations of prey and their predators is open to question. One thing is clear: The patterns of such waves are complex. They may be triggered at odd intervals, perhaps by unusual weather, and the interrelation of such a random driving "force" and the normal prey–predator interaction is not obvious. There may be several epicenters, so that what is observed is a mish-mash of overlapping waves. Epizootic waves may obscure prey–predator cycles. On the other hand, it may be reasonable to suggest that the 10-year cycles (and/or perhaps the 3 to 4 year cycles) we observe are not prey–predator cycles at all but the compounding of numerous epizootic waves of a purely migratory nature, each triggered in characteristic areas by unusual weather, perhaps at random intervals, and each traveling outward with some characteristic speed. If this is so, the existence of 10-year cycles may be a fortuitous consequence of the size and shape of the Canadian boreal forest. This would explain the apparent lack of such cycles in Europe (and perhaps the Soviet Union—see part 11.1).

Note in conclusion that not everyone accepts the notion of epizootic waves. Murray (1965) points to the outbreak of *Microtus montanus* in 1957 and 1958 which occurred simultaneously over several areas in California, central Oregon, and parts of Nevada. No wave was observed. Of course, waves may occur in some areas, and involve some species but not others. Butler (1953) feels that the lynx does not follow the hare or show progressive, wavelike increases; but that several nonspreading and apparently isolated centers react to the geographically expanding hare population. This is not an argument against epizootic waves, but only the nature of their propagation. Here, hare populations spread outward and promote a discrete rather than continuous wave of response by the lynx populations.

# 15
# Species Diversity I

It has been suggested that one goal of an ecologist is to go into an area, measure a minimal number of parameters, and on the basis of those measurements be able to predict with reasonable accuracy how many kinds of organism are present, how many of each, and how they interact. We have discussed the last two points in Section III; in this and the next chapter we examine the question of how many kinds.

There are a large number of problems facing the ecologist interested in species diversity. Some will be discussed specifically, others will become clear as the following discussion proceeds. We begin with one. Because all species are not equally common and evenly distributed in space, it is found that, as a sample area is increased in size and as the number of individuals in a sample increases, the number of different species encountered grows. Diversity varies with sample size and a good measure of diversity must compensate for this. To speak of number of species present is all right in theory, but in practice is valid only after exhaustive resampling fails to turn up significant numbers of new species.

## 15.1 THE DIVERSITY PROBLEM

### 15.1.1 Meanings and Measures of Diversity

Let the probability with which species $i$ is encountered in a complete sampling of a unit area be $\bar{q}_i$. Then the expected probability with which it is found in a complete sample of an area, $A$, assuming it is randomly distributed, is given by

$$1 - (1 - \bar{q}_i)^A = 1 - e^{A \ln(1 - \bar{q}_i)},$$

which, for $\bar{q}_i$ small (unit areas very small), approaches

$$1 - e^{-\bar{q}_i A}. \tag{15.1}$$

The expected number of species, $s^*$, found on the area, $A$, is thus given by

$$s^* = \sum_{i=1}^{s} (1 - e^{-\bar{q}_i A}), \tag{15.2}$$

where $s$ is the total number of species which at one time or another enter the area.

This expression can be expanded to give

$$s^* = \sum_{i=1}^{s} \{1 - [1 + (-\bar{q}_i A) + \tfrac{1}{2}(-\bar{q}_i A)^2 + \tfrac{1}{6}(-\bar{q}_i A)^3 + \cdots]\}$$

$$= \sum_{i=1}^{s} [(\bar{q}_i A) - (\tfrac{1}{2}\bar{q}_i^2 A^2) + (\tfrac{1}{6}\bar{q}_i^3 A^3) - \cdots].$$

Now if $\bar{q}_i A < 2$, this series converges—that is, each term is smaller in value than the preceding term—so that to a rough approximation over small ranges of $A$, $s^*$, it can be written as increasing with something less than the sum of the first power of $\bar{q}_i A \ldots$ . We thus write

$$s^* \approx cA^x \tag{15.3}$$

where $c$ is some constant and $x < 1$. This is the so-called "species-area" curve, which has been found to describe the increase in $s^*$ with sample size reasonably well over small ranges of $A$.

The simplest measure of species diversity is $s^*$, for a given, prescribed area. But this measure gives equal weight to very rare and very common species. It might be useful to have a measure which is greater when all species are equally common than when one species is common and the others all rare. Suppose, as our measure, we let diversity be proportional to the instantaneous rate of change in $s^*$ with the number of individuals, over some prescribed area size $A$. Then

$$\text{Diversity } (Dv) \propto \frac{ds^*}{dn} = \frac{ds^*/dA}{dn/dA} = \frac{\dfrac{d}{dA}\left(\sum\limits_{i=1}^{s} 1 - e^{-\bar{q}_i A}\right)}{\rho}$$

$$= \frac{1}{\rho} \sum_{i=1}^{s} \bar{q}_i e^{-\bar{q}_i A}.$$

But $\bar{q}_i$, to fit this formulation, is the probability of finding species $i$ over infinitesimally small unit areas, which is equivalent to the average number of species $i$ individuals per infinitesimal unit area, $n_i^* \to 0$. Thus

$$Dv \propto \frac{1}{\rho} \sum_{i=1}^{s} n_i^* e^{-n_i^* A}.$$

We now let our prescribed area size be $1/\Sigma n_i^*$ so that

$$Dv \propto \frac{1}{\rho A} \sum_{i=1}^{s} \left[\frac{n_i^*}{\Sigma n_i^*} \exp\left(-\frac{n_i^*}{\Sigma n_i^*}\right)\right] = \frac{1}{\rho A} \sum_{i=1}^{s} p_i e^{-p_i},$$

where $p_i$ is the relative abundance of species $i$ in the area. But $A = 1/\Sigma n_i^* \propto 1/\rho$, where $\rho$ is the density of all individuals over the area. Thus we define

$$Dv = \sum_{i=1}^{s} p_i e^{-p_i}. \tag{15.4}$$

This measure is practical to use, it is affected minimally by rare species ($p_i$ very small), and thus does not require exhaustive sampling. It also gives greater values when species are of equal commonnesss than when great differences in abundance exist.

The usually used measure of diversity, $H'$, has roughly the same qualities as $Dv$. It is the Shannon and Weaver information theoretic expression (Shannon and Weaver, 1963)

$$H' = -\sum_{i=1}^{s} p_i \log p_i. \qquad (15.5)$$

$H'$ measures the uncertainty with which we can predict the species of the next individual encountered; $Dv$ measures the increase in the number of species per individual. Table 15.1 gives values of $p_i$ for a two-species system ($s = 2$) and compares the corresponding values $H'$ and $Dv$.

**Table 15.1.** Relative species' abundances and diversity index values

| $p_1$ | $p_2$ | $Dv$ | $H'$ |
|-------|-------|------|------|
| 0.0 | 1.0 | 0.37 | 0.00 |
| 0.1 | 0.9 | 0.46 | 0.46 |
| 0.2 | 0.8 | 0.52 | 0.61 |
| 0.3 | 0.7 | 0.57 | 0.88 |
| 0.4 | 0.6 | 0.60 | 0.97 |
| 0.5 | 0.5 | 0.67 | 1.00 |

Sampling techniques, discussion of sampling problems, and alternative forms of Eq. (15.5) can be found in Pielou's (1969) book and earlier papers of Pielou cited therein.

Another diversity index commonly used is $1/\Sigma p_i^2$. When the number of individuals per species plotted against number of species fits a certain curve, still other appropriate diversity measures may be used. For a discussion of some of these other measures see Fisher *et al.*, 1943; Preston, 1948.

### 15.1.2 Practical Problems

$Dv$ and $H'$ have the advantage that all species are not scored equally in the calculation of diversity. But most communities over which we may measure diversity will undergo changes as their constituent species increase and decrease seasonally and as prey–predator cycling and random fluctuations in populations occur. The value of $s^*$, though, may remain constant, suggesting this as the best measure of diversity. Diversity changes seasonally as species migrate, hibernate, or aestivate and thus absent themselves or become nonfunctional, noninteracting citizens. Measures of diversity must therefore be specified as applying to a given portion of a year or the entire year.

Most environments are obviously patchy to the human eye, and probably all vary from place to place in subtle ways. If we could define the patch types and their boundaries we might measure diversity within each patch type. But patches are invariably overlapping or nondiscrete, so that the environment is a collage of varying habitats. Furthermore, there is dispersal between different areas in an environment, considerable movement by some species, little or none by others. Are we to include "accidentals" which appear in an area by accident of wandering in our diversity calculations? And how do we determine whether they are of the patch type under consideration? Theoretically, of course, these difficulties are surmountable. It is the practical aspects of the problem which are frustrating. About the best that can be done with actual measurements is to pick an area, regardless of its micro-environmental features and count species and individuals present. That is, we simply ignore what we cannot overcome.

In calculating a diversity index of the $Dv$ or $H'$ variety, we might use something other than relative abundance for $p_i$. If diversity is to reflect interactions of species, we might use, in place of the above meaning of $p_i$, the relative amount of energy passing through the $i$th species, or the relative biomass, or within one trophic level (where that can be defined) some measure of the competition coefficients, $\alpha$. For example we might write

$$p_i = \left( \sum_j \alpha_{ji} n_i \right) \Big/ \left( \sum_{i,j} \alpha_{ji} n_i \right). \tag{15.6}$$

This plethora of problems and alternative measures has led Hurlbert (1971) to speak of species diversity as a "nonconcept." This seems a rather strong view, however.

Finally, what do we mean by a species? Ecologically speaking we refer to a coherent group of interbreeding individuals. But ecological properties vary between individuals and age classes within species, and spatial distribution patterns and nonrandom mating systems tend to subdivide a population into semi-discrete, not necessarily interacting groups. Should we base $p_i$ not on population size but rather on effective population size (see Discussion, Section I)? Where competition occurs not only between different species and between individuals within a group, but also between groups of the same species, should we count the species only once in our calculation of diversity?

In this chapter and the next, we use $p_i$ as the proportion of the total number of individuals which belong to species $i$ and measure diversity over some habitat type, defined however we like, at some defined season. "Species" is taken to be phenetic species (not as used in Chapter 4). In practice, phenetic species are also biological species (Mayr, 1963), correspond to groupings based on DNA hybridization, and of necessity are cladistic species. These simplifications do not get us around the above problems, but are in keeping with general practice and offer a base from which to discuss the factors affecting diversity.

## 15.2 NEW SPECIES

The species diversity of a given area at any time depends on the rate at which new species arrive or arrived in the past, and the rate at which extinction occurs or occurred in the past. Part 15.2 deals with the arrival of new species, part 15.3 discusses the maintenance or local extinction of existing species.

When a large area undergoes a major disturbance such as glaciation, the local flora and fauna is either changed markedly or simply wiped out. When the climate returns to normal, it takes time for the original species to re-invade. If the life of the area is largely destroyed by the disturbance, then this re-invasion process is one of gradually increasing diversity which levels off when all the original species have returned. Thus recently disturbed areas may be *impoverished* and species diversity of a given taxon depends on time elapsed since the disturbance. This hypothesis of increasing diversity has been called the *ecological time theory* (Pianka, 1966). Its importance is hard to assess and largely untested.

Of overriding concern is the *geological time theory* (Pianka, 1966) championed by Simpson (1964). According to this approach, the older an area—that is, the longer it has existed in its present physical and climatic state—the longer have speciation and colonization by new species been operating. A long-since glaciated area gains species not only by reinvasion of its original fauna and flora, but through the arrival and arising of new species. Geologically older areas should, theoretically, be more diverse. In keeping with this hypothesis are the data of Southwood (1961). Southwood used the data of Godwin (1956) and calculated the "cumulative abundance" of various tree species—that is, the number of individuals integrated over the evolutionary time of existence of the species. He found a correlation coefficient of $+0.85$ ($p < 0.001$) between the cumulative abundance of a tree and the number of insect species associated with it.

We discussed dispersal and colonization in Chapter 14. We now briefly discuss the appearance of new forms through speciation.

Speciation may occur in the following way. Two populations of a single species somehow become geographically separated from one another and subsequently experience divergent evolution or genetic drift, or both. When, after considerable time, contact between the populations is re-established, hybrids, if they occur at all, tend to show low viability or low fecundity. Selection then favors the development of behavioral, physiological, or ecological barriers to interbreeding and we say that we have two distinct species. In practice such species are distinguished phenetically but it is assumed that they are truly noninterbreeding.

On the basis of the above arguments, species diversity might be expected to be high not only in geologically old areas, but also in geologically diverse areas, where mountains, lakes, and rivers form a number of potential barriers between a population and the occasional splinter group which manages to disperse across them. Of course, the areas of concern here are clearly very large, and there are no large areas of the world which seem remarkably rich or poor in barriers (archipelagos and areas with many isolated lakes—both island systems—are exceptions but are not comparable with continental areas—see part 16.3). There are, however, reasons for believing that geological variations are effective as barriers more often in some places than others.  Janzen (1967) suggests that in areas with little climatic variation, such as the tropics, species do not need to adapt to a variety of weather situations and become very narrow in their weather tolerances.  Slight climatic variations due to small differences in altitude may then become effective barriers.  On the other hand, such variations would be well within the tolerance limits of temperate species.  This may be a partial explanation for the large species diversity of tropical as opposed to temperate regions (see 16.1).

While the need for geographic isolation in speciation is generally accepted, there is some evidence that it is not necessary.  As Ehrlich and Raven (1969) point out, gene flow even over small distances in areas populated without clear breaks is often remarkably slow.  The marvel is that so little evolutionary divergence takes place at the two extremes of a population's range.  Where selective pressures differ over space, distant individuals may become incapable of interbreeding (Bazykin, 1969).

If disruptive selection (see 4.2.3) is sufficiently strong it may lead to positive-assortative mating and the genetic divergence of two or more groups.  The frugivorous fly, *Rhagoletis pomonella*, consists of two sympatric races.  One mates and lays its eggs on hawthorn, *Crataegus*, the other on apple, *Malus*.  Selection has favored a shift in the timing of emergence of the two races to correspond with the ripening of their respective fruits.  Where further selection pressures differ because of the timing difference or where interbreeding leads to errors in timing, barriers to interbreeding between the races should evolve.  The situation seems ripe for sympatric speciation (Bush, 1969).

Pimentel and his coworkers have attempted to demonstrate the possibility of sympatric speciation by connecting three or four small cages in a row with tubes, placing different larval media in either end cage and releasing flies into the apparatus.  Disruptive selection over an ecotone was provided by giving the flies in one end cage nine vials of fishmeal and one of banana in which to lay their eggs, and then destroying all larvae in the nine fishmeal vials.  In the other end cage the procedure was reversed.  Flies at one end of the apparatus thus experienced 90% selection for use of fishmeal, flies at the other end, 90% selection for use of banana.  After 18 weeks, flies hatched in the end cages were tested in the laboratory for medium preference.  Of the banana-selected flies, 71.2% chose banana, of the fishmeal-selected flies, 65.2% chose fishmeal.  After 44 weeks, the figures rose to 75% and 67%.  Preferences tested inside the cages were even more marked: 67.4% and 82.2% at 18 weeks, 88.8% and 84.1% at 70 weeks ($p < 0.01$).  Hybrids showed no preferences.  Migration rate between the end cages was 4% to 5%.  With higher migration rates (about 30%), the fishmeal strain evolved a significant preference by 36 weeks and the banana strain died out (Pimentel, Smith, and Soans, 1967).  Positive-assortative mating is the next phase expected in these cages, eventually followed by speciation.

If we accept the hypothesis of speciation through disruptive selection, then we may state that in spatially heterogeneous environments there is more opportunity for disruptive selection and hence speciation.  Spatially heterogeneous areas should on this basis be richer in species (see also 15.3.2).

Isolation by geographic barriers or distance has greater effect on nonmobile species, and mobility is linked with niche breadth (among other things).  For example, ground-living monkeys move about more than species restricted strictly to tree patches.  There is thus less isolation between subpopulations, more gene flow, and less opportunity for speciation.  According to DeVore and Washburn (1963), there are only one or two species of generalized ground-living monkeys (baboons), while there are several species of the somewhat more forest-dependent macaques.  *Cercopithecus* monkeys are quite restricted to forests though they may do some ground foraging; there are more such monkeys in Africa than baboons and macaques in the whole world.  The langurs are strictly tree dwelling, and open areas present effective barriers to movement.  There are more langurs in southeast Asia than other monkeys over the entire world.  Predators and omnivores tend to be generalized in diet more than herbivores (part 7.3), and tend therefore to wander across more diverse areas to get food.  Predator and omnivore populations are thus less isolated from one another and speciation less frequent.  There are only a very few large dogs, cats, bears, and oppossum, and many ungulates.  In North America one dog, *Canis lupus*, and one cat, *Felis concolor*, prey on the deer species *Odocoileus virginiana* and *O. hemionus*; the elk, *Cervus canadensis*; the caribou, *Rangifer caribou*; the moose, *Alces americana*; the bison, *Bison bison*; the prong-horn, *Antelocapra americana*; and mountain sheep and goats, *Ovis canadensis* and *Oreamnos americanus*.  In general there are more species of herbivore, fewer carnivores and omnivores, in any one area, fewer generalists, and more specialists.

## 15.3 THE MAINTENANCE OF DIVERSITY

### 15.3.1 Spatial Heterogeneity

If an environment possesses a diverse flora there is greater food variety and perhaps greater variety of microhabitats for animals—hence more animal species can avoid competitive exclusion.  Plant species diversity might have several causes.  First, diversity in the physical environment promotes variety in plant species.  Even very small differences in soil surface texture may encourage different plants to grow.  Harper *et al.* (1965) sowed a mixture of seeds in planters with different soil texture and compaction and found germination of quite different plants.  Thus where plants die and rot, or fall, or where animals dig or trample, soil is physically altered and a variety of plants is maintained.  Plant species diversity may also be maintained by the other mechanisms discussed here (15.3).

Where the environment is varied in its topography, flora, or in some other way, it becomes theoretically possible for animals to subdivide it in more ways.  Greater subdivision of the environment—and hence the niche space—allows for the co-existence of more animal species.  We assume here that population is limited by some resource.  If not, subdivision of the environment will not affect competition, competitive exclusion, and the number of coexisting species.

Let us examine some of the ways in which animals partition their environment. One way is to utilize different resources. Another is to exploit all patches of a certain type, using all resources contained within them. A special case of the latter involves dividing the environment by zones. MacArthur and MacArthur (1961) discuss these possibilities for insectivorous woodland birds. Such birds might avoid competition by eating different insect species. MacArthur and MacArthur reason that this practice would require birds to spend time and energy bypassing potential food items. Except where insects possess poisons, they are pretty much similar in being crunchy bags of haemolymph. Thus the basis for differential specialization is not clear in any case. Patch subdivision might involve eating all the insects in one kind of tree. This also appears wasteful since it would involve bypassing all sorts of trees, many containing insects eaten when on the appropriate tree species. Specialization on conifers as opposed to deciduous trees might be a possibility since foraging techniques might differ on such divergent growth forms, but could not occur within a purely coniferous or a purely deciduous forest. MacArthur and MacArthur thought the most efficient subdivision would be by vertical zones. In this case, all points lying on a straight line between any two feeding points would also be potential feeding points—the foraging area would be convex—and a minimum of time and energy would be wasted. A rough approximation to this scheme, only slightly more complex, was found by MacArthur in his (1958) study of warblers in northeastern coniferous forests. We return to the MacArthurs' further exploration of this matter in a moment. First, let's take a general look at habitat separation by resources.

The following discussion, with some additions and deletions, is taken from MacArthur and Levins (1964). Suppose species $x$ and $y$ compete for resources 1 and 2, occurring in abundances $R_1$ and $R_2$. We denote the efficiency with which $x$ uses $R_1$ and $R_2$ by $i_1$ and $i_2$ so that the total amount of resources obtained by $x$ is $i_1 R_1 + i_2 R_2$. For species $y$, we use the symbols $j_1, j_2$. Suppose that $x$ increases in number if, and only if, there is sufficient food, in combination, such that $i_1 R_1 + i_2 R_2 > K_1$. Similarly, we suppose that $y$ increases if and only if $j_1 R_1 + j_2 R_2 > K_2$. We may then depict the changes in $R_1$, $R_2$ with the following graphs (Fig. 15.1). When the point $(R_1, R_2)$ lies above the line denoted by $x$ in Fig. 15.1(a), both $x$ and $y$ are increasing and are using up the resources. Thus $R_1$ and $R_2$ fall. Between the lines $x$ and $y$ in Fig. 15.1(a), $y$ is still increasing and will continue to do so until $R_1$ and $R_2$ fall to the line $y$. But $x$ cannot survive at these low resource levels and hence dies out. Competitive exclusion has occurred. In Fig. 15.1(b) the changes, by similar argument, are as shown by the arrows. Between the lines, in the areas marked by question marks, however, the direction of the arrows may vary.

Consider the upper left one of these two areas. Species $x$ is declining, $y$ is growing. Species $y$ needs less of resource 1 than resource 2 to survive (the intercept along the $R_2$ axis is closer to the origin), and is thus more efficient at utilizing resource 1 and probably shows a preference for resource 1. Species $y$, by the same token, should prefer resource 2. If the preferences are not marked (that is, the use of the two resources by the two species is very similar) both $R_1$ and $R_2$ may be expected to decline and the arrows in this region will point down and to the left. If, however, the preferences are marked (the two species are sufficiently different in their utilization of the resources), $R_2$, by virtue of $x$'s decline, may actually increase. The arrows now point down and to the right. In the lower right member of the two areas, the arrows, by similar argument, point up and to the left. Clearly, an equilibrium situation exists in which $R_1$ and $R_2$ are given by the point of intersection of the two lines, and $x$ and $y$ coexist.

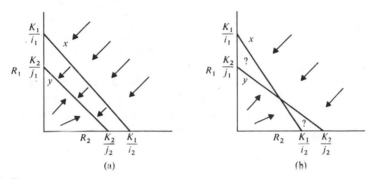

Fig. 15.1. The conditions for population increase in competing species limited by two resources.

So far this has been merely a restatement of the argument of Gause and Witt (part 12.2), but there is an added feature in this approach. Imagine a third line representing the needs of a third competitor. Unless this line intersects both of the other two at exactly the same point—a negligible possibility—equilibrium cannot occur with all three species present. Two resources can support, at most, two species *which are competing for them*. The addition of a third axis, of course, turns the lines of Fig. 15.1 into planes which may now intersect a third plane at a single point. Three resources may support, at most, three competing species. The argument may be extended for any number of resources.

At first glance, the notion that $n$ resources will support only $n$ species seems very useful. But one should never expect such simplicity in nature. What constitutes a resource? If a dandelion can be subdivided into two resources in the form of heads and stems (12.3.1), can we neglect the possibility that heads or stems may be further broken down? Do different morphs of a single prey species constitute different resources? What about different size classes within a species—and how are these size classes delineated? In terms of space and shelter, does a rocky fell-field constitute a single resource, or do rock bottoms, rock crannies, large boulders, and smaller rocks constitute different resources? Unfortunately, a resource is what its user makes it. A quantitative application of the above conclusion, then, is circular. The conclusion is nonetheless very useful in a qualitative sense. It seems reasonable that, for a given group of animals or plants, the number of resources interpreted by these creatures bears a consistently increasing relation to the number of resources as measured by any number of man-made criteria. This being so, the conclusion that *n* resources may support up to *n* species supports the idea that more complex environments support more diverse fauna and flora. The very weakness of the conclusion is useful. Those species most capable of breaking "resources" down into larger numbers of resources are the small species. A bird may cover miles of forest, an insect may live its entire life on one plant individual or even one leaf. Small organisms are thus presented with a larger array of resources and are represented by more species.

We return now to the subdivision of habitats by zoning and the data of MacArthur and MacArthur (1961) on insectivorous woodland birds. The approach taken by these authors was that if birds partition themselves by vertical zones, the amount of information required to specify the birds present should be directly related to the amount of information needed to describe the vertical profile of the forest. Accordingly, MacArthur and MacArthur divided the forest into what seemed obvious layers: litter and ground foliage, shrub zone, and tree zone. An obvious clue as to the zone in which a bird finds itself is foliage density. Foliage density in each of these layers was thus measured, converted into a relative density value,

$$p_i = \frac{\text{density of layer } i}{\Sigma \text{ density of layer } i}, \tag{15.7}$$

and used to calculate the information content according to the Shannon–Weaver equation: Foliage height diversity (FHD) $= -\sum_i p_i \log_2 p_i$. Plotting bird species diversities (BSD) against the values obtained above over 13 forest areas in the northeastern United States, a rather nice graph was obtained (Fig. 15.2). The regression equation derived was: BSD = 2.01 FHD + 0.46.

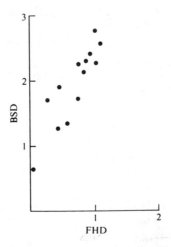

**Fig. 15.2.** The relation between foliage height diversity and bird species diversity.

Although plant species diversity was a reasonably good predictor of bird species diversity, it accounted for none of the remaining variance in BSD after the partial regression on FHD.

Recher (1969), working in Australia, gathered similar data for five areas—wet sclerophyll, dry sclerophyll, dry sclerophyll–heath transition, and two coastal heath areas. Of 74 bird families in Australia only 20 are common to North America, yet Recher obtained the same curve, with the same slope, as MacArthur and MacArthur. MacArthur, Recher, and Cody (1966) also found good fits in Puerto Rico (using two rather than three vertical layers) and in Panama (four layers), with almost exactly the same regression lines. It is not clear what meaning can be ascribed to the number of layers into which the forests of an area must be divided to achieve a fit between FHD and BSD. However, the goodness of fit and the similarity in slope of the regression lines over several areas indicate that this number is not without meaning. MacArthur, Recher, and Cody also performed the following analysis: Let $p_i$, $q_i$ be the proportion of bird species $i$ (or foliage density in layer $i$) in two areas, $P$ and $Q$. Then a measure of the difference between $P$ and $Q$ is:

$$H_{P \wedge Q} - \tfrac{1}{2}(H_P + H_Q) = \tfrac{1}{2} \sum_i (p_i \log_2 p_i + q_i \log_2 q_i) - \sum_i \left( \frac{p_i + q_i}{2} \log_2 \frac{p_i + q_i}{2} \right).$$

$$(15.8)$$

Applying this formula to FHD and BSD in many different areas, and plotting the former on the abscissa, the latter on the ordinate, the authors generated a curve rising from the origin and leveling off to the right. Points from Panama lay above the curve, implying that the birds here not only recognized more layers (four) but also were more specific in their choice of layers. The points from Puerto Rico lay below the curve implying a sloppier subdivision of (two) layers.

A different approach has been used by Cody (1968) with respect to grassland birds. Cody reasoned that the birds might divide their environment by height, horizontal spacing, or resources (food in this case). In low grass, niche separation by resources seems the most likely. Horizontal separation is inefficient for the same reason that specialization by tree species is inefficient for insectivorous, forest birds. In tall grass vertical zonation becomes possible and by the argument (above) of MacArthur and MacArthur, the most likely means of niche separation. Thus we can depict partitioning of the environment by grassland birds as shown in Fig. 15.3. The circles represent bird species.

Suppose that the niche of each species is separated from those of its nearest two competitors by equal amounts—the circles in Fig. 15.3 are evenly spaced. Now divide all grasslands into three types, one in which grass height varies from 0 to $\tfrac{1}{3}$ maximum, one in which grass varies from $\tfrac{1}{3}$ to $\tfrac{2}{3}$ maximum, and one in which grass is $\tfrac{2}{3}$ of maximum to maximum (horizontal lines in Fig. 15.3). Species number is least in habitats of medium height grass. A more accurate way to put this: For grasslands with equal variance in grass height (a measure of range in height) those with the shortest and the longest grass possess the largest number of bird species.

The number of bird species present in an area increases with the variance in grass height in the area. Although the uniform spacing of species along the partitioning gradients seems rather arbitrary, Cody's predictions are supported by observation.

It has often been noted that number of individuals as well as species number is high at interfaces between two types of habitat—a forest and a field, for example. It may be that productivity of one or both of two abutting habitats is enhanced by being peripheral and thus not surrounded by the usual competitors. This argument seems forced, but accounts for the increased number of individuals. The large number of species may be due to the mixing of species from both habitats as well as the presence of some adapted to the narrow ecotone between the two.

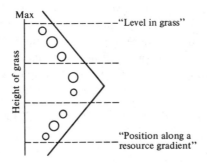

**Fig. 15.3.** The partitioning of niche space by grassland birds (modified from Cody, 1968).

## 15.3.2 The Competition Theory

Svärdson (1949), Dobzhansky (1950), and Williams (1964) have suggested that where competition is severe, niches will narrow and that more species can thus fit into an environment without their niches overlapping enough to lead to exclusion. This hypothesis has been called the *competition theory*.

First we show a way in which niche overlap can be easily visualized. We suppose that the resources in contention between a number of species can be ordered so that we can draw the niches of the species (after the Discussion, Section II) as in

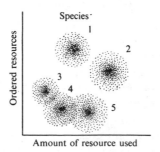

**Fig. 15.4.** Depiction of the niches of 5 species.

Fig. 15.4. Such ordering will be possible if partitioning of the environment is along a spatial (horizontal or vertical) gradient or by (say) food species size. We now view the projection of these niches on the ordinate and, recalling that each niche, as drawn, is a probability cloud representing population density, note that Fig. 15.4 can be depicted as in Fig. 15.5. If we let the distance between the modal point of two successive peaks (say of species $i$ and $j$) be $D_{ij}$ and the respective standard deviations of the curves be $\sigma_i$ and $\sigma_j$, the competition theory may be restated as follows: Strong interspecific competition decreases the diameters of the $R = 1$ isoclines and thus also the values of the $\sigma$'s resulting in less niche overlap and therefore less chance of competitive exclusion. Interspecific competition therefore allows for greater numbers of species. I believe this theory can be totally disposed of.

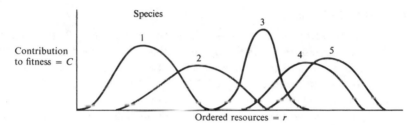

**Fig. 15.5.** Depiction of the niches of 5 species.

Increased interspecific competition comes about in several ways. We suppose that competition intensity is related to the number of individuals per (shared) food unit. Then, if there is a sudden shortage of resources, interspecific competition increases. Intraspecific competition also increases to the same degree, unless the shortage is less in the region of niche overlap. If the shortage is general, since niche breadth is a balance between intra and interspecific competition, we would not necessarily expect a change in $\sigma_i$ and $\sigma_j$ relative to $D_{ij}$. If the shortage is least in the region of overlap, each species will use resources in the overlap region more heavily, out of necessity, and while $\sigma_i$ and $\sigma_j$ may decrease, so will $D_{ij}$. If competition is less at some point along the resource gradient, individuals will shift their utilization so that this difference is obliterated. Reshuffling of preferences and use patterns will very quickly equalize competition pressure along the gradient, and there will be no place of entry for a competitor who was not already present. The only difference is that competition is now more intense, making colonization even more unlikely. Eventually, of course, populations will fall to new, lower carrying capacities, and competition intensity will fall to its initial value. Competition intensity will increase (momentarily) with the successful invasion of a new species. But note that while a new species implies increased competition, and perhaps subsequently narrowed niches, competition decidedly does not narrow niches in such a way as to make coexistence or the colonization by new species more likely.

It is probably true that niche overlap represents a balance between intra and interspecific competitive pressures. Niche overlap also depends on such things as genetic variability in habitat utilization and the degree to which an increase in niche breadth diminishes fitness within the old habitats—generalists are usually less efficient at utilizing a variety of resources than specialists are at using only a few. It is thus impossible at this stage in our knowledge to describe the shape of the niche (Fig. 15.5), or to predict evolved overlap patterns on a theoretical basis. For this reason I have not included the well-known paper by MacArthur and Levins (1967). It seems sufficient merely to say that in a community at equilibrium, in which certain competitors approach their carrying capacities (corrected for competitive effects), successful colonization of a new species depends on a use of the available resources at least as efficient as that shown by the established species. Since the established species have evolved *in situ*, this seems an unlikely situation.

### 15.3.3 Predation Theory

It is clear from the above arguments as well as those in part 14.2 that in a community where species approach their carrying capacities (corrected for effects by competitors), colonization is a possible but unlikely event unless the community is impoverished and some resources are left unused, or is disturbed. (Note that an environment can be unsaturated only if a disturbance has eliminated species recently enough that niche overlap patterns have not readjusted to their optimum values.) But what of communities, or subsets of communities, in which populations are held (most of the time) below the point where resources become limiting? We have discussed the role of predation on competitive exclusion (12.2.7). We now examine this role in more detail.

Suppose that $r_0$ denotes intrinsic rate of natural increase in a population before predation, and that the death rate is augmented by this predation by an amount $mn_p$ (where $n_p$ is the predator population and $m$ the predation rate). Then to an approximation

$$
\begin{cases}
\dfrac{dn_1}{dt} = \dfrac{(r_1)_0 n_1}{K_1}(K_1 - n_1 - \alpha_{21} n_2) - m_1 n_1 n_p, \\[2em]
\dfrac{dn_2}{dt} = \dfrac{(r_2)_0 n_2}{K_2}(K_2 - \alpha_{12} n_1 - n_2) - m_2 n_2 n_p,
\end{cases}
\tag{15.9}
$$

where $n_1$ and $n_2$ are the population sizes of two competitors. At equilibrium we have

$$(r_1)_0 (K_1 - \hat{n}_1 - \alpha_{21}\hat{n}_2) - m_1 K_1 \hat{n}_p = 0$$
$$(r_2)_0 (K_2 - \alpha_{12}\hat{n}_1 - \hat{n}_2) - m_2 K_2 \hat{n}_p = 0,$$

so that

$$\hat{n}_1 = \{(r_2)_0 K_1[(r_1)_0 - m_1\hat{n}_p] - (r_1)_0 K_2\alpha_{21}[(r_2)_0 - m_2\hat{n}_p]\}/(1 - \alpha_{12}\alpha_{21})(r_1)_0(r_2)_0$$
$$\hat{n}_2 = \{(r_1)_0 K_2[(r_2)_0 - m_2\hat{n}_p] - (r_2)_0 K_1\alpha_{12}[(r_1)_0 - m_1\hat{n}_p]\}/(1 - \alpha_{12}\alpha_{21})(r_1)_0(r_2)_0.$$

Both competitors coexist when $\hat{n}_1 > 0$, $\hat{n}_2 > 0$. That is

$$\frac{K_1}{\alpha_{21} K_2} > \frac{[(r_2)_0 - m_2 \hat{n}_p](r_1)_0}{[(r_1)_0 - m_1 \hat{n}_p](r_2)_0},$$

$$\frac{\alpha_{12} K_1}{K_2} < \frac{[(r_2)_0 - m_2 \hat{n}_p](r_1)_0}{[(r_1)_0 - m_1 \hat{n}_p](r_2)_0}.$$

Now, in the absence of predation ($m_1 = m_2 = 0$) species 2 is always excluded if $\alpha_{12} K_1 / K_2 > 1$. Thus predation allows for coexistence when in its absence exclusion would occur, when

$$\frac{K_1}{\alpha_{21} K_2} > \frac{[(r_2)_0 - m_2 \hat{n}_p](r_1)_0}{[(r_1)_0 - m_1 \hat{n}_p](r_2)_0} > \frac{\alpha_{12} K_1}{K_2} > 1. \qquad (15.10)$$

If $\hat{n}_p$ and $m_1/m_2$ are too high, predation turns the tables and species 2 excludes species 1. If these values are too low, species 1 wins. For intermediate values, coexistence occurs. Note that in nature, $m_1$ and $m_2$ need not be constants but may vary with the relative and absolute values of $n_1$ and $n_2$. If $m_1/m_2$ varies in the same direction as $n_1/n_2$, then the chances are enhanced that the conditions above are met. Of course, predation can also lead to extinction when in its absence coexistence would occur. The conditions for this event are

$$\frac{K_1}{\alpha_{21} K_2} > 1 > \frac{\alpha_{12} K_1}{K_2} > \frac{[(r_2)_0 - m_2 \hat{n}_p](r_1)_0}{[(r_1)_0 - m_1 \hat{n}_p](r_2)_0}.$$

The chances of predator-induced extinction are increased when $m_1/m_2$ varies inversely with $n_1/n_2$. Whether predation generally promotes species diversity or inhibits it, then, depends on how $m_1/m_2$ varies with $n_1/n_2$.

Predators *do* tend to switch preferences to common foods (7.1.2), but often only after a time lag. Furthermore, a novelty effect, in which animals eat more of a scarce food simply by virtue of its being scarce, is known to exist. These facts suggest that $m_1/m_2$ will decrease with an increase in $n_1/n_2$. However, not all predators display significant time lags in food preference changes—and during an increase phase of scarce food, a lag would favor its successful recovery. The novelty effect may be an artifact of domesticated or closely human-associated species. There are also reasons to believe that the relative resistance of a prey species to predation decreases as its relative number decreases. If a predator approaches an optimal choice strategy in food preference, the relative probability of being eaten by that predator declines with decreased relative population size (Emlen, 1966b, 1968). Except under certain conditions (Emlen, 1968) a preferred food is still eaten preferentially when scarce, but the degree of preference drops; that is, $m_1/m_2$ drops as $n_1/n_2$ drops.

In addition to this convenient change in food preference predictable from optimal choice models, there are also the effective changes in preference due to search images. As Holling (1965) envisions search images, a predator does not learn and remember the worth of a food unless it reaches some threshold abundance (both relative and absolute). Thus foods below this density threshold are eaten proportionally less than when common. Tinbergen's (1960) view of the search image is more what that name implies—a preference for items for which the predator has some kind of mental image resulting from past experience. Both viewpoints lead to the same conclusion: Scarce food species are taken proportionately less often than common ones, all else being equal. Existence of search images in predators may, in the same sense as preference changes under optimal choice, lead to the coexistence of competing prey species.

As the abundance of one prey species falls, it is less and less likely to occur in large, dense patches. If, as would seem likely, predators concentrate their attentions on rich food areas, then there is a built-in buffer to the elimination of scarce foods; $m_1/m_2$ declines with $n_1/n_2$. Actual information on predation intensity as a function of local concentrations of food is surprisingly scarce. However, as noted in Chapter 7, very good data have been gathered by Gibb (1962, 1966) on the intensity of predation by tits on the *Ernarmonia* larvae inhabiting Scots pine cones. Trees and areas with high larval density, as expected, experienced the greatest concentration of the birds' predatory activity. Tinbergen *et al.* (1967) tried an experimental procedure in which eggs were spread in a grid over a field. When interegg distance was 0.5 m, the proportion predated was almost always higher than when this distance was increased to 8 m.

Another factor which should buffer extinction of scarce species is the change in prey population age structure with population growth rate. In a prey species increasing at the expense of its competitors, individuals of pre-reproductive age are more prevalent than in its competitors' populations. Young individuals are more prone to predation than mature animals and their increased relative numbers should feed back negatively on the rate of population increase. (But this could detrimentally affect a sparse, recovering population.) It can be argued that this factor works both ways since declining populations are likely to have increased numbers of very old, also very vulnerable, individuals. This is probably seldom a valid argument, since few individuals approach senility in the wild, and middle-aged animals may well make up in experience what they lack in speed.

Finally, it can be argued that as a predated species dwindles in number, selection acts strongly in favor of those individuals possessing superior escape abilities. At worst, such selection will buffer extinction, and, at best, promote coexistence.

In conclusion, then, predation which keeps competitor populations below their respective (competitor-affected) carrying capacities more often promotes than discourages coexistence and thus leads to higher species diversity. This is true except in the case of extremely heavy or extremely light predation. We use, as a measure of *predation intensity* (or *predation pressure*), the value

$$\sum_i (K_i - n_i) \Big/ \sum_i K_i,$$

summed over all prey species, *i*. Then the manner in which predation affects species diversity is (qualitatively) as shown in Fig. 15.6.

The idea for the predation theory must be credited to Paine (1966) who took a somewhat different tack than that presented above. Consider a situation in which competitor populations are decimated not continuously, but by periodic catastrophes which wipe out local subpopulations. Examples of the causes of such catastrophes include small forest or grass fires, ground heaving by frost or falling trees, local feeding on *Opuntia* by *Cactobastis* (part 11.2), and the feeding on mussels or barnacles on the American west coast by the sea star, *Pisaster ochraceus*. Where a successful competitor is suddenly eliminated, other species with more rapid dispersal powers or rates of population increase will replace it, at least temporarily. Thus a high number of species may be maintained in patches over an area because of recurrent crises followed by rapid dispersal and colonization. *Pisaster* feeds rapidly on whole clumps of mussels or barnacles and thus leaves open spaces where rapidly colonizing plants and animals grab a temporary foothold. When the predator is removed spaces fail to open and competition takes its usual course, eliminating a number of species. Paine (1966) furnishes data showing that when *Pisaster* is removed from an area, the numbers of both plant and animal species decline. Paine and Vadas (1969) obtained similar results by removing urchins (*Strongylocentrotus franciscanus* and *S. purpuratus*) from west coast tide pools.

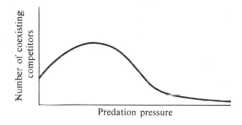

Predation pressure

**Figure 15.6**

It is tempting to make an oversimplified interpretation of the predation theory —that at least nonselective predation (or decimation by other means), by holding competitor populations at low levels, acts to alleviate competition and thereby "allow" coexistence. This is not quite correct. If no competition for resources occurs, but the combined populations of prey species is held to a finite level, then that species with the highest $r_0$ value will still displace other prey. Predation does not "allow" for coexistence, it changes the degree to which competitors depress each other's fitness ($R$) values with the result that new population equilibria, more often involving coexistence, are brought about.

It would also be a mistake to apply the predation theory as a hard and fast rule. Species number may rise and then fall with increasing predation intensity, all else being equal. But all else is seldom equal. The work of plant ecologists on the effects of grazing on the number of plant species (Harper, 1969, and papers cited therein) shows mixed effects. Species diversity should be encouraged if a herbivore prefers the major dominant species but will clearly be depressed if the plants preferred by the herbivore are the scarce ones. Harper (1969) believes that the effects of grazing on plant species diversity cannot be generalized. This seems too hesitant, however. It is probably safe to state that, over large numbers of different environments, predation intensity from very low to some intermediate level is correlated significantly with the number of prey species, but that at high predation intensities prey species number falls.

A different kind of predator theory altogether is presented by Janzen (1970). Where seed parasites are host-specific, the chances of successful germination and growth increase with increasing distance from a parent plant. On the other hand, the number of seeds dispersing a distance $x$ from a parent plant decreases with $x$. The product of the two curves described gives the expected number of offspring as a function of $x$. This value rises, reaches a peak (at $\hat{x}$) and then falls with $x$. In the tropics, extremely high parasitism selects for high dispersal distances, increasing $\hat{x}$. Also, host-specificity of seed parasites is high in the tropics, increasing $\hat{x}$ still more. This means that individuals of the same species grow at more widely spaced intervals in the tropics, allowing space for the establishment of other types, and thus a larger number of coexisting species.

### 15.3.4 Stability Theory

This approach includes that of the *climatic stability theory* (Pianka, 1966; see also Klopfer, 1959; Fischer, 1960; Dunbar, 1960). The arguments are of several basic sorts. First, instability (great change or uncertainty or both) in the environment (physical or biological) may keep populations low much of the time. Decimation by an unstable environment can be written in form analogous to decimation by predation—Eqs. (15.10). If decimation is seasonal, we simply write

$$\Delta n_1 = (R_1 - 1)\, n_1 - M_1 n_1 : R_1 \to \overline{e^{r_1}} = \exp \left[ \overline{\frac{(r_1)_0 n_1}{K_1} (K_1 - n_1 - \alpha_{21} n_2)} \right],$$

$$\Delta n_2 = (R_2 - 1)\, n_2 - M_2 n_2 : R_2 \to \overline{e^{r_2}} = \exp \left[ \overline{\frac{(r_2)_0 n_2}{K_2} (K_2 - \alpha_{12} n_1 - n_2)} \right],$$

where $M_1$, $M_2$ are the death rates due to environmental instability. But there is no clear manner in which $M_1/M_2$ varies with $n_1/n_2$. Thus we can make no clear statement on this basis as to whether slight diminishing of populations by an unstable environment will increase or decrease species number. Heavy decimation, like heavy predation, will act to lower species number.

Another argument (Klopfer and MacArthur, 1961) is that a stable environment allows for the evolution of specialization—narrow niches (see part 7.3), and thus for more species. We emphasize again that this tendency toward specialization can only occur when the resources in question are common—that is, not limiting. But if they are not limiting they are not contended for by species which, if in competition, must therefore be competing for something else. Thus narrow partitioning of these resources does not allow for added species and the argument is invalid. Of course, the addition of new species may force others into very narrow niches. When this occurs, stability encourages diversity, since it decreases the probability that the narrow range of resources used will fail and result in the stenotopic species' extinctions.

Where populations are held to low levels by unstable climate, food is, by definition, plentiful. There should accordingly be a tendency toward dietary specialization (see 7.3) *even though some food species may go uneaten.* This makes sympatric speciation more likely and makes it easy for new species to colonize. This is *not* the same argument as that of Klopfer and MacArthur. One might thus expect unstable climate to speed the increase in species number over geologic time, so long as the instability is not so great as to lead also to high extinction rates. Climate is generally unstable with respect to insects (10.1.1, 10.1.2) so we might in general expect to find insects quite specialized and diverse.

An argument which runs counter to the above is offered by Sanders and Hessler (1969) and Slobodkin and Sanders (1969). Consider two environments, the first fluctuating but predictable, the second constant. Species living in the first, to survive, must have adapted to the changes which occur either through eurytopy, aestivation or withdrawal of another sort, or regular migration. These species, by their generalized nature, may not often compete successfully with the species of the constant environment, but they *can* invade that environment and survive the physical rigors. Species of the constant environment are not equipped to deal with change, having never experienced selection for that ability, and thus cannot cope with the physical conditions of the fluctuating environment. Ignoring for the moment the problems presented to immigrants by biological competition and predation, we must conclude that physically constant areas will act as "species sinks", gradually accumulating more and more varieties of organisms. The same argument is presented for unpredictable as opposed to predictable environments. An unpredictably changing area requires eurytopic species, while a predictably changing area can be settled by either eurytopic species or stenotopic species which respond to regular cues to begin their periods of aestivation or whatever. On the basis of these arguments, constant, predictable areas should accumulate species much more rapidly—or much less slowly—than erratic, fluctuating environments.

In support of this hypothesis, Sanders (1968) and Sanders and Hessler (1969) point to the fact that the stable ocean floor off the eastern coast of the United States possesses a greater diversity of animal species than the shallower, less stable bottom of the continental shelf. Buzas and Gibson (1969) find dips in diversity of benthonic foraminiferans, along an otherwise increasing trend with depth, at 45 to 100 m, and 200–2500 m. Diversity peaks occur at 35–45 m, 100–200 m, and over 2500 m. The troughs correspond to the areas just above the wave base, and the drop over the continental shelf, both unpredictable areas. In marine benthic areas —and in old, deep lakes such as Baikal and Tanganyika—the high species diversity *may* well be accounted for by the geological time theory, working in a species sink. On the oher hand Paine (personal communication) has complained that while deep benthic climate may be constant, the influx of food from above may be quite erratic and unpredictable. If so, this throws some doubt on the meaning of the deep-sea data mentioned above. There are other difficulties besetting the hypothesis of Sanders and his coworkers. Janzen (personal communication) points out that a species which rhythmically adapts behaviorally or physiologically to a regularly fluctuating environment, or responds in the same ways to subtle environmental cues in a changing, unpredictable environment, may continue to change in this manner if placed in a constant, predictable environment. Such responses, adaptive in its native environment, could then prove disastrous. Furthermore, species specialized to a constant environment are likely to be extremely difficult to outcompete on their home ground. The species sink is probably never the less stable of two areas, but the validity of the species sink hypothesis is severely weakened as a *general* hypothesis by these considerations. Both Sanders' species sink hypothesis and the opposing specialization hypothesis may be valid, at least under some circumstances, and working at odds. Which is the more important cannot be determined at present, and will undoubtedly vary from situation to situation.

We write for a predator population

$$\frac{dn}{dt} = r_0 n \left(1 - \frac{n}{K}\right)$$

so that over time, a population in equilibrium can be expressed by

$$0 = \overline{r_0 n(1 - n/K)} = \overline{r_0 n} - \frac{\overline{r_0}}{K} n.$$

Now, if the environment is uncertain with respect to energy carried up through the food web to this predator, $K$ varies. Thus

$$0 = \overline{r_0 n} - \overline{r_0 n^2 \left(\frac{1}{K}\right)} = r_0 \left[\bar{n} - \overline{n^2 \left(\frac{1}{K}\right)}\right]$$

$$= r_0 \left\{\bar{n} - \left[\overline{n^2} \left(\frac{\overline{1}}{K}\right) + \mathrm{cov}\left(n^2, \frac{1}{K}\right)\right]\right\}$$

$$= r_0 \left[\bar{n} - (\overline{n^2} - \bar{n}^2)\left(\frac{\overline{1}}{K}\right) - \bar{n}^2\right) \frac{\overline{1}}{K}) - \mathrm{cov}\left(n^2, \frac{1}{K}\right)\right]$$

$$= r_0 \left\{\bar{n} - \left[\left(\frac{\overline{1}}{K}\right)(\bar{n}^2 + \mathrm{var}\, n) + \mathrm{cov}\left(\frac{1}{K}, n^2\right)\right]\right\},$$

and, solving the quadratic, we obtain

$$\bar{n} = \frac{1}{2(\overline{1/K})}\left\{1 + \sqrt{1 - 4\left(\frac{\overline{1}}{K}\right)\left[\frac{\overline{1}}{K}\, \mathrm{var}\, n + \mathrm{cov}\left(\frac{1}{K}, n^2\right)\right]}\right\}.$$

But the expression under the square root sign is less than one so that the expression in brackets must be less than two (call it $c$). Thus

$$\bar{n} = \frac{c}{2(1/K)} < \frac{1}{(1/K)}$$

which, in turn, is less than $\overline{K}$. In stable environments $\bar{n}$ is less depressed in this manner, meaning that predator populations remain at higher levels relative to the average of their prey populations ($\propto \overline{K}$), and the predation theory comes into more forceful play.

We conclude this section in essentially the same manner as 15.3.3. If environmental instability (degree and/or uncertainty of change) is very low, species number will be expected to increase with instability. Great instability will lead to extinctions and hence lowered numbers of species.

### 15.3.5 Productivity Theory

One idea which may have grown from the fact that there are more species in the tropics than in temperate zones, and perhaps also from the intuitive notion that high productivity implies lushness which implies more species, is that species diversity increases with productivity. Margalef (1963) mixes this idea with the concept of stability, suggesting that where environments are stable, more energy is available for productivity and hence there are more species. Connell and Orias (1964) clarify this matter somewhat with a complicated block diagram showing the interrelationship of such parameters as energy input, energy for reproduction, energy needed for (community) homeostasis, population sizes, stability, nutrient cycling, and number of species. While all the interrelationships proposed may be correct, some of the arguments seem a bit inconclusive. At least in terms of rigorous theoretical arguments, the role of productivity in promoting species diversity remains obscure.

## 15.4 SEASONALITY

Species may experience different spatial heterogeneity, different competitors, different predators, and different environmental stability in different seasons. We assume, then, that the above arguments apply only within a specified season. But the complexities are tremendous. Limitation of populations may occur in a larval stage; do the arguments in part 15.3 apply to adults? How does one predict trends in species diversity of adult organisms if their larval stages occurred at different seasons from each other? The problem of applying the arguments of 15.3 to seasonally varying environments (and almost all environments vary seasonally) are monumental, and theoretical work has virtually ignored these problems. The theoretical ecologist with a love for complexity can find here an almost untapped field.

## 15.5 EQUITABILITY

It has become popular to analyze diversity data looking at both species number and *equitability* value. If all species are equally frequent, then

$$H' = - \sum_{i=1}^{s} p_i \log p_i = - \log p_i = \log (1/p_i) = \log s,$$

so that

$$e^{H'} = e^{\log s} \propto s.$$

When species are *not* equally frequent, then $e^{H'}$ gives a measure of the *equivalent number of equally common species*. Thus equitability, which is defined by $e^{H'}/s$, approaches zero when species are maximally divergent in frequency, and $e^{\log s}/s$ (=1 when log base is $e$) when all species are equally frequent.

It is not yet clear what the ecological implications are of high or low equitability. However, we can make some theoretical stabs in this direction. Using the logistic equation as an approximation, we note that for two species in competition:

$$\begin{cases} r_1 \propto 1 - \dfrac{n_1}{K_1} - \dfrac{\alpha_{21} n_2}{K_1} - C_1 \\[2ex] r_2 \propto 1 - \dfrac{\alpha_{12} n_1}{K_2} - \dfrac{n_2}{K_2} - C_2 \text{ (see Horn's model, 10.1.2).} \end{cases}$$

Recall (Discussion, Section III) that selection pressure on a trait of value $x$ is given by

$$I'_x = \left| \frac{1}{R} \frac{\partial R}{\partial x} \right|.$$

Thus, since $r = \ln R$, we can write

$$\begin{cases} I'_{K_1} = \left| \dfrac{1}{R_1} \dfrac{\partial R_1}{\partial K_1} \right| = \left| \dfrac{\partial \ln R_1}{\partial K_1} \right| = \left| \dfrac{\partial r_1}{\partial K_1} \right| = (r_1)_0 \dfrac{n_1 + \alpha_{21} n_2}{K_1^2} \\[2ex] I'_{K_2} = \left| \dfrac{1}{R_2} \dfrac{\partial R_2}{\partial K_2} \right| = \left| \dfrac{\partial \ln R_2}{\partial K_2} \right| = \left| \dfrac{\partial r_2}{\partial K_2} \right| = (r_2)_0 \dfrac{\alpha_{12} n_1 + n_2}{K_2^2}. \end{cases}$$

But at equilibrium, $(r_1)_0 (n_1 + \alpha_{21} n_2) = K_1[(r_1)_0 - C_1]$, $(r_2)_0 (n_2 + \alpha_{12} n_1) = K_2[(r_2)_0 - C_1]$, so that

$$I'_{K_1} = \frac{(r_1)_0 - C_1}{K_1}$$

$$I'_{K_2} = \frac{(r_2)_0 - C_2}{K_2}.$$

Suppose now that these two species coexist over a long period in a stable environment so that equilibrium obtains and $C$'s approach zero. Then

$$I'_{K_1} \rightarrow \frac{(r_1)_0}{K_1}, \qquad I'_{K_2} \rightarrow \frac{(r_2)_0}{K_2}.$$

The intensity with which selection acts to increase $K$ for each species is inversely proportional to the size of $K$. Thus, where two species possess very nearly the same $r_0$ and very nearly the same additive genetic variance with respect to $K$, and coexist over long periods of time in a stable environment (so that a stable equilibrium obtains), the rate at which $K$ increases with selection is fastest for that species with the lowest $K$, and $K_1 \rightarrow K_2$. This is not a trivial argument since similar $r_0$ and $V_A(K)$ values do not necessarily imply close ecological similarity.

Thus high species number can be attributed to a number of causes, and high equitability might be expected to come about over long periods of time in highly stable environments. It is not surprising, therefore, that the increasing species diversity ($H'$) of benthonic foraminiferans found with increasing ocean depth by Buzas and Gibson (1969) was attributable largely to increasing species number at shallow depths, but in deep water (increasingly old, stable communities) was due primarily to an increase in equitability.

A word of caution is needed. The above argument lends meaning to the concept of equitability in species with similar $r_0$, $V_A(K)$ values. It is obscure what meaning, if any, can be placed on differences in equitability of species dissimilar in these respects over various environments.

# Species Diversity II

## 16.1 LATITUDINAL GRADIENTS IN DIVERSITY

In 1968, Johnson, Mason, and Raven published a curvilinear regression equation relating log number of plant species on islands or mainland areas to logs of land area, elevation, and latitude. They found land area to be the most important variable in accounting for species number (regression coefficient = 0.7212). Elevation was the least important variable (regression coefficient = 0.0143), with latitude accounting for an intermediate $0.1001/(0.1001 + 0.7212 + 0.0143) \times 100 = 12\%$ of the explained variance. Table 16.1 gives the results of the application of their equation. The variable in which we are interested here is latitude. These authors found that the number of plant species varied with the $-0.1001$ power of the latitude.

**Table 16.1.** Predicted and observed numbers of plant species in areas of California (from Johnson *et al.,* 1968)

| Area | Observed number of native plant species | Predicted number of plant species |
|---|---|---|
| California coast | 2525 | 2432 |
| Baja California | 1480 | 1450 |
| San Diego County | 1450 | 1316 |
| Monterey County | 1400 | 1473 |
| Marin County | 1060 | 1058 |
| Santa Cruz Mountains | 1200 | 1271 |
| Santa Monica Mountains | 640 | 772 |
| Santa Barbara area | 680 | 620 |
| San Francisco area | 640 | 590 |
| Tiburon Peninsula | 370 | 375 |

It is not only among plants that species diversity increases with decreasing latitude. Some of the documented latitudinal species gradients are illustrated in Tables 16.2, 16.3, and 16.4.

**Table 16.2.** Number of ant species as a function of latitude
(from Kusnezov, 1957)

|  | Area | Number of ant species |
|---|---|---|
| (North) | Alaska (Arctic) | 3 |
|  | Alaska (all) | 7 |
|  | Iowa | 73 |
|  | Utah | 63 |
|  | Cuba | 101 |
| (Equator) | Trinidad | 134 |
|  | Sao Paulo | 222 |
|  | Missiones | 191 |
|  | Tucuman | 139 |
|  | Buenos Aires | 103 |
|  | Patagonia (all) | 59 |
|  | Patagonia (western side) | 19 |
| (South) | Tierra del Fuego | 2 |

**Table 16.3.** Number of snake species as a function of latitude (from
Serie, 1936)

|  | Area (province in Argentina) | Number of snake species |
|---|---|---|
| (North, equatorial) | Missiones | 55 |
|  | Corrientes | 51 |
|  | Entre Rios | 34 |
|  | La Pampa | 15 |
|  | Buenos Aires | 22 |
|  | Rio Negro | 5 |
|  | Chub. | 5 |
|  | Santa Cruz | 1 |
| (South) | Tierra del Fuego | 0 |

Similar gradients are found also in tunicates, corals, crabs, nudibranchs, and amphipods. Epifaunal gastropods increase in variety toward the tropics, but burrowing snails (such as *Naticids*) show much less pronounced tendencies in this direction. Possibly most burrowers—cephalaspids, cumaceans, ophiuroids, holothurids—do not show significant latitudinal trends in diversity (Thorson, 1952, 1957), although this conclusion is disputed by Sanders (1968).

**Table 16.4.** Number of breeding bird species as a function of latitude (from Dobzhansky, 1950)

|                    | Area         | Number of breeding bird species | (Authority: cited in Dobzhansky, 1950) |
|--------------------|--------------|:-------------------------------:|----------------------------------------|
| (North)            | Greenland    | 56                              | Salomonsen                             |
|                    | Labrador     | 81                              | Peters                                 |
|                    | Newfoundland | 118                             | Peters                                 |
|                    | New York     | 195                             | Parkes                                 |
|                    | Florida      | 143                             | Grimes                                 |
|                    | Guatemala    | 469                             | Griscom                                |
|                    | Panama       | 1100                            | Griscom                                |
|                    | Colombia     | 1395                            | de Schauensee                          |
| (South, equatorial) |             |                                 |                                        |

## 16.1.1 Speciation Rates and Time

Diversity gradients with latitude are apparent even in the fossil record. For example, if we plot the percentage of Cretaceous planktonic foraminiferans against the latitude at which they were found, we obtain a curve increasing in height toward the equator—this in spite of the fact that, during the Cretaceous, climatic belts were less well defined and there was a milder world-wide climate than today (Stehli, Douglas, and Newell, 1969). The same is true, at the family level, for Permian brachiopods in the northern hemisphere. Stehli, Douglas, and Newell go on to note that, at least among bivalves and mammals, most "cold" area assemblages are cosmopolitan while many tropical assemblages are endemic, and also that the cosmopolitan families, as we might expect, are the older. These facts suggest higher rates of speciation in the tropics, with many (most?) higher taxa (families) arriving in temperate areas only with dispersal. It is tempting to believe that the faster rate in the arrival of new families—and presumably species—at lower latitudes is responsible for maintaining latitudinal diversity gradients. But why should evolution be faster in warm areas? The answer is clearly not related to body temperature since both bivalves and mammals (homeotherms) are affected. Stehli *et al.* suggest that evolution increases with the rate of capture of solar energy. But by what mechanism— higher mutation rates, perhaps? It seems more likely, in view of the discussion in part 15.2, that the relative physical constancy of lower latitudes encourages stenotopy, and increases the effectiveness of geographic barriers responsible for the higher speciation rates. There are also, undoubtedly, other factors working to that end.

Since the beginning of the Cenozoic and the period in which our modern day continents approached their present distribution, the currently existing temperate areas of the world have several times been glaciated. The tropics have not. Thus species of higher latitudes were driven out or driven extinct and may still be in the process of returning. The older tropics, having been disturbed less recently, have been accumulating species without interruption for a longer time and should display greater numbers of species. The argument is that of the *time theory* (15.2.2) and is sound, but as Pianka (1966) points out, we should expect, on the basis of this approach, that diversity gradients be steepest over the glaciated areas. In fact, gradients may be quite gradual over the glaciated areas and much steeper toward the equator (see Simpson, 1964; Table 16.4, for example). (Author's note: Table 16.4 may actually be misleading since Florida is a peninsula—see part 16.3. A plot of species number against latitude, deleting islands and peninsulas, leaves very few points, so that any conclusion must be taken with skepticism. The data in Table 16.2 are quite consistent with the time theory). The time theory is inconsistent with the possibility that volant forms show steeper gradients than burrowing forms which, we assume, disperse less rapidly.

### 16.1.2 Species Maintenance

Gillett (1962) gives the following data on tree species number in British Guiana. Each pair of numbers represents one 1.5-hectare sample. In Para, Brazil, 1482 individuals reflected 179 species over 3.5 hectares. These figures are typical for tropical rain forest. In the northeastern or midwestern United States the number of tree species expected over 1.5 (or 3.5) hectares would be closer to a dozen. Thus tree species diversity clearly increases toward the tropics. Gillett suggests that the reason may be increased seed parasitism in more stable climates (predation theory, 15.3.3). Seed parasitism was indeed found to be higher in tropical than temperate areas. At least two reasons exist for this increased rate of parasitism. Greater stability in climate implies greater stability in host populations, which means higher average parasite populations relative to host numbers (15.3.4). The stability of energy flow in the tropics also may mean more predator species per number of prey species (15.3.4). If this is so (and it seems to be true) then prey will have to evolve defense mechanisms against a larger variety of predators. This will tend to result in a slightly greater overall vulnerability to predation (4.2.2) and thus greater predation pressure. Of course, these same arguments apply to animal parasites and predators as well, implying that animal species should be numerous in stable, tropical areas.

| | Number of individuals | Number of species |
|---|---|---|
| | 310 | 60 |
| | 309 | 71 |
| | 432 | 91 |
| | 519 | 95 |
| | 617 | 74 |
| Total | 2187 | 120 |

Productivity increases toward the tropics (see Table 16.5), and many have been tempted to explain latitudinal diversity gradients by the productivity theory.

**Table 16.5.** Litter fall as a function of latitude (from Bray and Gorham, 1964).

| Area | Latitude | Litter fall in metric tons/ha/yr ≈ productivity | | |
|---|---|---|---|---|
| Congo | 1°N | 12.3 | | |
| | | 14.9 | | |
| | | 15.3 | | |
| | | 12.4 | | |
| | | 12.3 | | |
| Malaya | 3°N | 7.2 | 14.4 | |
| | | 5.5 | 9.3 | mean |
| | | 6.3 | 10.9 | 10.7 |
| | | 8.3 | 7.7 | |
| | | 10.5 | 14.8 | |
| | | | 10.2 | |
| Colombia | 4°S | 10.2 | | |
| Ghana | 6°N | 10.5 | | |
| Australia | 30°S | 6.0 | | |
| | 33°S | 2.4 | 2.6 | |
| | | 3.1 | 5.7 | |
| United States | 35°N | 6.3 | 4.6 | |
| | | 4.6 | 5.4 | mean |
| Australia | 37°S | 8.1 | | 5.2 |
| | | 8.1 | | |
| | | 6.9 | | |
| United States | 38°N | 6.7 | | |
| | 39°N | 2.9 | | |
| Canada | 44°N | 3.8 | | |
| | | 4.3 | | mean |
| Austria | 48°N | 3.5 | | 3.8 |
| Canada | 49°N | 3.6 | | |
| Denmark | 56°N | 2.6 | | |
| Finland | 61°N | 1.9 | 2.4 | |
| | | 2.3 | 2.0 | |
| | | 2.8 | 2.0 | |
| | | 3.2 | | |
| | 62°N | 1.6 | 2.2 | |
| | | 1.8 | | mean |
| Norway | 60°N | 2.8 | 2.0 | 2.1 |
| | | 2.1 | 2.0 | |
| | | 3.4 | 3.2 | |
| | 61°N | 2.8 | | |
| | 62°N | 1.2 | | |
| USSR | 67°N | 0.6 | | |
| | | 1.0 | | |

However, as noted in 15.3.5, the arguments of the productivity theory are somewhat hazy. The approach may well be valid but at this point should be viewed
with skepticism.

To look at communities with the assumption that either all niches are filled
or not tends to restrict one's attitude toward the flexibility of nature. Niches
don't just exist to be filled; they are in part created by their occupants and the
interaction of their occupants with other species in the community (see 15.3.2,
for example). This frame of mind may next lead to the notion that species fit into
their niches like books into a bookcase—that only so many fit in before the
community is saturated. One difficulty with this idea is given in 15.3.2.
Another is that as seasons change, some books are removed—through hibernation, migration, etc.—and others may replace them. Let us examine this notion
of replacement a little further.

Suppose that the populations of a group of animals are limited by food shortage
at the time when young must be fed. This is probably true for many birds
(Lack, 1954). Then we can view these creatures as differentiating their niches
not along some resource or vertical gradient—although this may occur too—but
along a time gradient. Figure 16.1 (Fig. 15.5 of the preceding chapter) now
depicts date of nesting against time, and the argument in Chapter 15.3.2 can be
applied. Were the situation this simple we would have to conclude that variance
in nesting date will decrease only with the addition of new species, but will not
provide open niche space for the arrival of new species (15.3.2). A longer
potential nesting season in the tropics will not necessarily result in the greater
separation of nesting dates and provide for greater numbers of nesting species.
However, suppose for the moment that greater niche separation occurs where
nesting can be widely spread over time. Niche separation then increases over its
optimal value as dictated by competition and there may exist spaces in time over
which competition is low and successful nesting by newly colonizing species is
likely. This could explain the higher number of species found in the tropics, at
least for many bird taxa. MacArthur (1964) tested this hypothesis, using data
from Skutch (1950) and Lack (1950). Table 16.6 gives the numbers of breeding
pairs viewed for all birds, by month (date of initiation of nest building).
MacArthur calculated an information theoretic index of diversity of nesting

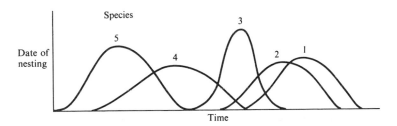

**Fig. 16.1.** Depiction of the niches of 5 species.

**Table 16.6.** Number of breeding bird pairs on a month by month basis

| Month | J | F | M | A | M | J | J | A | S | O | N | D |
|---|---|---|---|---|---|---|---|---|---|---|---|---|
| Costa Rica (Skutch) | 28 | 68 | 155 | 401 | 344 | 148 | 91 | 68 | 15 | 11 | 7 | 21 |
| England (Lack) | 0 | 1 | 285 | 1637 | 1413 | 726 | 163 | 56 | 4 | 0 | 0 | 0 |

months for each species ($H_i'$ for the $i$th species) and the same value for all species together ($H'$). Where $m$ species were present,

$$\bar{H}' = \frac{1}{m} \sum_{i=1}^{m} H_i'.$$

Now, since $e^{\bar{H}'}$ gives the equitability in months (number of equally good months) used for nesting by each species, on average, and $e^{H'}$ gives the equitability in months used by all species together, $e^{H'}/e^{\bar{H}'}$ is a measure of the turnover rate, or seasonal replacement of nesting bird species. MacArthur's calculations show seasonal replacements of 1.48 in Costa Rica, 1.37 in England, a ratio between the two areas of 1.08 : 1.00. If we make the intuitive (but not rigorously supported) supposition that ratio of seasonal replacement should equal the ratio in number of equally common bird species, then Costa Rica should possess 1.08 times as many equally common bird species as England. Since Costa Rica is comparatively much richer than that, MacArthur concluded that temporal separation was not terribly important as a cause of differences in species number.

**Table 16.7.** Temporal turnover rate in breeding birds and the number of bird species (see text for explanation)

| Area | Seasonal replacement | Number of nesting bird species with at least 20 observed nests |
|---|---|---|
| Costa Rica | 1.48 | 37 |
| E. Java | 1.43 | 71 |
| Kansas | 1.39 | 58 |
| Lapland | 1.37 | 40 |
| England | 1.37 | 15 |
| Uganda | 1.35 | 112 |
| Trinidad | 1.35 | 123 |
| W. Java (1) | 1.31 | 74 |
| S.E. Ecuador | 1.27 | 41 |
| Mid Java | 1.26 | 84 |
| W. Java (2) | 1.26 | 100 |

Ricklefs (1966) follows the same argument, and presents more data (Table 16.7). The similarity of seasonal replacement values and the total lack of correlation with either latitude or number of nesting bird species led him to the same conclusion as MacArthur: that temporal displacement in nesting was not an important factor in explaining temperate–tropical differences in number of bird species.

## 16.2 DIVERSITY AND SUCCESSION

### 16.2.1 General Information

There is a general belief that the number of species of any particular taxon increases with succession, reaching its peak at or slightly before climax. For the most part this belief seems to be true, but some qualifications and elaborations are in order.

Golley and Gentry (1964) plotted the total number of plant species/total number of individuals in a one-year-old and a twelve-year-old abandoned field, over each of eight months. They found, as expected, a higher relative number of species in the older field—from May through October. But in March and April, the young field showed greater diversity. This latter fact may be due to the influx into that field of opportunistic species still in the early stages of succession, species which later in the year die out. But it is clear that when comparing areas for species diversity we must guard against the temporary invader which, perhaps, should not be included in our calculations. Because of difficulties of this sort it is often convenient to stick to large perennials above some arbitrary age (or size) and regularly breeding animals.

Monk and McGinnis (1966) provide data on the changes in tree species diversity with succession in North Carolina and Florida. They found that within the areas examined, they could write, with some accuracy:

$$\log \text{ (number of individuals)} = \alpha + \beta \times \text{(number of species)}.$$

**Table 16.8.** Change in plant species diversity with succession (from Monk and McGinnis, 1966)

|  | Habitat type | $\beta$ |
|---|---|---|
| Successional | Cypress heads | 0.2090 |
|  | Sandhills and dry, mixed hardwood | 0.2262 |
| Fire disclimax should be intermediate in diversity | Flatwoods | 0.1275 |
|  | Bay heads | 0.1223 |
|  | Mixed swamps | 0.1378 |
| Climax southern mixed hardwood | Mesic, calcareous | 0.0677 |
|  | Mesic, noncalcareous | 0.0819 |
|  | West, calcareous | 0.0554 |
|  | Dry, calcareous | 0.0994 |
|  | Dry, noncalcareous | 0.1160 |

Thus $\beta$ was used as a measure of (lack of) diversity (Table 16.8). It is clear that diversity increases with succession here.

Pielou (1966) gives two other sets of data from six forest plots in eastern Canada. The first set of data was taken from stands which had been burned over 20, 22, 40, 40, 50, and 73 years previously. The second set was taken from the same stands five years later (Table 16.9). Reading across the table, there is no consistent increase either in species number or diversity ($H'$). Further, looking down the table, there is no trend over five years in the six areas examined. Pielou concludes that there is no significant change in these values with succession. This, however, may be too severe. The apparent rise and then fall in $H'$ with succession is too obvious to ignore. Equitability seems to increase with succession. This may be due to the evolutionary process described in part 15.5, but support for the operation of this process here would be difficult to obtain.

**Table 16.9.** Change in plant species diversity with succession (from Pielou, 1966)

|                   | Age of plot (years) | | | | | |
|                   | 20    | 22    | 40    | 40    | 50    | 73    |
|-------------------|-------|-------|-------|-------|-------|-------|
| Number of species | 4     | 7     | 8     | 13    | 9     | 6     |
| $H'$              | 1.235 | 2.168 | 2.048 | 2.709 | 1.668 | 1.383 |
| Equitability      | 0.83  | 0.25  | 0.97  | 1.15  | 1.60  | 1.81  |
| Number of species | 4     | 7     | 8     | 12    | 9     | 6     |
| $H'$              | 1.199 | 2.189 | 2.071 | 2.661 | 1.548 | 1.445 |
| Equitability      | 0.83  | 1.28  | 0.99  | 1.20  | 1.42  | 1.86  |
|                   | 25    | 27    | 45    | 45    | 55    | 78    |

Johnston and Odum (1956) examined the diversity of breeding bird species over a series of old fields, in various stages of succession, in Georgia. Their data are shown in Table 16.10. In view of the species–area relation and the limited area (about 20 acres) of the fields, it is difficult to know whether the increase in number of species reflects the age of the field or is the result of a sampling artifact reflecting the increased numbers of individuals. For the moment we accept the authors' view that, in fact, species number increases with succession. Whether the upward trend turns down again in late succession cannot be determined. The trend in winter residents in these fields seems to be one of increasing and then decreasing species number in older and older fields.

## 16.2.2  Some Speculation

It is clear that during primary succession the number of species increases. The data given above suggest that species number continues to increase through at least the early stages of secondary succession. Why this should be gives endless room for speculation but little to get our teeth into. Two rather obvious facts bear mentioning. First, species number tends to increase with spatial heterogeneity. As succession proceeds, community biomass increases, and with increasing biomass may come more structure to be subdivided. With the advent of great stability, late in secondary succession, fewer species fail to reach their competitive carrying capacities, so that competitive exclusion may become more common. Thus species number should increase until the approach of climax and then, perhaps, decline. Second, as succession proceeds toward an equilibrium (climax), there should be less fluctuating both of physical and biological parameters, less variability from place to place, and less variability in the number of different successional stages occurring simultaneously over a large area. A community which is in the early stages of succession will be a hodgepodge of a variety of stages, containing the species characteristic of all. This argument is countered by the fact that, later in succession, a small number of different stages may nevertheless run the gamut from very early to very late. Divergent stages share fewer species so that, in fact, the number of different species over locally differing areas may increase with succession. Where the successional process involves few steps and few major changes in life form (prairies, for example), the first argument may be of more importance. In areas with long, complex succession (forests) the latter may be the more important.

Table 16.10. The increase in bird species diversity with succession (from Johnston and Odum, 1956)

| Age of community (years) | Number of breeding species | Number of breeding pairs |
|---|---|---|
| 1–2 | 2 | 15 |
| 3 | 2 | 40 |
| 15 | 8 | 105 |
| 20 | 13 | 127 |
| 25 | 10 | 83 |
| 35 | 12 | 95 |
| 60 | 24 | 163 |
| 100 | 23 | 237 |
| 150 | 22 | 224 |

## 16.3 ISLANDS

One of the questions which has fascinated biogeographers and, more recently, other sorts of ecologists, is why islands—and to a smaller extent peninsulas—have so few species by comparison with equal-sized areas of mainland. Much study and speculation has gone into attempts to answer this question; a brief, partial review of these attempts follows.

## 16.3.1 Ecological Arguments

Lack (1969) suggests that where two species meet along a common border, they will gradually evolve tolerances for the physical environment occupied by the other and for the presence of the other species. This will allow for range extension and increasing range overlap. Over a large area there will be many abutting or slightly overlapping ranges and thus much opportunity for the evolution of further overlap. Any subarea will, in time, come to contain at least peripheral populations of many species. On small land areas, such as islands, there will be fewer abutting ranges, less opportunity for overlap in ranges, and over any given area, fewer species. Lack feels this may be one "minor reason" why islands have few species.

Another way of viewing the question is this: we suppose the first colonizer reaches an island. Where before this species' niche was of a given breadth, representing an equilibrium between intra and interspecific competition, its niche is now no longer constrained by competitors, and it broadens. This same argument applies to the second and third species, and so on, until all resources are utilized and all populations approach their competitor-affected carrying capacities. Readjustment of niches with each succeeding invasion keeps niche separation at its optimum value. Thus it should be about as difficult to invade this system as one with many more species, but in which niche separation is the same. By this argument, islands have fewer species because the invasion process has been shorter—islands are geologically younger than mainlands, or have experienced more recent geologic or climatological disturbance over their entire surface. Islands are likely to be almost universally younger than mainlands in this respect, simply because disturbances of the sort decimating large numbers of species are much less likely to occur continent-wide as they are over smaller areas. The fact that niches are generally broader on islands is fairly well accepted; recall that MacArthur et al. found Puerto Rican birds to distinguish less finely between fewer vertical zones than birds of either tropical or temperate mainland (15.3.1). Indirect evidence comes in the form of character displacement. On islands, if niches are broader, character displacement should be greater than on the mainland. Grant (1966) notes this to be true in island hummingbirds, and Selander (1966) finds marked character displacement between the sexes among island woodpeckers.

It may be only partly true that similar niche overlap values on island and mainland make them about equally difficult to invade, because islands seem to be more susceptible to invasion than mainlands (14.2.1). Of course, this may be due to the easy disturbance of islands, with their smaller areas, by man. Furthermore, there are data to support the contention that if one of a pair of species coexisting on the mainland establishes itself on an island, the other may have great difficulty invading, just as if it were a new invader to a mainland. Cameron (1964) finds that on the northeastern American mainland, the two mice *Microtus pennsylvanicus* and *Clethrionomys gapperi* live side by side, the former primarily in grassland, the latter in coniferous forests and occasionally scrubby areas. There are usually sharp boundaries between the two species, but in peak years

*M. pennsylvanicus* may sometimes occur in small numbers in grassy pockets in more open forest land. On Newfoundland, there are no *Clethrionomys*, and *Microtus* occurs throughout the forests as well as in grassland. Newfoundland is not far removed from the Canadian mainland and one presumes that *Clethrionomys* has had ample opportunity to invade the island. Nevertheless this hasn't happened. Grant (1966) notes that if established species can block colonization by competitors successful on the mainland, then the number of closely related species, being generally similar ecologically, should be less on small than on large islands. This is apparently the case.

### 16.3.2 Correlations

A more practical approach to the question of island species diversity is to gather data from many islands and calculate partial regression coefficients on several parameters. Such a study is that of Hamilton and Rubinoff (1963), who analyzed the distribution of Galapagos finches. They found the number of finch species on an island to be uncorrelated with the number of insular plant species and with island area. This latter finding is definitely not consistent with the results of other similar analyses. Darlington (1943) calculates the regression coefficient for carabid beetles on island area in the West Indies to be $+0.34$. For reptiles and amphibians in the West Indies, Hamilton *et al.* (1966) report a regression coefficient of $+0.301$, and for freshwater birds in the same area, $+0.237$. Preston (1962) gives a value of $+0.325$ for land plants in the Galapagos archipelago, and $+0.30$ for Ponerine ants in Melanesia (data of E. O. Wilson, cited in MacArthur and Wilson, 1967). Hamilton and Rubinoff (1963) *did* find a significant correlation between the number of Galapagos finch species and the distance from the nearest neighboring island, $D$, as well as the distance, $G$, from the archipelago center at Indefatigable Island:

$$\text{(number of finch species} = 9.8 - 0.10D - 0.02G).$$

With islands along the California coast, Johnson *et al.* (1968) find the regression coefficient of the number of plant species on the distance from the mainland to be $-0.0960$.

### 16.3.3 The Theory of Island Biogeography

Observation of recently formed islands tells us that the accumulation of species on islands is primarily one of dispersal and colonization, at least in the short run. Let us consider the colonization of islands. If species diversity on an island is to be predictable, then colonization must lead to the accumulation of species at a rate which falls and eventually reaches or approaches zero. For a given island, the rate should fall to zero when a characteristic number of species is present. If we clear the island of all life, we must, each time, regain the same (or nearly the same) species number at equilibrium. The classic studies by Patrick indicate that this may be so. For example, Patrick (1968) allowed diatoms, drifting through her laboratory in a natural stream in Pennsylvania, to colonize glass slides.   In one year's experiments, in eight "communities", with roughly 4509 individuals, only 4.5% of the total population was made up of species not common to all eight communities.   In the second year, 1% to 2% of 14,226 individuals were not common to four communities.   The diversity values for the two years were $H' = 2.2908$, and $H' = 2.5266$, and the number of species was 105 to 114, and 111 to 124.   The number of individuals varied, but the number and diversity of species and, to a large extent, the specific species were quite constant.   She concluded that under similar ecological conditions, at the same time, what and how many species will colonize similar substrates is highly predictable.   In another experiment, Cairns *et al.* (1969) floated two series of sponges in four to five feet of water in Douglas Lake, Michigan.   They checked the number of protozoan species colonizing and becoming extinct per unit time, and found that although the specific species differed in the two series, the rates of colonization and extinction were remarkably similar.   Data derived from experimental islands off the Florida coast (more fully discussed below) showed rather consistent trends toward predictable insect species number (Simberloff and Wilson, 1969).

If we accept the idea that the number of species on an island is due largely to colonization and extinction and is predictable, then by far the most satisfying theory to account for insular diversity is that of MacArthur and Wilson (1963, 1967).   These authors also considered the question of speciation, but this is not covered here.   The heading for this section is taken from these authors' (1967) book, *The Theory of Island Biogeography*, and the following discussion is largely condensed from their treatment.

For any island we can plot the expected number of new species immigrating and the expected number of species becoming extinct against the number of species on the island.   It is clear that immigration of new species will decline as the number already present increases, and it seems likely that if there are more species present, there will be more becoming extinct.   Also we might expect immigration to be higher for islands nearer the mainland and extinction lower on large islands where there is more opportunity for intra-island replacement of dying populations and perhaps more latitude for niche adjustment.   Thus we can draw curves as in Fig. 16.2.   Where immigration and extinction are equal—the curves cross—there will be an equilibrium.   On the basis of the above diagram, distant and small islands should have the fewest species.   As we have seen (part 16.2) this is true.

In the mathematically simplest case, the curves will be linear.   Let

$p$ = species pool on the mainland

$s$ = species on the island

$\lambda$ = immigration rate

$\mu$ = extinction rate

Then the immigration curve is given by

$$I = ds/dt = \lambda(p - s),$$

and the extinction curve by

$$ds/dt = \mu s.$$

When the two are equal,

$$\lambda(p - \hat{s}) = \mu\hat{s},$$

where $\hat{s}$ is the equilibrium number of species, so that

$$\hat{s} = \lambda p/(\lambda + \mu). \qquad (16.1)$$

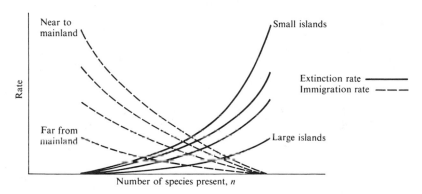

**Fig. 16.2.** Immigration and extinction rates for islands (after MacArthur and Wilson, 1963)

While this expression is a woeful oversimplification, it can nevertheless tell us something about the kinds of correlations we might expect to find in the study of islands (see 16.2). It is already clear in Fig. 16.1 that species number is positively correlated with island area and negatively correlated with distance from the mainland. Where $D$ is the distance from the mainland, we can write

$$\frac{d\hat{s}}{dD} = \frac{d\hat{s}}{d\lambda}\frac{d\lambda}{dD} = \frac{d}{d\lambda}\left(\frac{\lambda p}{\lambda + \mu}\right)\frac{d\lambda}{dD} = \frac{\mu p}{(\lambda + \mu)^2}\frac{d\lambda}{dD}.$$

This is small when $\mu$ is small and $\lambda$ large, conditions which exist when an island is large. Thus the distance effect ($d\hat{s}/dD$) should be smaller on large than small islands.

As a simple attempt to make the above expressions more specific, we suppose that the rate at which species arrive on an island is proportional to the angle the island subtends to some point on the mainland $\propto A^{\frac{1}{2}}/D$. Extinction rate should vary with some negative power of $A$, say $(\frac{1}{2} - C_3)$ where $C_3 > \frac{1}{2}$ is a constant. Then

$$\lambda = C_1\frac{A^{\frac{1}{2}}}{D}, \qquad \mu = C_2 A^{\frac{1}{2} - C_3},$$

and

$$\hat{s} = \frac{C_1 A^{\frac{1}{2}} p/D}{C_1 A^{\frac{1}{2}}/D + C_2 A^{\frac{1}{2}-C_3}} = \frac{p}{H(C_2/C_1)A^{-C_3}}.$$

Thus

$$d\hat{s}/dA = \frac{C_2}{C_1} DC_3 A^{-(C_3+1)} p \Big/ \left(1 + \frac{C_2}{C_1} DA^{-C_3}\right)^2.$$

Since the above values decline as $D$ increases we conclude that the area effect ($d\hat{s}/dA$) is less marked for distant islands.

Note that we may write

$$ds/dt = \lambda(p - s) - \mu s,$$

so that, where $\lambda$, $\mu$ are independent of $s$,

$$\int \frac{ds}{\lambda(p - s) - \mu s} = \int dt, \quad \text{or} \quad -\left(\frac{1}{\lambda + \mu}\right) \log \left[\lambda(p - s) - \mu s\right]_0^s = t.$$

Thus

$$\log \left[\frac{\lambda(p - s) - \mu s}{\lambda p}\right] = -(\mu + \lambda)t, \quad \lambda(p - s) - \mu s = \lambda p\, e^{-(\mu + \lambda)t},$$

and

$$s = \frac{\lambda p}{\lambda + \mu} (1 - e^{-(\lambda + \mu)t}) = \hat{s}(1 - e^{-(\lambda + \mu)t}). \tag{16.2}$$

Thus when an island is one-half of the way to saturation,

$$\tfrac{1}{2}\hat{s} = \hat{s}(1 - e^{-(\lambda + \mu)t_{\frac{1}{2}}}), \quad \text{and} \quad t_{\frac{1}{2}} = \ln 2/(\lambda + \mu).$$

Since $\lambda$ decreases with $D$, large $D$ implies large $t_{\frac{1}{2}}$. Distant islands reach equilibrium more slowly than near islands.

For further exploration the reader is referred to the original source (MacArthur and Wilson, 1967).

Since the publication of MacArthur and Wilson's book there have been a number of experiments designed to measure immigration and extinction curves. The work by Cairns *et al.* mentioned above was one. Another is that of Wilson (1969) who covered small mangrove islands in the Florida keys with plastic nets and killed all the arthropods by spraying poison. One island, 500 meters from land, refilled slowly; the other, nearer islands more rapidly. Those closest to shore refilled the fastest (as expected). Within about 280 days all islands except the most distant reached species number peaks above their original values, and then slumped back roughly to those original values. This is to be expected since there must be a time lag between the arrival of new competitors and the exclusion of competitors. That is, the extinction curve shows a slight lag. As Wilson points out, immigration was fast enough that by 280 days populations were still low and competitive interactions had barely made themselves felt. Table 16.11 is constructed from Wilson's data on island E9.

**Table 16.11.** Colonization by insects of a small mangrove island near Florida (data from Wilson, 1969)

| Time after island was reopened | Number of new species | Number of species disappearing | Total number of species |
|---|---|---|---|
| 24 | 1 | — | 1 |
| 45 | 8 | 0 | 9 |
| 62 | 5 | 5 | 9 |
| 84 | 9 | 2 | 16 |
| 101 | 6 | 4 | 18 |
| 117 | 3 | 4 | 17 |
| 136 | 2 | 1 | 18 |
| 153 | 6 | 1 | 23 |
| 171 | 7 | 4 | 26 |
| 193 | 7 | 4 | 29 |
| 210 | 8 | 5 | 32 |
| 229 | 8 | 6 | 34 |
| 247 | 7 | 7 | 34 |
| 266 | 7 | 9 | 32 |
| | Total 84 | 52 | |

From Table 16.11 data we can estimate, for the time between each census, the number of new species colonizing per day $(= \lambda(p - \bar{s}))$, where $\bar{s}$ is the estimated average number of species present over that period (mean of the number in the preceding and the following census), and the number becoming extinct per day $(= \mu \bar{s})$. Plotting these immigration and extinction values against $\bar{s}$, we find both curves to be extremely erratic. The value of $\mu$ appears to be roughly 0.0118, but the immigration curve rises ($\lambda \approx -0.0019$). (This may be an artifact of the linear approximation.) These results seem rather discouraging as support for the notion of smoothly rising extinction and smoothly falling immigration curves, but of course do not affect the basic hypothesis that $\hat{s}$ is an equilibrium between immigration and extinction. To those who consider that these results indicate low predictability of island diversity, it should be pointed out that large statistical error might be expected (Simberloff, 1969) especially when dealing with very small islands and very rapid immigration rates in a seasonally changing area.

# Discussion IV and Epilog

Unlike the material explored in the previous sections, that presented in Chapters 13 through 16 is too far removed from the laws of natural selection for any rigorous formulation of general axioms. This should be evident from the restraint with which hypotheses have been treated in this section, and the occasional references to rigorous thought or lack of it. Further clarification or general theoretical constructs or additions to theory of community ecology are therefore not attempted here. Since community ecology is the most pressing, practical aspect of ecology, however, it deserves further comment of a less technical nature.

The literature abounds in examples of the misuse of ecology and of misled attempts to improve man's environment. In Malaysia, for example, the United States has sprayed vast quantities of DDT in a program to kill mosquitoes and control malaria. The result was that cockroaches also ingested the pesticide and died. The lizards which ate the cockroaches followed suit. Cats, which ate both rats and lizards, were thus deprived of part of their diet and their numbers dwindled, allowing a temporary increase in the rat population. The immediate result of this was an increase in the number of fleas and the rapid spread of bubonic plague. Furthermore, the decreased lizard population allowed the population of one of their prey species, a straw-eating insect, to explode. The people, saved from malaria, died of the plague as their thatched roofs caved in around them (Gunter, 1970). Other painful examples need not be repeated here.

Two factors seem paramount in causing the sort of problem described above. One is clearly political. Government and industry are conservative institutions which resist any change even remotely threatening their continuing existence. The suggestion that DDT production and use be curbed brings immediate cries from the government which sees a danger to foreign relations, and from business which depends on the chemical for a part of its profits. Business interests, needless to say, have a very powerful influence on government policy. Efforts to curb pollution and control pesticides have often been opposed in subtle but significant ways by the regulatory agencies which, although designed to regulate, have evolved into promoting agencies. Successes in battles to save our environment and ourselves have occurred primarily when concerned people abandoned some of their efforts at opposing moneyed pressure groups in Congress and went to the less politically influenced, more open-minded forums provided by the courts. Yet, even in the case of DDT where the scientific facts are thoroughly understood and documented, the battle is uphill.

The second difficulty is a matter of awareness and knowledge. The recent upsurge in public concern for the environment is encouraging but not, so far, very significant. People still prefer ease to aesthetics, and the excitement of motor-boating seems to override the joy of quietly watching a loon not yet frightened away by noise and oil slicks. Most people would rather enjoy the convenience of tossing aluminum beer cans from their car windows than returning bottles.

Scientific knowledge of ecology is growing rapidly and the empirical data are beginning to provide the needed awareness of the subtleties of ecological inter-actions. But empirical data only show what happens, not why, and eventually, without the insights that theory alone can provide, must bog down as a back-ground for the intelligent management of the environment. Efforts of the sort made by such scientific–legal action groups as the Environmental Defense Fund can continue only when backed with good science, and too little data—and less understanding of the important questions—exist at the present time.

I end on a note of some pessimism. I am optimistic when it comes to the eventual awakening of the intellectual to problems of survival and his influence on political and economic change. But his direction of the change, especially in a world changing very rapidly, must be solidly based on the kind of understanding we are only beginning to glean. We cannot afford another panacea like DDT. Yet the number of wild areas, those places where evolutionary studies of ecology should be made, is rapidly dwindling and many habitat types have either disappeared or are represented by areas too small to be even remotely immune from devastating human influence. DDT occurs even in the Antarctic Ocean, thousands of miles from its nearest site of application, and lead from automobile exhaust is found in polar glaciers. Even in such seemingly wild areas as the Labrador peninsula, lumbering and other activities of man may have changed conditions significantly from those in which the ecological interactions we observe evolved. "Even on the extensive ranges now being provided by public-spirited conservationists, the bison remains a restricted and a typical offshoot of the freely migrating animal that once roamed our fenceless western plains. As a complete creature, the American bison is extinct. A unique experiment of nature has been terminated before its end product was adequately described or understood."

"There is no need to belabor the point that an animal removed from its natural environment or forced to live in a modified environment no longer provides us with the material needed for the final understanding of its real nature. It is no longer a whole animal. Like an excised frog's heart, it has lost its significance as a vital member of the organization to which it belongs.

"But far more serious to our cherished and dwindling heritage of unblemished wild creatures than extinction within its natural range is the sweeping and often drastic alteration of environments which modern mechanized man produces wherever he goes. In all but a few parts of the world, the ground has already been drastically transformed under the very feet of the animal inhabitants. Most species have adjusted to these changes—in some cases the adjustment has been achieved only with a struggle and a loss, in others it has been accomplished with overwhelming success.  The squirrels on our lawns, the swallows in our barns, the larks in our pastures, the weevils in our bean fields are examples of animals which have materially benefited by man's crude reworking of the environmental substrate.  Pheasants and house wrens have been deliberately managed through planned manipulations of the environment.  Such animals are of particular interest to the student of behavioral adjustments, but they are lost to us as subjects for the study of natural adaptations.  Like an urbanized aborigine, their actions and responses can only suggest, and with little reliability, the natural context of their lives." (Emlen, 1964).

How many of our conclusions are incorrect because they are based on measurements in man-altered environments?  Where in the United States, or over most of the world, could we see the sort of event described by Cox (1936) and viewed as recently as the early 1900's:

"One morning (in Glacier park), dark streaks were observed extending downward at various angles from the saddles or gaps on the mountains to the east of us.  Later in the day these streaks appeared to be much longer and at the lower end of each there could be discerned a dark speck.  Through binoculars these spots were seen to be animals floundering downward in the deep, soft snow. As they reached lower levels not so far distant, they proved to be porcupines. From every little gap there poured forth a dozen or twenty, or in one case actually fifty five, of these animals, wallowing down to timberline on the west side.  Hundreds of porcupines were crossing the main range of the Rockies."

## BIBLIOGRAPHY, SECTION IV

Agnew, A. D. Q., and J. E. C. Flux, 1970, Plant dispersal by hares (*Lepus capensis* L.) in Kenya, *Ecol.* **51**: 735–737.

Bazykin, A. D., 1969, Hypothetical mechanism of speciation, *Evol.* **23**: 685–687.

Belyea, R. M., 1952, Death and deterioration of balsam fir weakened by spruce budworm defoliation in Ontario, *J. Forestry* **50**: 729–738.

Bourliere, F., 1955, *The Natural History of Mammals,* Knopf, New York,

Bradt, G. W., 1938, A study of beaver colonies in Michigan, *J. Mammal.* **19**: 139–162.

Bray, J. R., and E. Gorham, 1964, Litter production in forests of the world, in J. B. Cragg (Ed.) *Advances in Ecological Research,* Academic Press, London and New York.

Brown, R. G. B., 1967, Species isolation between the herring gull, *Larus argentatus,* and lesser black-backed gull, *Larus fuscus, Ibis* **109**: 310–317.

Bush, G. L., 1969, Sympatric host race formation and speciation in frugivorous flies of the genus *Rhagoletis* (Diptera, Tephritidae), *Evol.* **23**: 237–251.

Butler, L., 1953, The nature of cycles in populations of Canadian mammals, *Can. J. Zool.* **31**: 242–262.

Buxton, P. A., 1923, *Animal Life in Deserts,* Edward Arnold, London.

Buzas, M. A., and T. G. Gibson, 1969, Species diversity: benthonic Foraminifera in western North Atlantic, *Science* **163**: 72–75.

Cairns, J., Jr., M. L. Dahlberg, K. L. Dickson, N. Smith, and W. T. Walker, 1969, The relation of freshwater protozoan communities to the MacArthur–Wilson equilibrium model, *Amer. Natur.* **103**: 439–454.

Cameron, A. W., 1964, Competitive exclusion between the rodent genera *Microtus* and *Clethrionomys, Evol.* **18**: 630–634.

Carlquist, S., 1966, The biota of long-distance dispersal, I, Principles of dispersal and evolution, *Quart. Rev. Biol.* **41**: 247–270.

Caughley, G., 1970, Eruption of ungulate populations, with emphasis on Himalayan thar in New Zealand, *Ecol.* **51**: 53–72.

Cody, M., 1968, On the methods of resource division in grassland bird communities, *Amer. Natur.* **102**: 107–147.

Connell, J., and E. Orias, 1964, The ecological regulation of species diversity, *Amer. Natur.* **98**: 399–414.

Cooper, W. S., 1913, The climax forest of Isle Royale, Lake Superior, and its development, *Bot. Gazette,* **58**: 1–44, 115–140, 183–235.

Cox, W. T., 1936, Snowshoe rabbit migration, tick infestation, and weather cycles, *J. Mammal.* **17**: 216–221.

Cruden, R. W., Birds as agents of long-distance dispersal for disjunct plant groups of the temperate Western Hemisphere, *Evol.* **20**: 517–532.

Cushing, D. H., 1968, *Fisheries Biology: A Study in Population Dynamics*, University of Wisconsin Press, Madison.

Dahlberg, B. L., and R. C. Guettinger, 1956, The white-tailed deer in Wisconsin, *Wisc. Conserv. Dept. Tech. Wildlife Bull.* No. 14.

Darlington, P., J. 1943, Carabidae of mountains and islands: data on the evolution of isolated faunas, and on atrophy of wings, *Ecol. Monog.* **13**: 37–61.

DeBach, P., 1965, Some biological and ecological phenomena associated with colonizing entomophagous insects, in H. G. Baker and G. L. Stebbins (Eds.), *The Genetics of Colonizing Species,* Academic Press, New York.

DeVore, J., and S. L. Washburn, 1963, Baboon ecology and human behavior, in F. C. Howell and F. Bourliere (Eds.), *African Ecology and Human Evolution,* Wenner–Gren Foundation, Aldine, Chicago.

Dobzhansky, T., 1950, Evolution in the tropics, *Amer. Sci.* **38**: 209–221.

Donnelly, U., 1947, Seasonal breeding and migrations of the desert locust (*Schistocerca gregaria* Forskal) in western and northwestern Africa, *Anti-locust Mem. (Brit. Mus. Nat. Hist.)* **3**: 1–43.

Dunbar, M. J., 1960, The evolution of stability in marine environments; natural selection at the level of the ecosystem, *Amer. Natur.* **94**: 129–136.

Ehrlich, P. R., and P. H. Raven, 1969, Differentiation of populations, *Science* **165**: 1228–1232.

Elton, C. S., 1958, *The Ecology of Invasions by Animals and Plants,* Methuen, London.

Emlen, J. M., 1966a, Time, energy, and risk in two species of carnivorous gastropods, Ph.D. thesis, University of Washington, Seattle.

Emlen, J. M., 1966b, The role of time and energy in food preference, *Amer. Natur.* **100**: 611–617.

Emlen, J. M., 1968, Optimal choice in animals, *Amer. Natur.* **102**: 385–389.

Emlen, J. T., 1964, Wilderness and behavior research, *Biosc.* **14**: 32–33.

Fischer, A. G., 1960, Latitudinal variation in organic diversity, *Evol.* **14**: 64–81.

Fisher, R. A., A. S. Corbet, and C. B. Williams 1943, The relation between the number of species and the number of individuals in a random sample of an animal population, *J. Anim. Ecol.* **12**: 42–58.

Gibb, J. A., 1962, L. Tinbergen's hypothesis of the role of specific search images, *Ibis* **104**: 106–111.

Gibb, J. A., 1966, Tit predation and the abundance of *Ernarmonia conicolana* (Heyl) on Weeting Heath, Norfolk, 1962–1963, *J. Anim. Ecol.* **35**: 43–53.

Gillett, J. B., 1962, Pest pressure, an underestimated factor in evolution, *Syst. Assoc. Publ.* No. 4 (Taxonomy and Geography) pp. 37–46.

Godwin, H., 1956, *The History of the British Flora,* Cambridge Univ. Press, Cambridge.

Golley, F. B., 1968, Secondary productivity in terrestrial communities, *Amer. Zool.* **8:** 53–59.

Golley, F. B., and J. B. Gentry, 1964, A comparison of variety and standing crop of vegetation on a 1-year and a 12-year abandoned field, *Oikos* **15:** 185–199.

Grant, P. R., 1966, Ecological compatability of bird species on islands, *Amer. Natur.* **100:** 451–462.

Gross, A. O., 1940, The migration of Kent Island herring gulls, *Bird Banding* **11:** 129–155.

Gunter, P. A., 1970, Mental inertia and environmental decay: the end of an era, *The Living Wilderness* **34:** 3–7.

Hamilton, T. H., R. H. Barth, Jr., and I. Rubinoff, 1966, The environmental control of insular variation in bird species abundance, *Proc. Nat. Acad. Sci.* **52:** 132–140.

Hamilton, T. H., and I. Rubinoff, 1963, Isolation, endemism, and multiplication of species in the Darwin finches, *Evol.* **17:** 388–403.

Harper, J. L., 1969, The role of predation in vegetational diversity, in *Diversity and Stability in Ecological Systems,* Brookhaven Symp. in Biology No. 22 pp. 48–61.

Harper, J. L., J. T. Williams, and G. R. Sagar, 1965, The behavior of seeds in soil, I, The heterogeneity of soil surfaces and its role in determining the establishment of plants from seed. *J. Ecol.* **53:** 273–286.

Harvey, H. W., 1950, On the production of living matter in the sea off Plymouth, *J. Mar. Biol. Assoc. U.K., * **29:** 97–137.

Hayne, D. W., and R. C. Ball, 1956, Benthic productivity as influenced by fish predation, *Limnol & Ocean* **1:** 162–175.

Hershkowitz, P., 1962, The evolution of neotropical cricetine rodents, *Fieldiana, Zool.* No. 46.

Holling, C. S., 1965, The functional response of predators to prey density and its role in mimicry and population regulation, *Mem. Ent. Soc. Canada,* **45:** 5–60.

Hope-Simpson, J. F., 1940, Studies of the vegetation of the English chalk, VI, Late stages in succession leading to chalk grassland, *J. Ecol.* **28:** 386–402.

Horn, H., 1970, *Forest Succession: the Form, Function, and Ecology of Trees,* Princeton University Press, Princeton.

Howard, W. E., 1965, Interaction of behavior, ecology, and genetics of introduced mammals, in H. G. Baker and C. and L. Stebbins, (Eds.), *The Genetics of Colonizing Species,* Academic Press, New York.

Hurlbert, S. H., 1971, The nonconcept of species diversity: A critique and alternative parameters, *Ecol.* **52:** 577–586.

Hutchinson, G. E., 1951, Copepodology for the ornithologist, *Ecol.* **32:** 571–577

Janzen, D. H., 1967, Why mountain passes are higher in the tropics, *Amer. Natur.* **101:** 233–250.

Janzen, D. H., 1970, Herbivores and the number of tree species in tropical forests, *Amer. Natur.* **104:** 501–528.

Johnson, C. 1966, Environmental modification of habitat selection in adult damselflies, *Ecol.* **47:** 674–676.

Johnson, M. P., L. G. Mason, and P. H. Raven, 1968, Ecological parameters and plant species diversity, *Amer. Natur.* **102:** 297–306.

Johnston, D. W., and E. P. Odum, 1956, Breeding bird populations in relation to plant succession on the piedmont of Georgia, *Ecol.* **37:** 50–62.

Keith, L. B., 1963, *Wildlife's 10-year Cycles,* University of Wisconsin Press, Madison.

Kennedy, J. S., 1956, Phase transformation in locust biology, *Biol. Rev.* **31:** 349–370.

Klopfer, P. H., 1959, Environmental determinates of faunal diversity, *Amer. Natur.* **93:** 337–342.

Klopfer, P. H., and R. H. MacArthur, 1961, On the causes of tropical species diversity: niche overlap, *Amer. Natur.* **95:** 223–226.

Kurten, B., 1969, Continental drift and evolution, *Sci. Amer.* **220**: 54–64.

Kusnezov, M., 1957, Numbers of species of ants in fauna of different latitudes, *Evol.* **11**: 298–299.

Lack, D., 1933, Habitat selection in birds with special reference to the effects of afforestation on the Breckland avifauna. *J. Anim. Ecol.* **2**: 239–262.

Lack, D., 1938, The psychological factor in bird distribution, *Brit. Birds.* **31**: 130–136.

Lack, D., 1940, Habitat selection and speciation in birds, *Brit. Birds.* **34**: 80–84.

Lack, D., 1950, The breeding seasons of European birds, *Ibis,* **92**: 288–316.

Lack, D., 1954, *The Natural Regulation of Animal Numbers,* Oxford University Press, London.

Lack D., 1969, Tit niches in two worlds; or homage to Evelyn Hutchinson, *Amer. Natur.* **103**: 43–49.

Leigh, E. G., Jr., 1965, On the relation between productivity, biomass, diversity, and stability of a community, *Proc. Nat. Acad. Sci.* **53**: 777–782.

Lewontin, R. C., 1969, The meaning of stability, in *Diversity and Stability in Ecological Systems,* Brookhaven Symposium in Biol. No. 22.

Lindeman, R. L., 1942, The trophic–dynamic aspect of ecology, *Ecol.* **23**: 399–418.

Lindsey, C. C., 1966, Body sizes of poikilotherm vertebrates at different latitudes, *Evol.* **20**: 456–465.

MacArthur, R. H., 1955, Fluctuations of animal populations and a measure of community stability, *Ecol.* **36**: 533–536.

MacArthur, R. H., 1958, Population ecology of some warblers of north-eastern coniferous forests, *Ecol.* **39**: 599–619.

MacArthur, R. H., 1959, On the breeding distribution pattern of North American migrant birds, *Auk.* **76**: 318–325.

MacArthur, R. H., 1960, On the relation between reproduction value and optimal predation, *Proc. Nat. Acad. Sci.* **46**: 143–145.

MacArthur, R. H., 1964, Environmental factors affecting bird species diversity, *Amer. Natur.* **98**: 387–397.

MacArthur, R. H., and J. Connell, 1966, *The Biology of Populations,* Princeton University Press, Princeton, New Jersey.

MacArthur, R. H., and R. Levins, 1964, Competition, habitat selection, and character displacement in a patchy environment, *Proc. Nat. Acad. Sci.* **51**: 1207–1210.

MacArthur, R. H., and R. Levins, 1967, The limiting similarity of convergence and divergence of coexisting species, *Amer. Natur.* **101**: 377 385.

MacArthur, R. H., and J. W. MacArthur, 1961, On bird species diversity *Ecol.* **42**: 594–598.

MacArthur, R. H., H. Recher, and M. S. Cody, 1966, On the relation between habitat selection and species diversity, *Amer. Natur.* **100**: 319–325.

MacArthur, R. H., and E. O. Wilson, 1963, An equilibrium model of insular zoogeography, *Evol.* **17**: 373–387.

MacArthur, R. H., and E. O. Wilson, 1967, *The Theory of Island Biogeography* Princeton University Press, Princeton, New Jersey.

Malone, C. M., 1965, Killdeer, (*Charadrius vociferus*) as a means of dispersal for aquatic gastropods, *Ecol.* **46**: 551–555.

Margalev, R., 1958, Temporal succession and spatial heterogeneity in phytoplankton, in A. A. Buzzati–Traverso (Ed.), Perspectives in Marine Biology, *Union Internationale des Sciences Biologiques* No. 27.

Margalev, R., 1963, On certain unifying principles in ecology, *Amer. Natur.* **97**: 357–374.

Margalev, R., 1968, *Perspectives in Ecological Theory,* University of Chicago Press, Chicago.

Mayr, E., 1963, *Animal Species and Evolution,* Belknap Press, Harvard.

Mayr, E., 1965, The Nature of Colonization in Birds, in H. G. Baker and G. L. Stebbins (Eds.), *The Genetics of Colonizing Species,* Academic Press, New York.

Moir, W. H., 1969, The lodgepole pine zone in Colorado, *Amer. Midl. Natur.* **81**: 87–98.

Monk, C. D., and J. T. McGinnis, 1966, Tree species diversity in six forest types in North Carolina and Florida, *J. Ecol.* **54**: 341–344.

Murray, K. F., 1965, Population changes during the 1957–58 vole (*Microtus*) outbreaks in California, *Ecol.* **46**: 163–171.

Nicholson, A. J., 1933, The balance of animal populations, *J. Anim. Ecol.* **2**: 132–178.

Odum, E. P., 1960, Organic production and turnover in old field succession, *Ecol.* **41**: 34–49.

Odum, E. P., 1969, The strategy of ecosystem development, *Science* **164**: 262–269.

Odum, H. T., 1957, Trophic structure and productivity of Silver Springs, Florida, *Ecol. Monog.* **27**: 55–112.

Olson, J. S., 1963, Energy storage and the balance of procedure and decomposers in ecological systems, *Ecol.* **44**: 322–331.

Oosting, H. J., 1958, *The Study of Plant Communities,* second edition, Freeman, San Francisco.

Paine, R. T., 1966, Food web complexity and species diversity, *Amer. Natur.* **100**: 65–76.

Paine, R. T., 1969, A note on trophic complexity and community stability, *Amer. Natur.* **103**: 91–93.

Paine, R. T., and R. L. Vadas, 1969, The effects of grazing by sea urchins, *Strongylocentrotus sp.,* on benthic algal populations, *Limnol and oceanog.* **14**: 710–719.

Patrick, R., 1968, The structure of diatom communities in similar ecological conditions, *Amer. Natur.* **102**: 173–183.

Pelseneer, P., 1935, Essai d'ethologie zoologique d'apres l'etude des mollusques, *Acad. R. Belge. Cl. Sci. publ.* Foundation Agathon de Potter, **1**: 1–662.

Pianka, E. R., 1966, Latitudinal gradients in species diversity: a review of concepts, *Amer. Natur.* **100**: 33–46.

Pielou, E. C., 1966, Species-diversity and pattern diversity in the study of ecological succession, *J. Theoret. Biol.* **10**: 370–383.

Pielou, E. C., 1969, *An Introduction to Mathematical Ecology,* Wiley-Interscience, New York.

Pimentel, D., G. J. C. Smith, and J. Soans, 1967, A population model of sympatric speciation, *Amer. Natur.* **101**: 493–504.

Preston, F. W., 1948, The commonness and rarity of species, *Ecol.* **29**: 254–283.

Preston, F. W., 1962, The canonical distribution of commonness and rarity, Part I, *Ecol.* **43**: 185–215. Part II, *Ecol.* **43**: 410–432.

Ramirez, W. B., 1969, Fig wasps: mechanism of pollen transfer, *Science* **163**: 580–581.

Ramirez, W. B., 1970, Host specificity of fig wasps (*Agaonidae*) *Evol.* **24**: 680–691.

Recher, H. F., 1969, Bird species diversity and habitat diversity in Australia and North America, *Amer. Natur.* **103**: 75–80.

Revill, D. L., K. W. Stewart, and H. E. Schlichting, Jr., 1967, Passive dispersal of viable algae and Protozoa by certain craneflies and midges, *Ecol.* **48**: 1023–1027.

Ricklefs, R. E., 1966, The temporal component of diversity among species of birds, *Evol.* **20**: 235–242.

Riley, G. A., 1956, Oceanography of Long Island Sound, IX, Production and utilization of organic matter, *Bull. Bingham Oceanogr. Coll.* **15**: 324–344.

Sanders, H. L., 1968, Marine benthic diversity: a comparative study, *Amer. Natur.* **102**: 243–282.

Sanders, H. L., and R. R. Hessler, 1969, Ecology of the deep-sea benthos, *Science* **163**: 1419–1424.

Savely, H. E., 1939, Ecological relations of certain animals in dead pine and oak logs, *Ecol. Monog.* **9**: 321–385.

Schildmacher, H., 1929, Uber den Warmehaushelt Kleiner Kornfresser, *Ornithol. Monat.* **37**: 102–106.

Selander, R. K., 1966, Sexual dimorphism and differential niche utilization in birds, *Condor* **68**: 113–151.

Serie, P., 1936, Distribution Geographica de los Ofidios Argentinos, *Obra Cincuentenario Museo de la Plata* **2**: 33–61.

Shannon, C. E., and W. Weaver, 1963, *The Mathematical Theory of Communication,* University of Illinois Press, Urbana.

Shelford, V. E., 1945, The relation of snowy owl migration to the abundance of the collared lemming, *Auk,* **62**: 592–596.

Simberloff, D. S., 1969, Experimental zoogeography of islands: a model for insular colonization, *Ecol.* **50**: 296–314.

Simberloff, D. S., and E. O. Wilson, 1969, Experimental zoogeography and islands: The colonization of empty islands, *Ecol.* **50**: 278–296.

Simpson, G. G., 1964, Species density of North American recent mammals, *Syst. Zool.* **13**: 57–73.

Skutch, A. F., 1950, The nesting seasons of Central American birds in relation to climate and food supply, *Ibis,* **92**: 185–222.

Slobodkin, L. B., 1959, Energetics in *Daphnia pulex* populations, *Ecol.* **40**: 232–243.

Slobodkin, L. B., 1961, *The Growth and Regulation of Animal Numbers,* Holt, Rinehart, and Winston, New York.

Slobodkin, L. B., 1968, How to be a predator, *Amer. Zool.* **8**: 43–51.

Slobodkin, L. B., and H. L. Sanders, 1969, On the contribution of environmental predictability to species diversity, in *Diversity and Stability in Ecological Systems,* Brookhaven Symp. in Biology No. 22.

Smith, A. D., 1940, A discussion of the application of a climatological diagram, the hythergraph, to the distribution of natural vegetation types, *Ecol.* **21**: 184–191.

Southwood, T. R. E., 1961, The number of species of insect associated with various trees, *J. Anim. Ecol.* **30**: 1 8.

Stehli, F. G., R. G. Douglas, and N. D. Newell, 1969, Generation and maintenance of gradients in taxonomic diversity, *Science* **164**: 947–949.

Svärdson, G., 1949, Competition and habitat selection in birds, *Oikos* **1**: 157–174.

Teal, J. M., 1962, Energy flow in the salt marsh ecosystems of Georgia, *Ecol.* **43**: 614–624.

Thorson, G., 1952, Zur Jatzigen Lage der Marinen Bodentier-Ökologie, *Verk. de Deutsch. Zool Ges.: Zool. Anzeiger Suppl. Bd.* **16**: 276–327.

Thorson, G., 1957, Bottom communities (sublittoral or shallow shelf), in J. W. Hedgpeth (Ed.), *Treatise on Marine Ecology and Palaecology,* Vol. 1, Geol. Soc. Amer. Mem. No. 67.

Tinbergen, L., 1960, The natural control of insects in pine woods, I, Factors influencing the intensity of predation by songbirds, *Archiv. Neerl. de Zool.* **XIII**: 265–343.

Tinbergen, N., M. Impekoven, and D. Franck, 1967, An experiment on spacing-out as a defense against predation, *Beh.* **28**: 307–321.

Turner, F. B., 1970, The ecological efficiency of consumer populations, *Ecol.* **51**: 741–742.

Van Valen, L., 1965, Morphological variation and width of ecological niche, *Amer. Natur.* **99**: 377–390.

Warner, R. E., 1968, The role of introduced diseases in the extinction of the endemic Hawaiian avifauna, *Condor* **70**: 101–120.

Watt, K. E. F., 1955, Studies on population productivity, I, Three approaches to the optimal yield problem in populations of *Tribolium confusum, Ecol. Monog.* **25**: 269–290.

Watt, K. E. F., 1968, *Ecology and Resource Management (A Quantitative Approach),* McGraw-Hill, New York.

Wecker, S.C., 1963, The role of early experience in habitat selection by the prairie deer-mouse, *Peromyscus maniculatus bairdi, Ecol. Monog.* **33**: 307–325.

Welty, J. C., 1962, *The Life of Birds,* Saunders, Philadelphia, and London.

Williams, C. B., 1964, *Patterns in the Balance of Nature,* Academic Press, New York.

Wilson, E. O., 1965, The challenge from related species, in H. G. Baker and G. L. Stebbins (Eds.), *The Genetics of Colonizing Species,* Academic Press, New York.

Wilson, E. O., 1969, The species equilibrium, in *Diversity and Stability in Ecological Systems,* Brookhaven Symp. in Biology No. 22.

Winston, P. W., 1956, The acorn microsere with special reference to arthropods, *Ecol.* **37**: 120–132.

Wolfenbarger, D. O., 1946, Dispersion of small organisms, *Amer. Midl. Natur.* **35**: 1–152.

Woodwell, G. M., and R. H. Whittaker, 1968, Primary productivity in terrestrial ecosystems, *Amer. Zool.* **8**: 19–30.

# Appendices

Appendices

# Appendix I
# Probability and the
# Binomial Distribution

The binomial distribution describes the probability of obtaining some given number of items of one category when drawing randomly from a collection of items of more than one category. It is useful when dealing with statistical problems in genetics such as those described under genetic drift in Chapter 1.

Suppose that the proportion of all items that are of category 1 (say, the proportion of $A_1$ genes in the gene pool) is given by $p$. Then the probability of obtaining an item of that category in one drawing is $p$. The probability of doing so twice in two drawings is $p \times p = p^2$. The probability of drawing $k$ such items in $k$ specified drawings is $p^k$, and the probability of drawing none is $(1 - p)^k$. This being the case, it is clear that the chance of drawing one such item in one *specified* drawing out of $n$ is the probability of drawing the one item in that one drawing multiplied by the probability of drawing something else in the other $n - 1$ drawings:

$$p^1(1 - p)^{n-1}.$$

Since there are $n$ drawings, any one of which might be the specified one, the probability of obtaining one such item in $n$ drawings is

$$np(1 - p)^{n-1}.$$

The probability of obtaining $k$ items of category 1 in $k$ *specified* drawings out of $n$ is, by the same argument,

$$p^k(1 - p)^{n-k}.$$

To calculate the probability of obtaining the $k$ items in *any* manner from $n$ drawings, we must multiply this by the number of distinct ways in which $k$ items can be chosen from $n$. This latter value can be found as follows.

There are $n$ different drawings in which one of the $k$ items might appear. The next can be chosen from any of the remaining $n - 1$, and the $k$th may appear in any of $n - (k - 1)$ remaining drawings. Thus there are $n(n - 1)(n - 2) \ldots (n - (k - 1)) = n!/(n - k)!$ ways in which $k$ items can be chosen from $n$. But if we had labeled each of the $k$ items drawn we would immediately see that item number one might have been drawn in the $i$th drawing, number two in the $j$th drawing, or vice versa. Thus to find the number of *distinct* ways in which $k$ items can be drawn from $n$ we must divide the quantity $n!/(n - k)!$, by the number

433

of ways in which the $k$ items can be ordered. This is given by $k!$ (The first can appear in any of $k$ positions, the second in any of the remaining $k - 1$, etc.) Thus we can write

$$\left. \begin{array}{l} \text{the number of } \textit{distinct} \text{ ways of drawing } k \\ \text{similar items from } n \end{array} \right\} = \frac{n!}{(n - k)!\, k!}$$

and,

$$\left. \begin{array}{l} \text{the probability of drawing } k \text{ items of one} \\ \text{category from } n \text{ items of more than one} \\ \text{category when the proportion of the } n \text{ belong-} \\ \text{ing to the category in question is } p \end{array} \right\} = \frac{n!}{(n - k)!\, k!} p^k (1 - p)^{n - k}.$$

This description may be written $p(k|n, p)$, and the quantity

$$\frac{n!}{(n - k)!\, k!} \qquad \text{is usually written} \qquad \binom{n}{k}.$$

Thus

$$p(k|n, p) = \binom{n}{k} p^k (1 - p)^{n - k}.$$

We return now to the discussion in part 1.1. If the frequency of the allele, $A_1$, is initially 0.4, and the population size is 5 (10 genes in the gene pool at the locus in question), then the probability that the frequency will be 0.3 in the next generation is

$$p(3|10, 0.4) = \binom{10}{3} (0.4)^3 (0.6)^7 = \frac{10!}{3!\, 7!} (0.064)(0.028) = 0.215.$$

The probability that the frequency will be equal to *or less than* 0.3 in the above example, is the probability that the number of $A_1$ alleles drawn is 3, 2, 1, or 0. That is

$$\sum_{i=0}^{3} \binom{10}{i} (0.4)^i (0.6)^{10 - i} = 0.215 + 0.105 + 0.035 + 0.005 = 0.360.$$

Clearly

$$\sum_{i=0}^{n} \binom{n}{i} p^i (1 - p)^{n - i} = 1.$$

Therefore, the expected number drawn (the mean) is

$$\sum_{i=0}^{n} i \binom{n}{i} p^i (1 - p)^{n - i} = \sum_{i=0}^{n} i \frac{n!}{(n - i)!\, i!} p^i (1 - p)^{n - i}$$

$$= \sum_{i=1}^{n} i \frac{n!}{(n - i)!\, i!} p^i (1 - p)^{n - i} = \sum_{i=1}^{n} \frac{n!}{(n - i)!\, (i - 1)!} p^i (1 - p)^{n - i}$$

$$= \sum_{i=1}^{n} \frac{n(n-1)!}{[(n-1)-(i-1)]!\,(i-1)!} p \cdot p^{i-1}(1-p)^{(n-1)-(i-1)}$$

$$= \sum_{(i-1)=0}^{(n-1)} np \frac{(n-1)!}{[(n-1)-(i-1)]!\,(i-1)!} p^{(i-1)}(1-p)^{(n-1)-(i-1)}$$

$$= np(1) = np.$$

We begin with the observation that an equation, $y = f(x)$, can be graphed, one axis denoting the value of $x$, the other the corresponding value of $y$. Differentiation is simply a process by which the slope of that graph, at any value of $x$, can be ascertained.

Slope is defined as the rise in $y$ with an increase in $x$. Where a change in $x$ is denoted $\Delta x$, a change in $y$, $\Delta y$, we write

$$\text{slope} = \Delta y / \Delta x.$$

Given the equation $y = 4x + 1$, for example, if $x$ increases by an amount $\Delta x$, the new value of $x$ is $x + \Delta x$ and the corresponding value of $y$ is

$$4(x + \Delta x) + 1 = 4x + 1 + 4\Delta x.$$

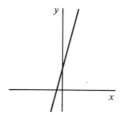

The change in $y$ is thus clearly

$$(4x + 1 + 4\Delta x) - (4x + 1) = 4\Delta x.$$

For any *linear* equation (of the form $y = mx + b$) the slope of the defined graph is

$$\Delta y / \Delta x = \frac{[m(x + \Delta x) + b] - [mx + b]}{\Delta x} = \frac{m\Delta x}{\Delta x} = m.$$

The slope of a straight line is clearly constant. But what of the situation where (say) $y = x^2$? Where $x < 0$ the curve slopes downward (negative slope), where $x > 0$, the slope is positive, and where $x = 0$, the slope is level ($=0$). Using the above method for calculating slope, we write

$$\Delta y / \Delta x = \frac{(x + \Delta x)^2 - (x)^2}{\Delta x} = \frac{x^2 + 2x\Delta x + \Delta x^2 - x^2}{\Delta x} = 2x + \Delta x.$$

In this case the slope changes not only with $x$, but with $\Delta x$. It is fairly obvious that this must be the case. If $\Delta x$ is taken between points 1 and 2 in the above figure, the slope defined is quite different than if the points 1 and 3, or 2 and 3 are used.

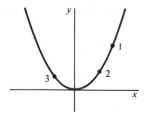

But what we wish to know is the slope *at some value*, $x$. To find this we must clearly define $\Delta x$ as ranging between two points infinitesimally close together —that is, we let $\Delta x$ approach zero and write

$$\text{slope at } x = \lim_{\Delta x \to 0} (\Delta y/\Delta x) = \lim_{\Delta x \to 0} (2x + \Delta x) = 2x.$$

Another example:
Let $y = 3x^2 - x$.

$$\text{slope at } x = \lim_{\Delta x \to 0} (\Delta y/\Delta x) = \lim_{\Delta x \to 0} \frac{[3(x + \Delta x)^2 - (x + \Delta x)] - [3x^2 - x]}{\Delta x}$$

$$= \lim_{\Delta x \to 0} \frac{6x\Delta x + 3\Delta x^2 - \Delta x}{\Delta x}$$

$$= \lim_{\Delta x \to 0} (6x + 3\Delta x - 1) = 6x - 1.$$

The values $\Delta y$, $\Delta x$, when $\Delta x \to 0$, are written $dy$, $dx$; the slope is known as the *derivative*, and the process of taking (finding) the derivative is *differentiation*.

$$\text{Derivative} = \lim_{\Delta x \to 0} (\Delta y/\Delta x) = \lim_{\Delta x \to 0} \frac{f(x + \Delta x) - f(x)}{\Delta x}$$

$$= \frac{f(x + dx) - f(x)}{dx} = \frac{df(x)}{dx}.$$

The slope, or derivative, of some value $y$ with respect to $x$, clearly is the rate of change in $y$ with respect to $x$. Thus if we write, for example,

$$s = vt,$$

where $s$ is distance, $t$ is time, the rate of change in $s$ with $t$ (that is, velocity) is given by

$$\frac{ds}{dt} = \frac{(v)(t + dt) - (v)(t)}{dt} = \frac{v\,dt}{dt} = v.$$

In this case the equation is linear and $v$ is constant. But $v$ is not constant if acceleration occurs. Suppose

$$s = at^2/2.$$

Then

$$\text{velocity } (=v) = ds/dt = \frac{(a)(t + dt)^2/2 - at^2/2}{dt} = at,$$

and

$$dv/dt \text{ (rate of change in } v \text{ with } t = \text{acceleration)}$$

$$= \frac{a(t + dt) - at}{dt} = a.$$

It is impossible in this space to derive the various rules for differentiation. We therefore simply list a few.

1.  $y = k = \text{constant.}$

$$dy/dx = dk/dx \left( \text{or} \quad \frac{d}{dx} k \right) = 0.$$

2.  $y = x^n.$

$$dy/dx = \frac{d}{dx} x^n = nx^{n-1}.$$

Example: $y = x^3.$

$$dy/dx = dx^3/dx = 3x^2.$$

3.  $y = kf(x).$

$$dy/dx = \frac{d}{dx} kf(x) = k \frac{d}{dx} f(x).$$

Example: $y = 3x^2.$

$$dy/dx = d(3x^2)/dx = 3\,dx^2/dx = 3(2x^1) = 6x.$$

4.  $y = f(x) + g(x).$

$$dy/dx = \frac{d}{dx} [f(x) + g(x)] = \frac{d}{dx} f(x) + \frac{d}{dx} g(x).$$

Example: $y = 3 + x^2.$

$$dy/dx = \left[ \frac{d}{dx} 3 + \frac{d}{dx} x^2 \right] = [0 + 2x^1] = 2x.$$

Example: $y = x - 3x^4.$

$$dy/dx = \left[ \frac{d}{dx} x + \frac{d}{dx} (-3x^4) \right] = [1 - 3dx^4/dx] = 1 - 12x^3.$$

5.   $y = f(x)g(x)$.

$$dy/dx = \frac{d}{dx}[f(x)g(x)] = f(x)\frac{d}{dx}g(x) + g(x)\frac{d}{dx}f(x).$$

Example: $y = (3x)(1 - x)$.

$$dy/dx = \frac{d}{dx}[(3x)(1 - x)]$$

$$= \left[(3x)\frac{d}{dx}(1 - x) + (1 - x)\frac{d}{dx}(3x)\right]$$

$$= \{(3x)[d1/dx - dx/dx] + (1 - x)3\,dx/dx\}$$

$$= [(3x)(0 - 1) + (1 - x)3] = [-3x + 3(1 - x)] = 3 - 6x.$$

6.   $y = g[f(x)]$.

$$dy/dx = \frac{d}{df(x)}g[f(x)]\frac{df(x)}{dx}.$$

Example: $y = (1 - x)^2$.

$$dy/dx = \frac{d}{dx}(1 - x)^2 = \frac{d}{d(1 - x)}(1 - x)^2\,\frac{d}{dx}(1 - x)$$

$$= 2(1 - x).\left(\frac{d1}{dx} - \frac{dx}{dx}\right) = 2(1 - x)(0 - 1) = -2(1 - x).$$

Example: $y = (x - 2)^{16}$.

$$dy/dx = \frac{d}{dx}(x - 2)^{16}$$

$$= \frac{d}{d(x - 2)}(x - 2)^{16}.\frac{d}{dx}(x - 2)$$

$$= 16(x - 2)^{15}(1 - 0) = 16(x - 2)^{15}.$$

Example: $y = (1 + x)/(1 - x)$.

$$dy/dx = \frac{d}{dx}[(1 + x)/(1 - x)] = \left(\frac{1}{1 - x}\right)\frac{d(1 + x)}{dx} + (1 + x)\frac{d}{dx}\left(\frac{1}{1 - x}\right)$$

$$= \left[(1 - x)^{-1}\frac{d}{dx}(1 + x)\right] + \left[(1 + x)\frac{d(1 - x)^{-1}}{d(1 - x)}\frac{d(1 - x)}{dx}\right]$$

$$= [(1 - x)^{-1}(0 + 1)] + [(1 + x)(-1)(1 - x)^{-2}(0 - 1)]$$

$$= \frac{1}{(1 - x)} + \frac{(1 + x)}{(1 - x)^2} = \frac{2}{(1 - x)^2}.$$

7.  $y = e^x$ (where $e$ is the base of the natural logarithm).

$$dy/dx = \frac{d}{dx} e^x = e^x.$$

8.  $y = \ln x$.

$$dy/dx = \frac{d}{dx} \ln x = 1/x.$$

Example: $y = x^2/(1 - x) - e^x \ln x$.

$$\frac{dy}{dx} = \frac{d}{dx} \left[ \frac{x^2}{1-x} - e^x \ln x \right] - \left[ \frac{d}{dx} \left( \frac{x^2}{1-x} \right) - \frac{d}{dx} e^x \ln x \right]$$

$$= \left[ (1-x)^{-1} \frac{d}{dx} x^2 + x^2 \frac{d}{dx} (1-x)^{-1} \right]$$

$$- \left[ e^x \frac{d}{dx} \ln x + \ln x \frac{d}{dx} e^x \right]$$

$$= \left[ \frac{2x}{(1-x)} - \frac{x^2}{(1-x)^2} \right] - \left[ \frac{e^x}{x} + e^x \ln x \right].$$

Example: $n = n_0 e^{rt}$.

$$\frac{dn}{dt} = \frac{d}{dt} n_0 e^{rt} = n_0 \frac{d}{dt} e^{rt}$$

$$= n_0 \frac{d}{d(rt)} e^{rt} \cdot \frac{d}{dt} (rt)$$

$$= n_0 e^{rt} \cdot r = nr.$$

From the above example it can be seen that given the expression, $dn/dt = rn$, it is possible to work backward and derive the solution $n = n_0 e^{rt}$. To explain the technique for doing this we digress a moment. Suppose we define the new variables $X$, $Y$, such that as the value along the horizontal axis varies from 0 to $x$, $X$ varies from $X_0$ to $X_x$ and $Y$ from $Y_0$ to $Y_x$. Now suppose we are given the equation

$$dy/dx = 3x/y.$$

Write:

$$dy = 3 \frac{x}{y} dx,$$

and then

$$y \, dy = 3x \, dx,$$

and define $dX = x\,dx$, $dY = y\,dy$. Then $dX/dx = x$, $dY/dy = y$: $X$, $Y$ are those expressions which, when differentiated, yield $x$, $y$. The above equation can now be rewritten:

$$dY = 3\,dX.$$

But $dX$, $dY$ are pieces of the graph, $X$ versus $Y$, which ranges from $X_0$, $Y_0$ to $X_x$, $Y_x$. Thus if we add up all of the pieces, $dX$, $dY$, in this interval we should obtain

$$X_x - X_0, \qquad Y_x - Y_0.$$

We write this

$$\int_0^x dX = X_x - X_0, \qquad \int_{y(0)}^{y(x)} dY = Y_x - Y_0,$$

where $\int$ represents summation.
Substituting back into the equation, then

$$\int_{y(0)}^{y(x)} dY = 3\int_0^x dX, \qquad \text{or} \qquad (Y_x - Y_0) = 3(X_x - X_0).$$

But $X$, $Y$ are the expressions which yield $x$, $y$ upon differentiation, so $X = \frac{1}{2}x^2$, $Y = \frac{1}{2}y^2$. Thus

$$X_x - X_0 = \tfrac{1}{2}x^2 - \tfrac{1}{2}0^2, \qquad Y_x - Y_0 = \tfrac{1}{2}y(x)^2 - \tfrac{1}{2}y(0)^2,$$

and

$$\tfrac{1}{2}y(x)^2 - y(0)^2 = (3)\tfrac{1}{2}x^2,$$

or

$$y(x)^2 = y(0)^2 + 3x^2.$$

This is the solution of the equation $dy/dx = 3x/y$.

In general, the procedure is to move all $x$'s to one side of the equation, all $y$'s to the other, obtaining an expression of the form:

$$\int_0^x f(x)\,dx = \int_{y(0)}^{y(x)} g(y)\,dy,$$

and then to find the functions of $x$, $y$, which, when differentiated with respect to $x$, $y$, yield $f(x)$, $g(y)$.

Example: $dy/dx = 2/3y$.
Then

$$3y\,dy = 2\,dx,$$

so

$$\int_{y(0)}^{y(x)} 3y\,dy = \int_0^x 2\,dx.$$

The expression which when differentiated by $y$ gives $3y$, is $3y^2/2$. The expression which when differentiated by $x$ gives $2$, is $2x$. Thus the solution is

$$\tfrac{3}{2}y(x)^2 - \tfrac{3}{2}y(0)^2 = 2x - 2\cdot 0 = 2x$$

or
$$y^2 = y(0)^2 + 4x/3.$$

Check: $y(0)$ is a constant, so its derivative is zero and the derivative of the right side of the equation is $(d/dx)\,4x/3 = 4/3$. The derivative of the left side (with respect to $x$) is $dy^2/dx = (dy^2/dy)(dy/dx) = 2y\,dy/dx$. Thus

$$2y\frac{dy}{dx} = 4/3,$$

or
$$dy/dx = 2/3y.$$

Example: $dn/dt = rn$.

Then

$$\frac{1}{n}\,dn = r\,dt,$$

so

$$\int_{n(0)}^{n(t)} \frac{1}{n}\,dn = \int_0^t r\,dt,$$

so

$$\ln n(t) - \ln n(0) = rt - r \cdot 0 = rt,$$

or

$$\ln \frac{n(t)}{n(0)} = rt,$$

so

$$n(t) = n(0)e^{rt}.$$

This process of reversing differentiation is called *integration*; and equations involving derivatives are known as *differential equations*.

Example: $dn/dt = rn(1 - n/K)$.

Then

$$\int_{n(0)}^{n(t)} \frac{dn}{n(1 - n/K)} = \int_0^t r\,dt.$$

The expression on the left is not easy to integrate, but tables of integrals such as this are easily obtainable. In this case the solution is given by

$$n = n(0)\,\frac{K}{n(0) + (K - n(0))\,e^{-rt}}.$$

Check: $dn/dt = n(0)K \left\{ \dfrac{d[n(0) + (K - n(0))\,e^{-rt}]^{-1}}{d[n(0) + (K - n(0))\,e^{-rt}]} \right\}$

$$\times \left\{ \frac{d[n(0) + (K - n(0))\,e^{-rt}]}{dt} \right\}$$

$$= -n(0)K[n(0) + (K - n(0))e^{-rt}]^{-2}\left[0 + (K - n(0))\frac{de^{-rt}}{dt}\right]$$

$$= \frac{-n(0)K}{[n(0) + (K - n(0))e^{-rt}]^2}\{[K - n(0)](-r)e^{-rt}\}.$$

But, from the above equation,

$$\begin{cases} \dfrac{n(0)K}{[n(0) + (K - n(0))e^{-rt}]^2} = n^2/n(0)K, \\ (K - n(0))e^{-rt} = n(0)(K - n)/n. \end{cases}$$

Thus

$$dn/dt = n^2\left[\frac{1}{n(0)K}\right]\left[\frac{n(0)(K - n)}{n}\right]r = rn(1 - n/K).$$

It is necessary to make two additional points.

**1.** Suppose $\overline{W} = ap(1 - p)^2$, where $\overline{W}$ is fitness, and $p$ is gene frequency. Selection acts to increase $\overline{W}$, and we should like to know at what value of $p$ it is that $\overline{W}$ reaches its maximum and selection stops. Graphing the expression, we obtain a curve of the general form shown here.

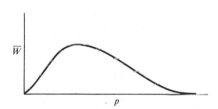

Clearly $W$ reaches its peak at the top of the curve, when the slope $d\overline{W}/dp$ is zero. To find the appropriate value of $p$, then, we simply write

$$0 = d\overline{W}/dp = \frac{d}{dp}ap(1 - p)^2 = a\left[p\frac{d}{dp}(1 - p)^2 + (1 - p)^2\,dp/dp\right]$$

$$= a\left[p\frac{d}{(1 - p)}(1 - p)^2\frac{d(1 - p)}{dp} + (1 - p)^2\right]$$

$$= a[(p)\,2(1 - p)(-1) + (1 - p)^2]$$

$$= a[(1 - p)^2 - 2(1 - p)p],$$

so

$$p = 1/3,$$

and

$$\overline{W}(\text{max}) = a(1/3)(2/3)^2 = 4a/27.$$

Of course, it is possible that in some instances the zero slope of an equation represents a minimum rather than a maximum, or, where a curve has several turning points, both maxima and minima. When the general shape of the curve is known this problem is eliminated and the procedure of setting the derivative equal to zero is useful in finding maxima and minima.

**2.** Often we encounter expressions of the form

$$U = 3x^2 + 2xy - y$$

(that is, equations with two independent variables). Suppose we wish to know how $U$ varies with $x$; we can write

$$dU/dx = d(3x^2)/dx + d(2xy)/dx - dy/dx$$

$$= 3 \cdot 2x + 2\left(x\frac{dy}{dx} + y\frac{dx}{dx}\right) - dy/dx$$

$$= 6x + 2x\,dy/dx + 2y - dy/dx.$$

If we wish to know how $U$ would vary with $x$ *were $y$ held constant*, we can write

$$\partial U/\partial x = \partial(3x^2)/\partial x + \partial(2xy)/\partial x - \partial y/\partial x$$

$$= 3 \cdot 2x + 2(x\,\partial y/\partial x + y\,\partial x/\partial x) - \partial y/\partial x$$

$$= 6x + 2(x \cdot 0 + y) - 0 = 6x + 2y.$$

$\partial U/\partial x$ is called the *partial derivative* of $U$ with respect to $x$.

It can be shown that $dU/dx = (\partial U/\partial x) + (\partial U/\partial y)\,dy/dx$,

or

$$dU = (\partial U/\partial x)\,dx + (\partial U/\partial y)\,dy.$$

Note, then, that if we wish to find values of $x$, $y$ which maximize (or minimize) $U$, we can write

$$0 = dU/dx = (\partial U/\partial x) + (\partial U/\partial y)\,dy/dx,$$

and

$$0 = dU/dy = (\partial U/\partial x)\,dx/dy + (\partial U/\partial y).$$

Both reduce to

$$0 = dU = (\partial U/\partial x)\,dx + (\partial U/\partial y)\,dy.$$

Thus

$$dy/dx = -\frac{\partial U/\partial x}{\partial U/\partial y}.$$

Example: Suppose fitness $= \overline{W} = c_1 x + c_2 y$, where $x$ is the amount of meat eaten, $y$ the number of berries eaten per day, and $c_1$ and $c_2$ are the relative "values" of these foods. Then $\overline{W}$ is maximum when

$$dy/dx = -\frac{\partial \overline{W}/\partial x}{\partial \overline{W}/\partial y} = -\frac{c_1}{c_2}.$$

Solving, we obtain

$$dy = -(c_1/c_2)\,dx, \qquad \int_{y(0)}^{y(x)} dy = \int_0^x -(c_1/c_2)\,dx, \qquad y(x) - y(0) = -c_1 x/c_2,$$

so

$$y = y(0) - c_1 x/c_2.$$

Substituting,  we  have

$$\overline{W} = c_1 x + c_2[y(0) - c_1 x/c_2] = x(c_1 - c_1) + c_2 y(0) = c_2 y(0).$$

Since selection acts to increase $\overline{W}$, it clearly acts:

1.  to maximize that number of berries which can be eaten in the absence of meat, $y(0)$,
2.  to regulate food preferences such that $y - y(0) - c_1 x/c_2$.

# Appendix III
# Determinants and
# Inverses

It is not possible in this space to give the reader any feel for matrices or to proceed rigorously with derivations. We thus simply define determinants and inverses and show briefly their relationship.

We begin by stating that every square matrix, $A$, has associated with it a number, the *determinant*, written $|A|$, or det $A$. In the case of a matrix with a single element, the value of the determinant is the same as the element

$$|a| = a.$$

For larger matrices we write

$$|A| = |\{a_{ij}\}| = \sum_j (-1)^{i+j}(a_{ij})|\phi_{ij}|, \tag{3.1}$$

where $a_{ij}$ is the element in the $i$th row, $j$th column, and $|\phi_{ij}|$ is the determinant of the matrix obtained from $A$ by deleting the $i$th row and $j$th column. Thus (example):

$$\begin{vmatrix} a_{11} & a_{12} \\ a_{21} & a_{22} \end{vmatrix} = (-1)^{1+1}(a_{11})|a_{22}| + (-1)^{1+2}(a_{12})|a_{21}| = + a_{11}a_{22} - a_{12}a_{21}.$$

When dealing with $2 \times 2$ matrices, the determinant is found most simply by multiplying the upper left and lower right elements and subtracting from this the product of the lower left and upper right elements.

$$\begin{vmatrix} 3 & 1 \\ 0 & 2 \end{vmatrix} = 3 \times 2 - 0 \times 1 = 6$$

$$\begin{vmatrix} 5 & 2 \\ 7 & 6 \end{vmatrix} = 5 \times 6 - 7 \times 2 = 16.$$

A $3 \times 3$ matrix can be dealt with in the following manner

$$\begin{vmatrix} a_{11} & a_{12} & a_{13} \\ a_{21} & a_{22} & a_{23} \\ a_{31} & a_{32} & a_{33} \end{vmatrix} = (-1)^{1+1}(a_{11})\begin{vmatrix} a_{22} & a_{23} \\ a_{32} & a_{33} \end{vmatrix} + (-1)^{1+2}(a_{12})\begin{vmatrix} a_{21} & a_{23} \\ a_{31} & a_{33} \end{vmatrix}$$

$$+ (-1)^{1+3}(a_{13})\begin{vmatrix} a_{21} & a_{22} \\ a_{31} & a_{32} \end{vmatrix}.$$

$$= + a_{11}(a_{22}a_{33} - a_{32}a_{23}) - a_{12}(a_{21}a_{33} - a_{31}a_{23})$$
$$+ a_{13}(a_{21}a_{32} - a_{31}a_{22}).$$

Notice that this can also be obtained by rewriting the first and second columns of the matrix, as shown below, and then adding the products of all upper left to lower right diagonal elements, and subtracting the sum of the products along the opposite diagonals:

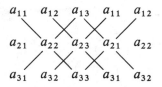

Determinant =

$$(a_{11}a_{22}a_{33} + a_{12}a_{23}a_{31} + a_{13}a_{21}a_{32}) - (a_{31}a_{22}a_{13} + a_{32}a_{23}a_{11} + a_{33}a_{21}a_{12}).$$

Both techniques are used in calculating determinants of $3 \times 3$ matrices. Some examples follow.

1. $A = \begin{pmatrix} 2 & 1 & 3 \\ 0 & 1 & 1 \\ 2 & 0 & 2 \end{pmatrix}$, $|A| = \begin{matrix} 2 & 1 & 3 & 2 & 1 \\ 0 & 1 & 1 & 0 & 1 \\ 2 & 0 & 2 & 2 & 0 \end{matrix} = (2 \times 1 \times 2 + 1 \times 1 \times 2 + 3 \times 0 \times 0)$

$$- (2 \times 1 \times 3 + 0 \times 1 \times 2 + 2 \times 0 \times 1) = (4 + 2 + 0) - (6 + 0 + 0) = 0,$$

or

$$|A| = (-1)^{1+1}(2) \begin{vmatrix} 1 & 1 \\ 0 & 2 \end{vmatrix} + (-1)^{1+2}(1) \begin{vmatrix} 0 & 1 \\ 2 & 2 \end{vmatrix} + (-1)^{1+3}(3) \begin{vmatrix} 0 & 1 \\ 2 & 0 \end{vmatrix}$$

$$= 2(2 - 0) - 1(0 - 2) + 3(0 - 2) = 4 + 2 - 6 = 0.$$

2. $A = \begin{pmatrix} 5 & 0 & 1 \\ 2 & 2 & 1 \\ 2 & 1 & 1 \end{pmatrix}$, $|A| = \begin{matrix} 5 & 0 & 1 & 5 & 0 \\ 2 & 2 & 1 & 2 & 2 \\ 2 & 1 & 1 & 2 & 1 \end{matrix} = (5 \times 2 \times 1 + 0 \times 1 \times 2 + 1 \times 2 \times 1)$

$$- (2 \times 2 \times 1 + 1 \times 1 \times 5 + 1 \times 2 \times 0) = (10 + 0 + 2) - (4 + 5 + 0) = 3,$$

or

$$|A| = (-1)^{1+1}(5) \begin{vmatrix} 2 & 1 \\ 1 & 1 \end{vmatrix} + (-1)^{1+2}(0) \begin{vmatrix} 2 & 1 \\ 2 & 1 \end{vmatrix} + (-1)^{1+3}(1) \begin{vmatrix} 2 & 2 \\ 2 & 1 \end{vmatrix}$$

$$= 5(2 - 1) - 0(2 - 2) + 1(2 - 4) = 5 - 0 - 2 = 3.$$

When we deal with matrices larger than $3 \times 3$, it is simplest to use Eq. (3.1).
Example:

$$\begin{vmatrix} 2 & 1 & 0 & 2 \\ 3 & 3 & 3 & 1 \\ 2 & 1 & 1 & 1 \\ 0 & 0 & 1 & 1 \end{vmatrix} = (-1)^{1+1}(2)\begin{vmatrix} 3 & 3 & 1 \\ 1 & 1 & 1 \\ 0 & 1 & 1 \end{vmatrix} + (-1)^{1+2}(1)\begin{vmatrix} 3 & 3 & 1 \\ 2 & 1 & 1 \\ 0 & 1 & 1 \end{vmatrix}$$

$$+ (-1)^{1+3}(0)\begin{vmatrix} 3 & 3 & 1 \\ 2 & 1 & 1 \\ 0 & 0 & 1 \end{vmatrix} + (-1)^{1+4}(2)\begin{vmatrix} 3 & 3 & 3 \\ 2 & 1 & 1 \\ 0 & 0 & 1 \end{vmatrix}$$

$$= 2\left(\begin{array}{ccccc} 3 & 3 & 1 & 3 & 3 \\ 1 & 1 & 1 & 1 & 1 \\ 0 & 1 & 1 & 0 & 1 \end{array}\right) - 1\left(\begin{array}{ccccc} 3 & 3 & 1 & 3 & 3 \\ 2 & 1 & 1 & 2 & 1 \\ 0 & 1 & 1 & 0 & 1 \end{array}\right)$$

$$+ 0 - 2\left(\begin{array}{ccccc} 3 & 3 & 3 & 3 & 3 \\ 2 & 1 & 1 & 2 & 1 \\ 0 & 0 & 1 & 0 & 1 \end{array}\right)$$

$$= 2[(3 + 0 + 1) - (0 + 3 + 3)] - [(3 + 0 + 2) - (0 + 3 + 6)]$$

$$- 2[(3 + 0 + 0) - (0 + 0 + 6)]$$

$$= 2(-2) - (-4) - 2(-3)$$

$$= -4 + 4 + 6 = 6.$$

Example:

$$\begin{vmatrix} 1 & 0 & 2 & 2 \\ 2 & 1 & 3 & 1 \\ 0 & 0 & 1 & 1 \\ 2 & 3 & 3 & 2 \end{vmatrix} = (-1)^{1+1}(1)\begin{vmatrix} 1 & 3 & 1 \\ 0 & 1 & 1 \\ 3 & 3 & 2 \end{vmatrix} + (-1)^{1+2}(0)\begin{vmatrix} 2 & 3 & 1 \\ 0 & 1 & 1 \\ 2 & 3 & 2 \end{vmatrix}$$

$$+ (-1)^{1+3}(2)\begin{vmatrix} 2 & 1 & 1 \\ 0 & 0 & 1 \\ 2 & 3 & 2 \end{vmatrix} + (-1)^{1+4}(2)\begin{vmatrix} 2 & 1 & 3 \\ 0 & 0 & 1 \\ 2 & 3 & 3 \end{vmatrix}$$

$$= 1 \begin{vmatrix} 1 & 3 & 1 & 1 & 3 \\ 0 & 1 & 1 & 0 & 1 \\ 3 & 3 & 2 & 3 & 3 \end{vmatrix} + 0 + 2 \begin{vmatrix} 2 & 1 & 1 & 2 & 1 \\ 0 & 0 & 1 & 0 & 0 \\ 2 & 3 & 2 & 2 & 3 \end{vmatrix}$$

$$- 2 \begin{vmatrix} 2 & 1 & 3 & 2 & 1 \\ 0 & 0 & 1 & 0 & 1 \\ 2 & 3 & 3 & 2 & 3 \end{vmatrix}$$

$$= [(2 + 9 + 0) - (3 + 3 + 0)] + 2[(0 + 2 + 0) - (0 + 6 + 0)]$$
$$- 2[(0 + 2 + 0) - (0 + 6 + 0)]$$
$$= (2) + 2(-4) - 2(-4) = 2.$$

Now that we have defined determinants we may proceed to inverses. Any square matrix, $A$, which possesses a nonzero determinant, has associated with it an inverse, $A^{-1}$, such that

$$AA^{-1} = A^{-1}A = I = \text{identity matrix.}$$

The method for calculating the inverse is given in the expression

$$A^{-1} = \frac{\text{Adj } A}{|A|},$$

where Adj $A$ is given by

$$\{(-1)^{i+j}|\phi_{ij}|\}'$$

and the prime denotes the switching of rows for columns.

We digress, and recall that $\{\phi_{ij}\}$ has been defined as the matrix obtained by deleting the $i$th row, $j$th column of $A$. Suppose

$$A = \begin{pmatrix} 2 & 1 & 0 \\ 3 & 2 & 1 \\ 2 & 2 & 1 \end{pmatrix}$$

Then

$$\phi_{11} = \begin{pmatrix} 2 & 1 \\ 2 & 1 \end{pmatrix} \quad \phi_{12} = \begin{pmatrix} 3 & 1 \\ 2 & 1 \end{pmatrix} \quad \phi_{13} = \begin{pmatrix} 3 & 2 \\ 2 & 2 \end{pmatrix}$$

$$\phi_{21} = \begin{pmatrix} 1 & 0 \\ 2 & 1 \end{pmatrix} \quad \phi_{22} = \begin{pmatrix} 2 & 0 \\ 2 & 1 \end{pmatrix} \quad \phi_{23} = \begin{pmatrix} 2 & 1 \\ 2 & 2 \end{pmatrix}$$

$$\phi_{31} = \begin{pmatrix} 1 & 0 \\ 2 & 1 \end{pmatrix} \quad \phi_{32} = \begin{pmatrix} 2 & 0 \\ 3 & 1 \end{pmatrix} \quad \phi_{33} = \begin{pmatrix} 2 & 1 \\ 3 & 2 \end{pmatrix}$$

so the matrix, $\{(-1)^{i+j}|\phi_{ij}|\}$, is given by

$$\left\{\begin{array}{ccc} (-1)^{1+1}(2-2) & (-1)^{1+2}(3-2) & (-1)^{1+3}(6-4) \\ (-1)^{2+1}(1-0) & (-1)^{2+2}(2-0) & (-1)^{2+3}(4-2) \\ (-1)^{3+1}(1-0) & (-1)^{3+2}(2-0) & (-1)^{3+3}(4-3) \end{array}\right\} = \left(\begin{array}{ccc} 0 & -1 & 2 \\ -1 & 2 & -2 \\ 1 & -2 & 1 \end{array}\right)$$

and the matrix $\{(-1)^{i+j}|\phi_{ij}|\}'$ is given by

$$\left(\begin{array}{ccc} 0 & -1 & 2 \\ -1 & 2 & -2 \\ 1 & -2 & 1 \end{array}\right)' = \left(\begin{array}{ccc} 0 & -1 & 1 \\ -1 & 2 & -2 \\ 2 & -2 & 1 \end{array}\right)$$

(Elements of the $i$th row, $j$th column are moved to the $j$th row, $i$th column).
The determinant of $A$ is

$$\left(\begin{array}{ccc} 2 & 1 & 0 & 2 & 1 \\ 3 & 2 & 1 & 3 & 2 \\ 2 & 2 & 1 & 2 & 2 \end{array}\right) = (4+2+0) - (0+4+3) = -1.$$

Thus

$$A^{-1} = \frac{\text{Adj}\,A}{|A|} = \frac{\{(-1)^{i+j}|\phi_{ij}|\}'}{|A|}$$

$$= \frac{\left(\begin{array}{ccc} 0 & -1 & 1 \\ -1 & 2 & -2 \\ 2 & -2 & 1 \end{array}\right)}{-1} = \left(\begin{array}{ccc} 0 & 1 & -1 \\ 1 & -2 & 2 \\ -2 & 2 & -1 \end{array}\right).$$

We can check the result by writing

$$A^{-1}A = \left(\begin{array}{ccc} 0 & 1 & -1 \\ 1 & -2 & 2 \\ -2 & 2 & -1 \end{array}\right)\left(\begin{array}{ccc} 2 & 1 & 0 \\ 3 & 2 & 1 \\ 2 & 2 & 1 \end{array}\right) = \left(\begin{array}{ccc} 1 & 0 & 0 \\ 0 & 1 & 0 \\ 0 & 0 & 1 \end{array}\right) = I$$

$$AA^{-1} = \left(\begin{array}{ccc} 2 & 1 & 0 \\ 3 & 2 & 1 \\ 2 & 2 & 1 \end{array}\right)\left(\begin{array}{ccc} 0 & 1 & -1 \\ 1 & -2 & 2 \\ -2 & 2 & -1 \end{array}\right) = \left(\begin{array}{ccc} 1 & 0 & 0 \\ 0 & 1 & 0 \\ 0 & 0 & 1 \end{array}\right) = I.$$

Suppose $A = \left(\begin{array}{ccc} 2 & 1 & 0 \\ 1 & 1 & 1 \\ 2 & 3 & 2 \end{array}\right) \ldots$ Then Adj $A = \{(-1)^{i+j}|\phi_{ij}|\}'$

$$= \left(\begin{array}{ccc} (-1)^{1+1}\begin{vmatrix} 1 & 1 \\ 3 & 2 \end{vmatrix} & (-1)^{1+2}\begin{vmatrix} 1 & 1 \\ 2 & 2 \end{vmatrix} & (-1)^{1+3}\begin{vmatrix} 1 & 1 \\ 2 & 3 \end{vmatrix} \\ (-1)^{2+1}\begin{vmatrix} 1 & 0 \\ 3 & 2 \end{vmatrix} & (-1)^{2+2}\begin{vmatrix} 2 & 0 \\ 2 & 2 \end{vmatrix} & (-1)^{2+3}\begin{vmatrix} 2 & 1 \\ 2 & 3 \end{vmatrix} \\ (-1)^{3+1}\begin{vmatrix} 1 & 0 \\ 1 & 1 \end{vmatrix} & (-1)^{3+2}\begin{vmatrix} 2 & 0 \\ 1 & 1 \end{vmatrix} & (-1)^{3+3}\begin{vmatrix} 2 & 1 \\ 1 & 1 \end{vmatrix} \end{array}\right)'$$

$$= \begin{pmatrix} -1 & 0 & 1 \\ -2 & 4 & -4 \\ 1 & -2 & 1 \end{pmatrix}' = \begin{pmatrix} -1 & -2 & 1 \\ 0 & 4 & -2 \\ 1 & -4 & 1 \end{pmatrix},$$

$$|A| = \begin{matrix} 2 & 1 & 0 & 2 & 1 \\ 1 & 1 & 1 & 1 & 1 \\ 2 & 3 & 2 & 2 & 3 \end{matrix} = (4 + 2 + 0) - (0 + 6 + 2) = -2,$$

so

$$A^{-1} = \frac{\text{Adj}\,A}{|A|} = \begin{pmatrix} -1 & -2 & 1 \\ 0 & 4 & -2 \\ 1 & -4 & 1 \end{pmatrix} \bigg/ (-2) = \begin{pmatrix} \frac{1}{2} & 1 & -\frac{1}{2} \\ 0 & -2 & 1 \\ -\frac{1}{2} & 2 & -\frac{1}{2} \end{pmatrix}.$$

Check:

$$A^{-1}A = \begin{pmatrix} \frac{1}{2} & 1 & -\frac{1}{2} \\ 0 & -2 & 1 \\ -\frac{1}{2} & 2 & -\frac{1}{2} \end{pmatrix} \begin{pmatrix} 2 & 1 & 0 \\ 1 & 1 & 1 \\ 2 & 3 & 2 \end{pmatrix} = \begin{pmatrix} 1 & 0 & 0 \\ 0 & 1 & 0 \\ 0 & 0 & 1 \end{pmatrix} = I$$

$$AA^{-1} = \begin{pmatrix} 2 & 1 & 0 \\ 1 & 1 & 1 \\ 2 & 3 & 2 \end{pmatrix} \begin{pmatrix} \frac{1}{2} & 1 & -\frac{1}{2} \\ 0 & -2 & 1 \\ -\frac{1}{2} & 2 & -\frac{1}{2} \end{pmatrix} = \begin{pmatrix} 1 & 0 & 0 \\ 0 & 1 & 0 \\ 0 & 0 & 1 \end{pmatrix} = I.$$

We begin with a topic that may initially seem quite unrelated to matrices or populations. Suppose we are given the numbers

$$a + ib, \qquad a - ib$$

where $i$ is the square root of minus one, and recalling that $i$ is imaginary, asked to describe their meaning in the real world. Specifically suppose we are told that the change over time of a real environmental parameter, $n$, can be described by the expression:

$$n(t) = \tfrac{1}{2}[(a + ib)^t + (a - ib)^t]\, n(0).$$

What does this mean?

To answer the question we first note that we can plot the value $a + ib$ on a graph, $a$ being the coordinate along one axis (the "real" axis) and $b$ the coordinate along the other ("imaginary") axis. When this is done it can be seen immediately that the point so designated can be described not only by $a$ and $b$, but also by its distance from the origin and the angle defined by the real axis and the line running from the origin through the point. The distance is given by $\sqrt{a^2 + b^2}$ (which we shall call $r$), and the angle, $\theta$, is given by $\cos^{-1}(a/r) = \sin^{-1}(b/r)$. Hence $b = r \sin\theta$, $a = r \cos\theta$, and

$$a + ib = r\,(\cos\theta + i \sin\theta), \qquad a - ib = r(\cos\theta - i \sin\theta).$$

Where the number $a - ib$ is real, that is $b = 0$, then $r = a$, and $\cos\theta = 1$.

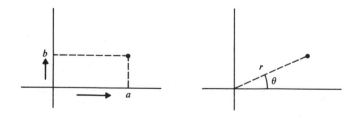

Note now that we can also write

$$(a + ib)^2 = a^2 + 2iab + (ib)^2 = (a^2 - b^2) + i(2ab),$$

which can be written

$$(r^2 \cos^2\theta - r^2 \sin^2\theta) + i(2r \cos\theta)(r \sin\theta)$$

$$= r^2(\cos^2\theta - \sin^2\theta) + ir(2 \cos\theta \sin\theta)$$

$$= r^2 \cos(2\theta) - ir^2 \sin(2\theta)$$

$$= r^2[\cos(2\theta) + i \sin(2\theta)].$$

In the general case

$$\tfrac{1}{2}[(a + ib)^t + (a - ib)^t] = \tfrac{1}{2}r^t[\cos(\theta t) + i \sin(\theta t)]$$

$$+ \tfrac{1}{2}r^t[\cos(\theta t) - i \sin(\theta t)] = r^t \cos\theta t$$

$$r = \sqrt{a^2 + b^2}$$

$$\cos\theta = a/r, \qquad \sin\theta = \frac{b}{r}. \tag{4.1}$$

We return now to the Leslie matrix. Recalling that, where $\vec{P}_j$ is some stable age distribution, we can write

$$L\vec{P}_j = \lambda_j\vec{P}_j. \tag{4.2}$$

Thus $(L - \lambda_j I)\vec{P}_j = 0$, so that (Chapter 9)

$$|L - \lambda_j I| = 0.$$

Now (Chapter 9) all but one $\lambda$ value will be either negative or *complex* (containing imaginary components). The corresponding $\vec{P}_j$ may thus also contain negative or complex elements. While this fact seems unrealistic biologically, it is useful as a mathematical abstraction. Writing all possible column vectors, representing stable age distributions, $\vec{P}_j$, together, we obtain

$$P = (\vec{P}_1\vec{P}_2\vec{P}_3 \dots \vec{P}_m)$$

and since $L\vec{P}_j = \lambda_j\vec{P}_j$, we can write

$$LP = \begin{pmatrix} \lambda_1 P_{11} & \lambda_2 P_{12}\dots \\ \lambda_1 P_{21} & \lambda_2 P_{22} \\ \vdots \end{pmatrix} = P\Lambda, \qquad \text{where } \Lambda = \begin{pmatrix} \lambda_1 & 0 & 0\dots \\ 0 & \lambda_2 & 0 \\ 0 & 0 & \lambda_3 \\ \vdots \end{pmatrix}.$$

But the characteristic equation of $L$, where there are $k$ age classes, has $k$ roots. Thus $m = k$, and $P$ is a square matrix. Thus, if $|P| \neq 0$, $P^{-1}$ exists and we write

$$LP = P\Lambda,$$

or

$$L = P\Lambda P^{-1}. \tag{4.3}$$

Now note that, where $\vec{N}^{(t)}$ is the observed age distribution at time $t$,

$$\vec{N}(t) = L^t \vec{N}^{(0)} = (P\Lambda P^{-1})^t \vec{N}^{(0)} = (P\Lambda P^{-1} P\Lambda P^{-1} P...) \vec{N}^{(0)}$$

$$= (P\Lambda^t P^{-1}) \vec{N}^{(0)}. \tag{4.4}$$

This equation gives us a tool to make use of the complex and negative roots of the characteristic equation and to see what they mean biologically.

Example: Suppose $L = \begin{pmatrix} 1 & 1 \\ \frac{1}{2} & 0 \end{pmatrix}$.

Then $\begin{vmatrix} 1 - \lambda & 1 \\ \frac{1}{2} & -\lambda \end{vmatrix} = 0 = -\lambda + \lambda^2 - \frac{1}{2}$ gives us the roots:

$$\lambda = \frac{1 \pm \sqrt{3}}{2}, \quad \text{so that} \quad \Lambda = \begin{pmatrix} \dfrac{1 + \sqrt{3}}{2} & 0 \\ 0 & \dfrac{1 - \sqrt{3}}{2} \end{pmatrix}.$$

To find $P$, we write (after Eq. 4.2),

$$L\vec{P}_j = \lambda_j \vec{P}_j,$$

or

for $j = 1$, $\begin{cases} (1)p_{11} + (1)p_{21} = \dfrac{1 + \sqrt{3}}{2} p_{11} \\ (\frac{1}{2})p_{11} + (0)p_{21} = \dfrac{1 + \sqrt{3}}{2} p_{21} \end{cases}$, $p_{11}:p_{21} = (1 + \sqrt{3}):1,$

for $j = 2$,
$$\left\{\begin{array}{l}(1)p_{12} + (1)p_{22} = \dfrac{1 - \sqrt{3}}{2}\, p_{12} \\[2mm] (\tfrac{1}{2})p_{12} + (0)p_{22} = \dfrac{1 - \sqrt{3}}{2}\, p_{22}\end{array}\right\}, \quad p_{12} : p_{22} = (1 - \sqrt{3}) : 1,$$

so

$$P = \begin{pmatrix} 1 + \sqrt{3} & 1 - \sqrt{3} \\ 1 & 1 \end{pmatrix},$$

$$P^{-1} = \frac{\operatorname{Adj} P}{|p|} = \begin{pmatrix} 1 & -1 \\ -1 + \sqrt{3} & 1 + \sqrt{3} \end{pmatrix}' \Big/ [(1 + \sqrt{3}) - (1 - \sqrt{3})]$$

$$= \begin{pmatrix} 1 & -1 + \sqrt{3} \\ -1 & 1 + \sqrt{3} \end{pmatrix} \Big/ 2\sqrt{3},$$

and

$$\vec{N}^{(t)} = \left[ \begin{pmatrix} 1 + \sqrt{3} & 1 - \sqrt{3} \\ 1 & 1 \end{pmatrix} \begin{pmatrix} \left(\dfrac{1 + \sqrt{3}}{2}\right)^t & 0 \\ 0 & \left(\dfrac{1 - \sqrt{3}}{2}\right)^t \end{pmatrix} \begin{pmatrix} 1 & -1 + \sqrt{3} \\ -1 & 1 + \sqrt{3} \end{pmatrix} \Big/ 2\sqrt{3} \right] \vec{N}^{(0)}$$

$$= \frac{1}{2\sqrt{3}} (\tfrac{1}{2})^t \begin{pmatrix} (1 + \sqrt{3})^{t+1} - (1 - \sqrt{3})^{t+1} & -(1 - \sqrt{3})(1 + \sqrt{3})^{t+1} \\ & + (1 + \sqrt{3})(1 - \sqrt{3})^{t+1} \\[2mm] (1 + \sqrt{3})^t - (1 - \sqrt{3})^t & -(1 - \sqrt{3})(1 + \sqrt{3})^t \\ & + (1 + \sqrt{3})(1 - \sqrt{3})^t \end{pmatrix} \vec{N}^{(0)}.$$

Since the negative components change sign every generation (if $-a^t$ is positive, $-a^{t+1}$ is negative, and vice versa), they clearly generate oscillations. Note also that as $t$ becomes larger, the absolute value of $(1 - \sqrt{3})^t$ is increasingly less than $(1 + \sqrt{3})^t$, so that its relative influence on $\vec{N}^{(t)}$ diminishes.

Example: Suppose $L = \begin{pmatrix} \tfrac{1}{2} & \tfrac{3}{8} & \tfrac{5}{4} \\ \tfrac{1}{2} & 0 & 0 \\ 0 & \tfrac{1}{2} & 0 \end{pmatrix}$, then

$$\lambda^3 - \tfrac{8}{16}\lambda^2 - \tfrac{3}{16}\lambda - \tfrac{5}{16} = 0$$

generates the roots:

$$\lambda = 1, \qquad -\tfrac{1}{4} \pm \tfrac{1}{2}i,$$

so

$$\Lambda = \begin{pmatrix} 1 & 0 & 0 \\ 0 & -\tfrac{1}{4} + \tfrac{1}{2}i & 0 \\ 0 & 0 & -\tfrac{1}{4} - \tfrac{1}{2}i \end{pmatrix},$$

and

$$L\vec{P}_j = \lambda_j \vec{P}_j,$$

or

for $j = 1$, $\left\{ \begin{aligned} \tfrac{1}{2}p_{11} + \tfrac{3}{8}p_{21} + \tfrac{5}{4}p_{31} &= p_{11} \\ \tfrac{1}{2}p_{11} &= p_{21} \\ \tfrac{1}{2}p_{21} &= p_{31} \end{aligned} \right\}$, $\quad p_{11}:p_{21}:p_{31} = 4:2:1$

for $j = 2$, $\left\{ \begin{aligned} \tfrac{1}{2}p_{12} + \tfrac{3}{8}p_{22} + \tfrac{5}{4}p_{32} &= (-\tfrac{1}{4} + \tfrac{1}{2}i)p_{12} \\ \tfrac{1}{2}p_{12} &= (-\tfrac{1}{4} + \tfrac{1}{2}i)p_{22} \\ \tfrac{1}{2}p_{22} &= (-\tfrac{1}{4} + \tfrac{1}{2}i)p_{32} \end{aligned} \right\}$, $\quad p_{12}:p_{22}:p_{32}$

$$= (-\tfrac{3}{4} - i):(-\tfrac{3}{4} + i):1,$$

for $j = 3$, $\left\{ \begin{aligned} \tfrac{1}{2}p_{13} + \tfrac{3}{8}p_{23} + \tfrac{5}{4}p_{33} &= (-\tfrac{1}{4} - \tfrac{1}{2}i)p_{13} \\ \tfrac{1}{2}p_{13} &= (-\tfrac{1}{4} - \tfrac{1}{2}i)p_{23} \\ \tfrac{1}{2}p_{23} &= (-\tfrac{1}{4} - \tfrac{1}{2}i)p_{33} \end{aligned} \right\}$, $\quad p_{13}:p_{23}:p_{33}$

$$= (-\tfrac{3}{4} + i):(-\tfrac{1}{2} - i):1,$$

so

$$P = \begin{pmatrix} 4 & -\tfrac{3}{4} - i & -\tfrac{3}{4} + i \\ 2 & -\tfrac{1}{2} + i & -\tfrac{1}{2} - i \\ 1 & 1 & 1 \end{pmatrix},$$

$$P^{-1} = \frac{\text{Adj } P}{|p|} = \frac{1}{29} \begin{pmatrix} 4 & 4 & 5 \\ -2 + 5i & -2 - \tfrac{19}{2}i & 12 - i \\ -2 - 5i & -2 + \tfrac{19}{2}i & 12 + i \end{pmatrix}.$$

Therefore, $P\Lambda P^{-1}$

$$\begin{pmatrix}[16+(\tfrac{13}{2}-\tfrac{7}{4}i)\lambda_2^t+(\tfrac{13}{2}+\tfrac{7}{4}i)\lambda_3^t] & [16+(-8+\tfrac{7}{8}i)\lambda_2^t+(-8-\tfrac{7}{8}i)\lambda_3^t] & [20+(-10+\tfrac{45}{4}i)\lambda_2^t+(-10-\tfrac{45}{4}i)\lambda_3^t]\\[2mm] [8+(-4-\tfrac{9}{2}i)\lambda_2^t+(-4+\tfrac{9}{2}i)\lambda_3^t] & [8+(\tfrac{21}{2}+\tfrac{11}{4}i)\lambda_2^t+(\tfrac{21}{2}-\tfrac{1}{4}i)\lambda_3^t] & [10+(-5+\tfrac{25}{2}i)\lambda_2^t+(-5-\tfrac{25}{2}i)\lambda_3^t]\\[2mm] [4+(-2+5i)\lambda_2^t+(-2-5i)\lambda_3^t] & [4+(-2+\tfrac{19}{2}i)\lambda_2^t+(-2-\tfrac{19}{2}i)\lambda_3^t] & [5+(12-i)\lambda_2^t+(12+i)\lambda_3^t]\end{pmatrix}.$$

But each element of the above matrix can be written in the form

$$k+(c+id)\lambda_2^t+(c-id)\lambda_3^t,$$

and $\lambda_2^t$, $\lambda_3^t$ can be written in the form

$$r^t(a-ib),\qquad r^t(a-ib),$$

so

$$k+(c+id)\lambda_2^t+(c-id)\lambda_3^t=k+(c+id)r^t(a+ib)+(c-id)r^t(a-ib)$$
$$=k+2r^t(ac-bd),$$

and the imaginary components cancel. In the above case

$$\begin{cases}\lambda_2^t=r^t(\cos\theta t+i\sin\theta t)=(\tfrac{1}{4}\sqrt{5})^2\,(\cos 2.9t+i\sin 2.9t),\\[1mm] \lambda_3^t=r^t(\cos\theta t-i\sin\theta t)=(\tfrac{1}{4}\sqrt{5})^2\,(\cos 2.9t-i\sin 2.9t),\end{cases}$$

where $(2.9t)$ is in radians. Thus $P\Lambda P^{-1}$ can be rewritten

$$\begin{pmatrix}[16+2(\tfrac{1}{4}\sqrt{5})^t(\tfrac{13}{2}\cos2.9t+\tfrac{7}{4}\sin2.9t)] & [16+2(\tfrac{1}{4}\sqrt{5})^t(-8\cos2.9t-\tfrac{7}{3}\sin2.9t)] & [20+2(\tfrac{1}{4}\sqrt{5})^t(-10\cos2.9t+\tfrac{45}{4}\sin2.9t)]\\[2mm] [8+2(\tfrac{1}{4}\sqrt{5})^t(-4\cos2.9t+\tfrac{9}{2}\sin2.9t)] & [8+2(\tfrac{1}{4}\sqrt{5})^t(\tfrac{21}{2}\cos2.9t-\tfrac{11}{4}\sin2.9t)] & [10+2(\tfrac{1}{4}\sqrt{5})^t(-5\cos2.9t-\tfrac{25}{2}\sin2.9t)]\\[2mm] [4+2(\tfrac{1}{4}\sqrt{5})^t(-2\cos2.9t-5\sin2.9t)] & [4+2(\tfrac{1}{4}\sqrt{5})^t(-2\cos2.9t-\tfrac{15}{2}\sin2.9t)] & [5+2(\tfrac{1}{4}\sqrt{5})^t(5\cos2.9t+\sin2.9t)]\end{pmatrix}.$$

The roots containing imaginary components clearly generate complicated oscillations.

Now, if the values of $r$ associated with all the $\lambda$'s except the one real, positive root are less than one (in the above example, $r = \sqrt{5}/4$), then the influence of the corresponding roots damps out in time,

$$\lim_{t \to \infty} (r)^t = 0, \qquad \text{for} \quad r < 1,$$

until only $\lambda_1$ remains. The population then ceases to oscillate, and

$$\vec{N}^{(t)} = (p)_{11}\lambda_1(p^{-1})_{11}\vec{N}^{(t-1)} = \lambda_1(pp^{-1})_{11}\vec{N}^{(t-1)} = \lambda_1(1)\vec{N}^{(t-1)} = \lambda\vec{N}^{(t-1)}.$$

This $\lambda_1$, the *dominant latent root*, is the one we call $R$ (Chapter 9).

When the first element of the first row of $L$ is nonzero (that is, individuals reproduce in the first reproductive season after birth, and if no breeding seasons are skipped by an age group as a whole, then it happens that $r_1(=\lambda_1) > r_j$ for all $j \neq 1$. Since, eventually, populations must stop growing ($\lambda_1 = r_1 = 1$), it follows that $r_j < 1$ for all $j \neq 1$. But it was noted above that when $r_j < 1$ the contribution of $\lambda_j$ gradually disappears over time. Thus, given sufficient time, all oscillations damp and

$$\vec{N}^{(t)} \to \lambda_1\vec{N}^{(t-1)}.$$

That is, the population approaches a stable age distribution.

# Indexes

# Author Index

Abramoff, P. (*see* R. M. Darnell)
Adams, J. (*see* R. W. Allard)
Agnew, A. D. Q.
  (and J. E. C. Flux)
    dispersal, 364
Alcock, J. (*see* L. M. Cook)
Allard, R. W.
  (and J. Adams)
    synergism in production, 329
Allee, W. C.
  (and A. E. Emerson, O. Park, T. Park,
    K. P. Schmidt)
    termite nests, 126
  (*see* B. Ginsburg)
Allen, D. L. (*see* P. A. Jordan)
Allen, J. A.
  (and B. Clarke)
    novelty, 172
Allison, A. C.
  sickle-cell anemia, 17
Alpers, F.
  food ration, 162
Ambrose, H. W., III (*see* S. T. Emlen)
Anderson, P. K.
  isolation of demes, 39
Anderson, W. W.
  (and C. E. King)
    age-specific selection, 326
  (*see* C. E. King)
Andrewartha, H. G.
  population regulation, 267
  (and L. C. Birch)
    population regulation, 268, 273, 314
  (and T. O. Browning)
    population regulation, 272
  (*see* J. Davidson)
Aneshansley, D. J.
  (and T. Eisner)
    toxic sprays, 101
Antonovics, J.
  genetic isolation, 81
  (*see* A. P. C. Seaton)

Armstrong, E. A.
  mating systems, 146
Armstrong, J. T.
  territory size, 200
Arnheim, N.
  (and C. E. Taylor)
    neutral alleles, 26
Aronson, L. R.
  reproductive cues, 127
Asdell, S. A.
  sperm storage, 133
Ashmole, N. P.
  clutch size, 138
  mobbing, 121
Aumann, G. D.
  (and J. T. Emlen)
    nutrients and population density, 280
Ayala, F. J.
  competitive exclusion, 314
  genetics and competition, 323, 328

Bacon, T. P., Jr. (*see* R. F. Inger)
Baerends, G. P.
  (and J. M. Baerends van Roon)
    startle effect, 118
Bailey, I. W.
  mutualism, 320
Bailey, V. A. (*see* A. J. Nicholson)
Bakken, A.
  food needs, 169
Ball, R. C. (*see* D. W. Hayne)
Barbieri, D. (*see* D. Mainardi)
Barelane, B., Jr. (*see* C. P. Richter)
Barnes, H.
  (and M. Barnes)
    egg size, 134
Barnes, M. (*see* H. Barnes)
Barnett, S. A.
  food choice, 170
  food ration, 162
  novelty, 172
Barth, R. A. (*see* T. H. Hamilton)

461

# Subject Index